Neurobiology

Edited by
Robert A. Meyers

Related Titles

Meyers, R. A. (Ed.)

Encyclopedia of Molecular Cell Biology and Molecular Medicine

16 Volume Set

2005

ISBN: 978-3-527-30542-1

von Bohlen und Halbach, O., Dermietzel, R.

Neurotransmitters and Neuromodulators

Handbook of Receptors and Biological Effects

2006

ISBN: 978-3-527-31307-5

Dermietzel, R., Spray, D. C., Nedergaard, M. (Eds.)

Blood-Brain Barriers

From Ontogeny to Artificial Interfaces

2006

ISBN: 978-3-527-31088-3

Nothwang, H., Pfeiffer, E. (Eds.)

Proteomics of the Nervous System

2008

ISBN: 978-3-527-31716-3

Müller, H. W. (Ed.)

Neural Degeneration and Repair Gene Expression Profiling, Proteomics and Systems Biology

2008

ISBN: 978-3-527-31707-3

Thiel, G. (Ed.)

Transcription Factors in the Nervous System

Development, Brain Function, and Diseases

2006

ISBN: 978-3-527-31285-6

Bähr, M. (Ed.)

Neuroprotection

Models, Mechanisms and Therapies

2004

ISBN: 978-3-527-30816-3

Neurobiology

From Molecular Basis to Disease

Edited by

Robert A. Meyers

Volume 2

WILEY-VCH

WILEY-VCH Verlag GmbH & Co. KGaA

The Editor

Dr. Robert A. Meyers
RAMTECH LIMITED
122, Escalle Lane
Larkspur, CA 94939
USA

Library of Congress Card No.: applied for

British Library Cataloguing-in-Publication Data: A catalogue record for this book is available from the British Library.

Bibliographic information published by the Deutsche Nationalbibliothek
Die Deutsche Nationalbibliothek lists this publication in the Deutsche Nationalbibliografie; detailed bibliographic data are available on the Internet at http://dnb.d-nb.de.

Composition: Laserwords Private Ltd., Chennai, India
Printing: Betz-Druck GmbH, Darmstadt
Bookbinding: Litges & Dopf GmbH, Heppenheim

Printed in the Federal Republic of Germany
Printed on acid-free paper

ISBN 978-3-527-32293-0

Contents

Neurobiology. From Molecular Basis to Disease. Edited by Robert A. Meyers.

ISBN: 978-3-527-32293-0

Preface

This treatise on cell and molecular neurobiology and neurological diseases was compiled from a selection of key articles from the recently published *Encyclopedia of Molecular Cell Biology and Molecular Medicine* (ISBN 978-3-527-30542-1, http://www.meyers-emcbmm.de/). This volume is composed of 21 detailed articles arranged in three sections covering Cell Biology and Biochemistry of the Nervous System, Neurodevelopmental Biology and Neurological Diseases. The articles were prepared by eminent researchers from many of the major neurobiology and neurology research institutions spanning the globe, including: the University of Cambridge; MRC Social, Genetic & Developmental Psychiatry Centre, King's College London; University College London; Charing Cross Hospital; the University of Leeds; the University of Salford; the University of Antwerp; Institut de Neurobiologie de la Méditerranée, Marseille; Institute of Clinical Neurobiology, Wuerzburg; Charité, Department of Experimental Neurology, Berlin; Humboldt University; National Institute of Neuroscience, Kodaira, Tokyo; the University of Pennsylvania; the University of California, San Francisco; Wesleyan University; Temple University; the University of Washington, Seattle; Washington University School of Medicine; The State University of New Jersey; and the University of California, Irvine.

Each article begins with a concise definition of the subject and its importance, followed by the body of the article and extensive references for further reading. The references are divided into secondary references (books and review articles) and primary research papers. Each subject is presented on a first-principle basis, including detailed figures, tables and drawings. Because of the self-contained nature of each article, some overlap among articles on related topics occurs. Extensive cross-referencing is provided to help the reader expand his or her range of inquiry.

The master publication, which is the basis of the *Neurobiology* set, is the *Encyclopedia of Molecular Cell Biology and Molecular Medicine*, which is the successor and second edition of the *Encyclopedia of Molecular Biology and Molecular Medicine*, covers the molecular and cellular basis of life at a university and professional researcher level. The first edition, published in 1996–1997 was utilized in libraries around the world. This second edition is double the first edition in length and comprises the most detailed treatment of both molecular and cell biology available today. The Board and I believe that there is a serious need for this publication, even in view of the vast amount of information available on the

Neurobiology. From Molecular Basis to Disease. Edited by Robert A. Meyers.

ISBN: 978-3-527-32293-0

World Wide Web and in text books and monographs. We feel that there is no substitute for our tightly organized and integrated approach to selection of articles and authors and implementation of peer review standards for providing an authoritative single-source reference for undergraduate and graduate students, faculty, librarians and researchers in industry and government.

Our purpose is to provide a comprehensive foundation for the expanding number of molecular biologists, cell biologists, pharmacologists, biophysicists, biotechnologists, biochemists and physicians as well as for those entering the field of molecular cell biology and molecular medicine from majors or careers in physics, chemistry, mathematics, computer science and engineering. For example there is an unprecedented demand for physicists, chemists and computer scientists who will work with biologists to define the genome, proteome and interactome through experimental and computational biology.

The Board and I first divided all of molecular cell biology and molecular medicine into primary topical categories and each of these was further defined into subtopics. The following is a summary of the topics and subtopics:

- Nucleic Acids: amplification, disease genetics overview, DNA structure, evolution, general genetics, nucleic acid processes, oligonucleotides, RNA structure, RNA replication and transcription.
- Structure Determination Technologies Applicable to Biomolecules: chromatography, labeling, large structures, mapping, mass spectrometry, microscopy, magnetic resonance, sequencing, spectroscopy, x-ray diffraction.
- Proteins, Peptides and Amino Acids: analysis, enzymes, folding, mechanisms, modeling, peptides, structural genomics (proteomics), structure, types.
- Biomolecular Interactions: cell properties, charge transfer, immunology, recognition, senses.
- Molecular Cell Biology of Specific Organisms: algae, amoeba, birds, fish, insects, mammals, microbes, nematodes, parasites, plants, viruses, yeasts.
- Molecular Cell Biology of Specific Organs or Systems: excretory, lymphatic, muscular, neurobiology, reproductive, skin.
- Molecular Cell Biology of Specific Diseases: cancer, circulatory, endocrine, environmental stress, immune, infectious diseases, neurological, radiation.
- Biotechnology: applications, diagnostics, gene altered animals, bacteria and fungi, laboratory techniques, legal, materials, process engineering, nanotechnology, production of classes or specific molecules, sensors, vaccine production.
- Biochemistry: carbohydrates, chirality, energetics, enzymes, biochemical genetics, inorganics, lipids, mechanisms, metabolism, neurology, vitamins.
- Pharmacology: chemistry, disease therapy, gene therapy, general molecular medicine, synthesis, toxicology.
- Cellular Biology: developmental cell biology, diseases, dynamics, fertilization, immunology, organelles and structures, senses, structural biology, techniques.

We then selected some 340 article titles and author or author teams to cover the above topics. Each article is designed as a self-contained treatment. Each article begins with a key word section, including definitions, to assist the scientist or student who is unfamiliar with the specific subject area. The encyclopedia includes more than 3000 key words, each defined within the context of the particular scientific field covered by the article. In addition to these definitions, the glossary of basic terms found at the back of each volume, defines the most commonly used terms in molecular and cell biology. These definitions, along with the reference materials printed at the back of each volume, should allow most readers to understand articles in the encyclopedia without referring to a dictionary, textbook or other reference work.

Larkspur, March 2008

Robert A. Meyers
Editor-in-Chief

List of Contributors

Arturo Alvarez-Buylla
University of California, San Francisco, CA, USA

Darren R. Brooks
School of Environment and Life Sciences, University of Salford, Salford, UK

Tanya Button
MRC Social, Genetic & Developmental Psychiatry Centre, Institute of Psychiatry, King's College London, De Crespigny Park, London, UK

Andrew J. Copp
Institute of Child Health, University College London, London, UK

Ulrich Dirnagl
Department of Experimental Neurology (Charité) Humboldt University, Berlin, Germany

Richard L. Doty
University of Pennsylvania Medical Center, Philadelphia, PA, USA

James Eberwine
University of Pennsylvania, Philadelphia, PA, USA

Charles J. Epstein
University of California, San Francisco, CA, USA

Bonnie L. Firestein
The State University of New Jersey, Piscataway, NJ, USA

Russell G. Foster
Charing Cross Hospital, London, UK

Ilse Gantois
University of Antwerp, Antwerp, Belgium

J. Jay Gargus
University of California, Irvine, CA, USA

Alison Goate
Department of Psychiatry, Washington University School of Medicine, St. Louis, MO, USA

Stephan J. Guyenet
University of Washington, Seattle, WA, USA

Georg Haase
Institut de Neurobiologie de la Méditerranée, Marseille, France

Tatsuya Haga
Gakushuin University, Tokyo, Japan

Silva Hecimovic
Department of Psychiatry, Washington University School of Medicine, St. Louis, MO, USA

Ian A. Hope
School of Biology, University of Leeds, Leeds, UK

Takayoshi Inoue
National Institute of Neuroscience, Ogawahigashi, Kodaira, Tokyo, Japan

R. Elwyn Isaac
School of Biology, University of Leeds, Leeds, UK

Erica L. Jackson
University of California, San Francisco, CA, USA

Kimihiko Kameyama
National Institute of Advanced Industrial Science and Technology, Tsukuta, Japan

R. Frank Kooy
University of Antwerp, Antwerp, Belgium

Albert R. La Spada
University of Washington, Seattle, WA, USA

Robert P. Lane
Wesleyan University, Middletown, CT, USA

Edward P. Masler
Nematology Laboratory Plant Sciences Institute, Beltsville, MD, USA

Peter McGuffin
MRC Social, Genetic & Developmental Psychiatry Centre, Institute of Psychiatry, King's College London, De Crespigny Park, London, UK

Peter A. McNaughton
University of Cambridge, Cambridge, UK

Andreas Meisel
Department of Experimental Neurology (Charité) Humboldt University, Berlin, Germany

Ben A. Oostra
Department of Clinical Genetics, Erasmus MC, Rotterdam, The Netherlands

Stuart Peirson
Charing Cross Hospital, London, UK

Konstantin Prass
Department of Experimental Neurology (Charité) Humboldt University, Berlin, Germany

Stanley B. Prusiner
University of California, San Francisco, CA, USA

Emily Rozak
University of Pennsylvania, Philadelphia, PA, USA

Michael Sendtner
Institute of Clinical Neurobiology, Wuerzburg, Germany

Gregory S. Smutzer
Temple University, Philadelphia, PA, USA and University of Pennsylvania Medical Center, Philadelphia, PA, USA

Jennifer Spaethling
University of Pennsylvania, Philadelphia, PA, USA

Jun Wang
Department of Psychiatry, Washington University School of Medicine, St. Louis, MO, USA

David Whitmore
University College London, London, UK

Tilo Wolf
Department of Experimental Neurology (Charité) Humboldt University, Berlin, Germany

Color Plates

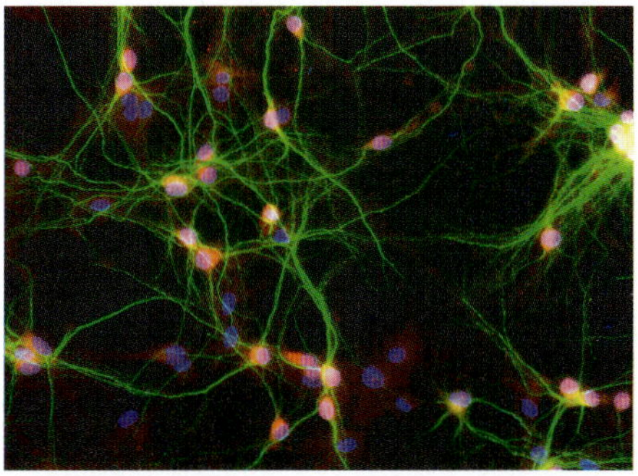

Fig. 2 (p. 12) Immunohistochemical localization of Map2 protein to the somatodendritic region of neurons – A rabbit polyclonal antibody raised against Map2 was applied to neurons of rat hippocampal neurons in primary culture. This first antibody was localized with a goat antirabbit secondary antibody that was fluorescein labeled (Green color). Green staining was observed in the fine hairlike structures that radiate from the cell soma. The nucleus is stained with DAPI. Glial cells in the primary cell culture system have a strong DAPI staining but no Map2 staining consistent with the absence of Map2 protein form glial cells.

Neurobiology. From Molecular Basis to Disease. Edited by Robert A. Meyers.

ISBN: 978-3-527-32293-0

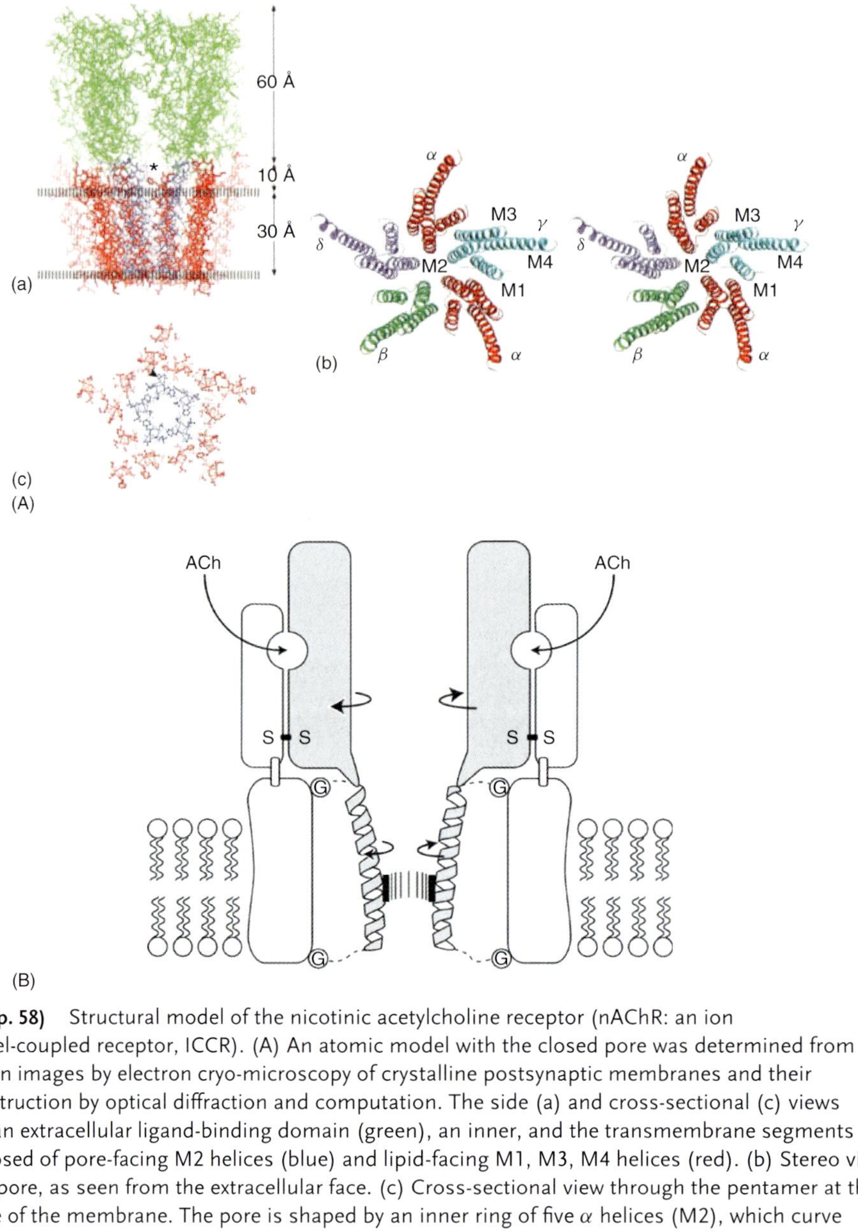

Fig. 1 (p. 58) Structural model of the nicotinic acetylcholine receptor (nAChR: an ion channel-coupled receptor, ICCR). (A) An atomic model with the closed pore was determined from electron images by electron cryo-microscopy of crystalline postsynaptic membranes and their reconstruction by optical diffraction and computation. The side (a) and cross-sectional (c) views show an extracellular ligand-binding domain (green), an inner, and the transmembrane segments composed of pore-facing M2 helices (blue) and lipid-facing M1, M3, M4 helices (red). (b) Stereo view of the pore, as seen from the extracellular face. (c) Cross-sectional view through the pentamer at the middle of the membrane. The pore is shaped by an inner ring of five α helices (M2), which curve radially to create a tapering path for ions, and an outer ring of 15 α-helices (M1, M3, M4), which coil around each other and shield the inner ring from the lipid layer. (B) A model for the gating mechanism. With binding of acetylcholine (ACh), two M2 α-helices of α-subunits rotate to opposite directions. This movement triggers the concerted movement of M2 α-helices of the other subunits resulting in the opening of the gate that consists of hydrophobic residues in the central part of five M2 segments. (From Miyazawa, A. et al. (2003) Structure and gating mechanism of the acetylcholine receptor pore, *Nature* **424**, 949–955).

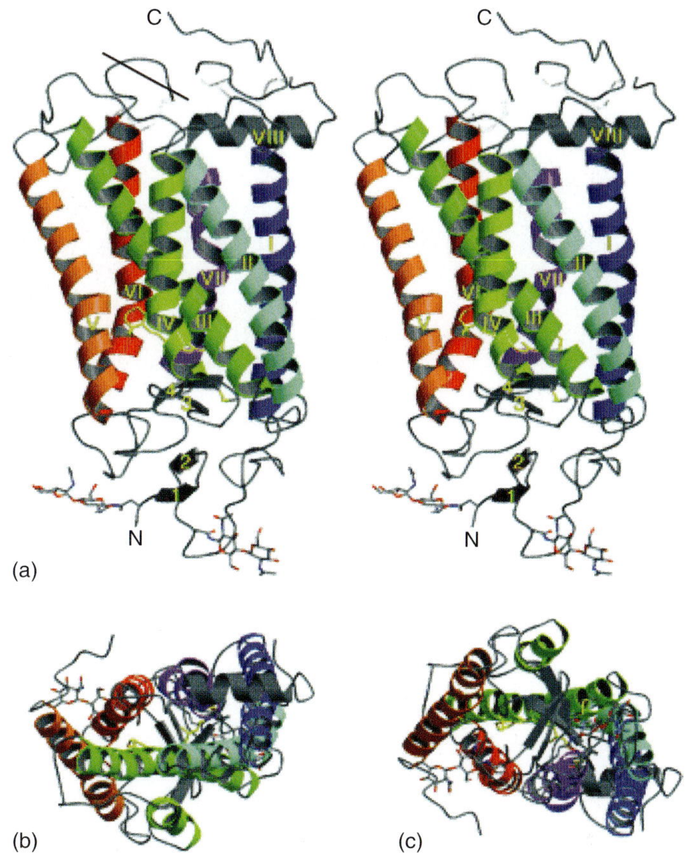

(A)

Fig. 2 (p. 67) Structures of rhodopsin and a heterotrimeric G-protein. (A) An atomic model for rhodopsin was obtained by X-ray analysis of the rhodopsin crystal. (a) A stereoview in parallel to the plane of the membrane. The upper and lower parts of this model correspond to the intra- and extracellular domains respectively. There are seven transmembrane α-helixes (I-VII), another α-helix in the C-terminal region (VIII), 11-cis-retinal (yellow) in the transmembrane region, and two β-sheets (1–2 and 3–4) and sugars in the N-terminal region. View of the receptor from the cytoplasmic (b) and intradiscal (extracellular) side (c) of the membrane. (From Palczewski, K. et al. (2000) Crystal structure of rhodopsin: A G protein-coupled receptor, *Science* **289**, 739–745). (B) Receptor-G protein interaction. (a) A structural model for the cytoplasmic surface of the M_1 muscarinic receptor. Residues important for interaction with G protein are shown in red, magenta (i3 loop), and yellow. The arrows indicates the outward movement of TMb and increased mobility of TM7 and helix 8, which are thought to accompany receptor activation. (b) A structural model for the interaction between G protein-coupled receptor (GPCR) and G-protein $\alpha\beta\gamma$ trimer. Structures are taken from crystal structures of rhodopsin and a chimera of transducin (Gt) and Gi. The N- and C-termini of the transducin α-subunit and the α4-β6 loop sequences, which are thought to be involved in the interaction with GPCR, are shown in red. The switch-II helix in the G-protein α-subunit is bound to the β-subunit when the α-subunit is bound to GDP, and is subject to conformational change and is released from the β-subunit when the α-subunit is bound with GTP. (From Lu, Z.L. et al. (2002) Seven-transmembrane receptors: crystals clarify, *Trends Pharmacol. Sci.* **23**, 140–146).

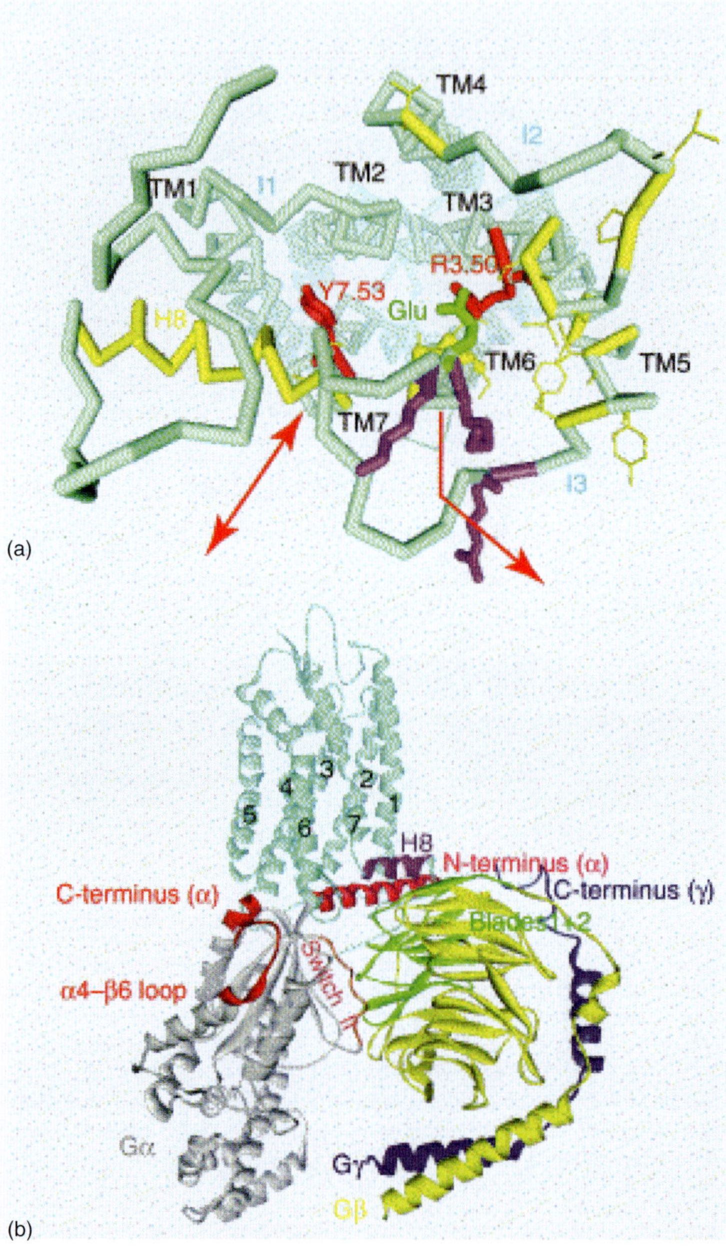

Fig. 2 (*Continued*)

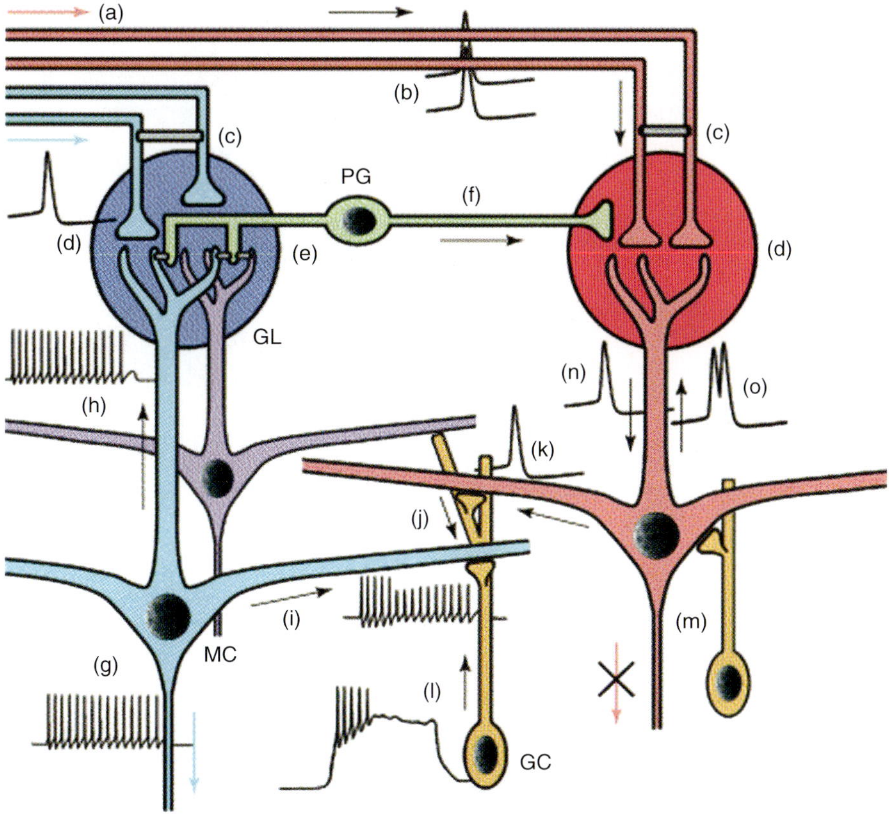

Fig. 5 (p. 178) Electrical signaling mechanisms in the olfactory bulb. (a) Action potentials propagate from olfactory receptor cells to glomeruli (GL), along axons segregated according to expressed odorant receptor genes (e.g. blue or red). Spikes may already be phase locked presynaptically (b) by ephaptic (point of contact where two or more nerve cell processes touch but do not form a typical synaptic contact), and (c) electrical coupling in nerve fascicles. (d) When action potentials invade olfactory nerve terminals, glutamate is released onto apical tufts of mitral cells (MC) and dendrites of periglomerular cells (PG). Tufts that share a glomerulus are coupled electrically by gap junctions, (e) perhaps through periglomerular cells, (f) which may relay timing information among glomeruli. (g) Glutamate evoked excitatory postsynaptic potentials in the tuft can depolarize the mitral soma and axon initial segment, which fires spike trains. (h) Spike output backpropagates up the primary dendrite to trigger glutamate release from tufts, and (e, f) coordinate firing in other mistral cells by electrical coupling. (I) Spikes also propagate laterally into secondary dendrites, in which they are subject to local inhibition by GABA release from granule cells (GC). (j) Lateral inhibition may spread locally among mitral cells through domains in the granule cell apical dendrite. (k) Domains are depolarized by glutamate released from other mitral cell dendrites with propagating spikes. (l) Strong depolarization drives ramped somatic spiking of granule cells, and these spikes back-propagate and release GABA throughout the granule dendrite. (m) Granule cell inhibition of the mitral soma can shift spike initiation to (n) the distal primary dendrite. (o) It can also veto mitral cell output, or generate compound waveforms composed of orthorhombic and back-propagating spikes. Reprinted from Lowe, G. (2003) Electrical signaling in the olfactory bulb. *Curr. Opin. Neurobiol.* **13**, 476–481. Copyright 2003 with permission from Elsevier.

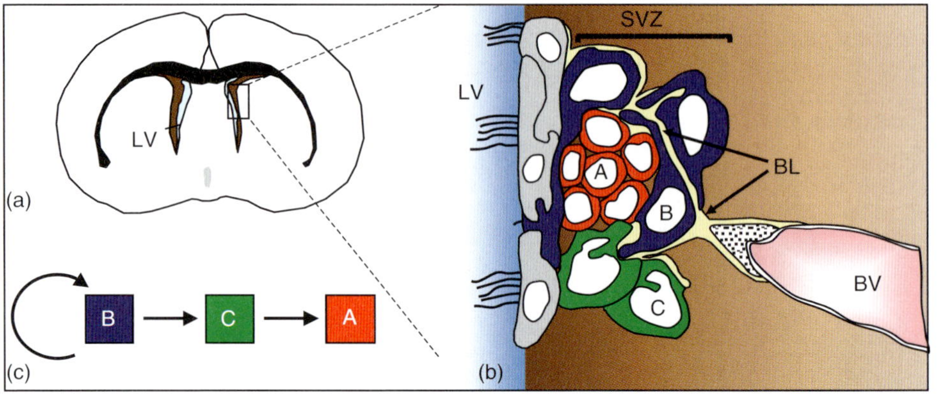

Fig. 6 (p. 180) Adult SVZ Neurogenesis in Rodents. (a) Coronal section through the adult mouse brain. Light blue shows the lateral ventricle (LV) space filled with cerebrospinal fluid. Boxed area is shown enlarged in B. (b) Architecture of the SVZ. B cells (dark blue) are the astrocytes that are stem cells of the SVZ, and also serve as niche cells. Some of the B cells contact the ventricle lumen and contain a single cilium (shown). C cells (green) are rapidly dividing transit-amplifying cells derived from the B cells. C cells give rise to A cells (red), neuroblasts that migrate to the olfactory bulb, where they become local interneurons. A blood vessel (BV, pink) is shown with a perivascular macrophage (dotted fill); a basal lamina (BL, yellow) extends from the BV and interdigitates extensively with the SVZ cells. Ciliated ependymal cells (gray) line the ventricle walls and have been shown to produce the organizer protein Noggin, which is important for this niche. Noggin, BMPs, Sonic hedgehog, Notch, TGFβ, ephrins, ephrin receptors, and vascular endothelial growth factor play roles in regulation of this neurogenic niche. (c) SVZ lineage. C cells can behave as stem cells under the influence of excess EGF signaling. Reprinted from Alvarez-Buylla, A., Lim, D.A. (2004) *Neuron*, **41**, 683–686. Copyright 2004 with permission from Elsevier.

(a)

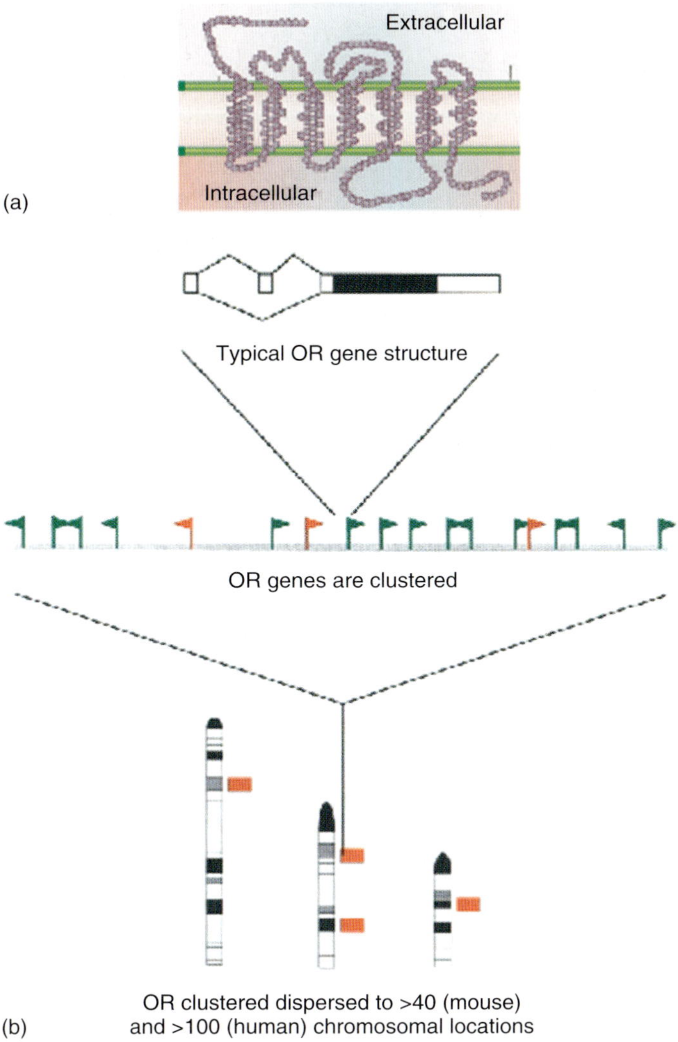

(b)

Fig. 9 (p. 188) OR genes. (a) Schematic drawing depicting the stereotyped structure of a GPCR protein, with seven membrane-spanning α-helices. Figure from Mombaerts, 2001. (b) Top: Gene structure and genome organization of OR genes. A typical OR gene consists of upstream noncoding exons that are commonly alternatively spliced, a single coding exon of approximately 1 kb (black), and a short 3′ untranslated region. Middle: OR genes are clustered in groups of 1 to over 100 genes (green arrows) and pseudogenes (red arrows) in both transcriptional orientations. Bottom: OR clusters are dispersed to more than 40 mouse and 100 human chromosomal locations. From Young, J.M., Trask, B.J. (2002) The sense of smell: genomics of vertebrate odorant receptors. *Hum. Mol. Genet.* **11**, 1153–60, by permission of Oxford University Press.

Fig. 10 (p. 190) Predicted odorant-binding pockets of OR proteins. (a) Schematic drawing depicting an overhead view of the seven transmembrane-spanning barrels of an OR protein modeled against the rhodopsin GPCR. Residues conserved among all GPCRs are shown in open circles. Colored squares and circles represent positions of conserved and variable residues, respectively, in OR proteins. Residues that align with ligand contact residues in other GPCRs are colored green and residues that do not align with such residues are colored red. The residues in each helix are numbered separately, according to the predicted TM boundaries. Hypervariable residues (putative odorant-binding residues) are indicated by asterisks. Area II denotes a hypervariable pocket that corresponds to the ligand binding pocket in other GPCRs Figure modified from Pilpel, Y., Lancet, D. (1999). The variable and conserved interfaces of modeled olfactory receptor proteins. *Prot. Sci.* **8**, 969–77. (b) Model depicting the predicted binding pocket for the hexanol ligand (purple) to the S25 OR protein (side view). Each TM and inter-TM loop is labeled. The membrane is represented in yellow. This computed model of the S25 OR protein atoms successfully predicts that relative affinities to a panel of odorants, including hexanol, which is predicted to interact with residues in TMs 3, 5, and 6. Figure modified from Floriano, W.B., Vaidehi, N., Goddard, W.A., Singer, M.S., Shepherd, G.M. (2000) Molecular mechanisms underlying differential odor responses of a mouse olfactory receptor. *Proc. Natl. Acad. Sci. U. S. A.* **97**, 10712–16. Copyright 2000 National Academy of Sciences, U.S.A.

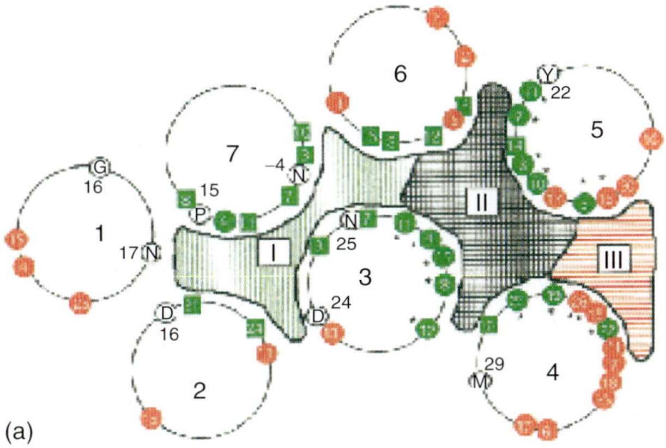
6
Y
22
5
7
-4
N
G
16
15
P
II
1
17
N
N
I
25
III
3
D
24
16
D
29
4
M
2

(a)

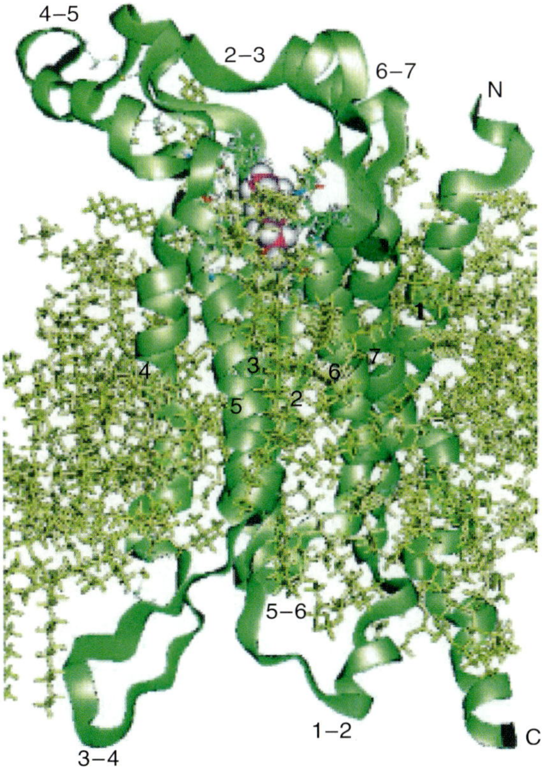
4–5
2–3
6–7
N
1
3
7
4
6
5
2
5–6
1–2
C
3–4

(b)

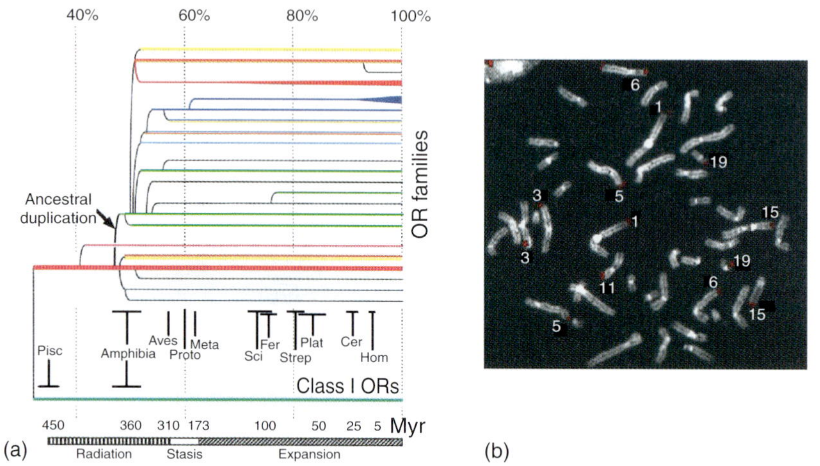

Fig. 13 (p. 201) OR evolution. (a) Application of a molecular clock based on subfamily sequence divergences (% divergence indicated above the tree) to infer periods of OR radiation and expansion during vertebrate evolution. According to this methodology, OR subfamily diversity arose predominantly during amphibian radiation, beginning ~370 million years ago (MYA). So-called "Class I" OR genes are described as *fish-like*, suggesting that these ORs are remnants of the preamphibian radiation. Taxonomy: Pisc = fish; Amphibia = frogs, salamander; Aves = chick; Proto = platypus; Meta = koala; Sci = marmot, mouse, rat; Fer: pig, dolphin, dog; Strep = lemurs, squirrel monkeys, Plat = marmoset; Cer = baboon, macaques; Hom = chimpanzee, gorilla. Figure modified from Glusman, G., Yanai, I., Rubin, I., Lancet, D. (2001) The complete human olfactory subgenome. *Genome Res.* **11**, 685–702. Copyright 2001 Cold Spring Harbor Laboratory Press. (b) Fluorescence *in situ* hybridization (FISH) on human chromosomes using a labeled probe (red) that recognizes a human subtelomeric OR sequence that is dispersed on seven human chromosomes (1, 3, 5, 6, 11, 15, 19) due to frequent duplication and translocation. The copy number and chromosomal distribution pattern of this OR locus is polymorphic in the human population, suggesting very recent duplications. Figure from Trask, B.J., Friedman, C., Martin-Gallardo, A., Rowen, L., Akinbami, C., Blankenship, J., Collins, C., Giorgi, D., Iadonato, S., Johnson, F., Kuo, W.-L., Massa, H., Morrish, T., Naylor, S., Nguyen, O.T.-H., Rouquier, S., Smith, T., Wong, D.J., Youngblom, J., van den Engh, G. (1998) Members of the olfactory receptor gene family are contained in large blocks of DNA duplicated polymorphically near the ends of human chromosomes. *Hum. Mol. Genet.* **7**, 13–26, by permission of Oxford University Press. (c) OR gene tree depicting evolutionary relationships between mouse (red) and human (blue) OR genes. Although the human OR repertoire is less than half the size of the mouse repertoire, the diversity of the human OR repertoire is comparable to that of the mouse repertoire. Therefore, it is expected that mice have better discriminatory power, but not necessarily a broader range of olfaction. Figure from Young et al., 2002 by permission of Oxford University Press. (d) Gene tree depicting species-specific expansion of a mouse and rat V1R subfamily. Most of the duplications occurred over a short period of evolution just after mouse and rat diverged. Figure modified from Lane, R.P., Young, J., Newman, T., Trask, B.J. (2004) Species specificity in rodent pheromone receptor repertoires. *Genome Res.* **14**, 603–608. Copyright 2002 Cold Spring Harbor Laboratory Press.

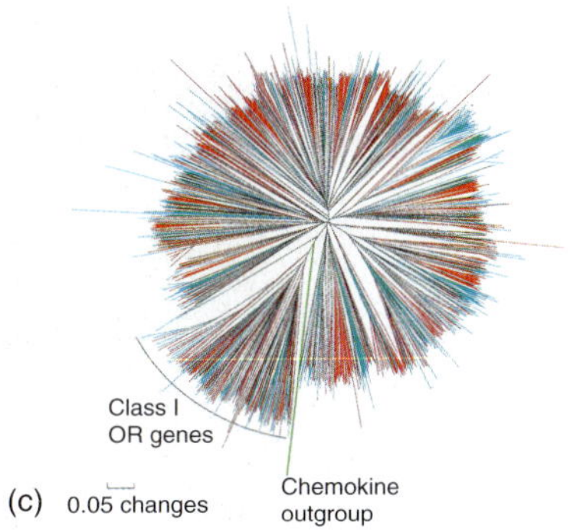

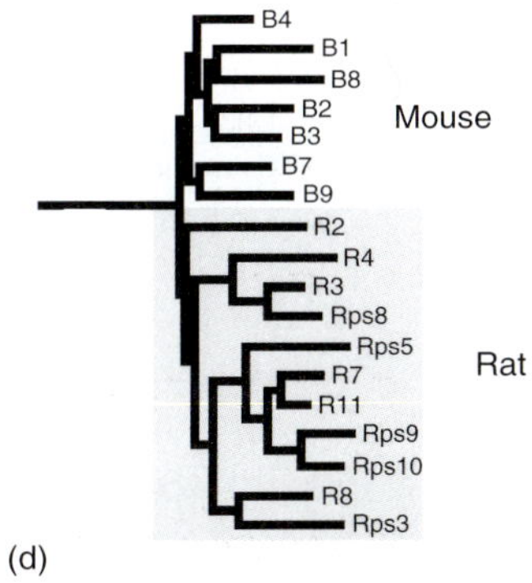

Fig. 13 (*Continued*)

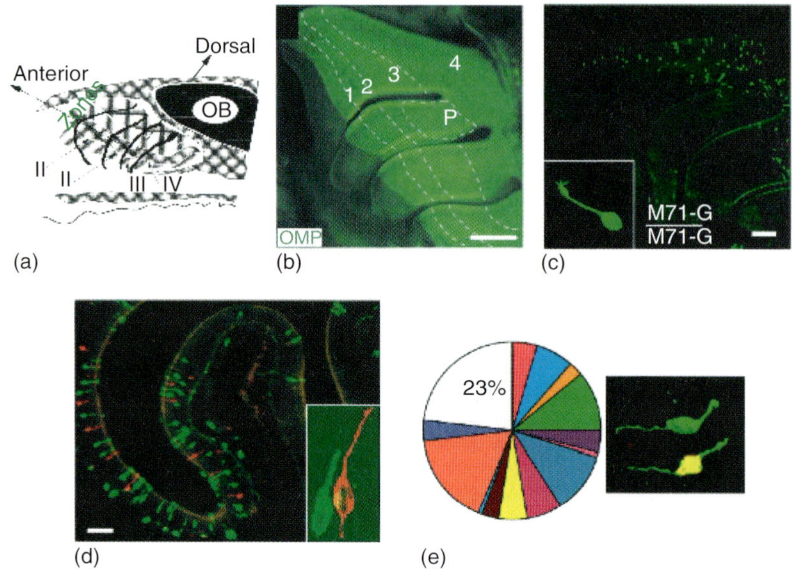

Fig. 14 (p. 205) OR expression patterns. (a) Schematic of the olfactory epithelium depicting four endoturbinate bones (II, II′, III, IV) and four topological OR expression zones along the dorsal axis (hatchings). Figure modified from Scott, J.W., Shannon, D.E., Charpentier, J., Davis, L.M., Kaplan, C. (1997) Spatially organized response zones in rat olfactory epithelium. *J. Neurophysiol.* **77**, 1950–1962. (b) Fully labeled olfactory epithelium using a construct that reports olfactory marker protein, OMP, expression within olfactory sensory neurons. The four topological zones are labeled (1–4); the patch, P, is a medial zone where some OR genes are expressed. Scale bar is 500 μm. Reprinted from Neuron, Vol. 4, Vassalli, A., Rothman, A., Feinstein, P., Zapotocky, M., Mombaerts, P. (2002). Minigenes impart odorant receptor-specific axon guidance in the olfactory bulb, pages 681–696. Copyright 2002, with permission from Elsevier. (c) Punctate expression of the M71 olfactory receptor gene using a construct in which M71 activation is reported by green fluorescence protein, GFP. Inset, fluorescence image of a single M71 marked olfactory sensory neuron. Scale bar is 250 μm. Figure from Bozza, T., Feinstein, P., Zheng, C., Mombaerts, P. (2002) Odorant receptor expression defines functional units in the mouse olfactory system. *J. Neurosci.* **22**, 3033–3043. Copyright 2002 by the Society for Neuroscience. (d) Coronal section showing mutually exclusive expression of a tagged MOR28 transgene (red) and the endogenous MOR28, tagged with GFP (green). Inset, higher magnification showing individual sensory neurons expressing one color or the other, but not both. Scale bar is 50 μm. Modified from Serizawa, S., Ishii, T., Nakatani, H., Tsuboi, A., Nagawa, F., Asano, M., Sudo, K., Sakagami, J., Sakano, H., Ijiri, T., Matsuda, Y., Suzuki, M., Yamanori, T., Iwakura, Y., Sakano, H. (2000) Mutually exclusive expression of odorant receptor transgenes. *Nat. Neurosci.* **3**, 687–693. Reprinted with permission by Nature Publishing Group. (e) Expression of an alternative OR gene in cells that select a GFP-marked transgene in which the MOR28 coding region is deleted. *In situ* hybridization using non-MOR28 probes (red) colabel the same sensory neuron with the MOR28-deleted transgene (green), producing a merged yellow color (coexpression of the red and green markers). The pie chart summarizes coexpression data for a panel of 13 different OR probes (different colors in chart); ~77% of the cells labeled with GFP (expressing the MOR28-deleted transgene) exhibit alternative expression of one of the OR probes tested. These data suggest that failure to express a functional OR gene results in the neuron choosing another OR locus to express. Reprinted with permissions from Serizawa, S., Miyamichi, K., Nakatani, H., Suzuki, M., Saito, M., Yoshihara, Y., Sakano, H. (2003) Negative feedback regulation ensures the one receptor-one olfactory neuron rule in mouse. *Science* **302**, 2088–2094. Copyright 2003 AAAS. Permission from AAAS required for all other uses.

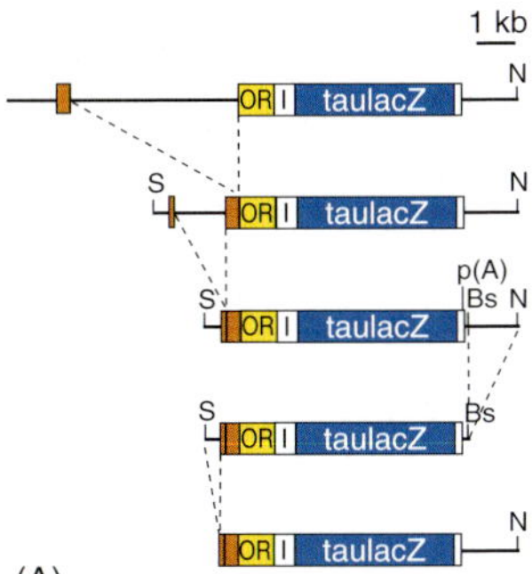

Expr. OSNs	Zone 4	Ecto. zone	Mut. excl.	Co-conv.	Ecto. glom.
+	+	–	+	+	–
+	+	–	+	+	–
+	+	+[b]	+	+	+
+	+	+	+	+	+
–[a]	NA	NA	NA	NA	NA

(A)

Fig. 17 (p. 215) Local and distant OR regulatory sequences. (A) mOR23 minigene constructs and resulting expression patterns. In each case, the OR coding region (yellow) is constructed with a downstream internal ribosome entry sequence (i) and reporter gene (*tauLacZ*). The wild-type construct (top) contains 400 bp upstream of the transcription start site, the two 5′ untranslated exons (orange) and introns, and >1 kb 3′ noncoding region. This construct reports the wild-type expression pattern of the mOR23 gene in olfactory sensory neurons. The table reports expression patterns for various deleted constructs; that is, whether the deleted transgene is expressed at detectable levels in olfactory sensory neurons (OSNs), whether its expression is within epithelial zone 4 (the normal zone of expression for mOR23), whether it expresses ectopically outside zone 4 (Ecto. zone), whether the mOR23 transgene exhibits mutually exclusive (Mut. excl.) expression with the endogenous mOR23 gene, whether the OSNs expressing the construct exhibit convergence onto the endogenous mOR23 glomerulus in the olfactory bulb (Co-conv.), and/or whether the OSNs expressing the construct extend axons to multiple glomeruli (+glom.). Deletion of the 400 bp just upstream of the transcription start site of mOR23 abolishes expression of the transgene (bottom construct, a). Deletion of 5′ introns results in transgene expression outside zone 4 (b). Deletion of sequences 3′ to the coding region do not seem to have an effect. Figure modified from Neuron, Vol. 4, Vassalli, A., Rothman, A., Feinstein, P., Zapotocky, M., Mombaerts, P. (2002) Minigenes impart odorant receptor-specific axon guidance in the olfactory bulb, pages 681–696. Copyright 2002, with permission from Elsevier. (B) Expression of OR genes from the MOR28 locus on mouse chromosome 14 requires a distant enhancer, the "H region." Various truncated constructs in which the H region is missing do not report MOR28 expression. The H region is situated ~75 kb upstream of the *MOR28* gene and between the MOR28 OR locus and the T-cell receptor (TCR) Vα gene segments. Inset, dot matrix plot showing conservation of the H region between mouse and human (diagonal trace within box). Reprinted with permission from Serizawa, S., Miyamichi, K., Nakatani, H., Suzuki, M., Saito, M., Yoshihara, Y., Sakano, H. (2003) Negative feedback regulation ensures the one receptor-one olfactory neuron rule in mouse. *Science* **302**, 2088–2094. Copyright 2003 AAAS.

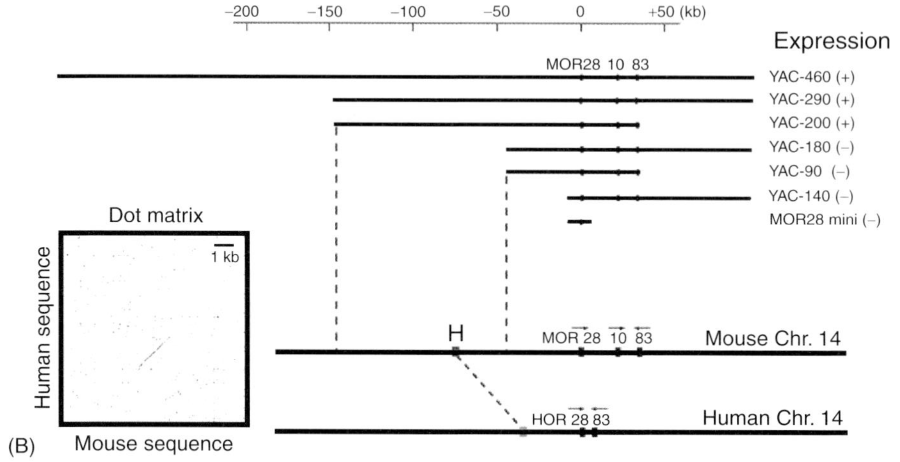

Fig. 17 (*Continued*)

Fig. 1 (p. 280) Gastrulation and neural induction in various vertebrate species. (a) Gastrulation and neural induction in an amphibian embryo. During gastrulation, bottle cells are observed around the blastpore, allowing the involution of cells (a dotted arrow). Amphibian neural induction events appear to be very dependent upon BMP (bone morphogenetic protein) signaling inhibition, in which BMP antagonists such as Chordin, Noggin, Cerberus, Follistatin, and so on secreted from the organizer play crucial roles. Regarding BMP signaling machineries, see Fig. 2(a) as BMP is a member of the TGF-β superfamily. (b) Gastrulation and neural induction in chicken embryo. During gastrulation, mesodermal cells detach from the epiblast layer and emigrate anterior-laterally (dotted arrows), forming the primitive streak. In the latest model, chicken neural tissue can be formed by FGF and Wnt signals earlier than gastrulation: At the blastula stage, only lateral epiblast (green) expresses Wnt3A/8C, demarcating medial epiblast as the prospective neural tissue (light pink region). The Wnt receptor, Frizzled8, is ubiquitously expressed by all epiblasts at this stage (regarding the Wnt signaling pathway, see Fig. 2(b)). a, anterior; N, neural tissue; n, future neural tissue, p, posterior. (c) Gastrulation, neural induction, and early anterior–posterior axis formation during mouse development. Blue arrows indicate the crucial movement of cells in the endodermal layer, determining anterior characters of the neuroectoderm. Expression of the secreted molecule, Cripto, is dynamically regulated at earlier stages, and this is thought to be critical to initiate A–P patterning events, including the endodermal cell movement as well as primitive-streak formation (purple region). During gastrulation, mesodermal cells detach from the epiblast layer (light green region) and emigrate just like those in the chicken embryo (dotted arrows). a, anterior; AVE, anterior visceral endoderm (dark blue region); E, embryonic day; EGO, early gastrula organizer (red region); ne, neuroectoderm (pink region); p, posterior; VE, visceral endoderm (light blue region). The light gray shading indicates Cripto expression, and the axial mesoderm is colored by orange.

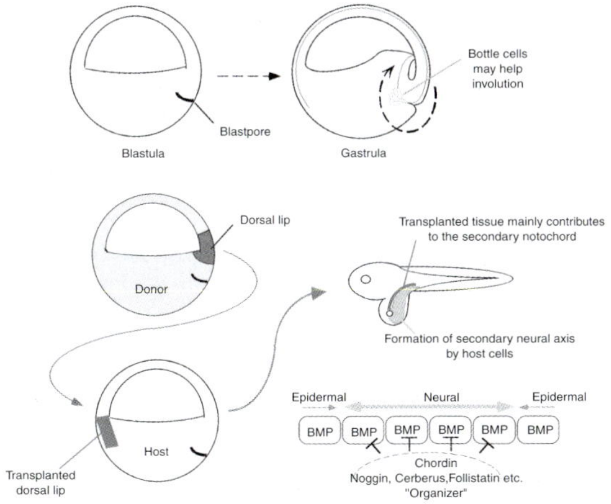

(a) Amphibian

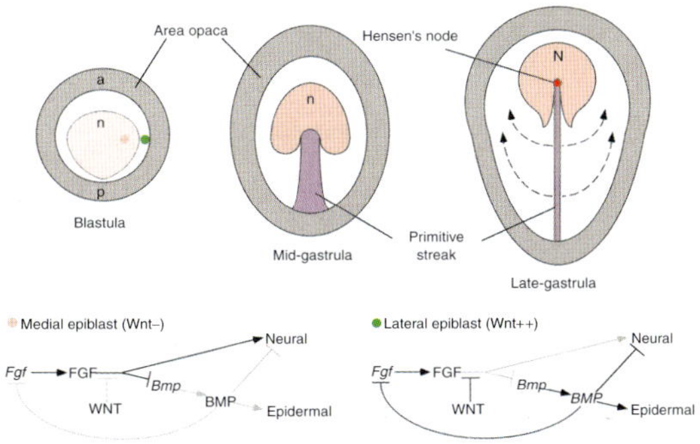

(b) Chicken

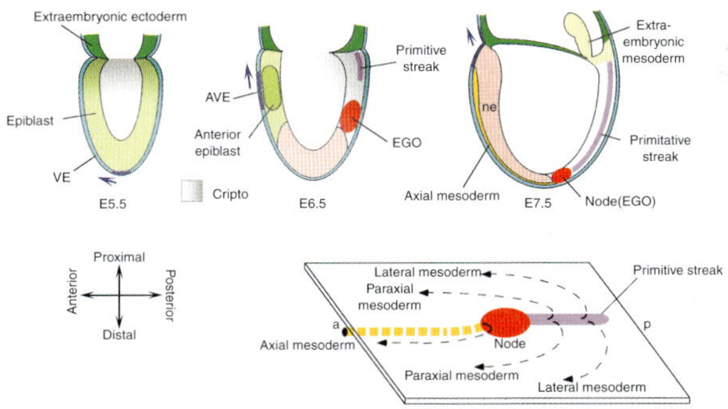

(c) Mouse

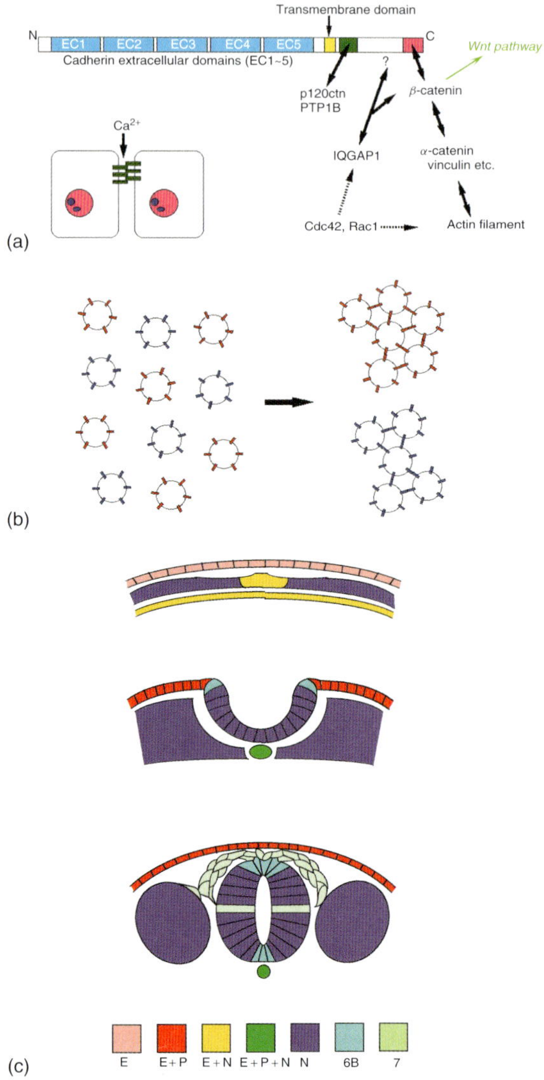

Fig. 5 (p. 285) Roles of cadherin cell adhesion molecules in embryogenesis (a) Cadherin is a transmembrane protein that confers rigid adhesiveness to cells in a Ca^{2+} dependent manner. Several molecules are known to interact with cadherin at the cytoplasmic domain, and they are important to tether cadherins to the actin cytoskeltons. Cadherins might be involved in the Wnt signaling pathway, as beta-catenin required for cadherin-mediated cell adhesion, is also a crucial element of Wnt signaling (Fig. 2b). (b) Cells with a particular subclass of cadherin can only adhere to cells with the same subclass of cadherin: when cells with different cadherin subclasses (red or blue) are mixed, they sort out from each other in *in vitro* aggregation assays, although some pairs of cadherins show heterotypic bindings. (c) During chicken neurulation, the expression pattern of cadherins is dynamically regulated. Note that the changes of the cadherin expression patterns are tightly linked to morphogenetic events, and it has been shown that overexpression of cadherins perturbs the processes of neurulation as well as neural crest emigrations.

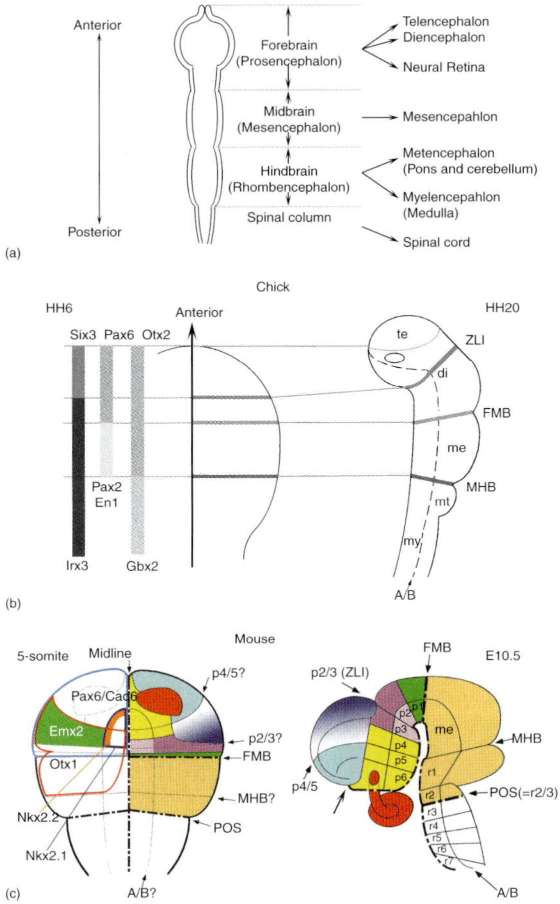

Fig. 6 (p. 288) A–P patterning of the neural tube/plate. (a) Anterior part of the neural/plate becomes segmented during development, generating brain regions with distinct structure and functions, (b) Early patterning of the chicken neural plate by the differential expression of transcription factors. Gene expression boundaries eventually form the anatomically distinguishable territories in the brain. A/B, alar/basal boundary; di, diencephalon; FMB, forebrain/midbrain boundary; HH, Hamburger Hamilton stages; me, mesencephalon; MHB, midbrain/hindbrain boundary; mt, metencephalon; my, myelencephalon; te, telencephalon; ZLI, zona limitans interthalamica. (c) A fate-mapping result of the mouse neural plate. The area painted by a given color on the right side of neural plate at the 5-somite stage (left) produces the brain region of the same color at E10.5 (right). Faint lines indicate prospective boundaries of brain regions. Cell lineage restricted compartment boundaries are indicated by bold dashed lines. Note that several gene expression boundaries demarcated by different colors on the left side of the neural plate coincide well with prospective boundaries of brain regions: blue line; Pax6/cadherin-6 expression boundary, red line; Otx1 expression boundary, orange line; Nkx2.2 expression boundary, purple line; Nkx2.1 expression boundary, and green area; Emx2 mRNA positive region in the neural plate. A/B; alar/basal boundary, FMB; forebrain/midbrain boundary, me; mesencephalon, MHB; midbrain/hindbrain boundary, p; prosomeres, POS; preotic sulcus, r; rhombomeres, ZLI; zona limitans interthalamica. An arrow indicates the anterior most part of the brain at the embryonic day (E) 10.5.

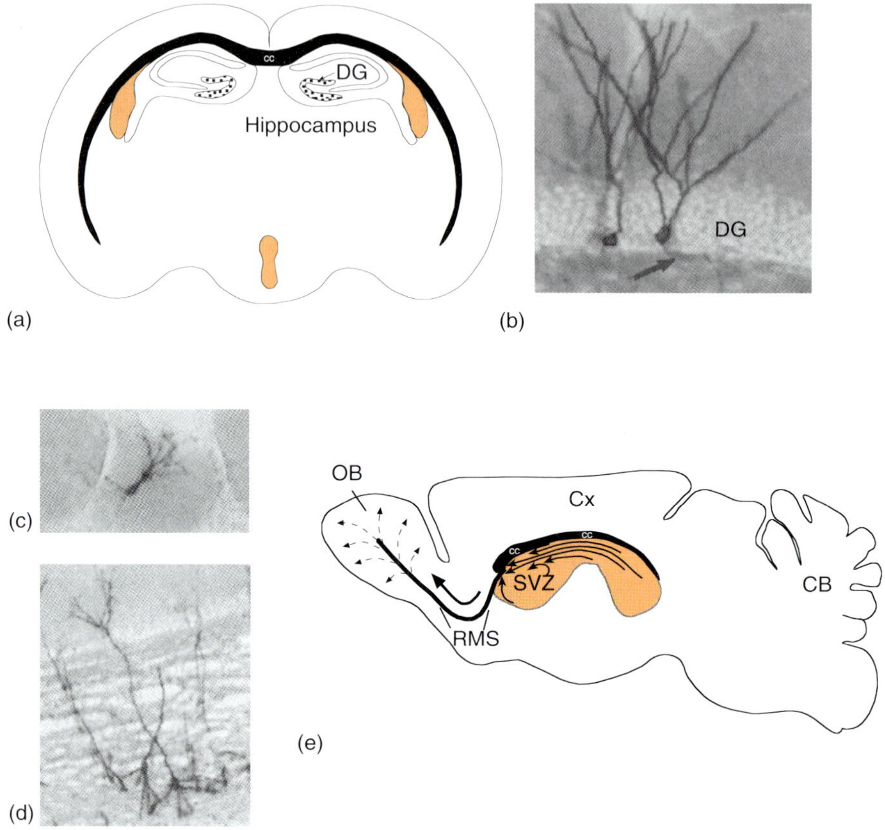

Fig. 1 (p. 326) Neurogenesis occurs in two regions of the adult brain. (a) Schematic cross-sectional view of the adult mouse brain at the level of the hippocampus. Retrovirally labeled progenitors in the subgranular layer of the dentate gyrus (DG) give rise to new neurons in the granule cell layer (b). Newly generated neurons, periglomerular cells (c) and granule cells (d) in the olfactory bulb are derived from labeled progenitors in the subventricular zone (SVZ). (e) The SVZ-OB system. Schematic sagittal view of the adult mouse brain with the OB to the left and cerebellum (CB) to the right. New neurons are produced throughout the SVZ (orange). The young neurons become aligned into long chains that form a series of interconnected paths throughout the SVZ. Once the chains reach the anterior SVZ, they join the rostral migratory stream (RMS), which leads the young neurons into the core of the olfactory bulb. Within the OB, cells disperse radially (dashed arrows) as individual cells and complete their differentiation into periglomerular cells (c) and granule cells (d). CX, Cortex; cc, corpus callosum.

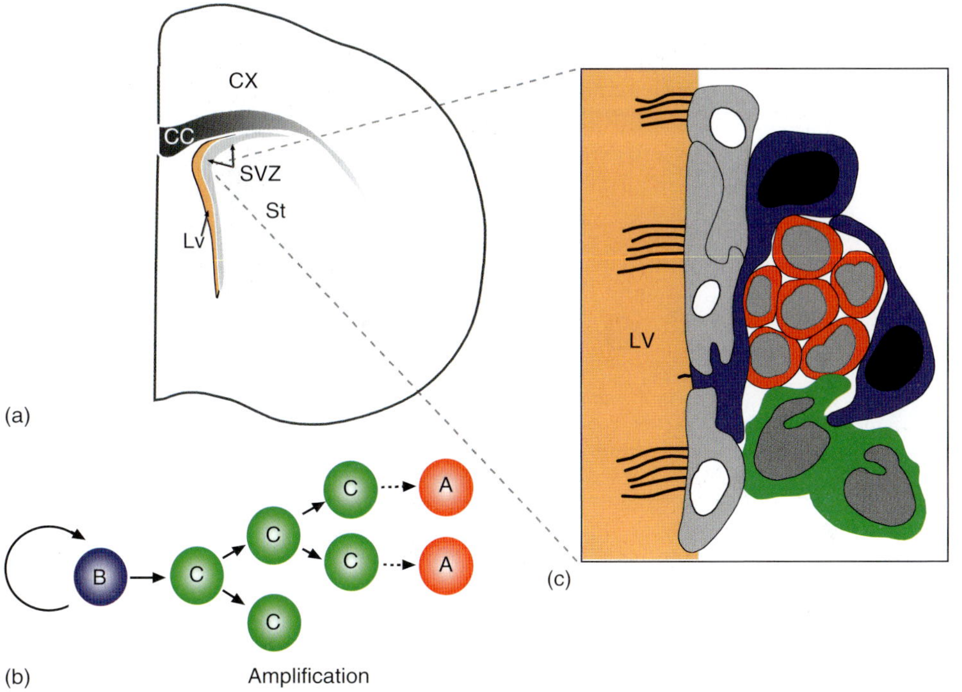

Fig. 2 (p. 331) Neurogenesis in the adult SVZ. (a) Cross-section of the adult mouse brain and (b) schematic illustration of SVZ composition and architecture; (c) expanded region. The SVZ lies adjacent to the lateral wall of the lateral ventricle (LV, tan). The chains of neuroblasts (A cells, red) destined for the olfactory bulb migrate as chains through glial tunnels formed by the processes of slowly dividing astrocytes (B cells, blue). Focal clusters of rapidly dividing precursors (C cells, green) are found scattered along the network of chains. Multiciliated ependymal cells line the wall of the ventricle (gray). Occasionally, an SVZ astrocyte extends a process with a single cilium to contact the ventricle. The lineage of olfactory bulb neurons has been described. SVZ astrocytes (B cells) act as neural stem cells in this region and divide to give rise to the rapidly dividing precursors (C cells) that in turn generate the neuroblasts that migrate to the olfactory bulb.

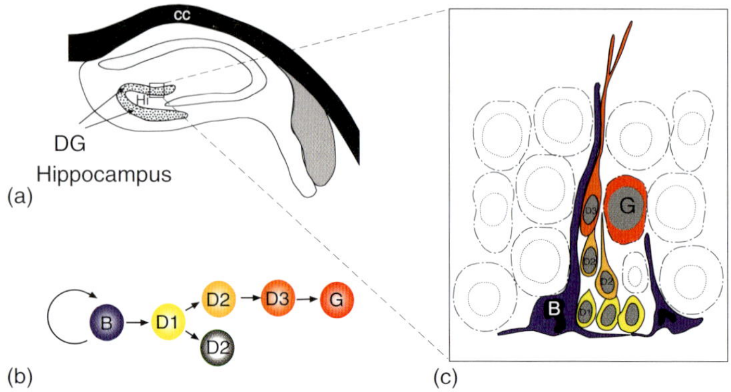

Fig. 3 (p. 333) Neurogenesis in the adult SGZ. (a) Cross-section of the adult mouse brain at the level of the hippocampus and (b) schematic illustration of composition and architecture of the SGZ and granule cell layer; (c) expanded region. SGZ astrocytes (B cells) have long radial processes that penetrate the granule cell layer and short tangential ones that run parallel to the SGZ. Dividing B cells generate immature, dark cells (D cells) that divide and mature into new granule neurons. D cells have been classified into three subtypes, D1, D2, and D3 that correspond to different stages in the maturation of the new neurons. The SGZ astrocytes (Blue cytoplasm, black nucleus); D cells (yellow or orange cytoplasm, gray nucleus), G: mature granule neuron (dotted, not filled).

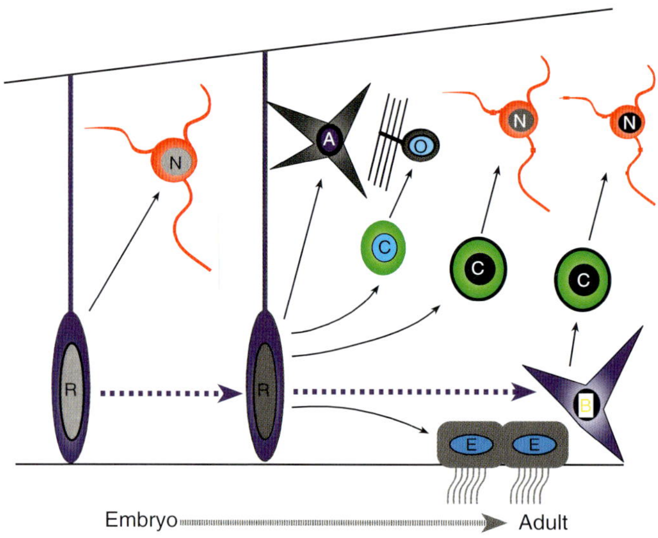

Fig. 4 (p. 338) Development of neural stem cells from the embryo to the adult. Primary progenitors in the embryo correspond to radial glia (R) in blue. These cells produce different generations of neurons (N, red) during development either directly, or indirectly through transit amplifying progenitors (C cells, green). Radial glia also generate oligodendrocytes (O, black) and ependymal cells (E, gray). It is not known whether single radial glia generate these different cell types or whether radial glia are lineage-restricted *in vivo*. Radial glia also turn into parenchymal astrocytes (A, black-blue) and germinal region astrocytes (B, blue-black). The latter continue to divide in the adult brain and generate neurons (N) and oligodendrocytes (not shown) through transit amplifying cells (C)]. The neural stem cell lineage is contained within what was classically considered an astrocytic lineage (Blue cells and arrows).

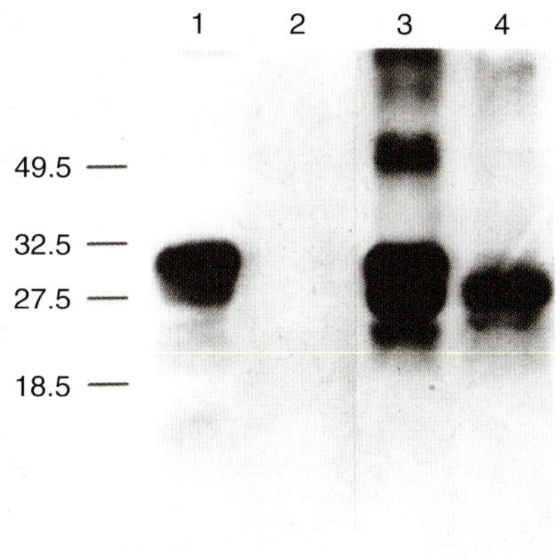

(a)

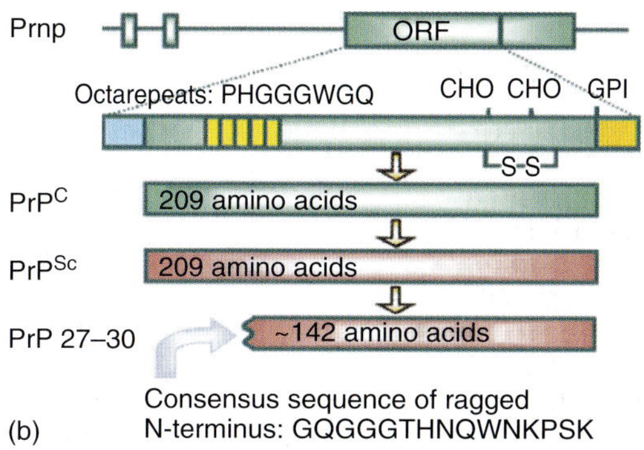

(b)

Fig. 1 (p. 408) Prion protein isoforms. Western immunoblot of brain homogenates from uninfected (lanes 1 and 2) and prion-infected (lanes 3 and 4) Syrian hamsters. Samples in lanes 2 and 4 were digested with 50 μg mL^{-1} of proteinase K for 30 min at 37 °C. PrP^C in lanes 2 and 4 was completely hydrolyzed under these conditions, whereas approximately 67 amino acids were digested from the N-terminus of PrP^{Sc} to generate PrP 27–30. After polyacrylamide gel electrophoresis (PAGE) and electrotransfer, the blot was developed with anti-SHaPrP R073 polyclonal rabbit antiserum. Molecular weight markers are depicted in kilodaltons (kDa). (b) Bar diagram of the SHaPrP gene that encodes a protein of 254 amino acids. After processing of the N- and C-termini, both PrP^C (green) and PrP^{Sc} (red) consist of 209 residues. After limited proteolysis, the N-terminus of PrP^{Sc} is truncated to form PrP 27–30, which is composed of approximately 142 amino acids, the N-terminal sequence of which was determined by Edman degradation.

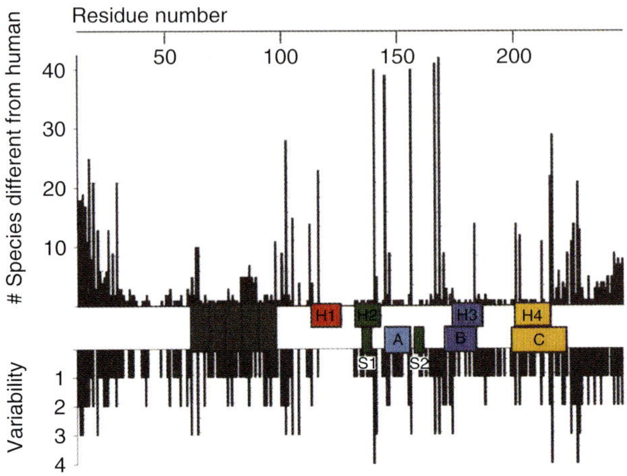

Fig. 5 (p. 420) Species variations and mutations of the prion protein gene. The *x*-axis represents the human PrP sequence, with the five octarepeats and H1–H4 regions of putative secondary structure shown as well as the three α-helices A, B, and C and the two β-strands S1 and S2 as determined by NMR. Vertical bars above the axis indicate the number of species that differ from the human sequence at each position. Below the axis, the length of the bars indicates the number of alternative amino acids at each position in the alignment. Data were compiled by Paul Bamborough and Fred E. Cohen.

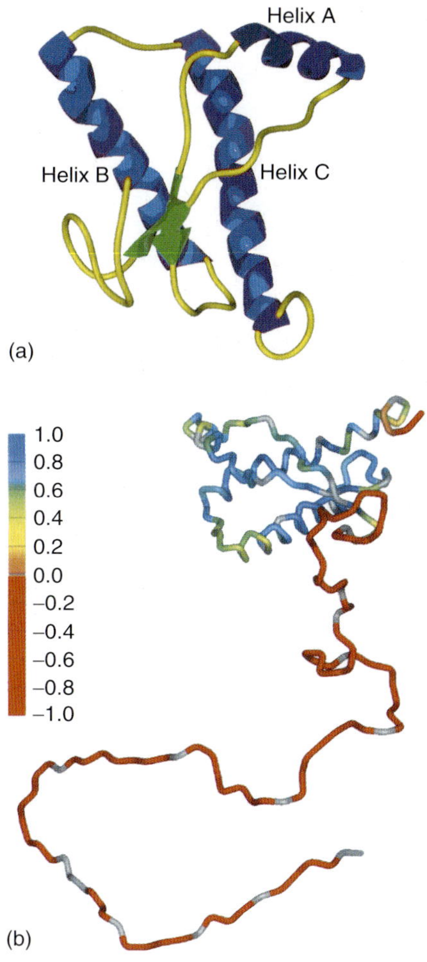

Fig. 7 (p. 424) Structures of PrP^C. (a) NMR structure of Syrian hamster (SHa) recombinant (rec) PrP(90–231). Presumably, the structure of the α-helical form of recPrP(90–231) resembles that of PrP^C. recPrP(90–231) is viewed from the interface where PrP^{Sc} is thought to bind to PrP^C. The color scheme is: α-helices A (residues 144–157), B (172–193), and C (200–227) in blue; loops in yellow; residues 129–134 encompassing strand S1 and residues 159–165 encompassing strand S2 in green. (b) Schematic diagram showing the flexibility of the polypeptide chain for PrP(29–231). The structure of the portion of the protein representing residues 90–231 was taken from the coordinates of PrP(90–231). The remainder of the sequence was hand-built for illustration purposes only. The color scale corresponds to the heteronuclear $[^1H]$-^{15}N NOE data: red for the lowest (most negative) values, where the polypeptide is most flexible, to blue for the highest (most positive) values in the most structured and rigid regions of the protein.

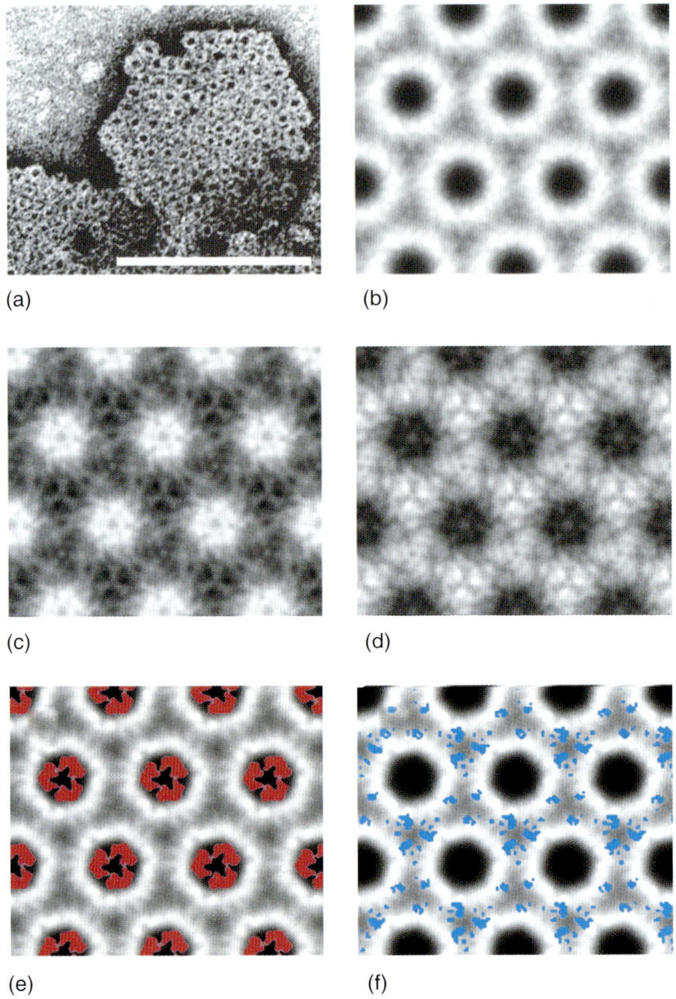

Fig. 8 (p. 426) Two-dimensional crystals of PrP^{Sc}. (a) A 2D crystal of $PrP^{Sc}106$ stained with uranyl acetate. Bar = 100 nm. (b) Image processing result after correlation-mapping and averaging followed by crystallographic averaging. (c and d) Subtraction maps between the averages of PrP 27–30 and $PrP^{Sc}106$ (panel B). (c) $PrP^{Sc}106$ minus PrP 27–30 and (d) PrP 27–30 minus $PrP^{Sc}106$, showing major differences in lighter shades. (e and f) The statistically significant differences between PrP 27–30 and $PrP^{Sc}106$ calculated from (c) and (d) in red and blue, respectively, overlaid onto the crystallographic average of PrP 27–30. Reprinted with permission, from Wille, H., Michelitsch, M.D., Guénebaut, V., Supattapone, S., Serban, A., Cohen, F.E., Agard, D.A. Prusiner, S.B. (2002) Structural studies of the scrapie prion protein by electron crystallography, *Proc. Natl. Acad. Sci. U. S. A.* **99**, 3563–3568 Wille et al., (2002), copyright 2002 National Academy of Sciences, USA.

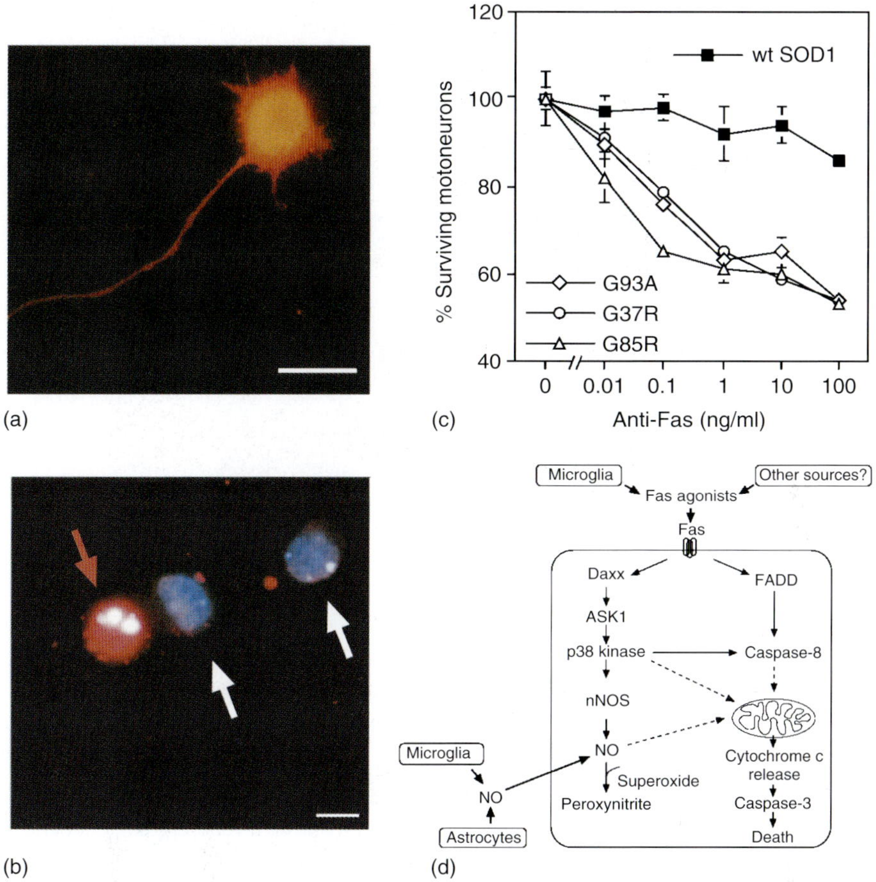

Fig. 1 (p. 486) Isolated motor neurons can be used to study mutant SOD1-linked motor neuron death. (a) Human superoxide SOD1 is expressed in cultured embryonic motor neurons from transgenic SOD1 G93A mice. (b) Treatment of SOD1 G85R motor neuron cultures with agonistic antibodies to the cell surface receptor Fas leads to increased apoptosis, as detected by staining with DAPI (in blue) and immunolabeling for activated caspase 3 (in red). An apoptotic motor neuron with strong caspase-3 activation and nuclear condensation c (red arrow) can be distinguished from healthy motor neurons displaying only weak caspase-3 activation and normal chromatin structure (white arrows). Scale bars: 25 μm. (c) Motor neurons from transgenic mutant SOD1 G93A, G85R, and G37R mice show higher susceptibility to Fas-triggered cell death than motor neurons from mice expressing wild-type SOD1. (d) Model of Fas-triggered motor neuron death. Cell death involves the classical FADD/caspase-8 pathway and a parallel pathway leading from Daxx, ASK1, and p38 activation to transcriptional upregulation of neuronal NO synthase (nNOS) and NO production. The presence of mutant SOD1 sensitizes motor neurons to Fas agonists and NO. Potential sources for these cell death triggers are astrocytes and microglia. See Fig. 2.

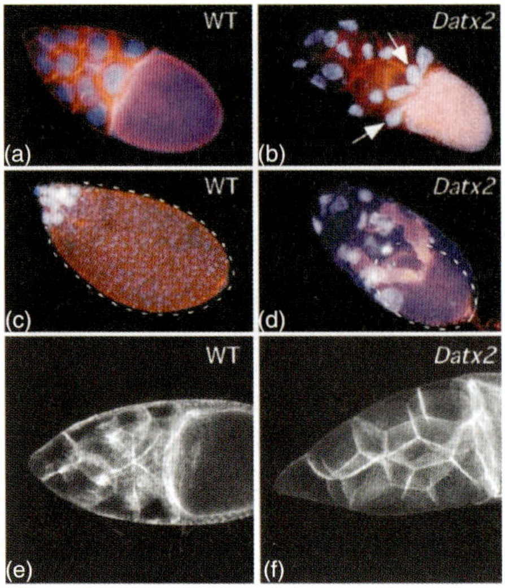

Fig. 4 (p. 608) Studies of the ataxin-2 ortholog in *Drosophila melanogaster* reveals a role for Drosophila ataxin-2 (*Datx2*) in actin filament formation during oogenesis. Egg chambers from normal (Wild-Type (WT)) and mutant flies with reduced expression (*Datx2*) of Drosophila ataxin-2 were stained with DAPI (blue) and phalloidin (red) to indicate nuclei and filamentous actin respectively (a–d). The egg chambers of WT flies prior to cytoplasmic transport (a) display well demarcated, separated but interconnected cells (blue) as expected, while the egg chambers of *Datx2* flies (b) contain irregularly arranged cells. After cytoplasmic transport, the egg chambers of WT flies (c) show one greatly enlarged oocyte (dashed line) with only a small section of compressed cells, while the egg chambers of *Datx2* flies (d) have failed to yield an enlarged oocyte, instead retaining dispersed and large adjacent cells. Confocal images of egg chambers prior to cytoplasmic transport stage reveal a prominent actin filament network in WT flies (e), but a remarkably transparent actin filament network in *Datx2* flies (f). The decreased density of the actin filament network underlies the cytoplasmic "dumping" defect in the *Datx2* flies. (From Satterfield, T.F., Jackson, S.M., Pallanck, L.J. (2002) A Drosophila homolog of the polyglutamine disease gene *SCA2* is a dosage-sensitive regulator of actin filament formation, *Genetics* **162**, 1687–1702, used with permission of *Genetics*.)

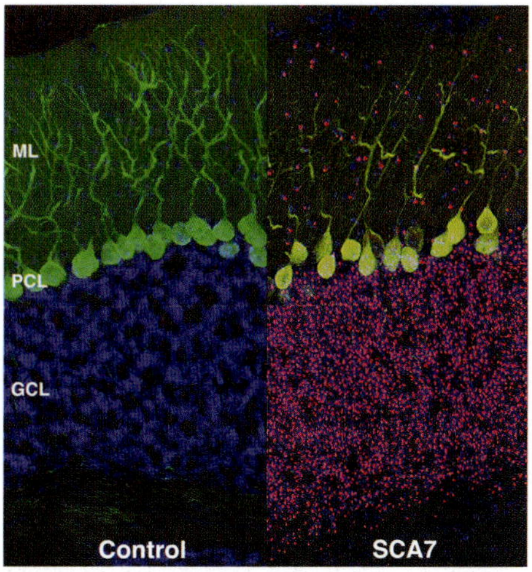

Fig. 5 (p. 612) Noncell autonomous Purkinje cell degeneration in a mouse model of spinocerebellar ataxia type 7 (SCA7). Confocal microscopy analysis of cerebellar sections from a SCA7 transgenic mouse (SCA7) created with an ataxin-7 CAG-92 containing murine prion protein expression vector and from an age- and sex-matched nontransgenic littermate (Control). Staining with an anti-ataxin-7 antibody (magenta), a calbindin antibody (green), and DAPI (blue) reveals a healthy, normal-appearing cerebellum characterized by properly oriented Purkinje cells with extensive dendritic arborization in the "Control" mice. However, SCA7 transgenic mice display pronounced Purkinje cell degeneration as evidenced by decreased dendritic arborization and displacement of Purkinje cell bodies. Interestingly, although numerous neurons in the granule cell layer (GCL) and the molecular layer (ML) display aggregates of ataxin-7, there is no accumulation of mutant ataxin-7 in the degenerating Purkinje cells due to lack of appreciable expression there. As the Purkinje cells degenerate without expressing the mutant protein, the degeneration is described as noncell autonomous. (Adapted from Garden, G.A., Libby, R.T., Fu, Y.H., Kinoshita, Y., Huang, J., Possin, D.E., Smith, A.C., Martinez, R.A., Fine, G.C., Grote, S.K., et al. (2002) Polyglutamine-expanded ataxin-7 promotes noncell-autonomous Purkinje cell degeneration and displays proteolytic cleavage in ataxic transgenic mice, *J. Neurosci.* **22**, 4897–4905, used with permission of the *Journal of Neuroscience*.)

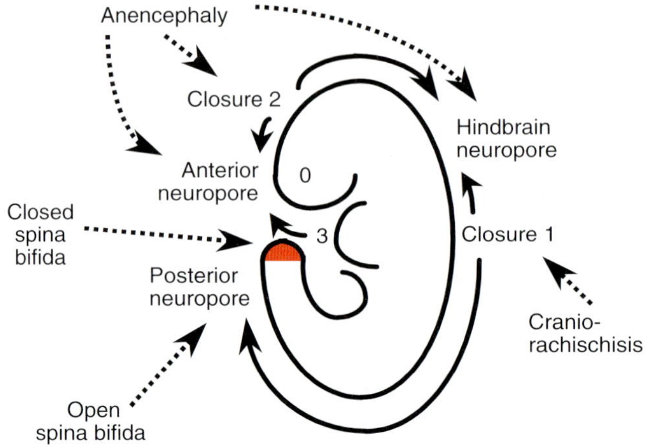

Fig. 2 (p. 653) Schematic summary of the successive initiation events of mouse neurulation (Closures 1, 2, and 3), the direction of spread of closure from the initiation sites (solid arrows), and the sites of completion of closure (anterior, hindbrain, and posterior neuropores). The tail bud region (red shading) is the site of secondary neurulation, which follows immediately after closure of the posterior neuropore. The main types of NTD that result from failure of the different components of neurulation are indicated by the dotted arrows. Modified from Copp, A.J., Bernfield, M. (1994) *Curr. Opin. Pediatr.* **6**, 624–631.

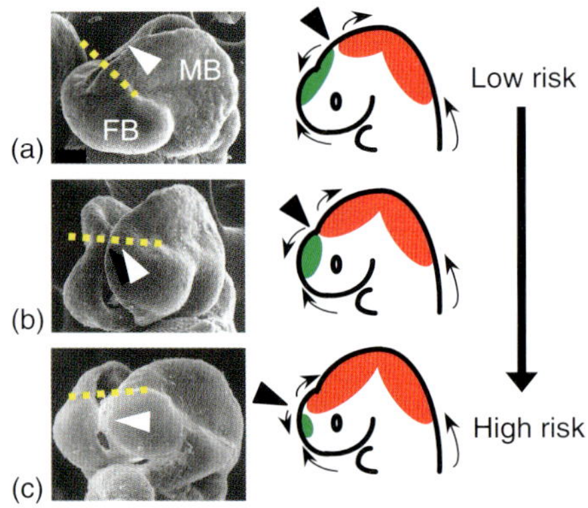

Fig. 3 (p. 657) Variations in position of Closure 2, in the future brain of different inbred mouse strains, in relation to the risk of exencephaly and anencephaly. Left panel shows scanning electron micrographs of embryonic brains from the left frontal view. Dotted lines indicate the forebrain/midbrain (FB/MB) boundary. Arrowheads point to the site of initiation of Closure, which is located caudal to the FB/MB boundary in the DBA/2 strain (a), at the FB/MB boundary in the CD1 strain (b), and rostral to the FB/MB boundary in the NZW strain (c). Middle panel summarizes diagrammatically this variation in Closure 2 morphology. Red shading indicates open neural folds caudal to the site of Closure 2, whereas green shading indicates the anterior neuropore, rostral to Closure 2. Arrows indicate direction of neural tube closure. Right panel shows an increasing risk of exencephaly correlated with a progressively more rostral position of Closure 2. The reason for this correlation is that the midbrain region, which lies at the apex of the cranial flexure, is mechanically the least favorable brain region for neural tube closure, and is aided in its closure by a caudal Closure 2 position (a) but hampered in its closure by a rostral Closure 2 position (c). Modified from Fleming, A., Copp, A.J. (2000) *Hum. Mol. Genet.* **9**, 575–581.

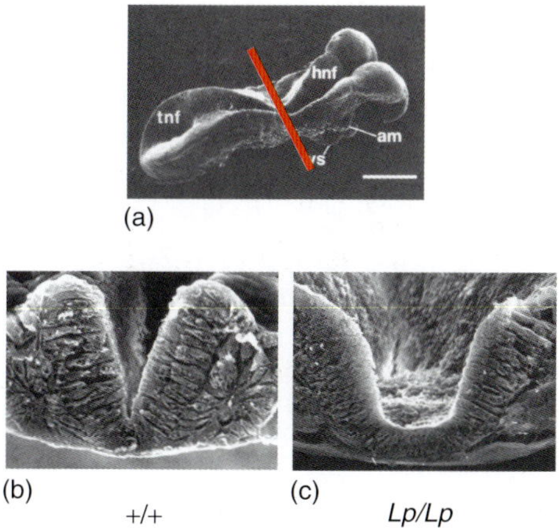

Fig. 5 (p. 659) Initiation of neurulation (Closure 1) in the mouse embryo, as studied by scanning electron microscopy. (a) Whole E8.5 embryo demonstrating the approaching neural folds approximately half way along the body axis, indicating the incipient Closure 1 event. This event occurs at the boundary of the hindbrain and cervical regions. (b, c) Sections transverse to the body axis as indicated by the red line in (a). The wild type embryo (b) shows "Mode 1" morphology with bending at a compact midline hinge point, and straight lateral neural folds. The homozygous *loop-tail* (*Lp/Lp*) embryo (c) shows a disturbance of the midline, which is broader than normal, without a compact bend in the neural plate. Although the neural folds elevate, laterally, they are not able to appose and fuse in the midline, leading to craniorachischisis in mutant embryos. Scale bar represents 0.25 mm (a) and 0.03 mm (b, c). Modified from Copp, A.J., Brook, F.A., Estibeiro, J.P., Shum, A.S.W., Cockroft, D.L. (1990) *Prog. Neurobiol.* **35**, 363–403; Greene, N.D.E., Gerrelli, D., Van Straaten, H.W.M., Copp, A.J. (1998) *Mech. Dev.* **73**, 59–72.

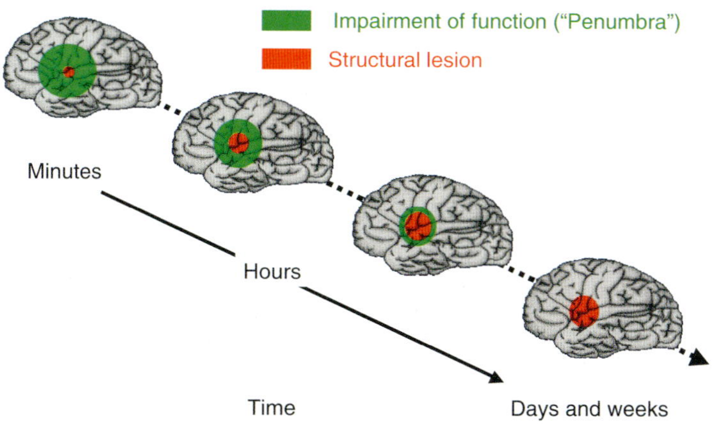

Fig. 1 (p. 749) Regression of the functional neurological deficit while the structural lesion grows. Early in the course of stroke, clinical symptoms mostly reflect an impairment of function (green) but not necessarily a structural lesion (blue). Over time, some areas either spontaneously, or because of therapy, recover function, which explains why symptoms in patients can regress, while the structural lesion actually grows (Reprinted from Dirnagl, U., Iadecola, C., Moskowitz, M.A. (1999) Pathobiology of ischaemic stroke: an integrated view, *Trends Neurosci.* **22**, 391–397, with permission from Elsevier).

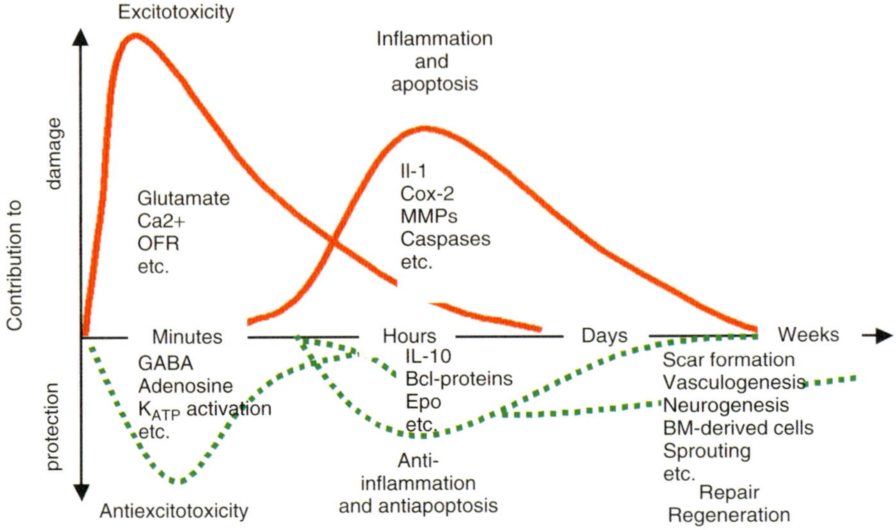

Fig. 4 (p. 766) Cell death–inducing and endogenous neuroprotective mechanisms in cerebral ischemia. Focal cerebral ischemia induces a complex series of mechanisms that damage brain cells. Brain tissue responds to most of the listed examples of noxious signals by inducing protective mechanisms. In infarcted tissue, destruction has overwhelmed protection, although tissue around the ischemic core might have been spared by restored substrate delivery (owing to delivery to collaterals, spontaneous or therapeutic reperfusion, or vasculogenesis) and cellular mechanisms of protection. In this cartoon, the *x*-axis reflects the evolution of the cascades over time and the *y*-axis illustrates the impact of each element of the destructive (top) and protective (bottom) cascades on the final outcome. The red dotted lines envelop major pathophysiological entities of tissue destruction in stroke (grouped for the acute mechanisms of excitotoxicity and the delayed mechanisms of inflammation and apoptosis) and the green broken lines envelop corresponding protective tissue responses. Abbreviations: BM, bone marrow; Cox-2, cyclooxygenase 2; Epo, erythropoietin; IL, interleukin; MMPs, matrix metalloproteinase; OFR, oxygen free radicals (Reprinted from Dirnagl, U., Simon, R.P., Hallenbeck, J.M. (2003) Ischemic tolerance and endogenous neuroprotection, *Trends Neurosci.* **26**, 248–254, with permission from Elsevier).

Part III
Neurological Diseases

Neurobiology. From Molecular Basis to Disease. Edited by Robert A. Meyers.

ISBN: 978-3-527-32293-0

12
Prions

Stanley B. Prusiner
University of California, San Francisco, CA, USA

Neurobiology. From Molecular Basis to Disease. Edited by Robert A. Meyers.

ISBN: 978-3-527-32293-0

Keywords

Prion
Proteinaceous infectious particle.

PrP^C
Normal isoform of the prion protein.

PrP^{Sc}
Disease-causing isoform of the prion protein.

Prions are infectious proteins that have been described in both mammals and fungi. That prions are composed solely of proteins makes them unprecedented infectious pathogens. Prions are proteins that can adopt at least two different conformations; they multiply by forcing the precursor protein to acquire a second conformation. Different conformations of proteins in the prion state encipher distinct strains and are prone to aggregation. In mammals, prions accumulate to high levels in the nervous system where they cause dysfunction and fatal degeneration.

Both mammalian and fungal prions have been produced in cell-free systems. Synthetic prion protein (PrP) peptides and recombinant PrP fragments have been used to form mammalian prions while N-terminal regions, called *prion domains* that are rich in glutamine and asparagine, have been used to form fungal prions.

Prions cause a group of invariably fatal, neurodegenerative diseases. Prion diseases may present as genetic, infectious, or sporadic disorders, all of which involve modification of PrP. The tertiary structure of PrP is profoundly altered as prions are formed, and as such, prion diseases represent disorders of protein conformation. Creutzfeldt-Jakob disease (CJD) generally presents as a progressive dementia in humans while scrapie of sheep and bovine spongiform encephalopathy (BSE) usually manifest as ataxic illnesses.

1
Prions

In mammals, prions reproduce by recruiting the normal, cellular isoform of the prion protein (PrP^C) and stimulating its conversion into the disease-causing isoform (PrP^{Sc}). PrP^C has a high α-helical content and little β-sheet structure, whereas PrP^{Sc} has less α-helical structure and a high β-sheet content. Comparisons of secondary structures of PrP^C and PrP^{Sc} were performed on proteins purified from Syrian hamster (SHa) brains. Limited proteolysis of PrP^{Sc} produces PrP 27–30, which retains prion infectivity; under these conditions, PrP^C is completely hydrolyzed (Fig. 1).

Solution structures of recombinant SHa and mouse (Mo) PrPs produced in bacteria showed three α-helices denoted A, B, and C as well as two short β-strands using nuclear magnetic resonance (NMR) imaging. The atomic structure of PrP^{Sc} has not been determined because of the insolubility of the protein. Two-dimensional crystals of PrP 27–30 have been used to constrain computational models of PrP^{Sc} suggesting it contains α-helix C and a portion of α-helix B while α-helix A, the two β-strands, and the surrounding segments form a β-helix. From these studies as well as from ionizing radiation inactivation analyses, it is likely that the infectious monomer is a trimer of PrP^{Sc} molecules. Recombinant

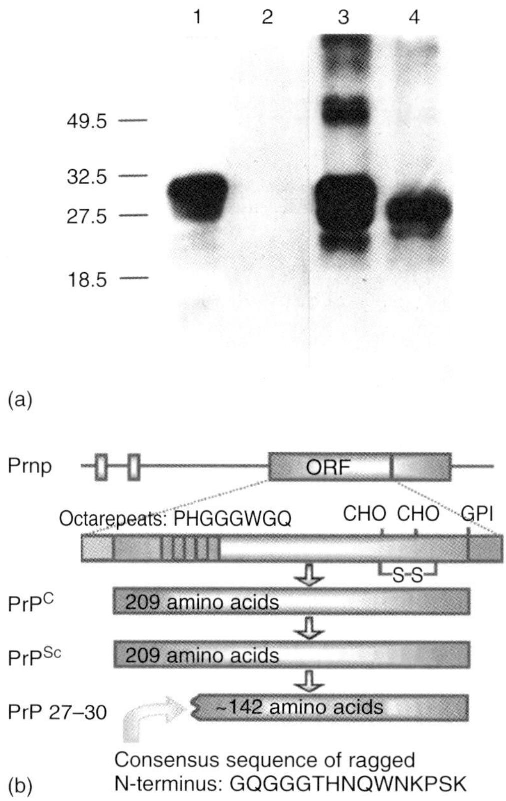

Fig. 1 Prion protein isoforms. Western immunoblot of brain homogenates from uninfected (lanes 1 and 2) and prion-infected (lanes 3 and 4) Syrian hamsters. Samples in lanes 2 and 4 were digested with 50 $\mu g\ mL^{-1}$ of proteinase K for 30 min at 37 °C. PrP^C in lanes 2 and 4 was completely hydrolyzed under these conditions, whereas approximately 67 amino acids were digested from the N-terminus of PrP^{Sc} to generate PrP 27–30. After polyacrylamide gel electrophoresis (PAGE) and electrotransfer, the blot was developed with anti-SHaPrP R073 polyclonal rabbit antiserum. Molecular weight markers are depicted in kilodaltons (kDa). (b) Bar diagram of the SHaPrP gene that encodes a protein of 254 amino acids. After processing of the N- and C-termini, both PrP^C (green) and PrP^{Sc} (red) consist of 209 residues. After limited proteolysis, the N-terminus of PrP^{Sc} is truncated to form PrP 27–30, which is composed of approximately 142 amino acids, the N-terminal sequence of which was determined by Edman degradation. (See color plate p. xxxiii).

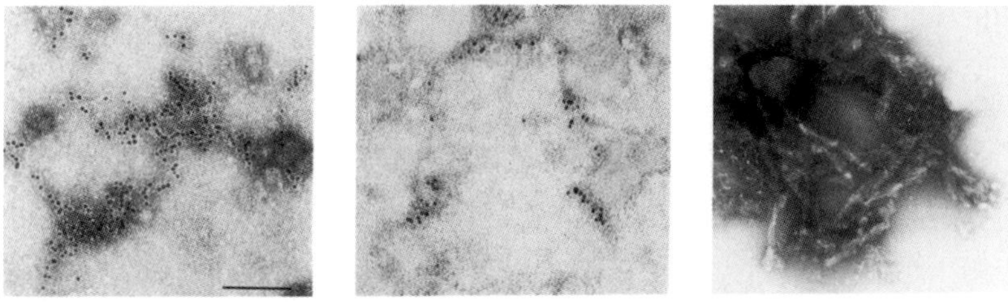

(a) (b) (c)

Fig. 2 Electron micrographs of negatively stained and immunogold-labeled prion proteins. (a) PrP^C and (b) PrP^{Sc}. Neither PrP^C nor PrP^{Sc} forms recognizable, ordered polymers. (c) Prion rods composed of PrP 27–30 were negatively stained. The prion rods are indistinguishable from many purified amyloids. Bar = 100 nm. Reprinted with permission, from (Pan, K-M., Baldwin, M., Nguyen, J., Gasset, M., Serban, A., Groth, D., Mehlhorn, I., Huang, Z., Fletterick, R.J., Cohen, F.E., Prusiner, S.B. (1993) Conversion of α-helices into β-sheets features in the formation of the scrapie prion proteins, *Proc. Natl. Acad. Sci. U. S. A.* **90**, 10962–10966), copyright 1993 National Academy of Sciences, USA.

antibody fragments have been used to map the surfaces of PrP^C and PrP^{Sc} and those results are consistent with the foregoing structural data. The structural transition from an α-helical protein into a β sheet–rich molecule seems to be the fundamental event underlying the formation of nascent prions.

Limited proteolysis truncates the N-terminus of PrP^{Sc} to produce PrP 27–30 consisting of the C-terminal ~142 amino acids. PrP 27–30 polymerizes into amyloid fibrils that are indistinguishable from fibrils found in amyloid plaques of the brains of mammals with prion disease. When full-length PrP^{Sc} was purified from SHa brain, only amorphous aggregates were seen by electron microscopy. While limited digestion of purified PrP^{Sc} with proteinase K produced PrP 27–30, the ultrastructural appearance of the samples remained unchanged (Fig. 2). Addition of a detergent, such as Sarkosyl, either before or after limited proteolysis provoked the assembly of PrP 27–30 into rod-shaped structures with the ultrastructural and tinctorial properties of amyloid.

2 Prions are Distinct from Viruses

Major features that distinguish prions from viruses are: (1) the ability to create prion infectivity by modifying the conformation of a polypeptide devoid of nucleic acid and (2) PrP is encoded by a chromosomal gene, designated *PRNP* in humans and *Prnp* in mice. The PrP gene is located on the short arm of chromosome 20 in humans and the syntenic region of chromosome 2 in mice.

Prions differ from viruses and viroids in that they lack a nucleic acid genome that directs the synthesis of their progeny. Many investigators argued for a nucleic acid genome within the infectious prion particle while others contended for a small, noncoding polynucleotide of either foreign or cellular origin. No nucleic acid has been found despite intensive searches using a wide variety of techniques and approaches. On the basis of a wealth of evidence, it is reasonable to assert that such a nucleic acid has not been found because it does not exist. Prions are composed of an alternative isoform of a cellular protein, whereas most viral proteins are encoded by a viral genome and viroids are devoid of protein.

In contrast to viruses, prions are nonimmunogenic. Prions do not elicit an immune response because the host is tolerant to PrP^{C}, which prevents the host from mounting an immune response to PrP^{Sc}. In contrast, foreign proteins of viruses that are encoded by the viral genome often elicit a profound immune response. Thus, it seems unlikely that vaccination, which has been so effective in preventing many viral illnesses, will be a useful strategy for preventing or treating prion diseases.

When prions and viruses are passaged from one host species to another, the consequences are very different. The passage of prions from one host to another is accompanied by the acquisition of a new PrP amino acid sequence encoded by the genome of the new host. The crossing of prions from one species to another is restricted by what has been called the "species barrier". The closer the evolutionary relationship is between the host in which the prions last replicated and the newly infected animal, the more likely replication will occur in the new host. However, particular strains of prions can change this rule. For example, variant CJD prions from humans replicate much more readily in transgenic (Tg) mice expressing bovine (Bo) PrP than in Tg mice expressing human or chimeric mouse-human PrP.

The issue of prion strains posed a profound conundrum for many years. How could an infectious pathogen composed only of protein encipher biological information? This riddle was solved when prion strains with different physical properties were isolated. Subsequently, new strains of prions have been isolated either by passage through mice expressing artificial PrP transgenes or by forming synthetic prions from PrP produced in *Escherichia coli*.

3 Disease Paradigms

Despite some similarities between prion and viral illnesses, these disorders are very different. Prion diseases are uniformly fatal. No human or animal has ever recovered from a prion disease once neurologic dysfunction is manifest. No host defenses are mounted in response to prion infection: no humoral immunity, no cellular immunity, and no interferons are elicited to the replicating prion.

While prions can be spread from one host to another, the most common form of prion disease is spontaneous or sporadic. In these illnesses, prions arise endogenously and replicate. Recent experiments with synthetic prions demonstrate that only PrP^{C} is required to generate prion infectivity. These findings contrast with viruses in which exogenous infection is required except in the case of latent retroviral genomes. For example, after infection with exogenous HIV, the virus may disappear but its RNA genome may be reverse-transcribed into DNA and the DNA copies may remain dormant for years.

The dramatically different principles that govern prion biology have often been poorly understood. This lack of understanding has led to some regrettable decisions of great economic, political, and possibly public health importance. For example, scrapie and BSE have different names, yet they are the same disease in two different species. Scrapie and BSE differ in only two respects: first, the PrP sequence in sheep differs from that of cattle at seven or eight positions of 270 amino acids, which results in different PrP^{Sc} molecules. Second, some aspects of each disease are determined by the particular prion strain that infects the respective host.

Understanding prion strains and the "species barrier" is of paramount importance with respect to the BSE epidemic in Britain, in which more than 180 000 cattle have died over the past decade. Brain extracts from eight cattle with BSE resulted in similar incubation times and patterns of vacuolation in the neuropil when inoculated into a variety of inbred mice. Incubation times and profiles of neuronal vacuolation have been used for three decades to study prion strains. Brain extracts prepared from three domestic cats, one nyala, and one kudu, all of which died of a neurologic illness, produced incubation times and lesion profiles indistinguishable from those found in the BSE cattle. Cats and exotic ungulates (such as the kudu) presumably developed prion disease from eating food containing bovine prions.

4 Nomenclature

Although the prions that cause transmissible mink encephalopathy (TME) and BSE are referred to as *TME prions* and *BSE prions*, this may be unjustified because both are thought to originate from the oral consumption of scrapie prions in sheep-derived foodstuffs and because many lines of evidence argue that the only difference among various prions is the sequence of PrP, which is dictated by the host and not the prion itself. Human (Hu) prions present a similar semantic conundrum. Transmission of Hu prions to laboratory animals produces prions carrying PrP molecules with sequences dictated by the PrP gene of the last host, not that of the inoculum.

To simplify the terminology, the generic term PrP^{Sc} was suggested in place of such terms as PrP^{CJD}, PrP^{BSE}, and PrP^{res}. To distinguish PrP^{Sc} found in humans and cattle from that found in other animals, "$HuPrP^{Sc}$" and "$BoPrP^{Sc}$" are suggested instead of PrP^{CJD} and PrP^{BSE}, respectively. Once human prions, or $HuPrP^{Sc}$ molecules, have been passaged into animals, the prions and PrP^{Sc} are no longer of the human species unless they were formed in an animal expressing a HuPrP transgene.

The "Sc" superscript of PrP^{Sc} was initially derived from the term "scrapie" because scrapie was the prototypic prion disease. Because all of the known prion diseases (Table 1) of mammals involve aberrant metabolism of PrP similar to that observed in scrapie, the "Sc" superscript was suggested for all abnormal, pathogenic PrP isoforms. In this context, the "Sc" superscript is used to designate the di<u>s</u>ease-<u>c</u>ausing isoform of PrP (Table 2). The development of the conformation-dependent immunoassay (CDI) for PrP^{Sc} led to the discovery of a protease-sensitive form of PrP^{Sc}, designated $sPrP^{Sc}$. The CDI is based on measuring antibody binding to an epitope that is exposed in PrP^{C} but buried in PrP^{Sc}. Under conditions of limited proteolysis in which the protease-resistant form of PrP^{Sc}, designated $rPrP^{Sc}$, is converted into PrP 27–30, $sPrP^{Sc}$ is completely hydrolyzed. Whether $sPrP^{Sc}$ is an intermediate in the formation of infectious prions remains to be established.

In the case of mutant PrP, mutations and polymorphisms can be denoted in parentheses following the PrP isoform.

Tab. 1 The prion diseases.

Disease	***Host***	***Mechanism of pathogenesis***
Kuru	Fore people	Infection through ritualistic cannibalism
Iatrogenic CJD	Humans	Infection from prion-contaminated HGH, dura mater grafts, etc.
Variant CJD	Humans	Infection from bovine prions?
Familial CJD	Humans	Germline mutations in PrP gene
GSS	Humans	Germline mutations in PrP gene
FFI	Humans	Germline mutations in PrP gene (D178N,M129)
Sporadic CJD	Humans	Somatic mutation or spontaneous conversion of PrP^C into PrP^{Sc}?
sFI		Somatic mutation or spontaneous conversion of PrP^C into PrP^{Sc}?
Scrapie	Sheep	Infection in genetically susceptible sheep
Bovine spongiform encephalopathy	Cattle	Infection with prion-contaminated MBM
Transmissible mink encephalopathy	Mink	Infection with prions from sheep or cattle
Chronic wasting disease	Mule deer, elk	Unknown
Feline spongiform encephalopathy	Cats	Infection with prion-contaminated bovine tissues or MBM
Exotic ungulate encephalopathy	Greater kudu, nyala, oryx	Infection with prion-contaminated MBM

Notes: CJD: Creutzfeldt–Jakob disease; FFI: fatal familial insomnia; sFI: sporadic fatal insomnia; GSS: Gerstmann–Sträussler–Scheinker disease; HGH: human growth hormone; MBM: meat and bone meal.

For fatal familial insomnia (FFI), in which it might be important to identify the mutation, the prions would be designated $HuPrP^{Sc}$(D178N,M129) (Table 3).

Parentheses following PrP^{Sc} can also be used to notate a particular prion strain. For example, prions from Syrian hamsters inoculated with Sc237 or 139H prion strains can be designated $SHaPrP^{Sc}$(Sc237) or $SHaPrP^{Sc}$(139H), respectively. Similarly, sheep inoculated with scrapie or BSE prions produce $OvPrP^{Sc}$(Sc) or $OvPrP^{Sc}$(BSE), respectively.

The terms "PrP^{res}" and "PrP-res" were derived to describe the protease resistance of PrP^{Sc} and have sometimes been used interchangeably with PrP^{Sc}. The use of "PrP^{res}" and "PrP-res" became particularly problematic with the discovery of $sPrP^{Sc}$. Protease resistance, insolubility, and high β-sheet content should be considered only as surrogate markers of PrP^{Sc} infectivity because not all characteristics may be present. For example, $MoPrP^{Sc}$(P101L) from Tg mice that express high levels of MoPrP(P101L) is transmissible to Tg(MoPrP,P101L)196/*Prnp*$^{0/0}$ mice, but these prions are sensitive to proteolytic digestion at 37 °C. When digestions were performed at 4 °C or the CDI was used to detect PrP^{Sc}, then $MoPrP^{Sc}$(P101L) was detected. In contrast, when Tg196 mice were inoculated with mouse RML prions, the resulting $MoPrP^{Sc}$(P101L) was resistant to proteolytic digestion at 37 °C. These

Tab. 2 Glossary of prion terminology.

Term	*Description*
Prion	A proteinaceous infectious particle that lacks nucleic acid. Prions are composed largely, if not entirely, of PrP^{Sc} molecules.
PrP^{Sc}	Abnormal, pathogenic isoform of the prion protein that causes illness. This protein is the only identifiable macromolecule in purified preparations of prions.
PrP^{C}	Cellular isoform of the prion protein.
PrP 27–30	N-terminally truncated PrP^{Sc}, generated by digestion with proteinase K.
PRNP	Human PrP gene located on chromosome 20.
Prnp	Mouse PrP gene located on syntenic chromosome 2. *Prnp* controls the length of the prion incubation time and is congruent with the incubation-time genes *Sinc* and *Prn-i*. PrP-deficient ($Prnp^{0/0}$) mice are resistant to prion infection.
PrP amyloid	Fibril of PrP fragments derived from PrP^{Sc} by proteolysis. Plaques containing PrP amyloid are found in the brains of some mammals with prion disease.
Prion rod	An amyloid polymer composed of PrP 27–30 molecules. Created by detergent extraction and limited proteolysis of PrP^{Sc}.
Protein X	A hypothetical macromolecule that is thought to act as a molecular chaperone in facilitating the conversion of PrP^{C} into PrP^{Sc}.

Tab. 3 Examples of human PrP gene mutations found in the inherited prion diseases.

Inherited prion disease	*PrP gene mutation*
Gerstmann–Sträussler–Scheinker disease	P102L[a]
Gerstmann–Sträussler–Scheinker disease	A117V
Familial Creutzfeldt–Jakob disease	D178N, V129
Fatal familial insomnia	D178N, M129[a]
Gerstmann-Sträussler-Scheinker disease	F198S[a]
Familial Creutzfeldt–Jakob disease	E200K[a]
Gerstmann–Sträussler–Scheinker disease	Q217R
Familial Creutzfeldt–Jakob disease	octarepeat insert[a]

[a] Signifies genetic linkage between the mutation and the inherited prion disease.

findings emphasize the ambiguities that may arise from simply assessing resistance to proteolytic digestion. Whether PrP^{res} is useful in denoting PrP molecules that have been subjected to procedures modifying resistance to proteolysis but that neither convey infectivity nor cause disease remains questionable. Perhaps PrP^{res} is best reserved for PrP molecules that exhibit resistance to limited proteolysis after binding to PrP^{Sc}.

The term "PrP*" has been used in two different ways. First, it has been used to identify a fraction of PrP^{Sc} molecules that are infectious. Such a designation is thought to be useful because there are $\sim 10^5$ PrP^{Sc} molecules per infectious (ID_{50}) unit. Second, PrP* has been used to designate a metastable intermediate of PrP^C that is bound to a putative conversion cofactor, provisionally designated protein X. It is noteworthy that neither a subset of biologically active PrP^{Sc} molecules nor a metastable intermediate of PrP^C has been identified, to date.

In mice, *Prnp* is now known to be identical to two genes, *Sinc* and *Prn-i*, that are known to control the length of the incubation time in mice inoculated with prions. A gene, designated *Pid-1*, on mouse chromosome 17 also appears to influence experimental CJD and scrapie incubation times but information on this locus is limited.

Distinguishing among CJD, FFI, and Gerstmann–Sträussler–Scheinker syndrome (GSS) has grown increasingly difficult with the recognition that familial (f) CJD, GSS, and FFI are autosomal-dominant diseases caused by mutations in *PRNP* (Table 3). Initially, it was thought that each *PRNP* mutation was associated with a particular cliniconeuropathologic phenotype, but more exceptions are being recognized. Multiple examples of variations in the cliniconeuropathologic phenotype have been recorded within a single family in which all affected members carry the same *PRNP* mutation. Most patients with a *PRNP* mutation at codon 102 present with ataxia and have PrP amyloid plaques; such patients are generally given the diagnosis of GSS, but some individuals present with dementia, a clinical characteristic that is usually associated with CJD. For most inherited prion disease, the disease is specified by the respective mutation, such as fCJD(E200K) and GSS(P101L). In the case of FFI, describing the D178N mutation and M129 polymorphism seems unnecessary because this is the only known mutation–polymorphism combination that results in the FFI phenotype. The sporadic form of fatal insomnia is denoted sFI.

5 Discovery of the Prion Protein

The discovery of PrP 27–30 in SHa brain fractions progressively enriched for scrapie infectivity transformed research on scrapie and related diseases. PrP 27–30 was so named because it has an apparent molecular weight (M_r) of 27 kDa to 30 kDa (Fig. 1). PrP 27–30 is derived from the larger PrP^{Sc} by N-terminal truncation; both PrP 27–30 and PrP^{Sc} are infectious. PrP 27–30 not only provided a molecular marker that is specific for prion disorders but it was later shown to be the sole component of the prion particle.

The molecular biology and genetics of prions began with the purification of PrP 27–30 that allowed determination of its N-terminal, amino acid sequence. Multiple signals in each cycle of the Edman degradation suggested that either multiple proteins were present in these "purified fractions" or a single protein with a ragged N-terminus was present. When the signals in each cycle were grouped according to their strong, intermediate, and weak intensities, it became clear that a single protein with a ragged N-terminus was being sequenced (Fig. 3). Determination of a single, unique sequence for the N-terminus of PrP 27–30 permitted the synthesis of isocoding mixtures of oligonucleotides that were subsequently used to identify incomplete PrP cDNA clones from hamster and mouse. cDNA clones encoding the entire open reading

Relative amount	Amino acid sequence*
1	G-Q-G-G-G-T-H-N-Q-W-N-K-P-S-K
0.4	X-X-X-T-H-N-X-W-X-K-P
0.2	X-X-P-W-X-Q-X-X-X-T-H-X-Q-W

*Single-letter amino acid code. X = amino acid not determined at that cycle.

Fig. 3 Interpreted sequence, shown by single-letter amino acid codes, of the N-terminus of PrP 27–30. The "ragged ends" of PrP 27–30 are shown. X = amino acid was not detected at that cycle of the Edman degradation.

frames (ORFs) of SHaPrP and MoPrP were eventually recovered.

6
Prion Protein Isoforms

In Syrian hamsters, PrP^C and PrP^{Sc} are 209-residue proteins that are anchored to the cell surface by a glycosylphosphatidyl inositol (GPI) moiety and have the same covalent structure (Fig. 1). The N-terminal sequencing, the deduced amino acid sequences from PrP cDNA, and immunoblotting studies argue that PrP 27–30 is a truncated protein of approximately 142 residues, which is derived from PrP^{Sc} by limited proteolysis of the N-terminus.

In general, $\sim 10^5$ PrP^{Sc} molecules correspond to one ID_{50} unit (U) using the most sensitive bioassay. PrP^{Sc} is probably best defined as the alternative or abnormal isoform of PrP, which stimulates conversion of PrP^C into nascent PrP^{Sc}, accumulates, and causes disease. Although resistance to limited proteolysis has proved to be a convenient tool for detecting PrP^{Sc}, not all PrP^{Sc} molecules possess protease resistance, as discussed above. Some investigators equate protease resistance with PrP^{Sc} and this erroneous view has been compounded by the use of the term "PrP-res".

Although insolubility and protease resistance were used in initial studies to differentiate PrP^{Sc} from PrP^C, subsequent investigations showed that these properties are only surrogate markers of infectivity, as are high β-sheet content and polymerization into amyloid. When these surrogate markers are present, they are useful, but their absence does not establish a lack of prion infectivity. PrP^{Sc} is usually not detected by Western immunoblotting if fewer than 10^5 ID_{50} U mL^{-1} are present in a sample. Furthermore, PrP^{Sc} from different species or prion strains may exhibit different degrees of protease resistance.

6.1
Cell Biology of PrP^{Sc} Formation

In scrapie-infected cells, PrP^C molecules destined to become PrP^{Sc} exit to the cell surface prior to conversion into PrP^{Sc}. Like other GPI-anchored proteins, PrP^C appears to reenter the cell through a subcellular compartment bounded by cholesterol-rich, detergent-insoluble membranes, which might be caveolae or early endosomes. Within this cholesterol-rich, nonacidic compartment, GPI-anchored PrP^C can be either converted into PrP^{Sc} or partially degraded. Subsequently, PrP^{Sc} is trimmed at the N-terminus in an acidic compartment in scrapie-infected cultured cells to form PrP 27–30. In contrast, N-terminal truncation of PrP^{Sc} is minimal in the brain, where little PrP 27–30 is found.

7 Rodent Models of Prion Disease

Mice and hamsters are commonly used in experimental studies of prion disease. The shortest incubation times are achieved with intracerebral inoculation of prions with a sequence identical to that of the host animal; under these conditions, all animals develop prion disease within a narrow interval for a particular dose. When the PrP sequence of the donor prion differs from that of the recipient host, the incubation time is prolonged, and can be quite variable; often, many of the inoculated animals do not develop disease. This phenomenon is generally called the "species barrier".

8 PrP Gene Structure and Organization

Prnp is a member of the *Prn* gene family. The second member of this family to be identified is the *Prnd* gene that lies approximately 19 kb downstream from the PrP locus and encodes the doppel (Dpl) protein. The respective genes that encode PrP and Dpl appear to represent ancient gene duplication that occurred an prior to the speciation of mammals. The sequences are approximately 25% identical but the structures of the two proteins are highly conserved (Fig. 4). In contrast to PrP, which is expressed in many different tissues, Dpl expression is confined to the testis. Both Dpl and PrP are found on the surface of sperm but their functions are unknown. In contrast to PrP-deficient ($Prnp^{0/0}$) mice, Dpl-deficient ($Prnd^{0/0}$) mice are sterile. The knockout of both PrP and Dpl genes resulted in a sterile phenotype.

The entire ORF of all known mammalian and avian PrP genes resides within a single exon, which eliminates the possibility that PrP^{Sc} arises from alternative RNA splicing. The two exons of the SHaPrP gene are separated by a 10-kb intron; exon 1 encodes a portion of the 5′ untranslated leader sequence. The PrP

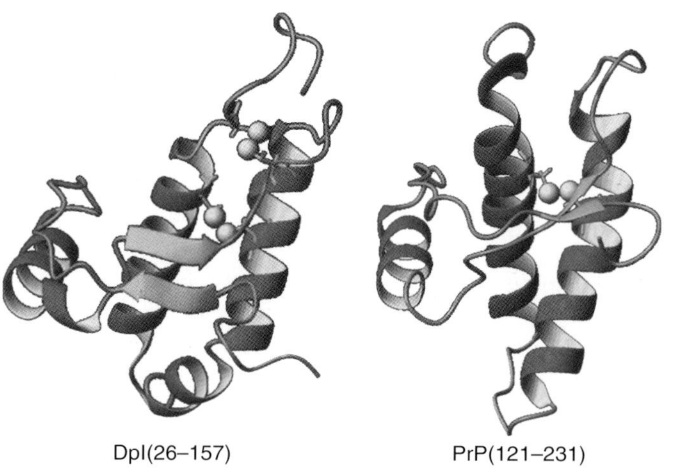

Fig. 4 Comparison of the NMR structures of mouse Dpl and PrP. Backbone topology of mouse Dpl(26–157) and mouse PrP(121–231). Figure prepared by Jane Dyson and Peter Wright.

genes of Syrian hamster, mouse, sheep, and rat contain three exons, with exon 3 encoding the ORF and 3′ untranslated region. The promoters of both the SHaPrP and MoPrP genes contain multiple copies of G-C–rich repeats and are devoid of TATA boxes. These G-C nonamers represent a motif that may function as a canonical binding site for the transcription factor Sp1.

Like the PrP gene, the entire ORF of the Dpl gene is encoded within a single exon. In some lines of *Prnp*$^{0/0}$ mice, high levels of Dpl expression were found in the brain. The expression of Dpl in the CNS was due to intergenic splicing of the nontranslated exons of the PrP gene with the translated exon of Dpl. That Dpl expression is neurotoxic was subsequently demonstrated by construction of Tg mice expressing Dpl in the brain.

8.1 Expression of the PrP Gene

Although PrP mRNA is constitutively expressed in the brains of adult animals, it is highly regulated during development. In the septum, levels of PrP mRNA and choline acetyltransferase were found to increase in parallel during development. In other brain regions, PrP gene expression occurs at an earlier age. *In situ* hybridization studies show that the highest levels of PrP mRNA are found in neurons.

Because no antibodies are currently available that clearly distinguish between PrP^{C} and PrP^{Sc}, PrP^{C} is generally measured in tissues from uninfected control animals, in which no PrP^{Sc} is found. PrP^{Sc} staining was minimal in the regions that intensely stained for PrP^{C}. A similar relationship between PrP^{C} and PrP^{Sc} was found in the amygdala. In contrast, PrP^{Sc} accumulated in the medial habenular nucleus, the medial septal nuclei, and the diagonal band of Broca; these areas were virtually devoid of PrP^{C}. In the white matter, bundles of myelinated axons contained PrP^{Sc} but were devoid of PrP^{C}. These findings suggest that prions are transported along axons, which is consistent with earlier findings in which scrapie infectivity migrated in a pattern consistent with retrograde transport. While the rate of PrP^{Sc} synthesis appears to be a function of the level of PrP^{C} expression in Tg mice, the level of PrP^{Sc} accumulation appears to be independent of PrP^{C} concentration.

8.2 Overexpression of Wild-type PrP Transgenes

Mice expressing different levels of the wild-type (wt) SHaPrP transgene were constructed. Inoculation of these Tg(SHaPrP) mice with SHa prions demonstrated abrogation of the species barrier, resulting in abbreviated incubation times due to a nonstochastic process. The length of the incubation time after inoculation with SHa prions was inversely proportional to the level of SHaPrP^{C} in the brains of Tg(SHaPrP) mice. Bioassays of brain extracts from clinically ill Tg(SHaPrP) mice inoculated with Mo prions revealed that Mo prions, but no SHa prions, were produced. Conversely, inoculation of Tg(SHaPrP) mice with SHa prions led to the synthesis of only SHa prions.

During transgenetic studies, uninoculated older mice harboring high copy numbers of wt PrP transgenes derived from Syrian hamsters, sheep, and *Prnp*^{b} mice spontaneously developed truncal ataxia, hind-limb paralysis, and tremors. These Tg mice exhibited profound necrotizing

myopathy involving skeletal muscle, demyelinating polyneuropathy, and focal vacuolation of the CNS. Development of disease was dependent on transgene dosage. For example, Tg(SHaPrP$^{+/+}$)7 mice homozygous for the SHaPrP transgene array regularly developed disease between 450 and 600 days of age, while hemizygous Tg(SHaPrP$^{+/0}$)7 mice developed disease after >650 days. Attempts to demonstrate spontaneous generation of SHa prions in Tg(SHaPrP$^{+/+}$)7 mice have been unsuccessful.

8.3 PrP Gene Dosage Controls the Incubation Time

Incubation times have been used to isolate prion strains inoculated into sheep, goats, mice, and hamsters. The *Sinc* gene is a major determinant of incubation periods in mice. Once molecular clones of *Prnp* became available, a study was performed to determine if control of the length of the incubation time is genetically linked to the PrP gene. Because the availability of VM mice with prolonged incubation times that were used to define *Sinc* was restricted, I/LnJ mice were used in the crosses. Indeed, the incubation-time locus, designated *Prn-i*, was found to be either congruent with or closely linked to *Prnp*.

Although the amino acid substitutions in PrP that distinguish *Prnp*^{a} from *Prnp*^{b} mice argued for the congruency of *Prnp* and *Prn-i*, experiments with *Prnp*^{a} mice expressing *Prnp*^{b} transgenes demonstrated a "paradoxical" shortening of incubation times. These Tg mice were predicted to exhibit a prolongation of the incubation time after inoculation with RML prions based on (*Prnp*^{a} × *Prnp*^{b}) F1 mice, which exhibit long incubation times. Those findings were described as a "paradoxical" shortening because we and others had believed for many years that long incubation times are dominant traits. From studies of congenic and transgenic mice expressing different numbers of the *a* and *b* alleles of *Prnp*, these findings were discovered not to be paradoxical; indeed, they resulted from increased PrP gene dosage. When the RML isolate was inoculated into congenic and transgenic mice, increasing the number of copies of the *a* allele was found to be the major determinant in reducing the incubation time; however, increasing the number of copies of the *b* allele also reduced the incubation time, but not to the same extent as that seen with the *a* allele. Gene-targeting studies established that the *Prnp* gene controls the incubation time and as such, is congruent with both *Prn-i* and *Sinc*.

8.4 PrP-deficient Mice

Ablation of the PrP gene in mice did not affect development of these animals. *Prnp*$^{0/0}$ mice remain healthy for more than two years. Acute suppression of PrP expression by addition of doxycycline to the drinking water of bigenic mice with an inducible PrP transgene under the control of the tetracycline promoter did not result in any untoward effects in adult mice. Similarly, bigenic mice, in which neuronal PrP expression was terminated in adulthood, showed no discernable deficits.

In two *Prnp*$^{0/0}$ lines, Purkinje cell loss was accompanied by ataxia beginning at approximately 70 weeks of age. Crossing one of these *Prnp*$^{0/0}$ lines with Tg mice overexpressing MoPrP rescued the ataxic phenotype. With the discovery of Dpl, it became clear, as described above, that Dpl

expression in the brains of these *Prnp*$^{0/0}$ mice provoked cerebellar degeneration.

Prnp$^{0/0}$ mice inoculated with prions are resistant to infection. *Prnp*$^{0/0}$ mice were sacrificed 5 days, 60 days, 120 days, and 315 days after inoculation with RML prions and brain extracts were bioassayed in Swiss CD-1 mice. Except for residual infectivity from the inoculum detected at 5 days after inoculation, no infectivity was found in the brain extracts, as the Swiss CD-1 mice did not develop disease.

Prnp$^{0/0}$ mice crossed with Tg(SHaPrP) mice were rendered susceptible to SHa prions but remained resistant to Mo prions. Because the absence of PrP^{C} expression does not provoke disease, it is likely that scrapie and other prion diseases are a consequence of PrP^{Sc} accumulation rather than inhibition of PrP^{C} function. Such an interpretation is consistent with the dominant inheritance of familial prion diseases.

Mice heterozygous (*Prnp*$^{0/+}$) for ablation of the PrP gene had prolonged incubation times when inoculated with Mo prions and developed signs of neurologic dysfunction at 400 to 460 days after inoculation. These findings are in accord with studies on Tg(SHaPrP) mice in which increased SHaPrP expression was accompanied with shortened incubation times.

Because *Prnp*$^{0/0}$ mice do not express PrP^{C}, we reasoned that they might more readily produce α-PrP antibodies. *Prnp*$^{0/0}$ mice immunized with Mo or SHa prion rods produced α-PrP antisera that bound MoPrP, SHaPrP, and HuPrP. These findings contrast with earlier studies in which α-MoPrP antibodies could not be produced in mice presumably because the mice had been rendered tolerant by the presence of MoPrP^{C}. That *Prnp*$^{0/0}$ mice readily produce α-PrP antibodies is consistent with the hypothesis that the lack of an immune response in prion diseases is due to the fact that PrP^{C} and PrP^{Sc} share many epitopes.

8.5 Species Variations in the PrP Sequence

PrP is posttranslationally processed to remove a 22-amino acid, N-terminal signal peptide. The C-terminal 120 amino acids contain two conserved disulfide-bonded cysteines and a sequence that beckons addition of a GPI anchor. Twenty-three residues are removed during the addition of this GPI moiety, which anchors the protein to the cell membrane. Contributing to the mass of the protein are two Asn side chains linked to large oligosaccharides with multiple structures that have been shown to be complex and diverse. Although many species variants of PrP have now been sequenced, only the chicken sequence has been found to differ greatly from the human sequence. The alignment of the translated sequences from more than 40 PrP genes shows a striking degree of conservation between the mammalian sequences and is suggestive of the retention of some important function through evolution (Fig. 5). Cross-species conservation of PrP sequences makes it difficult to draw conclusions about the functional importance of many of the individual residues in the protein.

8.6 N-terminal Sequence Repeats

The N-terminal domain of mammalian PrP contains five copies of a P(H/Q)GGG(G)WGQ octarepeat sequence, occasionally more, as in the case of one sequenced bovine allele, which has six copies. These repeats are remarkably conserved between species, which implies a functionally

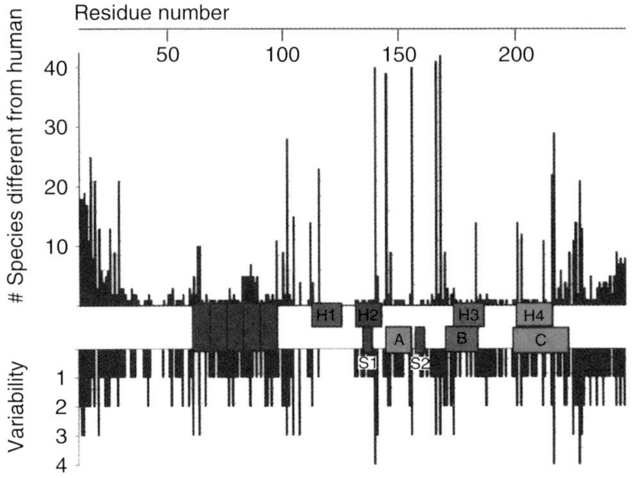

Fig. 5 Species variations and mutations of the prion protein gene. The *x*-axis represents the human PrP sequence, with the five octarepeats and H1–H4 regions of putative secondary structure shown as well as the three α-helices A, B, and C and the two β-strands S1 and S2 as determined by NMR. Vertical bars above the axis indicate the number of species that differ from the human sequence at each position. Below the axis, the length of the bars indicates the number of alternative amino acids at each position in the alignment. Data were compiled by Paul Bamborough and Fred E. Cohen (see color plate p. xxxiv).

important role. The chicken sequence contains a different repeat, PGYP(H/Q)N. Although insertions of extra repeats have been found in patients with familial prion disease, naturally occurring deletions of single octarepeats do not appear to cause disease, and deletion of all these repeats does not prevent PrP^C from undergoing a conformational transition into PrP^{Sc}.

It was suggested that the histidine residues in the octarepeats might bind metal ions and that immobilized metal ion affinity chromatography (IMAC) might facilitate the purification of PrP; indeed, IMAC did prove to be useful in the purification of PrP^C. The availability of purified PrP^C allowed studies comparing the secondary structures of PrP^C and PrP^{Sc}. The metal ion-peptide complexes were found to be much more soluble than the metal ions bound to full-length PrP. Using full-length recombinant (rec) PrP, Cu^{2+} was found to bind with a much higher avidity than any other metal ion, but the concentration for half-maximal binding for Cu^{2+} was 14 µM at pH 6.0, indicating a rather low affinity of PrP for Cu^{2+}. At neutral pH, Cu^{2+}–PrP complexes tended to form large aggregates and precipitate, so, synthetic peptides containing the octarepeats were used to study more extensively the interaction of Cu^{2+} with PrP. The concentration for half-maximal binding for Cu^{2+} to a peptide (51–75) containing two octarepeats was 6 µM at pH 7.5. When a peptide (58–91) containing four octarepeats was studied, the binding of Cu^{2+} was found to be cooperative and highly pH-dependent (Fig. 6). The midpoint of the pH-dependence transition was pH

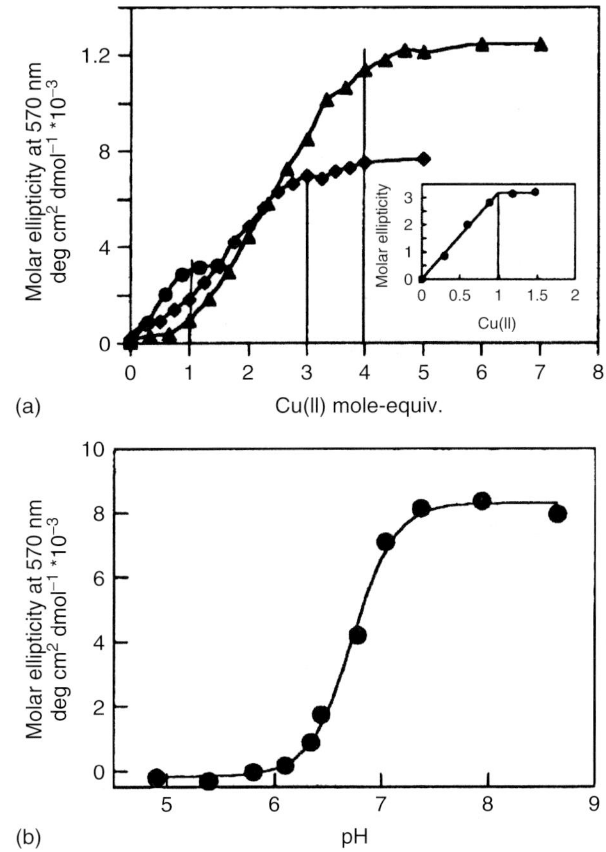

Fig. 6 Binding of copper ions to synthetic peptides containing the octarepeats of PrP. (a) Cu(II) binding curves: molar ellipticity at 570 nm with increasing amounts of Cu(II), pH 7.5. Filled circles, 2-His peptide, PrP(51–75) (0.34 mM). Filled diamonds, 3-His peptide, PrP(66–91) (0.021 mM). Filled triangles, 4-His peptide, PrP(58–91) (0.033 mM). (b) pH dependence of the ellipticity at 570 nm for PrP(58–91). The pH-dependence curve has been fitted to the following equation: $\Delta\varepsilon_{obs} = [\Delta\varepsilon_{acid}[H^+]^n + \Delta\varepsilon_{base}[H^+]K_a]/[H^+]^n + K_a]$, in which n = Hill coefficient and K_a = acid dissolution constant for the transition. The midpoint of the transition is pH 6.7. Reprinted with permission, from (Viles, J.H., Cohen, F.E., Prusiner, S.B., Goodin, D.B., Wright, P.E., Dyson, H.J. (1999) Copper binding to the prion protein: structural implications of four identical cooperative binding sites, *Proc. Natl. Acad. Sci. U. S. A.* **96**, 2042–2047), copyright 1999 National Academy of Sciences, USA.

6.7, suggesting that the binding of Cu^{2+} occurred through the imidazole nitrogens of histidine residues.

In studies of full-length recPrP, Cu^{2+} was found to catalyze the oxidation of the histidine residues within the octarepeats. These findings argue that PrP may function as Cu^{2+}-binding protein. Consistent with studies on recPrP, membranes isolated from the brains of *Prnp*$^{0/0}$ mice were reported to have substantially lower levels of Cu^{2+} than wt mice. However, attempts to confirm this relationship between brain Cu^{2+} and PrP levels were unsuccessful. Interestingly, the copper-chelating reagent cuprizone administered to rodents causes spongiform degeneration resembling that induced by prions.

8.7 Conserved Ala-Gly Region

In addition to the octarepeat, the other region of notable conservation is in the sequence at the C-terminal end of the last octarepeat. Here, an unusual glycine- and alanine-rich region from A113 to Y128 is found (Fig. 5). Although no differences have been found in this part of the sequence between species, a single point mutation A117V is linked to GSS. The conservation of structure suggests an important role in the function of PrP^C; in addition, this region is likely to be important in the conversion of PrP^C into PrP^{Sc}.

9 Structures of PrP Isoforms

Mass spectrometry and gas phase sequencing were used to search for posttranslational chemical modifications that might explain the differences in the properties of PrP^C and PrP^{Sc}. No modifications differentiating PrP^C from PrP^{Sc} were found. These observations forced the consideration of the possibility that a conformational change distinguishes the two PrP isoforms.

When the secondary structures of the PrP isoforms were compared by optical spectroscopy, they were found to be markedly different. Fourier transform infrared (FTIR) and circular dichroism (CD) spectroscopy studies showed that PrP^C contains approximately 40% α-helix and little β-sheet while PrP^{Sc} is composed of approximately 30% α-helix and 45% β-sheet. That the two PrP isoforms have the same amino acid sequence runs counter to the widely accepted view that the amino acid sequence specifies only one biologically active conformation of a protein. Like PrP^{Sc}, PrP 27–30 has a high β-sheet content, which is consistent with the earlier finding that PrP 27–30 polymerizes into amyloid fibrils. Denaturation of PrP 27–30 under conditions that reduced infectivity resulted in a concomitant diminution of β-sheet content.

Prior to comparative studies on the structures of PrP^C and PrP^{Sc}, metabolic labeling studies showed that the acquisition of protease resistance in PrP^{Sc} is a posttranslational process. In a search for chemical differences that would distinguish PrP^{Sc} from PrP^C, ethanolamine was identified in hydrolysates of PrP 27–30, which signaled the possibility that PrP might contain a GPI anchor. Both PrP isoforms were found to carry GPI anchors. PrP^C was found on the surface of cells, where it could be released by cleavage of the anchor. Subsequent studies showed that PrP^{Sc} formation occurs after PrP^C reaches the cell surface and localizes to caveolae-like domains (CLDs).

9.1 NMR Structures of recPrP

The NMR structures of more than 10 PrPs from different species have been determined to be quite similar (Table 4). The recPrP molecules were produced in *E. coli* and isotopically labeled. When refolded into proteins with a high α-helical content, these recPrPs appear to be representative of PrP^C. It is noteworthy that recPrPs produced in *E. coli* have neither N-linked sugar chains nor a GPI anchor, both of which are attached to PrP^C. However, synthetic prions produced from recMoPrP(89–230) demonstrated that neither N-linked sugar chains nor a GPI anchor is required for PrP^{Sc} formation. The structure of MoDpl

Tab. 4 Structures of PrP, Dpl, and Ure2p[a].

Species	***Protein***	***Determined by***	***PDB accession codes***
Mouse	recPrP(121–231)	NMR	1AG2
Mouse	recPrP(23–231)	NMR	data not released?
Syrian hamster	recPrP(90–231)	NMR	1B10, updated 1999
Syrian hamster	recPrP(29–231)	NMR	data not released?
Syrian hamster	synPrP(104–113) in complex with 3F4	NMR	1CU4 in complex with 3F4
Syrian hamster	recPrP(90–231)	NMR	used to update 1B10
Human	recPrP(23–230)	NMR	1QLX and 1QLZ
Human	recPrP(90–230)	NMR	1QM0 and 1QM1
Human	recPrP(121–230)	NMR	1QM2 and 1QM3
Cattle	recPrP(121–230)	NMR	1DWY and 1DWZ
Cattle	recPrP(23–230)	NMR	1DX0 and 1DX1
Human	recPrP(121–230), M166V	NMR	1E1G and 1E1J
Human	recPrP(121–230), S170N	NMR	1E1P and 1E1S
Human	recPrP(121–230), R220K	NMR	1E1U and 1E1W
Human	recPrP(90–231), E200K	NMR	1QM0, 1FKC, and 1FO7
Human	recPrP(90–231)	X ray	1I4M
Human	recPrP(121–230), M166C, E221C	NMR	1H0L
Sheep	synPrP(145–169)	NMR	1G04
Sheep	synPrP(145–169)	NMR	1M25
Mouse	recDpl(26–157)	NMR	1I17
Human	recDpl(24–152)	NMR	1LG4
S. cerevisiae	recUre2p(95–354)	X ray	1G6W and 1G6Y
S. cerevisiae	recUre2p(95–354) Glutathion complex	X ray	1JZR, 1K0B, and 1K0D
S. cerevisiae	recUre2p(95–354) S-P-Nitrobenzylglutathione complex	X ray	1K0C
S. cerevisiae	recUre2p(95–354) S-Hexylglutathione complex	X ray	1K0A
S. cerevisiae	recUre2p(97–354)	X ray	1HQO

[a] Data compiled by Holger Wille.

is similar to that of MoPrP even though the two proteins share only 25% sequence similarity (Fig. 4).

The NMR structure of recSHaPrP (90–231) was determined after the protein was purified and refolded (Fig. 7a). Residues 90–112 are not shown because marked conformational heterogeneity was found in this region while residues 113–126 constitute the conserved hydrophobic region that also displays some structural plasticity. Although some features of the structure of recSHaPrP(90–231) are similar to those reported earlier for the smaller recMoPrP(121–231) fragment, substantial differences were found. For example, the loop at the N-terminus of helix B is defined in recSHaPrP(90–231) but is disordered in recMoPrP(121–231); in addition, helix C is composed of residues 200–227 in recSHaPrP(90–231) but extends only from residues 200–217 in recMoPrP(121–231). The loop and the C-terminal portion of

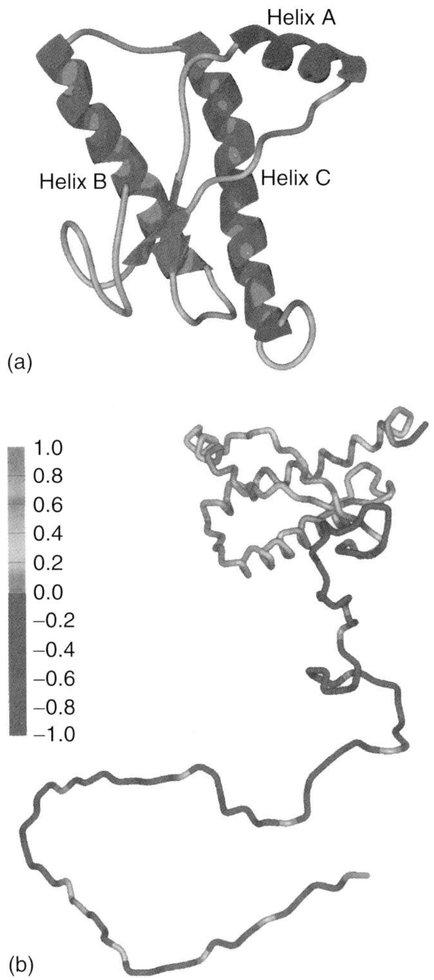

Fig. 7 Structures of PrP^{C}. (a) NMR structure of Syrian hamster (SHa) recombinant (rec) PrP(90–231). Presumably, the structure of the α-helical form of recPrP(90–231) resembles that of PrP^{C}. recPrP(90–231) is viewed from the interface where PrP^{Sc} is thought to bind to PrP^{C}. The color scheme is: α-helices A (residues 144–157), B (172–193), and C (200–227) in blue; loops in yellow; residues 129–134 encompassing strand S1 and residues 159–165 encompassing strand S2 in green. (b) Schematic diagram showing the flexibility of the polypeptide chain for PrP(29–231). The structure of the portion of the protein representing residues 90–231 was taken from the coordinates of PrP(90–231). The remainder of the sequence was hand-built for illustration purposes only. The color scale corresponds to the heteronuclear $[^{1}H]$-^{15}N NOE data: red for the lowest (most negative) values, where the polypeptide is most flexible, to blue for the highest (most positive) values in the most structured and rigid regions of the protein. (See color plate. p. xxxv)

helix C are particularly important, as described below. Whether the differences between the two recPrP fragments are due to (1) their different lengths, (2) species-specific differences in sequences, or (3) the conditions used for solving the structures remains to be determined.

NMR studies of recMoPrP(23–231) and recSHaPrP(29–231) showed that like the shorter fragments described above, full-length PrP is also a three helix–bundle protein with two short antiparallel β-strands. While the three helices form a globular C-terminal domain, the N-terminal domain is highly flexible and lacks identifiable secondary structure under the experimental conditions employed (Fig. 7b). Studies of recSHaPrP(29–231) indicate transient interactions between the C-terminal end of helix B and the highly flexible, N-terminal random-coil containing the octarepeats (residues 29–125).

9.2 Electron Crystallography of PrP^{Sc}

Because the insolubility of PrP^{Sc} has frustrated structural studies by X-ray crystallography or NMR spectroscopy, electron crystallography was used to characterize the structure of two infectious variants of PrP. Isomorphous, two-dimensional crystals of PrP 27–30 and a miniprion (PrP^{Sc}106) were identified by negative-stain electron microscopy. Image processing allowed the extraction of limited structural information to 7-Å resolution. Comparing projection maps of PrP 27–30 and PrP^{Sc}106, the 36-residue internal deletion of the miniprion was visualized and the N-linked sugars were localized (Fig. 8). The dimensions of the monomer and the locations of the deleted segment and sugars were used as constraints in the construction of models for PrP^{Sc}. Only models featuring parallel β-helices as the key element could satisfy the constraints.

A single antiparallel β-sheet as previously suggested was not consistent with the observed densities in the projection maps obtained from the 2D crystals. Specifically, the sheets were far too wide to fit into the observed hexameric arrangement. Efforts to adjust the sheet morphology to fit the density required the use of shorter strands. The amount of β-structure in these altered β-sheets was no longer compatible with the amounts of β-sheet observed by FTIR spectroscopy. Furthermore, antiparallel β-sheets typically have a twist of $\sim 20°$ per strand.

From a study of 119 all-β folds observed in globular proteins, it was determined that if PrP^{Sc} follows a known protein fold, it adopts either a β-sandwich or parallel β-helical architecture. With increasing evidence arguing for a parallel β-sheet organization in amyloids, it was argued that the sequence of PrP is compatible with a parallel left-handed β-helical fold. Left-handed β-helices readily form trimers, providing a natural template for a trimeric model of PrP^{Sc} (Fig. 9a). This trimeric model accommodates the PrP sequence from residues 89–175 in a β-helical conformation with the C-terminus (residues 176–227) retaining the disulfide-linked α-helical conformation observed in PrP^{C}. In addition, the proposed model matches the structural constraints of the crystals, positioning residues 141–176 and the N-linked sugars appropriately. The parallel left-handed β-helical model provides a coherent framework that is consistent with the stacking of trimeric units of PrP 27–30 to form PrP amyloid fibrils (Fig. 9b).

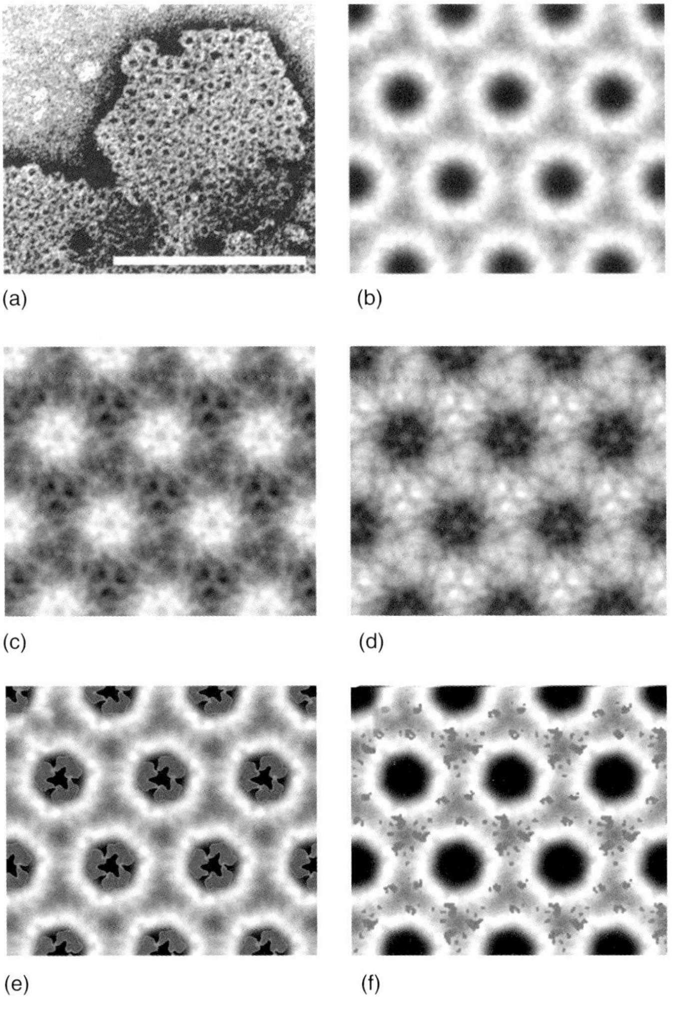

Fig. 8 Two-dimensional crystals of PrP^{Sc}. (a) A 2D crystal of PrP^{Sc}106 stained with uranyl acetate. Bar = 100 nm. (b) Image processing result after correlation-mapping and averaging followed by crystallographic averaging. (c and d) Subtraction maps between the averages of PrP 27–30 and PrP^{Sc}106 (panel B). (c) PrP^{Sc}106 minus PrP 27–30 and (d) PrP 27–30 minus PrP^{Sc}106, showing major differences in lighter shades. (e and f) The statistically significant differences between PrP 27–30 and PrP^{Sc}106 calculated from (c) and (d) in red and blue, respectively, overlaid onto the crystallographic average of PrP 27–30. Reprinted with permission, from Wille, H., Michelitsch, M.D., Guénebaut, V., Supattapone, S., Serban, A., Cohen, F.E., Agard, D.A. Prusiner, S.B. (2002) Structural studies of the scrapie prion protein by electron crystallography, *Proc. Natl. Acad. Sci. U. S. A.* **99**, 3563–3568 Wille et al., (2002), copyright 2002 National Academy of Sciences, USA. (See color plate. p. xxxvi)

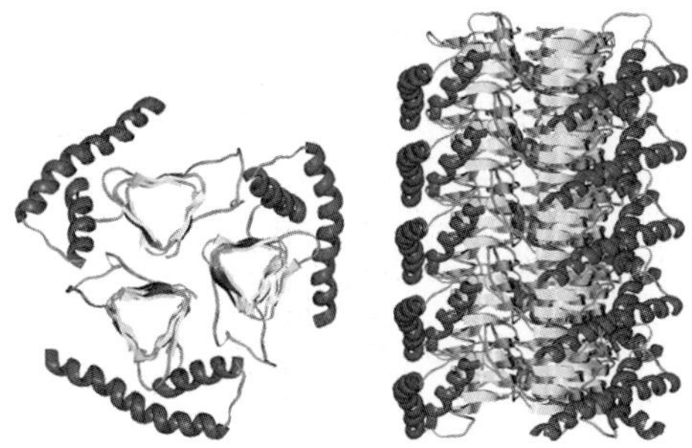
(a) (b)

Fig. 9 Left-handed β-helical models of PrP 27–30. (a) Trimeric model of PrP 27–30 built by superimposing three monomeric models onto the coordinates of the C'_αs of the structure of trimeric carbonic anhydrase from *Methanosarcina thermophila* (PDB ID 1THJ). (b) A model of the PrP 27–30 fiber, constructed by stacking five trimeric discs. Figure prepared by Cédric Govaerts.

10 Prion Replication

The mechanism of prion replication is not well understood. The formation of nascent prions requires that PrP^C be converted into PrP^{Sc}. Some investigators have argued that this process involves a nucleation–polymerization (NP) reaction while others have contended that a more likely mechanism employs a template-assisted process.

In an uninfected cell, PrP^C with the wt sequence is likely to exist in equilibrium in its monomeric α-helical, protease-sensitive state or bound to some other protein, such as protein X (Fig. 10). The conformation of PrP^C that is bound to protein X is denoted PrP*; this conformation is likely to be different from that determined under aqueous conditions for monomeric recPrP. The PrP*–protein X complex will bind PrP^{Sc}, thereby creating a replication-competent assembly. Order-of-addition experiments demonstrate that protein X binding to PrP^C precedes productive PrP^{Sc} interactions. A conformational change takes place wherein PrP*, in a shape competent for binding to protein X and PrP^{Sc}, represents the initial phase in the formation of infectious PrP^{Sc}.

Although many lines of evidence reviewed below argue for a template-assisted process in the formation of nascent PrP^{Sc}

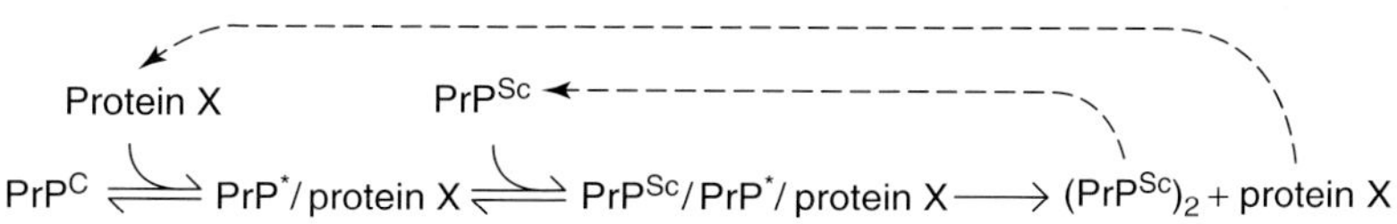

Fig. 10 Hypothetical scheme for template-assisted PrP^{Sc} formation. In the initial step, PrP^C binds to protein X to form the PrP*–protein X complex. Next, PrP^{Sc} binds to PrP* that has already formed a complex with protein X. When PrP* is transformed into a nascent molecule of PrP^{Sc}, protein X is released and a dimer of PrP^{Sc} remains. The inactivation target size of an infectious prion suggests that it is composed of a dimer of PrP^{Sc}. In the model depicted here, a fraction of infectious PrP^{Sc} dimers dissociates into uninfectious monomers as the replication cycle proceeds, while a majority of the dimers accumulates in accord with the increase in prion titer that occurs during the incubation period. Another fraction of PrP^{Sc} is cleared presumably by cellular proteases. The precise stoichiometry of the replication process remains uncertain. Reprinted with permission, from (Prusiner, S.B., Scott, M.R., DeArmond, S.J., Cohen, F.E. (1998) Prion protein biology, *Cell* **93**, 337–348), copyright 1998 Cell Press.

in mammalian cells, the formation of synthetic prions from PrP(89–143,P101L) as well as from recPrP(89–230) establishes that infectious prions can be generated *de novo* without the aid of auxiliary proteins.

10.1 Mechanism of Prion Propagation

In attempting to probe the mechanism of prion formation, it is reasonable to determine the rate-limiting step in prion formation. First, the impact of the concentration of PrP^{Sc} in the inoculum, which is inversely proportional to the length of the incubation time, must be considered. Second, the sequence of PrP^{Sc} that forms an interface with PrP^{C} must be taken into account. When the sequences of the two isoforms are identical, the shortest incubation times are observed. Third, the strain-specific conformation of PrP^{Sc} must be considered. Some prion strains exhibit longer incubation times than others; the mechanism underlying this phenomenon is not understood. From these considerations, a set of conditions exist under which initial PrP^{Sc} concentrations can be rate limiting. These effects presumably relate to the stability of the PrP^{Sc}, its targeting to the correct cells and subcellular compartments, and its ability to be cleared. Once infection in a cell is initiated and endogenous PrP^{Sc} production is operative, then the following discussion of PrP^{Sc} formation seems most applicable. If the assembly of PrP^{Sc} into a specific dimeric or multimeric arrangement is difficult, then an NP formalism would be relevant. In NP processes, nucleation is the rate-limiting step and elongation, or polymerization, is facile. These conditions are frequently observed in peptide models of aggregation phenomena; however, studies in ScN2a cells and with Tg mice expressing foreign PrP genes suggest that a different process occurs.

10.2 Template-assisted Prion Formation

From investigations with mice expressing both the SHaPrP transgene and the endogenous MoPrP gene, designated Tg(SHaPrP) mice, it is clear that PrP^{Sc} provides a template for directing prion replication; a template is defined as a catalyst that leaves its imprint on the product of the reaction. Inoculation of these mice with $SHaPrP^{Sc}$ leads to the production of nascent $SHaPrP^{Sc}$ and not $MoPrP^{Sc}$. Conversely, inoculation of the Tg(SHaPrP) mice with $MoPrP^{Sc}$ results in $MoPrP^{Sc}$ and not $SHaPrP^{Sc}$ formation.

Even stronger evidence for templating has emerged from studies of prion strains passaged in Tg(MHu2M)*Prnp*$^{0/0}$ mice, which express a chimeric mouse–human gene, as described in more detail below. Even though the conformational templates were initially generated with PrP^{Sc} molecules having different sequences from patients with inherited prion diseases, these templates are sufficient to direct replication of distinct PrP^{Sc} molecules when the amino acid sequences of the substrate PrPs are identical.

Another line of evidence for template-assisted prion replication comes from studies of FI. In FFI, the protease-resistant fragment of PrP^{Sc} after deglycosylation has an M_r of 19 kDa when measured by immunoblotting of brain extracts from humans with FFI or from Tg(MHu2M)*Prnp*$^{0/0}$ mice that were inoculated with FFI prions. In both humans and inoculated Tg(MHu2M)*Prnp*$^{0/0}$ mice on both first and second passage, PrP^{Sc} is confined largely to the thalamus. These findings argue that the conformation of

PrP^{Sc} that yields a 19-kDa polypeptide after deglycosylation is propagated both in humans and in mice expressing an artificial chimeric transgene; in addition, prion accumulation is confined to the thalamus. Additional evidence supporting these assertions comes from a patient who died after developing a clinical disease similar to FFI. Because both PrP alleles encoded the wt PrP sequence and methionine at position 129, we labeled this sFI. At autopsy, the spongiform degeneration, reactive astrocytic gliosis, and PrP^{Sc} deposition were confined to the thalamus. Moreover, the PrP^{Sc} both in the patient's brain and in the brains of inoculated Tg(MHu2M)$Prnp^{0/0}$ mice yielded a 19-kDa fragment after deglycosylation. These findings argue that the clinicopathologic phenotype is determined by the conformation and not the amino acid sequence of PrP^{Sc}. Additionally, FI prions can be replicated with mutant human PrP(D178N), wt human PrP, or chimeric human–mouse PrP. With these different PrPs as substrates, the conformation of PrP^{Sc} must be faithfully copied; template-assisted prion replication provides for such a mechanism.

10.3 Evidence for Protein X

Protein X was postulated to explain the results of the transmission of Hu prions to Tg mice (Table 5). Mice expressing both $MoPrP^C$ and $HuPrP^C$ were resistant while those expressing only $HuPrP^C$ were susceptible to Hu prions. These results argue that $MoPrP^C$ inhibited transmission of Hu prions, that is, the formation of nascent $HuPrP^{Sc}$. In contrast, mice expressing both $MoPrP^C$ and chimeric $MHu2MPrP^C$ were susceptible to Hu prions and mice expressing $MHu2MPrP^C$ alone were only slightly more susceptible. These findings contend that $MoPrP^C$ has only a minimal effect on the formation of chimeric $MHu2MPrP^{Sc}$.

When the data on Hu prion transmission to Tg mice were considered together, they suggested that $MoPrP^C$ prevented the conversion of $HuPrP^C$ into PrP^{Sc} by binding to another Mo protein (protein X) with a higher affinity than $HuPrP^C$ does. It was postulated that $MoPrP^C$ has little effect on the formation of PrP^{Sc} from $MHu2MPrP^C$ (Table 5) because $MoPrP^C$ and $MHu2MPrP^C$ share the same amino

Tab. 5 Influence of prion species and strain on transmission to Syrian hamsters and mice[a].

Inoculum	*Recipient*	*Incubation time [days ± SEM] (n/n_0)*	
		Sc237 prions	*139H prions*
SHa>SHa	SHa	77 ± 1 (48/48)	167 ± 1 (94/94)
SHa>SHa	non-Tg mice	>700 (0/9)	499 ± 15 (11/11)
SHa>SHa	Tg(SHaPrP)81 mice	75 ± 2 (22/22)	110 ± 2 (19/19)

[a]The species of prion is encoded by the primary structure of PrP^{Sc} and the strain of prion is enciphered by the tertiary structure of PrP^{Sc}. We recognize that the primary structure as well as posttranslational chemical modifications determine the tertiary structure of PrP^C, but we argue that the conformation of PrP^C is modified by PrP^{Sc} as it is refolded into a nascent PrP^{Sc} molecule.

acid sequence at the C-terminus. This also suggested that $MoPrP^C$ only weakly inhibited transmission of SHa prions to Tg(SHaPrP) mice because SHaPrP is more closely related to MoPrP than to HuPrP.

The search for protein X has been frustrating because many proteins are known to bind to PrP^C but none have been shown to participate in PrP^{Sc} formation. Whether protein X is one protein or a complex of proteins remains to be established. It is reasonable to expect that mice deficient in protein X would not replicate prions or do so extremely slowly, resulting in prolonged incubation times.

10.4 Dominant-negative Inhibition

To extend the foregoing findings in Tg mice, an expression vector with an insert encoding a chimeric mouse–hamster PrP, designated MHM2PrP, was transfected into ScN2a cells. MHM2PrP carries two Syrian hamster residues, which create an epitope that is recognized by the anti-SHaPrP 3F4 mAb that does not react with mouse PrP. Substitution of a Hu residue at position 214 or 218 in MHM2PrP prevented formation of $MHM2PrP^{Sc}$ in ScN2a cells. The side chains of these residues protrude from the same surface of the C-terminal α-helix, forming a discontinuous epitope with residues 167 and 171 in an adjacent loop. Like MHM2PrP(Q218K), substitution of a basic residue at position 167 or 171 prevented PrP^{Sc} formation. When MHM2PrP and MHM2PrP(Q218K) were coexpressed, the conversion of $MHM2PrP^C$ into PrP^{Sc} was inhibited, arguing that MHM2PrP(Q218K) was acting as a dominant negative. Similar results were obtained when studies were performed with cells expressing MHM2PrP(Q167R).

One interpretation of dominant-negative inhibition of prion propagation is that mutant PrP binds more avidly to protein X than does wt PrP. In this scenario, mutant PrP binds to protein X and prevents binding of wt PrP. This explanation is also consistent with the protective effects of basic polymorphic residues in PrP of humans and sheep. The E219K substitution seems to render humans resistant to sCJD and Q171R renders sheep resistant to scrapie. In MHM2, the Q218K mutation corresponds to E219K in humans and the Q167R mutation corresponds to Q171R in sheep (Fig. 11).

To determine whether dominant-negative inhibition of prion formation occurs *in vivo*, Tg mice expressing PrP with either the Q167R or Q218K substitution alone or in combination with wt PrP were produced. Tg(MoPrP,Q167R)$Prnp^{0/0}$ mice expressing mutant PrP at levels equal to non-Tg FVB mice were inoculated with prions and remained healthy for more than 550 days, indicating that inoculation did not initiate a process sufficient to cause disease. Immunoblots of brain homogenates and histologic analysis revealed no abnormalities. Tg(MoPrP,Q167R)$Prnp^{+/+}$ mice expressing both mutant and wt PrP exhibited neurologic dysfunction at ~450 days after inoculation; the brains of three of these mice that were sacrificed at 300 days revealed low levels of PrP^{Sc} as well as numerous vacuoles and severe astrocytic gliosis. Both Tg(MoPrP,Q218K)$Prnp^{0/0}$ and Tg(MoPrP,Q218K)$Prnp^{+/+}$ mice expressing the transgene product at 16× remained healthy for more than ~450 days after inoculation. Neither PrP^{Sc} deposition nor neuropathology was found. Tg mice expressing MoPrP(Q218K) at 32× developed spontaneous neurodegeneration at ~450 days of age. These studies

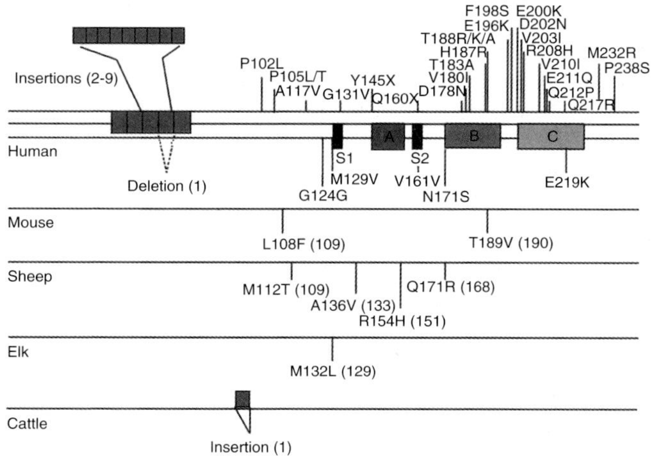

Fig. 11 Mutations and polymorphisms of the prion protein gene. PrP mutations causing inherited human prion disease and PrP polymorphisms found in humans, mice, sheep, elk, and cattle. Above the line of the human sequence are mutations that cause inherited prion disease. Below the lines are polymorphisms, some but not all of which are known to influence the onset as well as the phenotype of disease. Residue numbers in parentheses correspond to the human codons. Much of the data was compiled by J.-L. Laplanche.

demonstrate that although dominant-negative inhibition of wt PrP^{Sc} formation occurs, expression of dominant-negative PrP at the same level as wt PrP does not completely prevent prion formation. However, expression of dominant-negative PrP alone had no deleterious effects on the mice and did not support prion propagation.

11
Sporadic Human Prion Diseases

In most patients with CJD, there is neither an infectious nor heritable etiology. How prions arise in patients with sCJD is unknown; hypotheses include: (1) horizontal transmission from humans or animals; (2) somatic mutation of the *PRNP* ORF; (3) spontaneous conversion of PrP^C into PrP^{Sc}; and (4) the accumulation of PrP^{Sc}, which is normally present at very low levels. Numerous attempts to establish an infectious link between sCJD and a preexisting prion disease in animals or humans have been unrewarding. Studies demonstrating the formation of synthetic prions from PrP(89–143,P101L) as well as from recPrP(89–230) show that only PrP^C is needed for an animal for producing infectious prions. Thus, the spontaneous conversion of PrP^C to PrP^{Sc} seems reasonable and any mammal expressing PrP^C is capable of forming prions.

12
Heritable Human Prion Diseases

The recognition that 10 to 15% of CJD cases are familial led to the suspicion that genetics plays a role in this disease. As with scrapie, the relative contributions of genetic and infectious etiologies in the human prion diseases remain puzzling.

More than 30 different mutations of the PrP gene have been shown to segregate with the heritable human prion diseases. Five of these mutations have been genetically linked to the inherited human prion diseases (Table 3). Virtually all cases of GSS and FFI appear to be caused by germline mutations in the PrP gene. The brains of humans dying of inherited prion disease contain infectious prions that have been transmitted to experimental animals.

12.1 GSS and Genetic Linkage

The discovery that GSS, which is a familial disease, could be transmitted to apes and monkeys was first reported when many still thought that scrapie, CJD, and related disorders were caused by viruses. With the discovery that the P102L mutation of the human PrP gene was genetically linked to GSS, prion diseases were concluded to have both genetic and infectious etiologies. In that study, the P102L mutation was linked to development of GSS with a logarithm of the odds (LOD) score exceeding 3, demonstrating a tight association between the altered genotype and disease phenotype (Fig. 3B). This mutation may be caused by the deamination of a methylated CpG in a germline PrP gene, which results in the substitution of a leucine for proline. This mutation has been found in many families in numerous countries, including the first family identified to have GSS.

12.2 fCJD Caused by Octarepeat Inserts

An insert of 144 bp containing six octarepeats at codon 53, in addition to the five that are normally present, was described in patients with CJD from four families residing in southern England. Genealogic investigations showed that all four families are related, arguing for a single founder born more than two centuries ago. The LOD score for this extended pedigree exceeds 11. Studies from several laboratories have demonstrated that inserts of two, four, five, six, seven, eight, or nine octarepeats in addition to the normal five are found in individuals with inherited CJD (Fig. 11).

12.3 fCJD in Libyan Jews

The unusually high incidence of CJD among Israeli Jews of Libyan origin was thought to be due to the consumption of lightly cooked sheep brain or eyeballs. Molecular genetic investigations revealed that Libyan and Tunisian Jews with fCJD have a PrP gene point mutation at codon 200, resulting in a Glu-to-Lys substitution (Fig. 11). The E200K mutation has been genetically linked to fCJD, with an LOD score exceeding 3; the same mutation has also been found in patients from Orava in North Central Slovakia, in a cluster of familial cases in Chile, and in a large German family living in the United States.

Most patients are heterozygous for the mutation and thus, express both mutant and wt PrP^{C}. In the brains of patients who die of fCJD(E200K), mutant PrP^{Sc} is both insoluble and protease-resistant, whereas much of wt PrP differs from both PrP^{C} and PrP^{Sc} in that it is insoluble but readily digested by proteases. Whether this form of PrP is an intermediate in the conversion of PrP^{C} into PrP^{Sc} remains to be established.

12.4 Penetrance of fCJD

Life-table analyses of carriers harboring the codon 200 mutation exhibit complete penetrance. Therefore, if the carriers live long enough, they will all eventually develop prion disease. Some investigators have argued that the inherited prion diseases are not fully penetrant and thus an environmental factor, such as the ubiquitous "scrapie virus," is required for illness to be manifest, but no viral pathogen has been found to date.

12.5 Fatal Familial Insomnia

The D178N mutation has been linked to the development of FFI, with an LOD score exceeding 5. More than 30 families worldwide with FFI have been recorded. Studies of inherited human prion diseases demonstrate that the amino acid at polymorphic residue 129 with the D178N pathogenic mutation alters the clinical and neuropathologic phenotype. The D178N mutation combined with M129 results in FFI. In this disease, adults generally over age 50 years present with a progressive sleep disorder and usually die within one year. In their brains, deposition of PrP^{Sc} is confined largely within the anteroventral and the dorsal medial nuclei of the thalamus. In contrast, the same D178N mutation with V129 produces fCJD, in which the patients present with dementia and widespread deposition of PrP^{Sc} is found postmortem. The first family to be recognized with CJD was found to carry the D178N mutation.

12.6 Human PrP Gene Polymorphisms

Several polymorphisms in the PrP gene may affect the development and expression of prion disease. A Met/Val polymorphism at PrP codon 129 has been identified (Fig. 11). This polymorphism appears to influence expression of the inherited as well as the iatrogenic and sporadic forms of prion disease. A second polymorphism resulting in an amino acid substitution at codon 219 (Glu/Lys) has been reported to occur with a frequency of approximately 12% in the Japanese population but not in Caucasians. In people heterozygous for the lysine substitution, this expression seems to act as a dominant negative and protects against CJD by the avid binding of mutant PrP^{C} to protein X. By sequestering protein X, PrP^{C}(K219) prevents PrP^{C}(E219) from being converted into PrP^{Sc}. In people homozygous for the lysine substitution, it seems likely that PrP^{C}(K219) is not converted into PrP^{Sc}. A third polymorphism results in an amino acid substitution at codon 171 (Asn/Ser), which lies adjacent to the protein X binding site. This polymorphism has been found in Caucasians but it has not been studied extensively and it is not known to influence the binding of PrP^{C} to protein X. A fourth polymorphism is the deletion of a single octarepeat (24 bp), which has been found in 2.5% of Caucasians. In another study of over 700 individuals, this single octarepeat deletion was found in 1.0% of the population.

Studies of Caucasian patients with sCJD have shown that most are homozygous for Met or Val at codon 129. This contrasts with the general Caucasian population, in which frequencies for the codon 129 polymorphism are 12% Val/Val, 37% Met/Met, and 51% Met/Val. In contrast, the frequency of the valine allele in the Japanese population is much lower. Heterozygosity at codon 129 is more frequent in Japanese CJD patients (18%) than in the general Japanese population, in which the polymorphism frequencies are 0% Val/Val, 92% Met/Met, and 8% Met/Val.

While no specific mutations have been identified in the PrP gene of patients with sCJD, homozygosity at codon 129 is consistent with the results of Tg mouse studies. The finding that homozygosity at codon 129 predisposes to sCJD supports a model of prion production that favors PrP interactions between homologous proteins, as appears to occur in Tg(SHaPrP) mice inoculated with either hamster prions or mouse prions, as well as in Tg(MHM2) mice inoculated with "artificial" prions.

13 Infectious Human Prion Diseases

Prions from different sources have infected humans. Human prions have been transmitted to others both by ritualistic cannibalism and iatrogenic means. Kuru in the highlands of New Guinea was transmitted by ritualistic cannibalism, as people in the region attempted to immortalize their dead relatives by eating their brains. Iatrogenic transmissions include prion-tainted human growth hormone (HGH) and gonadotropin, dura mater grafts, and corneal transplants from people who died of CJD. In addition, CJD cases have been recorded after neurosurgical procedures in which ineffectively sterilized depth electrodes or instruments were used.

13.1 Human Growth Hormone

More than 165 young adults have been diagnosed with iatrogenic (i) CJD between 4 and 30 years after receiving HGH or gonadotropin from cadaveric pituitaries. The longest incubation periods (20 to 30 years) are similar to those associated with more recent cases of kuru. Since 1985, recombinant HGH produced in *E. coli* has been used in place of cadaveric HGH. With recombinant HGH, no cases of iCJD have been identified.

13.2 Variant Creutzfeldt–Jakob Disease (vCJD)

In 1994, the first cases of CJD in teenagers and young adults that were eventually labeled variant (v) CJD occurred in Great Britain. More than 150 teenagers and young adults have died of vCJD in Britain, Ireland, Italy, and France. While the average age of vCJD patients is 26 years of age, the youngest patient was 12 years old and the oldest was 74 years of age. The median duration of illness was 13 months, with the range from 6 to 69 months.

In addition to the young age of these patients, vCJD is characterized by numerous PrP amyloid plaques surrounded by a halo of intense spongiform degeneration in the brain. These unusual neuropathologic changes have not been seen in CJD cases in the United States, Australia, or Japan. Both macaque monkeys and marmosets developed neurologic disease several years after inoculation with bovine prions, but only the macaques exhibited numerous PrP plaques similar to those found in vCJD.

The majority of vCJD patients presented with psychiatric symptoms, including dysphoria, withdrawal, anxiety, insomnia, and loss of interest. Generally, neurologic deficits did not appear until at least four months later; these neurologic changes consisted of memory loss, paresthesias, sensory deficits, gait disturbances, and dysarthria. To date, all vCJD cases have been reported from Britain, except for six cases in France, one in Italy, one in Ireland, and one in the United States. The US case is a 23-year-old woman, who is thought to have been exposed to bovine prions while living in Britain during the first 12 years of her life. From both epidemiologic and experimental studies, evidence is now quite compelling that vCJD is the result of prions being transmitted from cattle with BSE to humans through consumption of contaminated beef products.

Studies of the prion diseases have taken on new significance with the identification of vCJD. The number of cases of vCJD caused by bovine prions that will occur in the years ahead is unknown. Until more time passes, it will be impossible to assess the magnitude of this problem. These tragic cases have generated a continuing discourse concerning mad cows, prions, and the safety of the human and animal food supplies throughout the world. Untangling politics and economics from the science of prions seems to have been difficult in disputes between Great Britain and other European countries over the safety of beef and lamb products.

13.3 Transmission of vCJD Prions by Blood Transfusion

vCJD has been identified in two patients who received blood transfusions from donors that later died of vCJD. In one case, the recipient was a 69-year-old male who was transfused 6.5 years before the onset of neurologic dysfunction. His PrP gene encoded Met/Met at position 129. Many details of the second case are not published but the patient is known to have died of a non-neurologic disease. While vCJD prions were found in the spleen and cervical lymph nodes of this patient, none were found in the brain. The patient's PrP gene encoded Met/Val at position 129 and as such, is the first heterozygous codon 129 human identified with vCJD prions.

A glimpse of future vCJD cases caused by prion-tainted transfused blood may come from a survey of tissues collected during appendectomies and tonsillectomies. This UK survey reports that of the 12 674 appendectomy specimens examined, three were positive for PrP^{Sc} by immunohistochemistry. This finding argues that as many as 3800 people in the United Kingdom may be replicating vCJD prions in their lymphoid tissues. Considering that immunohistochemistry is less sensitive than Western immunoblotting, the number of people with vCJD prions in their lymphoid tissues may be substantially greater than the above estimate.

14 Strains of Prions

That goats with scrapie can manifest two different syndromes, one in which the goats became hyperactive and the other in which they became drowsy, raised the possibility that strains of prions might exist. Subsequent studies with mice documented the existence of multiple strains through careful measurements of incubation times and the distribution of vacuoles in the CNS.

The search for an explanation of how biological information encrypting the disease phenotype could be enciphered within the prion posed a conundrum. Many investigators kept arguing for a small nucleic acid but none could be found. Others argued that this information must be enciphered within the structure of PrP^{Sc}. The first evidence supporting the hypothesis that strain-specific information is enciphered in PrP^{Sc} came from studies on prions causing TME, which were passaged into Syrian hamsters. On serial passaging, two strains emerged: one strain (HY) produced hyperactivity in Syrian hamsters and the other (DY) was manifest as a drowsy syndrome, as was seen with scrapie strains in goats. The HY strain was similar to the Sc237 prion strain with respect to protease resistance and sedimentation properties, while the DY strain showed minimal protease resistance and was less readily sedimented.

Because the HY and DY strains emerged after passaging TME prions in Syrian hamsters, concern over the origin of these strains made the conclusions somewhat questionable. Because of such concerns, a more convincing investigation came from the transmission of FFI and fCJD to Tg mice. In this study, two different strains of prions were generated *de novo* in patients with PrP gene mutations and propagated in mice expressing the chimeric MHu2MPrP transgene. Brain homogenates of FFI and fCJD(E200K) patients transmitted disease to Tg(MHu2M) mice approximately 200 days after inoculation. The FFI inoculum induced formation of a 19-kDa PrP^{Sc} fragment, as measured by SDS-PAGE after limited proteolysis and removal of Asn-linked carbohydrates; in contrast, fCJD(E200K) produced a 21-kDa PrP^{Sc} fragment. On second passage, Tg(MHu2M) mice inoculated with FFI prions showed an incubation time of ~130 days and a 19-kDa PrP^{Sc} fragment while those inoculated with fCJD(E200K) prions exhibited an incubation time of ~170 days and a 21-kDa PrP^{Sc} fragment. The experimental data demonstrated that $MHu2MPrP^{Sc}$ can exist in two different conformations, based on the sizes of the protease-resistant fragments, within an invariant amino acid sequence. The results of these studies argue that PrP^{Sc} acts as a template for the conversion of PrP^{C} into nascent PrP^{Sc}. Imparting the size of the protease-resistant fragment of PrP^{Sc} through conformational templating provides a mechanism for both the generation and propagation of prion strains.

In another set of studies, prions from cattle with BSE and humans with vCJD were used to demonstrate distinct prion strains in humans expressing wt PrP, in contrast to the studies described above with prions from patients harboring the E200K mutation. Both BSE and vCJD prions transmitted to Tg mice expressing BoPrP but not to mice expressing either HuPrP or MHu2MPrP. Moreover, sCJD, iCJD, and fCJD(E200K) prions failed to transmit disease to Tg(BoPrP) mice after more than 500 days. These findings demonstrate the prions that cause BSE and vCJD are similar, but distinct from those responsible for sCJD, iCJD, and fCJD(E200K).

14.1 Isolation of New Strains

Additional evidence implicating PrP in the phenomenon of prion strains comes from studies on the interspecies transmission of strains. Such studies were especially revealing when mice expressing chimeric hamster–mouse transgenes were used. Evidence was obtained showing that new strains pathogenic to Syrian hamsters could be obtained from prion strains that had been previously replicated by limiting dilution in mice. Both the generation and propagation of prion strains result from interactions between PrP^{Sc} and PrP^{C}. Additionally, strains that were isolated from different breeds of sheep with scrapie and thought to be distinct were shown to have identical properties. Such findings argue for the convergence of some strains and raise the issue of the limits of prion diversity.

Studies with different strains of prions propagated in Syrian hamsters and subsequently in Tg mice expressing MH2MPrP are most illustrative with respect to the isolation of new prion strains. These Tg mice were inoculated with the two prion strains, Sc237 and DY that had been serially passaged in Syrian hamsters. On first passage in Tg(MHM2)*Prnp*$^{0/0}$ mice, the SHa(Sc237) prions exhibited prolonged incubation times, typical of a species barrier. On subsequent passage in Tg(MHM2)*Prnp*$^{0/0}$ mice, the MH2M(Sc237) strain showed a profound shortening of incubation time. Moreover, PrP^{Sc} of the MH2M(Sc237) strain possesses different structural properties from PrP^{Sc} of the SHa(Sc237) strain, as demonstrated by relative conformational stability measurements. Conversely, transmission of SHa(DY) prions to Tg(MHM2)*Prnp*$^{0/0}$ mice did not encounter a species barrier and the MH2M(DY) strain retained the conformational and phenotypic properties of SHa(DY). These results extend the findings described above for FFI and fCJD(E200K) prions and contend that a change in PrP^{Sc} conformation is intimately associated with the emergence of a new prion strain.

14.2
Interplay between the Species and Strains of Prions

The recent advances in our understanding of the role of the primary and tertiary structures of PrP in the transmission of disease have given new insights into the pathogenesis of the prion diseases. The amino acid sequence of PrP encodes the species of the prion (Table 5), and the prion derives its PrP^{Sc} sequence from the last mammal in which it was passaged. While the primary structure of PrP is likely to be the most important or even sole determinant of the tertiary structure of PrP^{C}, existing PrP^{Sc} seems to function as a template in determining the tertiary structure of nascent PrP^{Sc} molecules as they are formed from PrP^{C}. In turn, prion diversity appears to be enciphered in the conformation of PrP^{Sc} and prion strains represent different conformers of PrP^{Sc}. The total number of prion strains, that is, different conformations of PrP^{Sc}, in existence remains to be established.

15
Prion Diseases of Animals

The prion diseases of animals include scrapie of sheep and goats, BSE, TME, chronic wasting disease (CWD) of mule deer and elk, feline spongiform encephalopathy, and exotic ungulate encephalopathy (Table 1).

15.1
PrP Polymorphisms in Sheep, Cattle, and Elk

In 1962, Parry argued that host genes were responsible for the development of scrapie in sheep. He was convinced that natural scrapie is a genetic disease that could be eradicated by proper breeding protocols. He considered its transmission by inoculation of importance primarily for laboratory studies and communicable infection of little consequence in nature. Other investigators viewed natural scrapie as an infectious disease and argued that host genetics only modulates susceptibility to an endemic infectious agent.

In sheep, polymorphisms at codons 136, 154, and 171 of the PrP gene that produce amino acid substitutions have been studied with respect to the incidence of scrapie (Fig. 11). Studies of natural scrapie in the United States have shown that 85% of afflicted sheep are of the Suffolk breed. Only those Suffolk sheep homozygous for Gln (Q) at codon 171 developed scrapie although healthy controls with Gln/Gln, Gln/Arg, and Arg/Arg genotypes were also found. These results argue that susceptibility in Suffolk sheep is governed by the PrP codon 171 polymorphism. As with the Suffolk breed, the PrP codon 171 polymorphism in Cheviot sheep has a profound influence on susceptibility to scrapie, and codon 136 also modulates susceptibility but less so than codon 171.

In contrast to sheep, different breeds of cattle have no specific PrP polymorphisms. The only polymorphism recorded in cattle is a variation in the number of octarepeats: most cattle, like humans, have five octarepeats but some have six. However, the presence of six octarepeats does not seem to be overrepresented in BSE.

In studies of CWD, the susceptibility of elk, but not deer, seems to be modulated by codon 132, which corresponds to codon 129 in humans (Fig. 11). Elk with CWD consistently express Met/Met at position 132; no elk with CWD expressing leucine at this residue have been found.

15.2 Bovine Spongiform Encephalopathy

Prion strains and the species barrier are of paramount importance in understanding the BSE epidemic in Britain, in which it is estimated that between one and two million cattle were infected with prions. The mean incubation time for BSE is approximately 5 years. Therefore, most cattle did not manifest disease because they were slaughtered between 2 and 3 years of age. Nevertheless, more than 180 000 cattle, primarily dairy cows, have died of BSE over the past decade. BSE is a massive common-source epidemic caused by meat and bone meal (MBM) fed primarily to dairy cows. MBM was prepared from the offal of sheep, cattle, pigs, and chickens as a high-protein nutritional supplement. In the late 1970s, the hydrocarbon-solvent extraction method used in the rendering of offal began to be abandoned, resulting in MBM with a much higher fat content. It is now thought that this change allowed scrapie prions from sheep to survive the rendering process and to be passed into cattle. Alternatively, bovine prions may have been present at low levels prior to modification of the rendering process and with the processing change, survived in sufficient numbers to initiate the BSE epidemic when reintroduced into cattle through ingestion of MBM. Perhaps, a particular conformation of $BoPrP^{Sc}$ selected for heat-resistance during the rendering process and then reselected multiple times as cattle infected by ingesting prion-contaminated MBM were slaughtered and their offal rendered into more MBM. Against the latter hypothesis is the widespread geographical distribution of the initial 17 cases of BSE throughout England, which occurred almost simultaneously.

In July 1988, the practice of feeding MBM to sheep and cattle was banned. Although statistical analyses demonstrate that the epidemic is disappearing as a result of this ruminant feed ban, reminiscent of the disappearance of kuru in the Fore people of New Guinea, it is unclear how many current cases of BSE are due to infection and how many arise spontaneously. In 2003, there were 1390 cases of BSE; how many were due to tainted feed is unknown. Although evidence of a preexisting prion disease of cattle, either in Great Britain or elsewhere, is scant, an outbreak of TME in Wisconsin has been cited as evidence for sporadic BSE. Moreover, the formation of synthetic prions from wt PrP argues that any mammal expressing PrP^{C} can produce prions spontaneously.

The origin of BSE prions cannot be determined by examining the amino acid sequence of PrP^{Sc} from cattle with BSE because it has the bovine sequence regardless of whether the initial prions originated from sheep or cattle. The bovine PrP sequence differs from that of sheep at seven or eight positions.

Brain extracts from cattle with BSE cause disease in cattle, sheep, mice, pigs, and mink after intracerebral inoculation, but prions in brain extracts from sheep with scrapie fed to cattle produced an illness substantially different from BSE. However, no exhaustive effort has been made to test different strains of sheep prions or to examine the disease following bovine-to-bovine passage.

15.3 Monitoring Cattle for BSE Prions

Although no reliable, specific test for prion disease in live animals is available, immunoassays for PrP^{Sc} in the brainstems of cattle currently provide the best available approach for preventing BSE prions from entering the human food supply. Determining how early in the incubation period PrP^{Sc} can be detected by immunological methods is now possible because a reliable bioassay is now available with the creation of Tg(BoPrP)*Prnp*$^{0/0}$ mice. Prior to the construction of Tg(BoPrP)*Prnp*$^{0/0}$ mice, cattle were used for bioassays of bovine prions. In a very limited study using cattle bioassays, bovine prions were undetectable in the obex of the brainstem until 26 months after oral inoculation. In these studies, prion infectivity was detected much earlier in the lymphoid tissue of the distal ileum.

Prior to development of Tg(BoPrP) *Prnp*$^{0/0}$ mice, non-Tg mice inoculated intracerebrally with BSE brain extracts required more than 300 days to develop disease. Generally, many of the inoculated mice fail to develop prion disease. It is noteworthy that a highly stable, mouse prion strain denoted 301V was derived by passage of BSE inocula into non-Tg *Prnp*$^{b/b}$ mice.

Both Western blots and ELISA tests are currently used in the postmortem evaluation of bovine brainstems. As configured, these tests are quite insensitive. The CDI, whose sensitivity is superior to conventional Western blots and ELISAs, offers a new approach to measuring PrP^{Sc} in tissues. The sensitivity of the CDI for BSE prions was found to be similar to that of bioassays using Tg(BoPrP)*Prnp*$^{0/0}$ mice. These studies suggest that the CDI may be able to detect BSE prions in asymptomatic cattle even when titers are quite low.

15.4 Compelling Evidence of Transmission of Bovine Prions to Humans

The restricted geographical occurrence and chronology of vCJD raise the possibility that BSE prions have been transmitted to humans. That only $\sim$150 vCJD cases have been recorded and the incidence has remained relatively constant made establishing the origin of vCJD difficult. No set of dietary habits distinguishes vCJD patients from apparently healthy people. Moreover, there is no explanation for the predilection of vCJD for teenagers and young adults. Why have older individuals not developed vCJD-based neuropathologic criteria? It is noteworthy that epidemiologic studies over the past three decades have failed to find evidence of transmission of sheep prions to humans. Attempts to predict the future number of cases of vCJD, assuming exposure to bovine prions prior to the offal ban, have been uninformative because so few cases of vCJD have occurred. Are we at the beginning of a prion disease epidemic in Britain, as seen for BSE and kuru, or will the number of vCJD cases remain small, as seen with iCJD caused by cadaveric HGH?

Recent studies of PrP^{Sc} from brains of patients who died of vCJD show a PrP glycoform pattern different from those found for sCJD and iCJD. The utility of measuring PrP glycoforms is questionable in trying to relate BSE to vCJD because PrP^{Sc} is formed after the protein is glycosylated and enzymatic deglycosylation of PrP^{Sc} requires denaturation. Moreover, synthetic prions have been formed from unglycosylated recMoPrP(89–230), demonstrating that glycosylation is unnecessary for prion formation.

Prior to the investigations recorded below laboratory studies were unconvincing in establishing a relationship between the conformations of PrP^{Sc} from cattle with BSE and those from humans with vCJD. Perhaps slightly more persuasive is the relationship demonstrated between vCJD and BSE based on similar incubation times in non-Tg RIII mice, of ~310 days after inoculation with human vCJD or bovine BSE prions. But such studies suffer from transmission of both BSE and vCJD prions to a heterologous host, that is, non-Tg mice expressing $MoPrP^{C}$. Using Tg mice, as was done for strains generated in the brains of patients with FFI or fCJD, compelling evidence for the transmission of bovine prions to humans was found. BSE prions transmitted to all inoculated Tg(BoPrP) mice after ~240 days but not to Tg mice expressing either HuPrP or MHu2MPrP. On second passage to Tg(BoPrP)*Prnp*$^{0/0}$ mice, the incubation time was unaltered, demonstrating the complete absence of a species barrier. Similar to BSE prions, vCJD prions transmitted readily to Tg(BoPrP)*Prnp*$^{0/0}$ mice, with a slightly longer incubation time of ~270 days, but poorly to Tg(HuPrP) and Tg(MHu2M) mice. Moreover, sCJD, iCJD, and fCJD(E200K) prions failed to transmit disease to Tg(BoPrP) mice after more than 500 days. On second passage of vCJD prions to Tg(BoPrP)*Prnp*$^{0/0}$ mice, the incubation time was reduced to ~225 days, demonstrating a small but expected species barrier. These findings argue that the strain-specific PrP^{Sc} conformations from BSE and vCJD prions are quite similar despite substantial differences in the amino acid sequences of BoPrP and HuPrP. Clearly, the conformation of $HuPrP^{Sc}$(vCJD) makes these prions much more readily transmittable to Tg(BoPrP)*Prnp*$^{0/0}$ mice than to either Tg(HuPrP)*Prnp*$^{0/0}$ or Tg(MHu2M)*Prnp*$^{0/0}$ mice.

15.5 Chronic Wasting Disease

Mule deer, white-tailed deer, and elk have been reported to develop CWD, which is unique among the prion diseases because it seems to be far more communicable than scrapie, BSE, or TME; moreover, it is the only prion disease known in free-ranging animals. CWD was first described in 1967 and reported to be a spongiform encephalopathy in 1978 based on histopathology in the brain. CWD has been found in the United States, Canada, and South Korea. In the United States,

CWD has been reported in Colorado, Wyoming, South Dakota, Nebraska, Oklahoma, Montana, New Mexico, Minnesota, and Wisconsin. In captive cervid herds, up to 90% of mule deer have been reported to be positive for prions and up to 60% of elk in Colorado and Wyoming develop CWD. Moreover, the incidence of CWD in cervids living in the wild has been estimated to be as high as 15%. The mode of transmission of CWD prion among mule deer, white-tailed deer, and elk is unresolved but contamination of grass with prions excreted in fecal matter seems to be a likely source. The high content of PrP^{Sc} in the intestinal lymphoid tissue of cervids with CWD supports such a scenario.

Brain homogenates from mule deer with CWD have transmitted disease to 4 of 13 cattle after intracerebral inoculation. These findings are particularly important because there is great concern that CWD prions might be transmitted to cattle grazing in contaminated pastures. In addition, CWD has been transmitted to ferrets, mink, squirrel monkeys, goats, and mice after intracerebral inoculation; however, only mule deer demonstrate efficient transmission of CWD prions by intracerebral inoculation. To date, endpoint titrations of CWD prions have not been performed in mule deer and Tg mice susceptible to CWD prions have not been developed, but the CDI has been adapted to measure CWD prions rapidly.

16 Fungal Prions

Although prions were originally defined in the context of an infectious pathogen, it is now becoming widely accepted that prions are elements that impart and propagate variability through multiple conformers of a normal, cellular protein. It is likely that such a mechanism will not be restricted to a single class of transmissible pathogens. Indeed, it is probable that this original definition will need to be extended to encompass other situations in which a similar mechanism of information transfer occurs. Two notable prion-like determinants, [*URE3*] and [*PSI*], have been described in yeast and another prion-like determinant has been reported in other fungi.

16.1 Yeast Prion Domains

In both [*PSI*] and [*URE3*], the "functional" determinants have been mapped to the C-terminal region of the protein, distinct from the "prion" domain, which comprises the N-terminal residues 65 and 114 of Ure2p and Sup35p, respectively. Although neither of the prion domains displays sequence identity either to each other or to PrP, the N-terminal regions of both Sup35p and mammalian PrP contain short repeated sequence elements: PQGGYQQYN in Sup35p and PHGGGWGQ in PrP. Interestingly, when the prion domains of both Sup35 and Ure2p are expressed in *E. coli* and purified, they polymerize spontaneously into amyloid-like fibrils. Polymerization of the prion domains of both Sup35 and Ure2p are able to create [*PSI*] and [*URE3*] states in yeast, respectively.

16.2 Dependence of Yeast Prions on Molecular Chaperones

The intrinsic power of the yeast genetic system has provided striking evidence of the involvement of chaperones in the propagation of yeast [*PSI*] "prions." A genetic screen for factors that suppress the [*PSI*] phenotype resulted in the isolation of a single suppressor plasmid, which was found to contain the chaperone Hsp104. Furthermore, propagation of [*PSI*] was eliminated by either overproduction or absence of Hsp104, and treatment of cells with guanidine or UV light led to induction of Hsp104. The significance of Hsp104 is unclear because no published data exist to indicate that [*URE3*] utilizes Hsp104; furthermore, overexpression of Sup35 at high levels can induce [*PSI*] in the absence of Hsp104.

16.3
Differences between Yeast and Mammalian Prions

A wealth of studies has established the concept of fungal prions. While fungal prions are produced in the cytoplasm, mammalian prions are produced within cholesterol-rich microdomains on the surface of cells called *rafts* or *CLDs*. Whether the process of fungal and mammalian prion replication are fundamentally different or quite similar remains to be established. It seems likely that many more similarities than differences will emerge because distinct strains of both fungal and mammalian prions seem to represent different conformations of a particular protein. This finding necessitates a mechanism by which a particular conformation can be templated and reproduced with a high degree of fidelity.

Interestingly, the prion state in yeast is proposed to be functionally inert in the case of both [*PSI*] and [*URE3*] and produces the same phenotype as inactivation of the maintenance gene. In contrast, prion diseases in mammals cannot be explained simply by the loss of function of PrP because ablation of the PrP gene does not as yet have a detectable deleterious effect.

17
Prion Diseases are Disorders of Protein Conformation

The study of prions has followed several unexpected directions over the past three decades. The discovery that prion diseases in humans are uniquely genetic and infectious has greatly strengthened and extended the prion concept. To date, more than 30 different mutations in the human PrP gene, all resulting in nonconservative substitutions have been found either to be linked genetically to or to segregate with the inherited prion diseases (Fig. 11).

Understanding how PrP^{C} unfolds and refolds into PrP^{Sc} will be of paramount importance in transferring advances in the prion diseases to studies of other degenerative illnesses. The mechanism by which PrP^{Sc} is formed must involve a templating process in which existing PrP^{Sc} directs the refolding of PrP^{C} into nascent PrP^{Sc} with the same conformation. Undoubtedly, molecular chaperones of some type participate in a process that appears to be associated with CLDs on the cell surface.

Studies of prions in fungi have been extremely helpful in establishing the prion concept. In the case of yeast, Sup35 and Ure2p fold into alternative conformations that create the new metabolic states [*PSI*] and [*URE3*], respectively. Whether PrP^{Sc} in mammals represents a misfolded protein or an alternatively folded protein, as with yeast prions, remains to be established.

18 Prevention and Therapeutics for Prion Diseases

Because people at risk for inherited prion diseases can now be identified decades before neurologic dysfunction is evident, the development of an effective therapy for these fully penetrant disorders is imperative. Although it is difficult to predict the number of individuals who may develop neurologic dysfunction from bovine prions in the future, seeking an effective therapy now seems most prudent. Interfering with the conversion of PrP^{C} into PrP^{Sc} would seem to be the most attractive therapeutic target.

Defining the pathogenesis of prion disease is an important issue with respect to developing an effective therapy. The issue of whether large aggregates of misprocessed proteins or misfolded monomers (or oligomers) cause CNS degeneration has been addressed in several studies of prion diseases in humans as well as in Tg mice. In humans, the frequency of PrP amyloid plaques varies from 100% in GSS and vCJD to ~70% in kuru and ~10% in sCJD, arguing that these plaques are a nonobligatory feature of the disease. In Tg mice expressing both MoPrP and SHaPrP, animals inoculated with hamster prions produced hamster prions and developed amyloid plaques composed of $SHaPrP^{Sc}$. In contrast, Tg mice inoculated with mouse prions did not develop plaques even though they produced mouse prions and died of prion disease.

18.1 Prion Therapeutics

Various compounds have been proposed as potential therapeutics for treatment of prion diseases; these include polysulfated anions, dextrans, Congo red dye, oligonucleotides, and cyclic tetrapyrroles, all of which have been shown to increase survival time when given prior to prion infection in rodents, but not when administered a month or more after infection has been established.

Besides studies in rodents, ScN2a cells chronically infected with scrapie prions have been used to identify several candidate antiprion drugs but none have been shown to be effective in halting prion diseases in either animals or humans. Structure-based drug design based on dominant-negative inhibition of prion formation has produced several lead compounds. Prion replication depends on protein–protein interactions and a subset of these interactions gives rise to dominant-negative phenotypes produced by single residue substitutions. A particularly interesting set of drugs is the branched polyamines, or dendrimers, which enhance the clearance of PrP^{Sc} from cells. While these compounds cure cultured cells of prion infection, they have not been successfully deployed in mice because of difficulties in delivering such highly charged compounds to the CNS.

18.1.1 Quinacrine and other Acridine Derivatives

Tricyclic derivatives of acridine exhibit half-maximal inhibition of PrP^{Sc} formation at effective concentrations (IC_{50}) between 0.3 μM and 3 μM in cultured ScN2a cells. The IC_{50} for chlorpromazine was 3 μM while quinacrine was 10 times more potent. A variety of 9-substituted, acridine-based analogs of quinacrine were synthesized, which demonstrated variable potencies similar to chlorpromazine and emphasized the importance of the side chain in mediating the inhibition of PrP^{Sc}

formation. These studies showed that tricyclic compounds with an aliphatic side chain at the middle ring moiety constitute a new class of antiprion agents. Because quinacrine and chlorpromazine have been used for many years in humans as antimalarial and antipsychotic drugs, respectively, and are known to penetrate the blood-brain barrier (BBB), these compounds became immediate candidates for the treatment of CJD and other human prion diseases.

An asymmetric carbon in the side chain of quinacrine creates two stereoisomers. (*S*)-Quinacrine is two to three times more potent than the (*R*)-quinacrine enantiomer in cultured cells. Whether the use of (*S*)-quinacrine in place of the racemic mixture will allow significantly more drug to be given remains to be established. Studies performed almost six decades ago in humans demonstrated that (*S*)-quinacrine is selectively metabolized and that (*R*)-quinacrine remains intact. Whether (*R*)-quinacrine or the metabolites of (*S*)-quinacrine are responsible for the side effects recorded in patients receiving the racemic mixture is unknown. Side effects of quinacrine include liver toxicity, cardiomyopathy, and toxic psychoses.

18.1.2 Anti-PrP Antibodies

A panel of recombinant antibody fragments (recFabs), recognizing different epitopes on PrP, was studied with respect to inhibition of prion propagation in cultured ScN2a cells. Recombinant Fabs binding to PrP^C on the cell surface inhibited PrP^{Sc} formation in a dose-dependent manner. In ScN2a cells treated with the most potent recFab D18, prion replication was completely abolished and preexisting PrP^{Sc} rapidly cleared, suggesting that this antibody may cure established infection. The activity of recFab D18 is associated with its ability to recognize more completely the total population of PrP^C molecules on the cell surface than other recFabs and with the location of its epitope on PrP^C. In other studies, a monoclonal antibody, pH4, which is thought to bind to the same region of PrP as recFab D18, was found to inhibit prion accumulation in ScN2a cells.

Whether antibodies or Fabs can be effectively administered to humans for the prevention and treatment of prion diseases is unclear. Neither antibodies nor Fabs cross the BBB in high concentration so that delivery of these proteins to the CNS remains a critical issue.

It is notable that anti-PrP antibodies delivered by transgenetics or by injection were successful in blocking prion replication in mice inoculated intraperitoneally. The same anti-PrP antibodies were ineffective in treating prion infection initiated by intracerebral inoculation.

18.2 Dominant-negative Sheep

The production of domestic animals that do not replicate prions may also prove to be a practical way to prevent prion disease. Sheep encoding Arg/Arg at polymorphic position 171 seem resistant to scrapie; presumably, this was the genetic basis of Parry's scrapie eradication program in Great Britain 30 years ago. A more effective approach applying dominant-negative inhibition for producing prion-resistant domestic animals, including sheep and cattle, is probably the expression of PrP transgenes encoding K219 and/or R171 (Fig. 11). Such an approach has been evaluated in Tg mice and shown to be effective. In fact, the replacement of all sheep with Arg/Arg at codon 171 is underway in the Netherlands. Such an approach can be instituted by artificial insemination using sperm from males homozygous for Arg/Arg at residue 171.

In comparison, a less practical approach is the production of PrP-deficient cattle and sheep. Although such animals would not be susceptible to prion disease, they might suffer some deleterious effects from the ablation of the PrP gene.

19 Some Principles of Prion Biology

Many principles of prion replication are clearly unprecedented in biology. As such, it is not surprising that some of these principles have been slow to be embraced. Although prion replication resembles viral replication superficially, the underlying principles are quite different. For example, in prion replication, the substrate is a host-encoded protein, PrP^{C}, which undergoes modification to form PrP^{Sc}, the only known component of the infectious prion particle. In contrast, viruses carry a DNA or RNA genome that is copied and directs the synthesis of most, if not all, of the viral proteins. The mature virion consists of a nucleic acid genome surrounded by a protein coat, whereas a prion appears to be composed of a dimer of PrP^{Sc}. When viruses pass from one species to another, they often replicate without any structural modification, whereas prions undergo a profound change. The prion adopts a new PrP sequence, which is encoded by the PrP gene of the current host. That change in amino acid sequence can result in a restriction of transmission for some species while making the new prion permissive to others. In viruses, different properties exhibited by distinct strains are encoded in the viral genome, whereas in prions, strain-specific properties are enciphered in the conformation of PrP^{Sc}.

19.1 Implications for Common Neurodegenerative Diseases

Understanding how PrP^{C} unfolds and refolds into PrP^{Sc} may also open new approaches to deciphering the causes of and to developing effective therapies for some common neurodegenerative diseases, including Alzheimer's disease, Parkinson's disease, and amyotrophic lateral sclerosis (ALS). Whether or not therapies designed to prevent the conversion of PrP^{C} into PrP^{Sc} will be effective in these more common neurodegenerative diseases is unknown. Alternatively, developing a therapy for the prion diseases might provide a blueprint for designing somewhat different drugs for these common disorders. Like the inherited prion diseases, subsets of Alzheimer's disease and ALS are caused by mutations that result in nonconservative amino acid substitutions in proteins expressed in the CNS.

As knowledge about prions continues to expand, our understanding of how prions replicate and cause disease will undoubtedly evolve. It is important to add that many of the basic principles of prion biology are becoming increasingly well understood.

See also Alzheimer's Disease; Motor Neuron Diseases: Cellular and Animal Models; Motor Neuron Diseases: Molecular Mechanism, Pathophysiology, and Treatments.

Bibliography

Books and Reviews

Aguzzi, A., Heppner, F.L., Heikenwalder, M., Prinz, M., Mertz, K., Seeger, H., Glatzel, M. (2003) Immune system and peripheral nerves in propagation of prions to CNS, *Br. Med. Bull.* **66**, 141–159.

Alpers, M.P. (1968) Kuru: Implications of its Transmissibility for the Interpretation of its Changing Epidemiological Pattern, in: Bailey, O.T., Smith, D.E. (Eds.) *The Central Nervous System: Some Experimental Models of Neurological Diseases*, Williams and Wilkins Company, Baltimore, MD, pp. 234–251.

Behrens, A. (2003) Physiological and pathological functions of the prion protein homologue Dpl., *Br. Med. Bull.* **66**, 35–42.

Brandner, S. (2003) CNS pathogenesis of prion diseases, *Br. Med. Bull.* **66**, 131–139.

Bruce, M., Chree, A., McConnell, I., Foster, J., Pearson, G., Fraser, H. (1994) Transmission of bovine spongiform encephalopathy and scrapie to mice: strain variation and the species barrier, *Philos. Trans. R. Soc. Lond. B. Biol. Sci.* **343**, 405–411.

Bruce, M.E. (1996) Strain Typing Studies of Scrapie and BSE, in: Baker, H.F., Ridley, R.M. (Eds.) *Methods in Molecular Medicine: Prion Diseases*, Humana Press, Totowa, NJ, pp. 223–236.

Bruce, M.E. (2003) TSE strain variation: An investigation into prion disease diversity, *Br. Med. Bull.* **66**, 99–108.

Budka, H. (2003) Neuropathology of prion diseases, *Br. Med. Bull.* **66**, 121–130.

Caughey, B. (2003) Prion protein conversions: insight into mechanisms, TSE transmission barriers and strains, *Br. Med. Bull.* **66**, 109–120.

Cervenáková, L., Brown, P., Piccardo, P., Cummings, J.L., Nagle, J., Vinters, H.V., Kaur, P., Ghetti, B., Chapman, J., Gajdusek, D.C., Goldfarb, L.G. (1996) 24-Nucleotide Deletion in the *PRNP* Gene: Analysis of Associated Phenotypes, in: Court, L., Dodet, B. (Eds.) *Transmissible Subacute Spongiform Encephalopathies: Prion Diseases*, Elsevier, Paris, France, pp. 433–444.

Chesebro, B. (2003) Introduction to the transmissible spongiform encephalopathies or prion diseases, *Br. Med. Bull.* **66**, 1–20.

Chien, P., Weissman, J.S., DePace, A.H. (2004) Emerging principles of conformation-based prion inheritance, *Annu. Rev. Biochem.* **73**, 617–656.

Collinge, J., Palmer, M.S. (1997) Human Prion Diseases, in: Collinge, J., Palmer, M.S. (Eds.) *Prion Diseases*, Oxford University Press, Oxford, UK, pp. 18–56.

DeArmond, S.J., Prusiner, S.B. (1997) Prion Diseases, in: Lantos, P., Graham, D. (Eds.) *Greenfield's Neuropathology*, 6th edition, Edward Arnold, London, UK, pp. 235–280.

Dickinson, A.G., Outram, G.W. (1988) Genetic Aspects of Unconventional Virus Infections: the Basis of the Virino Hypothesis, in: Bock, G., Marsh, J. (Eds.) *Novel Infectious Agents and the Central Nervous System. Ciba Foundation Symposium 135*, John Wiley and Sons, Chichester, UK, pp. 63–83.

Diringer, H. (1991) Transmissible spongiform encephalopathies (TSE) virus-induced amyloidoses of the central nervous system (CNS), *Eur. J. Epidemiol.* **7**, 562–566.

Dormont, D. (2003) Approaches to prophylaxis and therapy: An investigation into prion disease diversity, *Br. Med. Bull.* **66**, 281–292.

Erdtmann, R., Sivitz, L.B. (Eds.) (2004) *Advancing Prion Science: Guidance for the National Prion Research Program*, National Academies Press, Washington, DC

Gambetti, P., Kong, Q., Zou, W., Parchi, P., Chen, S.G. (2003) Sporadic and familial CJD: classification and characterisation, *Br. Med. Bull.* **66**, 213–239.

Gambetti, P., Parchi, P., Petersen, R.B., Chen, S.G., Lugaresi, E. (1995) Fatal familial insomnia and familial Creutzfeldt-Jakob disease: clinical, pathological and molecular features, *Brain Pathol.* **5**, 43–51.

Goodman, L.S., Gilman, A. (Eds.) (1970) *The Pharmacological Basis of Therapeutics; A Textbook of Pharmacology, Toxicology, and Therapeutics for Physicians and Medical Students*, Macmillan, New York.

Harris, D.A. (2003) Trafficking, turnover and membrane topology of PrP: protein function in prion disease, *Br. Med. Bull.* **66**, 71–85.

Harris, D.A. (Ed.) (2004) *Mad Cow Disease and Related Spongiform Encephalopathies*, Current Topics in Microbiology and Immunology, Springer-Verlag, Berlin, Germany.

Hill, A.F., Collinge, J. (2003) Subclinical prion infection in humans and animals, *Br. Med. Bull.* **66**, 161–170.

Hunter, N. (2003) Scrapie and experimental BSE in sheep, *Br. Med. Bull.* **66**, 171–183.

Kimberlin, R.H. (1996) Speculations on the Origin of BSE and the Epidemiology of CJD, in: Gibbs Jr., C.J. (Ed.) *Bovine Spongiform Encephalopathy: The BSE Dilemma,* Springer Verlag, New York, pp. 155–175.

Kübler, E., Oesch, B., Raeber, A.J. (2003) Diagnosis of prion diseases, *Br. Med. Bull.* **66**, 267–279.

Lasmézas, C.I. (2003) Putative functions of PrP^C, *Br. Med. Bull.* **66**, 61–70.

Malmgren, R., Kurland, L., Mokri, B., Kurtzke, J. (1979) The Epidemiology of Creutzfeldt-Jakob Disease, in: Prusiner, S.B. Hadlow, W.J. (Eds.) *Slow Transmissible Diseases of the Nervous System,* Vol. 1, Academic Press, New York, pp. 93–112.

Parry, H.B. (1983) *Scrapie Disease in Sheep,* Academic Press, New York.

Pattison, I.H. (1965) Experiments with Scrapie with Special Reference to the Nature of the Agent and the Pathology of the Disease, in: Gajdusek, D.C. Gibbs Jr., C.J., Alpers, M.P. (Eds.) *Slow, Latent and Temperate Virus Infections, NINDB Monograph 2,* U.S. Government Printing, Washington, DC, pp. 249–257.

Prusiner, S.B. (1989) Scrapie prions, *Annu. Rev. Microbiol.* **43**, 345–374.

Prusiner, S.B. (1991) Molecular biology of prion diseases, *Science* **252**, 1515–1522.

Prusiner, S.B. (1997) Prion diseases and the BSE crisis, *Science* **278**, 245–251.

Prusiner, S.B. (Ed.) (2004) *Prion Biology and Diseases,* Cold Spring Harbor Laboratory Press, Cold Spring Harbor, New York.

Prusiner, S.B., Scott, M.R., DeArmond, S.J., Cohen, F.E. (1998) Prion protein biology, *Cell* **93**, 337–348.

Prusiner, S.B., Baron, H., Carlson, G., Cohen, F.E., DeArmond, S.J., Gabizon, R., Gambetti, P., Hope, J., Kitamoto, T., Kretzschmar, H.A., Laplanche, J.-L., Tateishi, J., Telling, G., Will, R. (2000) Prions, in: van Regenmortel, M.H.V., Fauquet, C.M., Bishop, D.H.L., Carstens, E.B., Estes, M.K., Lemon, S.M., Maniloff, J., Mayo, M.A., McGeoch, D.J., Pringle, C.R., Wickner, R.B. et al., *Virus Taxonomy – Classification and Nomenclature of Viruses,* Academic Press, San Diego, CA, 1032–1039.

Riesner, D. (2003) Biochemistry and structure of PrP^C and PrP^{Sc}, *Br. Med. Bull.* **66**, 21–33.

Sigurdson, C.J., Miller, M.W. (2003) Other animal prion diseases, *Br. Med. Bull.* **66**, 199–212.

Smith, P.G., Bradley, R. (2003) Bovine spongiform encephalopathy (BSE) and its epidemiology, *Br. Med. Bull.* **66**, 185–198.

Solassol, J., Crozet, C., Lehmann, S. (2003) Prion propagation in cultured cells, *Br. Med. Bull.* **66**, 87–97.

Tabrizi, S.J., Elliott, C.L., Weissmann, C. (2003) Ethical issues in human prion diseases, *Br. Med. Bull.* **66**, 305–316.

Tateishi, J., Kitamoto, T. (1993) Developments in diagnosis for prion diseases, *Br. Med. Bull.* **49**, 971–979.

Taylor, D.M. (2003) Preventing accidental transmission of human transmissible spongiform encephalopathies, *Br. Med. Bull.* **66**, 293–303.

Tuite, M.F., Lindquist, S.L. (1996) Maintenance and inheritance of yeast prions, *Trends Genet.* **12**, 467–471.

Wadsworth, J.D., Hill, A.F., Beck, J.A., Collinge, J. (2003) Molecular and clinical classification of human prion disease, *Br. Med. Bull.* **66**, 241–254.

Weissmann, C. (1991) A "unified theory" of prion propagation, *Nature* **352**, 679–683.

Weissmann, C., Flechsig, E. (2003) PrP knockout and PrP transgenic mice in prion research, *Br. Med. Bull.* **66**, 43–60.

Wells, G.A.H., Wilesmith, J.W. (1995) The neuropathology and epidemiology of bovine spongiform encephalopathy, *Brain Pathol.* **5**, 91–103.

Wickner, R.B., Liebman, S.W., Saupe, S.J. (2004) Prions of Yeast and Filamentous Fungi: [URE3], [PSI^+], [PIN^+], and [Het-s], in: Prusiner, S.B. (Ed.) *Prion Biology and Diseases,* Cold Spring Harbor Laboratory Press, Cold Spring Harbor, pp. 305–372.

Will, R.G. (2003) Acquired prion disease: iatrogenic CJD, variant CJD, kuru, *Br. Med. Bull.* **66**, 255–265.

Will, R.G., Alpers, M.P., Dormont, D., Schonberger, L.B. (2004) Infectious and Sporadic Prion Diseases, in: Prusiner, S.B. (Ed.) *Prion Biology and Diseases,* Cold Spring Harbor Laboratory Press, Cold Spring Harbor, pp. 629–671.

Williams, E.S., Miller, M.W. (2002) Chronic wasting disease in deer and elk in North America, *Rev. Sci. Tech.* **21**, 305–316.

Primary Literature

Anderson, R.M., Donnelly, C.A., Ferguson, N.M., Woolhouse, M.E.J., Watt, C.J., Udy, H.J., MaWhinney, S., Dunstan, S.P., Southwood, T.R.E., Wilesmith, J.W., Ryan, J.B.M., Hoinville, L.J., Hillerton, J.E., Austin, A.R., Wells, G.A.H. (1996) Transmission dynamics and epidemiology of BSE in British cattle, *Nature* **382**, 779–788.

Anfinsen, C.B. (1973) Principles that govern the folding of protein chains, *Science* **181**, 223–230.

Asante, E.A., Linehan, J.M., Desbruslais, M., Joiner, S., Gowland, I., Wood, A.L., Welch, J., Hill, A.F., Lloyd, S.E., Wadsworth, J.D., Collinge, J. (2002) BSE prions propagate as either variant CJD-like or sporadic CJD-like prion strains in transgenic mice expressing human prion protein, *EMBO J.* **21**, 6358–6366.

Baker, H.F., Ridley, R.M., Wells, G.A.H. (1993) Experimental transmission of BSE and scrapie to the common marmoset, *Vet. Rec.* **132**, 403–406.

Barry, R.A., Prusiner, S.B. (1986) Monoclonal antibodies to the cellular and scrapie prion proteins, *J. Infect. Dis.* **154**, 518–521.

Basler, K., Oesch, B., Scott, M., Westaway, D., Wälchli, M., Groth, D.F., McKinley, M.P., Prusiner, S.B., Weissmann, C. (1986) Scrapie and cellular PrP isoforms are encoded by the same chromosomal gene, *Cell* **46**, 417–428.

Bastian, F.O., Foster, J.W. (2001) Spiroplasma sp. 16S rDNA in Creutzfeldt-Jakob disease and scrapie as shown by PCR and DNA sequence analysis, *J. Neuropathol. Exp. Neurol.* **60**, 613–620.

Bateman, D., Hilton, D., Love, S., Zeidler, M., Beck, J., Collinge, J. (1995) Sporadic Creutzfeldt-Jakob disease in a 18-year-old in the UK (Lett.), *Lancet* **346**, 1155–1156.

Behrens, A., Genoud, N., Naumann, H., Rülicke, T., Janett, F., Heppner, F.L., Ledermann, B., Aguzzi, A. (2002) Absence of the prion protein homologue Doppel causes male sterility, *EMBO J.* **21**, 3652–3658.

Bellinger-Kawahara, C.G., Kempner, E., Groth, D.F., Gabizon, R., Prusiner, S.B. (1988) Scrapie prion liposomes and rods exhibit target sizes of 55 000 Da, *Virology* **164**, 537–541.

Belt, P.B., Muileman, I.H., Schreuder, B.E.C., Ruijter, J.B., Gielkens, A.L.J., Smits, M.A. (1995) Identification of five allelic variants of the sheep PrP gene and their association with natural scrapie, *J. Gen. Virol.* **76**, 509–517.

Bertoni, J.M., Brown, P., Goldfarb, L., Gajdusek, D., Omaha, N.E. (1992) Familial Creutzfeldt-Jakob disease with the PRNP codon 200^{lys} mutation and supranuclear palsy but without myoclonus or periodic EEG complexes, *Neurology* **42**(4), Suppl. 3), 350 [Abstr].

Bessen, R.A., Marsh, R.F. (1994) Distinct PrP properties suggest the molecular basis of strain variation in transmissible mink encephalopathy, *J. Virol.* **68**, 7859–7868.

Billette de Villemeur, T., Deslys, J.-P., Pradel, A., Soubrié, C., Alpérovitch, A., Tardieu, M., Chaussain, J.-L., Hauw, J.-J., Dormont, D., Ruberg, M., Agid, Y. (1996) Creutzfeldt-Jakob disease from contaminated growth hormone extracts in France, *Neurology* **47**, 690–695.

Borchelt, D.R., Scott, M., Taraboulos, A., Stahl, N., Prusiner, S.B. (1990) Scrapie and cellular prion proteins differ in their kinetics of synthesis and topology in cultured cells, *J. Cell Biol.* **110**, 743–752.

Britton, T.C., Al-Sarraj, S., Shaw, C., Campbell, T., Collinge, J. (1995) Sporadic Creutzfeldt-Jakob disease in a 16-year-old in the UK (Lett.), *Lancet* **346**, 1155.

Brown, D.R., Qin, K., Herms, J.W., Madlung, A., Manson, J., Strome, R., Fraser, P.E., Kruck, T., von Bohlen, A., Schulz-Schaeffer, W., Giese, A., Westaway, D., Kretzschmar, H. (1997) The cellular prion protein binds copper *in vivo*, *Nature* **390**, 684–687.

Brown, P., Cathala, F., Raubertas, R.F., Gajdusek, D.C., Castaigne, P. (1987) The epidemiology of Creutzfeldt-Jakob disease: conclusion of a 15-year investigation in France and review of the world literature, *Neurology* **37**, 895–904.

Bruce, M., Chree, A., McConnell, I., Foster, J., Fraser, H. (1993) Transmissions of BSE, scrapie and related diseases to mice (Abstr.), *Proceedings of the IXth International Congress of Virology*, Glasgow, Scotland, Aug. 9–12, 93.

Bruce, M.E., Dickinson, A.G. (1987) Biological evidence that the scrapie agent has an independent genome, *J. Gen. Virol.* **68**, 79–89.

Bruce, M.E., McConnell, I., Fraser, H., Dickinson, A.G. (1991) The disease characteristics of different strains of scrapie in *Sinc* congenic mouse lines: implications for the nature of the agent and host control of pathogenesis, *J. Gen. Virol.* **72**, 595–603.

Bruce, M.E., Will, R.G., Ironside, J.W., McConnell, I., Drummond, D., Suttie, A., McCardle, L., Chree, A., Hope, J., Birkett, C., Cousens, S., Fraser, H., Bostock, C.J. (1997) Transmissions to mice indicate that 'new variant' CJD is caused by the BSE agent, *Nature* **389**, 498–501.

Büeler, H., Raeber, A., Sailer, A., Fischer, M., Aguzzi, A., Weissmann, C. (1994) High prion and PrP^{Sc} levels but delayed onset of disease in scrapie-inoculated mice heterozygous for a disrupted PrP gene, *Mol. Med.* **1**, 19–30.

Büeler, H., Aguzzi, A., Sailer, A., Greiner, R.-A., Autenried, P., Aguet, M., Weissmann, C. (1993) Mice devoid of PrP are resistant to scrapie, *Cell* **73**, 1339–1347.

Büeler, H., Fisher, M., Lang, Y., Bluethmann, H., Lipp, H.-P., DeArmond, S.J., Prusiner, S.B., Aguet, M., Weissmann, C. (1992) Normal development and behaviour of mice lacking the neuronal cell-surface PrP protein, *Nature* **356**, 577–582.

Burns, C.S., Aronoff-Spencer, E., Legname, G., Prusiner, S.B., Antholine, W.E., Gerfen, G.J., Peisach, J., Millhauser, G.L. (2003) Copper coordination in the full-length, recombinant prion protein, *Biochemistry* **42**, 6794–6803.

Buschmann, A., Pfaff, E., Reifenberg, K., Müller, H.M., Groschup, M.H. (2000) Detection of cattle-derived BSE prions using transgenic mice overexpressing bovine PrP^{C}, *Arch. Virol. [Suppl].* **16**, 75–86.

Carlson, G.A., Westaway, D., DeArmond, S.J., Peterson-Torchia, M., Prusiner, S.B. (1989) Primary structure of prion protein may modify scrapie isolate properties, *Proc. Natl. Acad. Sci. U. S. A.* **86**, 7475–7479.

Carlson, G.A., Kingsbury, D.T., Goodman, P.A., Coleman, S., Marshall, S.T., DeArmond, S., Westaway, D., Prusiner, S.B. (1986) Linkage of prion protein and scrapie incubation-time genes, *Cell* **46**, 503–511.

Carlson, G.A., Ebeling, C., Yang, S.-L., Telling, G., Torchia, M., Groth, D., Westaway, D., DeArmond, S.J., Prusiner, S.B. (1994) Prion isolate specified allotypic interactions between the cellular and scrapie prion proteins in congenic and transgenic mice, *Proc. Natl. Acad. Sci. U. S. A.* **91**, 5690–5694.

Caughey, B., Race, R.E. (1992) Potent inhibition of scrapie-associated PrP accumulation by Congo red, *J. Neurochem.* **59**, 768–771.

Caughey, B., Raymond, G.J. (1991) The scrapie-associated form of PrP is made from a cell-surface precursor that is both protease- and phospholipase-sensitive, *J. Biol. Chem.* **266**, 18217–18223.

Caughey, B., Kocisko, D.A., Raymond, G.J., Lansbury Jr., P.T. (1995) Aggregates of scrapie-associated prion protein induce the cell-free conversion of protease-sensitive prion protein to the protease-resistant state, *Chem. Biol.* **2**, 807–817.

Caughey, B., Raymond, G.J., Ernst, D., Race, R.E. (1991) N-terminal truncation of the scrapie-associated form of PrP by lysosomal protease(s): implications regarding the site of conversion of PrP to the protease-resistant state, *J. Virol.* **65**, 6597–6603.

Caughey, B., Neary, K., Butler, R., Ernst, D., Perry, L., Chesebro, B., Race, R.E. (1990) Normal and scrapie-associated forms of prion protein differ in their sensitivities to phospholipase and proteases in intact neuroblastoma cells, *J. Virol.* **64**, 1093–1101.

Caughey, B.W., Dong, A., Bhat, K.S., Ernst, D., Hayes, S.F., Caughey, W.S. (1991) Secondary structure analysis of the scrapie-associated protein PrP 27–30 in water by infrared spectroscopy, *Biochemistry* **30**, 7672–7680.

Centers for Disease Control (1996) Surveillance for Creutzfeldt-Jakob disease – United States, *MMWR Morb. Mortal. Wkly. Rep.* **45**, 665–668.

Chapman, J., Ben-Israel, J., Goldhammer, Y., Korczyn, A.D. (1994) The risk of developing Creutzfeldt-Jakob disease in subjects with the *PRNP* gene codon 200 point mutation, *Neurology* **44**, 1683–1686.

Chernoff, Y.O., Lindquist, S.L., Ono, B., Inge-Vechtomov, S.G., Liebman, S.W. (1995) Role of the chaperone protein Hsp104 in propagation of the yeast prion-like factor [psi^{+}], *Science* **268**, 880–884.

Chesebro, B. (1998) Prion diseases: BSE and prions: uncertainties about the agent, *Science* **279**, 42–43.

Chesebro, B., Race, R., Wehrly, K., Nishio, J., Bloom, M., Lechner, D., Bergstrom, S., Robbins, K., Mayer, L., Keith, J.M., Garon, C., Haase, A. (1985) Identification of scrapie prion protein-specific mRNA in scrapie-infected and uninfected brain, *Nature* **315**, 331–333.

Clousard, C., Beaudry, P., Elsen, J.M., Milan, D., Dussaucy, M., Bounneau, C., Schelcher, F., Chatelain, J., Launay, J.-M., Laplanche, J.-L. (1995) Different allelic effects of the codons 136 and 171 of the prion protein gene in sheep with natural scrapie, *J. Gen. Virol.* **76**, 2097–2101.

Cohen, F.E., Pan, K.-M., Huang, Z., Baldwin, M., Fletterick, R.J., Prusiner, S.B. (1994) Structural clues to prion replication, *Science* **264**, 530–531.

Collinge, J., Palmer, M.S., Dryden, A.J. (1991) Genetic predisposition to iatrogenic Creutzfeldt-Jakob disease, *Lancet* **337**, 1441–1442.

Collinge, J., Sidle, K.C.L., Meads, J., Ironside, J., Hill, A.F. (1996) Molecular analysis of prion strain variation and the aetiology of "new variant" CJD, *Nature* **383**, 685–690.

Collinge, J., Whittington, M.A., Sidle, K.C., Smith, C.J., Palmer, M.S., Clarke, A.R., Jefferys, J.G.R. (1994) Prion protein is necessary for normal synaptic function, *Nature* **370**, 295–297.

Collinge, J., Palmer, M.S., Sidle, K.C., Hill, A.F., Gowland, I., Meads, J., Asante, E., Bradley, R., Doey, L.J., Lantos, P.L. (1995) Unaltered susceptibility to BSE in transgenic mice expressing human prion protein, *Nature* **378**, 779–783.

Cousens, S.N., Vynnycky, E., Zeidler, M., Will, R.G., Smith, P.G. (1997) Predicting the CJD epidemic in humans, *Nature* **385**, 197–198.

Cousens, S.N., Harries-Jones, R., Knight, R., Will, R.G., Smith, P.G., Matthews, W.B. (1990) Geographical distribution of cases of Creutzfeldt-Jakob disease in England and Wales 1970–84, *J. Neurol. Neurosurg. Psychiatry* **53**, 459–465.

Dawson, M., Wells, G.A.H., Parker, B.N.J. (1990) Preliminary evidence of the experimental transmissibility of bovine spongiform encephalopathy to cattle, *Vet. Rec.* **126**, 112–113.

Dawson, M., Wells, G.A.H., Parker, B.N.J., Scott, A.C. (1990) Primary parenteral transmission of bovine spongiform encephalopathy to the pig, *Vet. Rec.* **127**, 338.

DeArmond, S.J., McKinley, M.P., Barry, R.A., Braunfeld, M.B., McColloch, J.R., Prusiner, S.B. (1985) Identification of prion amyloid filaments in scrapie-infected brain, *Cell* **41**, 221–235.

Derkatch, I.L., Chernoff, Y.O., Kushnirov, V.V., Inge-Vechtomov, S.G., Liebman, S.W. (1996) Genesis and variability of [*PSI*] prion factors in *Saccharomyces cerevisiae*, *Genetics* **144**, 1375–1386.

Dickinson, A.G., Fraser, H., Outram, G.W. (1975) Scrapie incubation time can exceed natural lifespan, *Nature* **256**, 732–733.

Dickinson, A.G., Meikle, V.M.H., Fraser, H. (1968) Identification of a gene which controls the incubation period of some strains of scrapie agent in mice, *J. Comp. Pathol.* **78**, 293–299.

Dickinson, A.G., Meikle, V.M., Fraser, H. (1969) Genetical control of the concentration of ME7 scrapie agent in the brain of mice, *J. Comp. Pathol.* **79**, 15–22.

Dickinson, A.G., Young, G.B., Stamp, J.T., Renwick, C.C. (1965) An analysis of natural scrapie in Suffolk sheep, *Heredity* **20**, 485–503.

Doel, S.M., McCready, S.J., Nierras, C.R., Cox, B.S. (1994) The dominant *PNM2*$^-$ mutation which eliminates the *ψ* factor of *Saccharomyces cerevisiae* is the result of a missense mutation in the *SUP35* gene, *Genetics* **137**, 659–670.

Doh-ura, K., Iwaki, T., Caughey, B. (2000) Lysosomotropic agents and cysteine protease inhibitors inhibit scrapie-associated prion protein accumulation, *J. Virol.* **74**, 4894–4897.

Doh-ura, K., Kitamoto, T., Sakaki, Y., Tateishi, J. (1991) CJD discrepancy, *Nature* **353**, 801–802.

Doh-ura, K., Tateishi, J., Sasaki, H., Kitamoto, T., Sakaki, Y. (1989) Pro→Leu change at position 102 of prion protein is the most common but not the sole mutation related to Gerstmann-Sträussler syndrome, *Biochem. Biophys. Res. Commun.* **163**, 974–979.

Donne, D.G., Viles, J.H., Groth, D., Mehlhorn, I., James, T.L., Cohen, F.E., Prusiner, S.B., Wright, P.E., Dyson, H.J. (1997) Structure of the recombinant full-length hamster prion protein PrP(29–231): the N terminus is highly flexible, *Proc. Natl. Acad. Sci. U. S. A.* **94**, 13452–13457.

Edenhofer, F., Rieger, R., Famulok, M., Wendler, W., Weiss, S., Winnacker, E.-L. (1996) Prion protein PrP^C interacts with molecular chaperones of the Hsp60 family, *J. Virol.* **70**, 4724–4728.

Ehlers, B., Diringer, H. (1984) Dextran sulphate 500 delays and prevents mouse scrapie by impairment of agent replication in spleen, *J. Gen. Virol.* **65**, 1325–1330.

Enari, M., Flechsig, E., Weissmann, C. (2001) Scrapie prion protein accumulation by scrapie-infected neuroblastoma cells abrogated by exposure to a prion protein antibody, *Proc. Natl. Acad. Sci. U. S. A.* **98**, 9295–9299.

Endo, T., Groth, D., Prusiner, S.B., Kobata, A. (1989) Diversity of oligosaccharide structures linked to asparagines of the scrapie prion protein, *Biochemistry* **28**, 8380–8388.

Fink, J.K., Peacock, M.L., Warren, J.T., Roses, A.D., Prusiner, S.B. (1994) Detecting prion protein gene mutations by denaturing gradient gel electrophoresis, *Hum. Mutat.* **4**, 42–50.

Fischer, M., Rülicke, T., Raeber, A., Sailer, A., Moser, M., Oesch, B., Brandner, S., Aguzzi, A., Weissmann, C. (1996) Prion protein (PrP) with amino-proximal deletions restoring susceptibility of PrP knockout mice to scrapie, *EMBO J.* **15**, 1255–1264.

Foster, J., Goldmann, W., Parnham, D., Chong, A., Hunter, N. (2001) Partial dissociation of PrP^{Sc} deposition and vacuolation in the brains of scrapie and BSE experimentally affected goats, *J. Gen. Virol.* **82**, 267–273.

Fraser, H., Dickinson, A.G. (1973) Scrapie in mice. Agent-strain differences in the distribution and intensity of grey matter vacuolation, *J. Comp. Pathol.* **83**, 29–40.

Fraser, H., McConnell, I., Wells, G.A.H., Dawson, M. (1988) Transmission of bovine spongiform encephalopathy to mice, *Vet. Rec.* **123**, 472.

Fraser, H., Bruce, M.E., Chree, A., McConnell, I., Wells, G.A.H. (1992) Transmission of bovine spongiform encephalopathy and scrapie to mice, *J. Gen. Virol.* **73**, 1891–1897.

Furukawa, H., Kitamoto, T., Tanaka, Y., Tateishi, J. (1995) New variant prion protein in a Japanese family with Gerstmann-Sträussler syndrome, *Mol. Brain Res.* **30**, 385–388.

Gabizon, R., Telling, G., Meiner, Z., Halimi, M., Kahana, I., Prusiner, S.B. (1996) Insoluble wild-type and protease-resistant mutant prion protein in brains of patients with inherited prion disease, *Nat. Med.* **2**, 59–64.

Gabizon, R., Rosenmann, H., Meiner, Z., Kahana, I., Kahana, E., Shugart, Y., Ott, J., Prusiner, S.B. (1993) Mutation and polymorphism of the prion protein gene in Libyan Jews with Creutzfeldt-Jakob disease (CJD), *Am. J. Hum. Genet.* **53**, 828–835.

Gabriel, J.-M., Oesch, B., Kretzschmar, H., Scott, M., Prusiner, S.B. (1992) Molecular cloning of a candidate chicken prion protein, *Proc. Natl. Acad. Sci. U. S. A.* **89**, 9097–9101.

Gajdusek, D.C. (1977) Unconventional viruses and the origin and disappearance of kuru, *Science* **197**, 943–960.

Gajdusek, D.C., Gibbs Jr., C.J., Asher, D.M., Brown, P., Diwan, A., Hoffman, P., Nemo, G., Rohwer, R., White, L. (1977) Precautions in medical care of, and in handling materials from, patients with transmissible virus dementia (Creutzfeldt-Jakob disease), *N. Engl. J. Med.* **297**, 1253–1258.

Gambetti, P., Parchi, P. (1999) Insomnia in prion diseases: sporadic and familial, *N. Engl. J. Med.* **340**(21), 1675–1677.

Gasset, M., Baldwin, M.A., Fletterick, R.J., Prusiner, S.B. (1993) Perturbation of the secondary structure of the scrapie prion protein under conditions that alter infectivity, *Proc. Natl. Acad. Sci. U. S. A.* **90**, 1–5.

Gauczynski, S., Peyrin, J.M., Haik, S., Leucht, C., Hundt, C., Rieger, R., Krasemann, S., Deslys, J.P., Dormont, D., Lasmézas, C.I., Weiss, S. (2001) The 37-kDa/67-kDa laminin receptor acts as the cell-surface receptor for the cellular prion protein, *EMBO J.* **20**, 5863–5875.

Ghani, A.C., Ferguson, N.M., Donnelly, C.A., Anderson, R.M. (2000) Predicted vCJD mortality in Great Britain, *Nature* **406**, 583–584.

Glasse, R.M. (1967) Cannibalism in the kuru region of New Guinea, *Trans. NY Acad. Sci. [Ser. 2]* **29**, 748–754.

Glover, J.R., Kowal, A.S., Schirmer, E.C., Patino, M.M., Liu, J.-J., Lindquist, S. (1997) Self-seeded fibers formed by Sup35, the protein determinant of $[PSI^+]$, a heritable prion-like factor of S. cerevisiae, *Cell* **89**, 811–819.

Goldfarb, L., Brown, P., Goldgaber, D., Garruto, R., Yanaghiara, R., Asher, D., Gajdusek, D.C. (1990) Identical mutation in unrelated patients with Creutzfeldt-Jakob disease, *Lancet* **336**, 174–175.

Goldfarb, L.G., Mitrova, E., Brown, P., Toh, B.H., Gajdusek, D.C. (1990) Mutation in codon 200 of scrapie amyloid protein gene in two clusters of Creutzfeldt-Jakob disease in Slovakia, *Lancet* **336**, 514–515.

Goldfarb, L.G., Haltia, M., Brown, P., Nieto, A., Kovanen, J., McCombie, W.R., Trapp, S., Gajdusek, D.C. (1991) New mutation in scrapie amyloid precursor gene (at codon 178) in Finnish Creutzfeldt-Jakob kindred, *Lancet* **337**, 425.

Goldfarb, L.G., Brown, P., Goldgaber, D., Asher, D.M., Rubenstein, R., Brown, W.T., Piccardo, P., Kascsak, R.J., Boellaard, J.W., Gajdusek, D.C. (1990) Creutzfeldt-Jakob disease and kuru patients lack a mutation consistently found in the Gerstmann-Sträussler-Scheinker syndrome, *Exp. Neurol.* **108**, 247–250.

Goldfarb, L.G., Brown, P., McCombie, W.R., Goldgaber, D., Swergold, G.D., Wills, P.R., Cervenakova, L., Baron, H., Gibbs, C.J.J., Gajdusek, D.C. (1991) Transmissible familial Creutzfeldt-Jakob disease associated with five, seven, and eight extra octapeptide coding repeats in the *PRNP* gene, *Proc. Natl. Acad. Sci. U. S. A.* **88**, 10926–10930.

Goldfarb, L.G., Petersen, R.B., Tabaton, M., Brown, P., LeBlanc, A.C., Montagna, P., Cortelli, P., Julien, J., Vital, C., Pendelbury, W.W. (1992) Fatal familial insomnia and familial Creutzfeldt-Jakob disease: disease phenotype determined by a DNA polymorphism, *Science* **258**(5083), 806–808.

Goldfarb, L.G., Brown, P., Mitrova, E., Cervenakova, L., Goldin, L., Korczyn, A.D., Chapman, J., Galvez, S., Cartier, L., Rubenstein, R., Gajdusek, D.C. (1991) Creutzfeldt-Jacob disease associated with the *PRNP* codon 200^{Lys} mutation: an analysis of 45 families, *Eur. J. Epidemiol.* **7**, 477–486.

Goldgaber, D., Goldfarb, L.G., Brown, P., Asher, D.M., Brown, W.T., Lin, S., Teener, J.W., Feinstone, S.M., Rubenstein, R., Kascsak, R.J., Boellaard, J.W., Gajdusek, D.C. (1989) Mutations in familial Creutzfeldt-Jakob disease and Gerstmann-Sträussler-Scheinker's syndrome, *Exp. Neurol.* **106**, 204–206.

Goldmann, W., Hunter, N., Manson, J. Hope, J. (1990) The PrP gene of the sheep, a natural host of scrapie, *Proceedings of the VIIIth International Congress of Virology*, Berlin, Germany, Aug. 26–31, 284.

Goldmann, W., Hunter, N., Benson, G., Foster, J.D., Hope, J. (1991) Different scrapie-associated fibril proteins (PrP) are encoded by lines of sheep selected for different alleles of the *Sip* gene, *J. Gen. Virol.* **72**, 2411–2417.

Goldmann, W., Hunter, N., Martin, T., Dawson, M., Hope, J. (1991) Different forms of the bovine PrP gene have five or six copies of a short, G-C-rich element within the protein-coding exon, *J. Gen. Virol.* **72**, 201–204.

Goldmann, W., Hunter, N., Smith, G., Foster, J., Hope, J. (1994) PrP genotype and agent effects in scrapie: change in allelic interaction with different isolates of agent in sheep, a natural host of scrapie, *J. Gen. Virol.* **75**, 989–995.

Goldmann, W., Hunter, N., Foster, J.D., Salbaum, J.M., Beyreuther, K., Hope, J. (1990) Two alleles of a neural protein gene linked to scrapie in sheep, *Proc. Natl. Acad. Sci. U. S. A.* **87**, 2476–2480.

Gorodinsky, A., Harris, D.A. (1995) Glycolipid-anchored proteins in neuroblastoma cells form detergent-resistant complexes without caveolin, *J. Cell Biol.* **129**, 619–627.

Govaerts, C., Wille, H., Prusiner, S.B., Cohen, F.E. (2004) Evidence for assembly of prions with left-handed β-helices into trimers, *Proc. Natl. Acad. Sci. U. S. A.* **101**, 8342–8347.

Graner, E., Mercadante, A.F., Zanata, S.M., Forlenza, O.V., Cabral, A.L.B., Veiga, S.S., Juliano, M.A., Roesler, R., Walz, R., Mineti, A., Izquierdo, I., Martins, V.R., Brentani, R.R. (2000) Cellular prion protein binds laminin and mediates neuritogenesis, *Mol. Brain Res.* **76**, 85–92.

Grassi, J., Comoy, E., Simon, S., Creminon, C., Frobert, Y., Trapmann, S., Schimmel, H., Hawkins, S.A., Moynagh, J., Deslys, J.P., Wells, G.A. (2001) Rapid test for the preclinical postmortem diagnosis of BSE in central nervous system tissue, *Vet. Rec.* **149**, 577–582.

Hamir, A.N., Cutlip, R.C., Miller, J.M., Williams, E.S., Stack, M.J., Miller, M.W., O'Rourke, K.I., Chaplin, M.J. (2001) Preliminary findings on the experimental transmission of chronic wasting disease agent of mule deer to cattle, *J. Vet. Diagn. Invest.* **13**, 91–96.

Hammick, D.L., Chambers, W.E. (1945) Optical activity of excreted mepacrine, *Nature* **155**, 141.

Haraguchi, T., Fisher, S., Olofsson, S., Endo, T., Groth, D., Tarantino, A., Borchelt, D.R., Teplow, D., Hood, L., Burlingame, A., Lycke, E., Kobata, A., Prusiner, S.B. (1989) Asparagine-linked glycosylation of the scrapie and cellular prion proteins, *Arch. Biochem. Biophys.* **274**, 1–13.

Harries-Jones, R., Knight, R., Will, R.G., Cousens, S., Smith, P.G., Matthews, W.B. (1988) Creutzfeldt-Jakob disease in England and Wales, 1980–1984: a case-control study of potential risk factors, *J. Neurol. Neurosurg. Psychiatry* **51**, 1113–1119.

Harris, D.A., Falls, D.L., Walsh, W., Fischbach, G.D. (1989) Molecular cloning of an acetylcholine receptor-inducing protein, *Soc. Neurosci.* **15**, 70. 7.

Heppner, F.L., Musahl, C., Arrighi, I., Klein, M.A., Rülicke, T., Oesch, B., Zinkernagel, R.M., Kalinke, U., Aguzzi, A. (2001) Prevention of scrapie pathogenesis by transgenic expression of antiprion protein antibodies, *Science* **294**, 178–182.

Hill, A.F., Desbruslais, M., Joiner, S., Sidle, K.C.L., Gowland, I., Collinge, J., Doey, L.J., Lantos, P. (1997) The same prion strain causes vCJD and BSE, *Nature* **389**, 448–450.

Hilton, D.A., Ghani, A.C., Conyers, L., Edwards, P., McCardle, L., Ritchie, D., Penney, M., Hegazy, D., Ironside, J.W. (2004) Prevalence of lymphoreticular prion protein accumulation in UK tissue samples, *J. Pathol.* **203**, 733–739.

Hornshaw, M.P., McDermott, J.R., Candy, J.M. (1995) Copper binding to the N-terminal tandem repeat regions of mammalian and avian prion protein, *Biochem. Biophys. Res. Commun.* **207**, 621–629.

Hornshaw, M.P., McDermott, J.R., Candy, J.M., Lakey, J.H. (1995) Copper binding to the N-terminal tandem repeat region of mammalian and avian prion protein: structural studies using synthetic peptides, *Biochem. Biophys. Res. Commun.* **214**, 993–999.

Hsiao, K., Baker, H.F., Crow, T.J., Poulter, M., Owen, F., Terwilliger, J.D., Westaway, D., Ott, J., Prusiner, S.B. (1989) Linkage of a prion protein missense variant to Gerstmann-Sträussler syndrome, *Nature* **338**, 342–345.

Hsiao, K., Meiner, Z., Kahana, E., Cass, C., Kahana, I., Avrahami, D., Scarlato, G., Abramsky, O., Prusiner, S.B., Gabizon, R. (1991) Mutation of the prion protein in Libyan Jews with Creutzfeldt-Jakob disease, *N. Engl. J. Med.* **324**, 1091–1097.

Hsiao, K.K., Cass, C., Schellenberg, G.D., Bird, T., Devine-Gage, E., Wisniewski, H., Prusiner, S.B. (1991) A prion protein variant in a family with the telencephalic form of Gerstmann-Sträussler-Scheinker syndrome, *Neurology* **41**, 681–684.

Hsiao, K.K., Groth, D., Scott, M., Yang, S.-L., Serban, H., Rapp, D., Foster, D., Torchia, M., DeArmond, S.J., Prusiner, S.B. (1994) Serial transmission in rodents of neurodegeneration from transgenic mice expressing mutant prion protein, *Proc. Natl. Acad. Sci. U. S. A.* **91**, 9126–9130.

Huang, Z., Prusiner, S.B., Cohen, F.E. (1995) Scrapie prions: a three-dimensional model of an infectious fragment, *Fold. Des.* **1**, 13–19.

Hundt, C., Peyrin, J.M., Haik, S., Gauczynski, S., Leucht, C., Rieger, R., Riley, M.L., Deslys, J.P., Dormont, D., Lasmézas, C.I., Weiss, S. (2001) Identification of interaction domains of the prion protein with its 37-kDa/67-kDa laminin receptor, *EMBO J.* **20**, 5876–5886.

Hunter, N., Foster, J.D., Benson, G., Hope, J. (1991) Restriction fragment length polymorphisms of the scrapie-associated fibril protein (PrP) gene and their association with susceptiblity to natural scrapie in British sheep, *J. Gen. Virol.* **72**, 1287–1292.

Hunter, N., Goldmann, W., Smith, G., Hope, J. (1994) Frequencies of PrP gene variants in healthy cattle and cattle with BSE in Scotland, *Vet. Rec.* **135**, 400–403.

Hunter, N., Goldmann, W., Benson, G., Foster, J.D., Hope, J. (1993) Swaledale sheep affected by natural scrapie differ significantly in PrP genotype frequencies from healthy sheep and those selected for reduced incidence of scrapie, *J. Gen. Virol.* **74**, 1025–1031.

Hunter, N., Moore, L., Hosie, B.D., Dingwall, W.S., Greig, A. (1997) Association between natural scrapie and PrP genotype in a flock of Suffolk sheep in Scotland, *Vet. Rec.* **140**, 59–63.

Hunter, N., Cairns, D., Foster, J.D., Smith, G., Goldmann, W., Donnelly, K. (1997) Is scrapie solely a genetic disease? *Nature* **386**, 137.

Ikeda, T., Horiuchi, M., Ishiguro, N., Muramatsu, Y., Kai-Uwe, G.D., Shinagawa, M. (1995) Amino acid polymorphisms of PrP with reference to onset of scrapie in Suffolk and Corriedale sheep in Japan, *J. Gen. Virol.* **76**, 2577–2581.

Ingrosso, L., Ladogana, A., Pocchiari, M. (1995) Congo red prolongs the incubation period in scrapie-infected hamsters, *J. Virol.* **69**, 506–508.

Ironside, J.W. (1997) The new variant form of Creutzfeldt-Jakob disease: a novel prion protein amyloid disorder (Editorial), *Amyloid: Int. J. Exp. Clin. Invest.* **4**, 66–69.

James, T.L., Liu, H., Ulyanov, N.B., Farr-Jones, S., Zhang, H., Donne, D.G., Kaneko, K., Groth, D., Mehlhorn, I., Prusiner, S.B., Cohen, F.E. (1997) Solution structure of a 142-residue recombinant prion protein corresponding to the infectious fragment of the scrapie isoform, *Proc. Natl. Acad. Sci. U. S. A.* **94**(19), 10086–10091.

Jeffrey, M., Wells, G.A.H. (1988) Spongiform encephalopathy in a nyala *(Tragelaphus angasi)*, *Vet. Pathol.* **25**, 398–399.

Jendroska, K., Heinzel, F.P., Torchia, M., Stowring, L., Kretzschmar, H.A., Kon, A.,

Stern, A., Prusiner, S.B., DeArmond, S.J. (1991) Proteinase-resistant prion protein accumulation in Syrian hamster brain correlates with regional pathology and scrapie infectivity, *Neurology* **41**, 1482–1490.

Kahana, E., Milton, A., Braham, J., Sofer, D. (1974) Creutzfeldt-Jakob disease: focus among Libyan Jews in Israel, *Science* **183**, 90–91.

Kaneko, K., Zulianello, L., Scott, M., Cooper, C.M., Wallace, A.C., James, T.L., Cohen, F.E., Prusiner, S.B. (1997) Evidence for protein X binding to a discontinuous epitope on the cellular prion protein during scrapie prion propagation, *Proc. Natl. Acad. Sci. U. S. A.* **94**, 10069–10074.

Kaneko, K., Ball, H.L., Wille, H., Zhang, H., Groth, D., Torchia, M., Tremblay, P., Safar, J., Prusiner, S.B., DeArmond, S.J., Baldwin, M.A., Cohen, F.E. (2000) A synthetic peptide initiates Gerstmann-Sträussler-Scheinker (GSS) disease in transgenic mice, *J. Mol. Biol.* **295**(4), 997–1007.

Kascsak, R.J., Rubenstein, R., Merz, P.A., Tonna-DeMasi, M., Fersko, R., Carp, R.I., Wisniewski, H.M., Diringer, H. (1987) Mouse polyclonal and monoclonal antibody to scrapie-associated fibril proteins, *J. Virol.* **61**, 3688–3693.

Kellings, K., Prusiner, S.B., Riesner, D. (1994) Nucleic acids in prion preparations: unspecific background or essential component? *Philos. Trans. R. Soc. Lond. B. Biol. Sci.* **343**, 425–430.

Kellings, K., Meyer, N., Mirenda, C., Prusiner, S.B., Riesner, D. (1992) Further analysis of nucleic acids in purified scrapie prion preparations by improved return refocussing gel electrophoresis (RRGE), *J. Gen. Virol.* **73**, 1025–1029.

Keshet, G.I., Bar-Peled, O., Yaffe, D., Nudel, U., Gabizon, R. (2000) The cellular prion protein colocalizes with the dystroglycan complex in the brain, *J. Neurochem.* **75**, 1889–1897.

Kimberlin, R.H. (1982) Reflections on the nature of the scrapie agent, *Trends Biochem. Sci.* **7**, 392–394.

Kimberlin, R.H. (1990) Scrapie and possible relationships with viroids, *Semin. Virol.* **1**, 153–162.

Kimberlin, R.H., Walker, C.A. (1983) The antiviral compound HPA-23 can prevent scrapie when administered at the time of infection, *Arch. Virol.* **78**, 9–18.

Kimberlin, R.H., Collis, S.C., Walker, C.A. (1976) Profiles of brain glycosidase activity in cuprizone-fed Syrian hamsters and in scrapie-affected mice, rats, Chinese hamsters and Syrian hamsters, *J. Comp. Pathol.* **86**, 135–142.

Kimberlin, R.H., Field, H.J., Walker, C.A. (1983) Pathogenesis of mouse scrapie: evidence for spread of infection from central to peripheral nervous system, *J. Gen. Virol.* **64**, 713–716.

Kimberlin, R.H., Walker, C.A., Fraser, H. (1989) The genomic identity of different strains of mouse scrapie is expressed in hamsters and preserved on reisolation in mice, *J. Gen. Virol.* **70**, 2017–2025.

King, C.-Y., Tittman, P., Gross, H., Gebert, R., Aebi, M., Wüthrich, K. (1997) Prion-inducing domain 2–114 of yeast Sup35 protein transforms *in vitro* into amyloid-like filaments, *Proc. Natl. Acad. Sci. U. S. A.* **94**, 6618–6622.

Kingsbury, D.T., Kasper, K.C., Stites, D.P., Watson, J.D., Hogan, R.N., Prusiner, S.B. (1983) Genetic control of scrapie and Creutzfeldt-Jakob disease in mice, *J. Immunol.* **131**, 491–496.

Kirkwood, J.K., Cunningham, A.A., Wells, G.A.H., Wilesmith, J.W., Barnett, J.E.F. (1993) Spongiform encephalopathy in a herd of greater kudu (*Tragelaphus strepsiceros*): epidemiological observations, *Vet. Rec.* **133**, 360–364.

Kitamoto, T., Tateishi, J. (1994) Human prion diseases with variant prion protein, *Philos. Trans. R. Soc. Lond. B. Biol. Sci.* **343**, 391–398.

Klatzo, I., Gajdusek, D.C., Zigas, V. (1959) Pathology of kuru, *Lab. Invest.* **8**, 799–847.

Klitzman, R.L., Alpers, M.P., Gajdusek, D.C. (1984) The natural incubation period of kuru and the episodes of transmission in three clusters of patients, *Neuroepidemiology* **3**, 3–20.

Koch, T.K., Berg, B.O., DeArmond, S.J., Gravina, R.F. (1985) Creutzfeldt-Jakob disease in a young adult with idiopathic hypopituitarism. Possible relation to the administration of cadaveric human growth hormone, *N. Engl. J. Med.* **313**, 731–733.

Kocisko, D.A., Come, J.H., Priola, S.A., Chesebro, B., Raymond, G.J., Lansbury Jr., P.T., Caughey, B. (1994) Cell-free formation of protease-resistant prion protein, *Nature* **370**, 471–474.

Korth, C., May, B.C.H., Cohen, F.E., Prusiner, S.B. (2001) Acridine and phenothiazine derivatives as pharmacotherapeutics for prion disease, *Proc. Natl. Acad. Sci. U. S. A.* **98**, 9836–9841.

Korth, C., Kaneko, K., Groth, D., Heye, N., Telling, G., Mastrianni, J., Parchi, P., Gambetti, P., Will, R., Ironside, J., Heinrich, C., Tremblay, P., DeArmond, S.J., Prusiner, S.B. (2003) Abbreviated incubation times for human prions in mice expressing a chimeric mouse – human prion protein transgene, *Proc. Natl. Acad. Sci. U. S. A.* **100**, 4784–4789.

Korth, C., Stierli, B., Streit, P., Moser, M., Schaller, O., Fischer, R., Schulz-Schaeffer, W., Kretzschmar, H., Raeber, A., Braun, U., Ehrensperger, F., Hornemann, S., Glockshuber, R., Riek, R., Billeter, M., Wuthrick, K., Oesch, B. (1997) Prion (PrP^{Sc})-specific epitope defined by a monoclonal antibody, *Nature* **389**, 74–77.

Kretzschmar, H.A., Neumann, M., Stavrou, D. (1995) Codon 178 mutation of the human prion protein gene in a German family (Backer family): sequencing data from 72 year-old celloidin-embedded brain tissue, *Acta Neuropathol.* **89**, 96–98.

Kretzschmar, H.A., Prusiner, S.B., Stowring, L.E., DeArmond, S.J. (1986) Scrapie prion proteins are synthesized in neurons, *Am. J. Pathol.* **122**, 1–5.

Kretzschmar, H.A., Honold, G., Seitelberger, F., Feucht, M., Wessely, P., Mehraein, P., Budka, H. (1991) Prion protein mutation in family first reported by Gerstmann, Sträussler, and Scheinker, *Lancet* **337**, 1160.

Kurschner, C., Morgan, J.I. (1995) The cellular prion protein (PrP) selectively binds to Bcl-2 in the yeast two-hybrid system, *Mol. Brain Res.* **30**, 165–168.

Laplanche, J.-L., Chatelain, J., Launay, J.-M., Gazengel, C., Vidaud, M. (1990) Deletion in prion protein gene in a Moroccan family, *Nucleic Acids. Res.* **18**, 6745.

Laplanche, J.-L., Chatelain, J., Beaudry, P., Dussaucy, M., Bounneau, C., Launay, J.-M. (1993) French autochthonous scrapied sheep without the 136Val PrP polymorphism, *Mamm. Genome* **4**, 463–464.

Lasmézas, C.I., Deslys, J.-P., Demaimay, R., Adjou, K.T., Hauw, J.-J., Dormont, D. (1996) Strain specific and common pathogenic events in murine models of scrapie and bovine spongiform encephalopathy, *J. Gen. Virol.* **77**, 1601–1609.

Lasmézas, C.I., Deslys, J.-P., Demaimay, R., Adjou, K.T., Lamoury, F., Dormont, D., Robain, O., Ironside, J., Hauw, J.-J. (1996) BSE transmission to macaques, *Nature* **381**, 743–744.

Lasmézas, C.I., Deslys, J.-P., Robain, O., Jaegly, A., Beringue, V., Peyrin, J.-M., Fournier, J.-G., Hauw, J.-J., Rossier, J., Dormont, D. (1997) Transmission of the BSE agent to mice in the absence of detectable abnormal prion protein, *Science* **275**(17), 402–405.

Legname, G., Baskakov, I.V., Nguyen, H.-O.B., Riesner, D., Cohen, F.E., DeArmond, S.J., Prusiner, S.B. (2004) Synthetic mammalian prions, *Science* **305**, 673–676.

Li, G., Bolton, D.C. (1997) A novel hamster prion protein mRNA contains an extra exon: increased expression in scrapie, *Brain Res.* **751**, 265–274.

Liu, H., Farr-Jones, S., Ulyanov, N.B., Llinas, M., Marqusee, S., Groth, D., Cohen, F.E., Prusiner, S.B., James, T.L. (1999) Solution structure of Syrian hamster prion protein rPrP(90–231), *Biochemistry* **38**(17), 5362–5377.

Lledo, P.-M., Tremblay, P., DeArmond, S.J., Prusiner, S.B., Nicoll, R.A. (1996) Mice deficient for prion protein exhibit normal neuronal excitability and synaptic transmission in the hippocampus, *Proc. Natl. Acad. Sci. U. S. A.* **93**, 2403–2407.

Llewelyn, C.A., Hewitt, P.E., Knight, R.S., Amar, K., Cousens, S., Mackenzie, J., Will, R.G. (2004) Possible transmission of variant Creutzfeldt-Jakob disease by blood transfusion, *Lancet* **363**, 417–421.

Locht, C., Chesebro, B., Race, R., Keith, J.M. (1986) Molecular cloning and complete sequence of prion protein cDNA from mouse brain infected with the scrapie agent, *Proc. Natl. Acad. Sci. U. S. A.* **83**, 6372–6376.

Lugaresi, E., Medori, R., Montagna, P., Baruzzi, A., Cortelli, P., Lugaresi, A., Tinuper, P., Zucconi, M., Gambetti, P. (1986) Fatal familial insomnia and dysautonomia with selective degeneration of thalamic nuclei, *N. Engl. J. Med.* **315**, 997–1003.

Maddelein, M.L., Dos Reis, S., Duvezin-Caubet, S., Coulary-Salin, B., Saupe, S.J. (2002) Amyloid aggregates of the HET-s prion protein are infectious, *Proc. Natl. Acad. Sci. U. S. A.* **99**, 7402–7407.

Mallucci, G., Dickinson, A., Linehan, J., Klohn, P.C., Brandner, S., Collinge, J. (2003) Depleting neuronal PrP in prion infection prevents disease and reverses spongiosis, *Science* **302**, 871–874.

Manson, J.C., Clarke, A.R., Hooper, M.L., Aitchison, L., McConnell, I., Hope, J. (1994) 129/Ola mice carrying a null mutation in PrP that abolishes mRNA production are developmentally normal, *Mol. Neurobiol.* **8**, 121–127.

Manson, J.C., Jameison, E., Baybutt, H., Tuzi, N.L., Barron, R., McConnell, I., Somerville, R., Ironside, J., Will, R., Sy, M.-S., Melton, D.W., Hope, J., Bostock, C. (1999) A single amino acid alteration (101L) introduced into murine PrP dramatically alters incubation time of transmissible spongiform encephalopathy, *EMBO J.* **18**, 6855–6864.

Marsh, R.F., Bessen, R.A., Lehmann, S., Hartsough, G.R. (1991) Epidemiological and experimental studies on a new incident of transmissible mink encephalopathy, *J. Gen. Virol.* **72**, 589–594.

Martins, V.R., Graner, E., Garcia-Abreu, J., de Souza, S.J., Mercadante, A.F., Veiga, S.S., Zanata, S.M., Neto, V.M., Brentani, R.R. (1997) Complementary hydropathy identifies a cellular prion protein receptor, *Nat. Med.* **3**, 1376–1382.

Masison, D.C., Wickner, R.B. (1995) Prion-inducing domain of yeast Ure2p and protease resistance of Ure2p in prion-containing cells, *Science* **270**, 93–95.

Masters, C.L., Gajdusek, D.C., Gibbs Jr., C.J. (1981) Creutzfeldt-Jakob disease virus isolations from the Gerstmann-Sträussler syndrome, *Brain* **104**, 559–588.

Masters, C.L., Harris, J.O., Gajdusek, D.C., Gibbs Jr., C.J., Bernouilli, C., Asher, D.M. (1978) Creutzfeldt-Jakob disease: patterns of worldwide occurrence and the significance of familial and sporadic clustering, *Ann. Neurol.* **5**, 177–188.

Mastrianni, J., Nixon, F., Layzer, R., DeArmond, S.J., Prusiner, S.B. (1997) Fatal sporadic insomnia: fatal familial insomnia phenotype without a mutation of the prion protein gene, *Neurology* **48** (Suppl.), A296.

Mastrianni, J.A., Nixon, R., Layzer, R., Telling, G.C., Han, D., DeArmond, S.J., Prusiner, S.B. (1999) Prion protein conformation in a patient with sporadic fatal insomnia, *N. Engl. J. Med.* **340**(21), 1630–1638.

McKinley, M.P., Meyer, R.K., Kenaga, L., Rahbar, F., Cotter, R., Serban, A., Prusiner, S.B. (1991) Scrapie prion rod formation *in vitro* requires both detergent extraction and limited proteolysis, *J. Virol.* **65**, 1340–1351.

McKnight, S., Tjian, R. (1986) Transcriptional selectivity of viral genes in mammalian cells, *Cell* **46**, 795–805.

Medori, R., Montagna, P., Tritschler, H.J., LeBlanc, A., Cortelli, P., Tinuper, P., Lugaresi, E., Gambetti, P. (1992) Fatal familial insomnia: a second kindred with mutation of prion protein gene at codon 178, *Neurology* **42**, 669–670.

Meggendorfer, F. (1930) Klinische und genealogische Beobachtungen bei einem Fall von spastischer Pseudosklerose Jakobs, *Z. Gesamte Neurol. Psychiatr.* **128**, 337–341.

Meyer, R.K., McKinley, M.P., Bowman, K.A., Braunfeld, M.B., Barry, R.A., Prusiner, S.B. (1986) Separation and properties of cellular and scrapie prion proteins, *Proc. Natl. Acad. Sci. U. S. A.* **83**, 2310–2314.

Miller, M.W., Wild, M.A., Williams, E.S. (1998) Epidemiology of chronic wasting disease in captive Rocky Mountain elk, *J. Wildl. Dis.* **34**, 532–538.

Miller, M.W., Williams, E.S., McCarty, C.W., Spraker, T.R., Kreeger, T.J., Larsen, C.T., Thorne, E.T. (2000) Epizootiology of chronic wasting disease in free-ranging cervids in Colorado and Wyoming, *J. Wildl. Dis.* **36**, 676–690.

Mills, J.L., Schonberger, L.B., Wysowski, D.K., Brown, P., Durako, S.J., Cox, C., Kong, F., Fradkin, J.E. (2004) Long-term mortality in the United States cohort of pituitary-derived growth hormone recipients, *J. Pediatr.* **144**, 430–436.

Minor, P., Newham, J., Jones, N., Bergeron, C., Gregori, L., Asher, D., Van Engelenburg, F., Stroebel, T., Vey, M., Barnard, G., Head, M. (2004) Standards for the assay of Creutzfeldt-Jakob disease specimens, *J. Gen. Virol.* **85**, 1777–1784.

Miyazono, M., Kitamoto, T., Doh-ura, K., Iwaki, T., Tateishi, J. (1992) Creutzfeldt-Jakob disease with codon 129 polymorphism (Valine): a comparative study of patients with codon 102 point mutation or without mutations, *Acta Neuropathol.* **84**, 349–354.

Mo, H., Moore, R.C., Cohen, F.E., Westaway, D., Prusiner, S.B., Wright, P.E., Dyson, H.J. (2001) Two different neurodegenerative diseases caused by proteins with similar structures, *Proc. Natl. Acad. Sci. U. S. A.* **98**, 2352–2357.

Mobley, W.C., Neve, R.L., Prusiner, S.B., McKinley, M.P. (1988) Nerve growth factor increases mRNA levels for the prion protein and the β-amyloid protein precursor in developing hamster brain, *Proc. Natl. Acad. Sci. U. S. A.* **85**, 9811–9815.

Moore, R.C., Hope, J., McBride, P.A., McConnell, I., Selfridge, J., Melton, D.W., Manson, J.C. (1998) Mice with gene targeted prion protein alterations show that *Prnp, Sinc* and *Prni* are congruent, *Nat. Genet.* **18**, 118–125.

Moore, R.C., Mastrangelo, P., Bouzamondo, E., Heinrich, C., Legname, G., Prusiner, S.B., Hood, L., Westaway, D., DeArmond, S.J., Tremblay, P. (2001) Doppel-induced cerebellar degeneration in transgenic mice, *Proc. Natl. Acad. Sci. U. S. A.* **98**, 15288–15293.

Moore, R.C., Lee, I.Y., Silverman, G.L., Harrison, P.M., Strome, R., Heinrich, C., Karunaratne, A., Pasternak, S.H., Chishti, M.A., Liang, Y., Mastrangelo, P., Wang, K., Smit, A.F.A., Katamine, S., Carlson, G.A., Cohen, F.E., Prusiner, S.B., Melton, D.W., Tremblay, P., Hood, L.E., Westaway, D. (1999) Ataxia in prion protein (PrP)-deficient mice is associated with upregulation of the novel PrP-like protein doppel, *J. Mol. Biol.* **292**(4), 797–817.

Mouillet-Richard, S., Ermonval, M., Chebassier, C., Laplanche, J.-L., Lehmann, S., Launay, J.M., Kellermann, O. (2000) Signal transduction through prion protein, *Science* **289**, 1925–1928.

Nathanson, N., Wilesmith, J., Griot, C. (1997) Bovine spongiform encephalopathy (BSE): cause and consequences of a common-source epidemic, *Am. J. Epidemiol.* **145**, 959–969.

Nishida, N., Tremblay, P., Sugimoto, T., Shigematsu, K., Shirabe, S., Petromilli, C., Erpel, S.P., Nakaoke, R., Atarashi, R., Houtani, T., Torchia, M., Sakaguchi, S., DeArmond, S.J., Prusiner, S.B., Katamine, S. (1999) A mouse prion protein transgene rescues mice deficient for the prion protein gene from Purkinje cell degeneration and demyelination, *Lab. Invest.* **79**(6), 689–697.

Oesch, B., Teplow, D.B., Stahl, N., Serban, D., Hood, L.E., Prusiner, S.B. (1990) Identification of cellular proteins binding to the scrapie prion protein, *Biochemistry* **29**, 5848–5855.

Oesch, B., Westaway, D., Wälchli, M., McKinley, M.P., Kent, S.B.H., Aebersold, R., Barry, R.A., Tempst, P., Teplow, D.B., Hood, L.E., Prusiner, S.B., Weissmann, C. (1985) A cellular gene encodes scrapie PrP 27–30 protein, *Cell* **40**, 735–746.

O'Rourke, K.I., Holyoak, G.R., Clark, W.W., Mickelson, J.R., Wang, S., Melco, R.P., Besser, T.E., Foote, W.C. (1997) PrP genotypes and experimental scrapie in orally inoculated Suffolk sheep in the United States, *J. Gen. Virol.* **78**, 975–978.

O'Rourke, K.I., Besser, T.E., Miller, M.W., Cline, T.F., Spraker, T.R., Jenny, A.L., Wild, M.A., Zebarth, G.L., Williams, E.S. (1999) PrP genotypes of captive and free-ranging Rocky Mountain elk (Cervus elaphus nelsoni) with chronic wasting disease, *J. Gen. Virol.* **80**, 2765–2769.

Owen, F., Poulter, M., Collinge, J., Crow, T.J. (1990) Codon 129 changes in the prion protein gene in Caucasians, *Am. J. Hum. Genet.* **46**, 1215–1216.

Owen, F., Poulter, M., Lofthouse, R., Collinge, J., Crow, T.J., Risby, D., Baker, H.F., Ridley, R.M., Hsiao, K., Prusiner, S.B. (1989) Insertion in prion protein gene in familial Creutzfeldt-Jakob disease, *Lancet* **1**, 51–52.

Paisley, D., Banks, S., Selfridge, J., McLennan, N.F., Ritchie, A.M., McEwan, C., Irvine, D.S., Saunders, P.T., Manson, J.C., Melton, D.W. (2004) Male infertility and DNA damage in Doppel knockout and prion protein/Doppel double-knockout mice, *Am. J. Pathol.* **164**, 2279–2288.

Palmer, M.S., Dryden, A.J., Hughes, J.T., Collinge, J. (1991) Homozygous prion protein genotype predisposes to sporadic Creutzfeldt-Jakob disease, *Nature* **352**, 340–342.

Palmer, M.S., Mahal, S.P., Campbell, T.A., Hill, A.F., Sidle, K.C.L., Laplanche, J.-L., Collinge, J. (1993) Deletions in the prion protein gene are not associated with CJD, *Hum. Mol. Genet.* **2**, 541–544.

Pan, K.-M., Stahl, N., Prusiner, S.B. (1992) Purification and properties of the cellular prion protein from Syrian hamster brain, *Protein Sci.* **1**, 1343–1352.

Pan, K.-M., Baldwin, M., Nguyen, J., Gasset, M., Serban, A., Groth, D., Mehlhorn, I., Huang, Z., Fletterick, R.J., Cohen, F.E., Prusiner, S.B. (1993) Conversion of α-helices into β-sheets features in the formation of the scrapie prion proteins, *Proc. Natl. Acad. Sci. U. S. A.* **90**, 10962–10966.

Paramithiotis, E., Pinard, M., Lawton, T., LaBoissiere, S., Leathers, V.L., Zou, W.Q., Estey, L.A., Lamontagne, J., Lehto, M.T., Kondejewski, L.H., Francoeur, G.P., Papadopoulos, M., Haghighat, A., Spatz, S.J.,

Head, M., Will, R., Ironside, J., O'Rourke, K., Tonelli, Q., Ledebur, H.C., Chakrabartty, A., Cashman, N.R. (2003) A prion protein epitope selective for the pathologically misfolded conformation, *Nat. Med.* **9**, 893–899.

Parchi, P., Capellari, S., Chin, S., Schwarz, H.B., Schecter, N.P., Butts, J.D., Hudkins, P., Burns, D.K., Powers, J.M., Gambetti, P. (1999) A subtype of sporadic prion disease mimicking fatal familial insomnia, *Neurology* **52**, 1757–1763.

Parchi, P., Capellari, S., Chen, S.G., Petersen, R.B., Gambetti, P., Kopp, P., Brown, P., Kitamoto, T., Tateishi, J., Giese, A., Kretzschmar, H. (1997) Typing prion isoforms (Lett.), *Nature* **386**, 232–233.

Parry, H.B. (1962) Scrapie: a transmissible and hereditary disease of sheep, *Heredity* **17**, 75–105.

Patino, M.M., Liu, J.-J., Glover, J.R., Lindquist, S. (1996) Support for the prion hypothesis for inheritance of a phenotypic trait in yeast, *Science* **273**, 622–626.

Pattison, I.H., Millson, G.C. (1961) Scrapie produced experimentally in goats with special reference to the clinical syndrome, *J. Comp. Pathol.* **71**, 101–108.

Pattison, I.H., Jebbett, J.N. (1971) Clinical and histological observations on cuprizone toxicity and scrapie in mice, *Res. Vet. Sci.* **12**, 378–380.

Peden, A.H., Head, M.W., Ritchie, D.L., Bell, J.E., Ironside, J.W. (2004) Preclinical vCJD after blood transfusion in a *PRNP* codon 129 heterozygous patient, *Lancet* **364**, 527–529.

Peretz, D., Scott, M., Groth, D., Williamson, A., Burton, D., Cohen, F.E., Prusiner, S.B. (2001) Strain-specified relative conformational stability of the scrapie prion protein, *Protein Sci.* **10**, 854–863.

Peretz, D., Williamson, R.A., Legname, G., Matsunaga, Y., Vergara, J., Burton, D., DeArmond, S.J., Prusiner, S.B., Scott, M.R. (2002) A change in the conformation of prions accompanies the emergence of a new prion strain, *Neuron* **34**, 921–932.

Peretz, D., Williamson, R.A., Matsunaga, Y., Serban, H., Pinilla, C., Bastidas, R.B., Rozenshteyn, R., James, T.L., Houghten, R.A., Cohen, F.E., Prusiner, S.B., Burton, D.R. (1997) A conformational transition at the N-terminus of the prion protein features in formation of the scrapie isoform, *J. Mol. Biol.* **273**, 614–622.

Peretz, D., Williamson, R.A., Kaneko, K., Vergara, J., Leclerc, E., Schmitt-Ulms, G., Mehlhorn, I.R., Legname, G., Wormald, M.R., Rudd, P.M., Dwek, R.A., Burton, D.R., Prusiner, S.B. (2001) Antibodies inhibit prion propagation and clear cell cultures of prion infectivity, *Nature* **412**, 739–743.

Perrier, V., Wallace, A.C., Kaneko, K., Safar, J., Prusiner, S.B., Cohen, F.E. (2000) Mimicking dominant-negative inhibition of prion replication through structure-based drug design, *Proc. Natl. Acad. Sci. U. S. A.* **97**, 6073–6078.

Perrier, V., Kaneko, K., Safar, J., Vergara, J., Tremblay, P., DeArmond, S.J., Cohen, F.E., Prusiner, S.B., Wallace, A.C. (2002) Dominant-negative inhibition of prion replication in transgenic mice, *Proc. Natl. Acad. Sci. U. S. A.* **99**, 13079–13084.

Peters, J., Miller, J.M., Jenny, A.L., Peterson, T.L., Carmichael, K.P. (2000) Immunohistochemical diagnosis of chronic wasting disease in preclinically affected elk from a captive herd, *J. Vet. Diagn. Invest.* **12**, 579–582.

Petersen, R.B., Tabaton, M., Berg, L., Schrank, B., Torack, R.M., Leal, S., Julien, J., Vital, C., Deleplanque, B., Pendlebury, W.W., Drachman, D., Smith, T.W., Martin, J.J., Oda, M., Montagna, P., Ott, J., Autilio-Gambetti, L., Lugaresi, E., Gambetti, P. (1992) Analysis of the prion protein gene in thalamic dementia, *Neurology* **42**, 1859–1863.

Phillips, N.A., Bridgeman, J. Ferguson-Smith, M. (2000) Findings & Conclusions. *The BSE Inquiry*, Vol. 1, Stationery Office, London, UK.

PHS (1997) Report on Human Growth Hormone and Creutzfeldt-Jakob Disease, Public Health Service Interagency Coordinating Committee. **14**: 1–11.

Poulter, M., Baker, H.F., Frith, C.D., Leach, M., Lofthouse, R., Ridley, R.M., Shah, T., Owen, F., Collinge, J., Brown, G., Hardy, J., Mullan, M.J., Harding, A.E., Bennett, C., Doshi, R., Crow, T.J. (1992) Inherited prion disease with 144 base pair gene insertion. 1. Genealogical and molecular studies, *Brain* **115**, 675–685.

Priola, S.A., Raines, A., Caughey, W.S. (2000) Porphyrin and phthalocyanine antiscrapie compounds, *Science* **287**, 1503–1506.

Proske, D., Gilch, S., Wopfner, F., Schätzl, H.M., Winnacker, E.L., Famulok, M. (2002) Prion-protein-specific aptamer reduces PrP^{Sc} formation, *Chembiochem* **3**, 717–725.

Prusiner, S.B. (1982) Novel proteinaceous infectious particles cause scrapie, *Science* **216**, 136–144.

Prusiner, S.B. (2004) Detecting mad cow disease, *Sci. Am.* **291**, 86–93.

Prusiner, S.B., Groth, D.F., Bolton, D.C., Kent, S.B., Hood, L.E. (1984) Purification and structural studies of a major scrapie prion protein, *Cell* **38**, 127–134.

Prusiner, S.B., Bolton, D.C., Groth, D.F., Bowman, K.A., Cochran, S.P., McKinley, M.P. (1982) Further purification and characterization of scrapie prions, *Biochemistry* **21**, 6942–6950.

Prusiner, S.B., McKinley, M.P., Bowman, K.A., Bolton, D.C., Bendheim, P.E., Groth, D.F., Glenner, G.G. (1983) Scrapie prions aggregate to form amyloid-like birefringent rods, *Cell* **35**, 349–358.

Prusiner, S.B., Groth, D., Serban, A., Koehler, R., Foster, D., Torchia, M., Burton, D., Yang, S.-L., DeArmond, S.J. (1993) Ablation of the prion protein (PrP) gene in mice prevents scrapie and facilitates production of anti-PrP antibodies, *Proc. Natl. Acad. Sci. U. S. A.* **90**, 10608–10612.

Prusiner, S.B., Fuzi, M., Scott, M., Serban, D., Serban, H., Taraboulos, A., Gabriel, J.-M., Wells, G., Wilesmith, J., Bradley, R., DeArmond, S.J., Kristensson, K. (1993) Immunologic and molecular biological studies of prion proteins in bovine spongiform encephalopathy, *J. Infect. Dis.* **167**, 602–613.

Prusiner, S.B., Scott, M., Foster, D., Pan, K.-M., Groth, D., Mirenda, C., Torchia, M., Yang, S.-L., Serban, D., Carlson, G.A., Hoppe, P.C., Westaway, D., DeArmond, S.J. (1990) Transgenetic studies implicate interactions between homologous PrP isoforms in scrapie prion replication, *Cell* **63**, 673–686.

Raymond, G.J., Hope, J., Kocisko, D.A., Priola, S.A., Raymond, L.D., Bossers, A., Ironside, J., Will, R.G., Chen, S.G., Petersen, R.B., Gambetti, P., Rubenstein, R., Smits, M.A., Lansbury Jr., P.T., Caughey, B. (1997) Molecular assessment of the potential transmissibilities of BSE and scrapie to humans, *Nature* **388**, 285–288.

Requena, J.R., Groth, D., Legname, G., Stadtman, E.R., Prusiner, S.B., Levine, R.L. (2001) Copper-catalyzed oxidation of the recombinant SHa(29–231) prion protein, *Proc. Natl. Acad. Sci. U. S. A.* **98**, 7170–7175.

Rieger, R., Edenhofer, F., Lasmézas, C.I., Weiss, S. (1997) The human 37-kDa laminin receptor precursor interacts with the prion protein in eukaryotic cells, *Nat. Med.* **3**, 1383–1388.

Riek, R., Hornemann, S., Wider, G., Billeter, M., Glockshuber, R., Wüthrich, K. (1996) NMR structure of the mouse prion protein domain PrP(121–231), *Nature* **382**, 180–182.

Robinson, M.M., Hadlow, W.J., Knowles, D.P., Huff, T.P., Lacy, P.A., Marsh, R.F., Gorham, J.R. (1995) Experimental infection of cattle with the agents of transmissible mink encephalopathy and scrapie, *J. Comp. Path.* **113**, 241–251.

Rogers, M., Yehiely, F., Scott, M., Prusiner, S.B. (1993) Conversion of truncated and elongated prion proteins into the scrapie isoform in cultured cells, *Proc. Natl. Acad. Sci. U. S. A.* **90**, 3182–3186.

Rogers, M., Serban, D., Gyuris, T., Scott, M., Torchia, T., Prusiner, S.B. (1991) Epitope mapping of the Syrian hamster prion protein utilizing chimeric and mutant genes in a vaccinia virus expression system, *J. Immunol.* **147**, 3568–3574.

Rossi, D., Cozzio, A., Flechsig, E., Klein, M.A., Rülicke, T., Aguzzi, A., Weissmann, C. (2001) Onset of ataxia and Purkinje cell loss in PrP null mice inversely correlated with Dpl level in brain, *EMBO J.* **20**, 694–702.

Ryou, C., Legname, G., Peretz, D., Craig, J.C., Baldwin, M.A., Prusiner, S.B. (2003) Differential inhibition of prion propagation by enantiomers of quinacrine, *Lab. Invest.* **83**, 837–843.

Saeki, K., Matsumoto, Y., Hirota, Y., Matsumoto, Y., Onodera, T. (1996) Three-exon structure of the gene encoding the rat prion protein and its expression in tissues, *Virus Genes* **12**, 15–20.

Safar, J., Roller, P.P., Gajdusek, D.C., Gibbs, C.J.J. (1993) Thermal-stability and conformational transitions of scrapie amyloid (prion) protein correlate with infectivity, *Protein Sci.* **2**, 2206–2216.

Safar, J., Wille, H., Itri, V., Groth, D., Serban, H., Torchia, M., Cohen, F.E., Prusiner, S.B. (1998) Eight prion strains have PrP^{Sc} molecules with different conformations, *Nat. Med.* **4**(10), 1157–1165.

Safar, J.G., Scott, M., Monaghan, J., Deering, C., Didorenko, S., Vergara, J., Ball, H., Legname, G., Leclerc, E., Solforosi, L., Serban, H., Groth, D., Burton, D.R., Prusiner, S.B., Williamson, R.A. (2002) Measuring prions causing bovine spongiform encephalopathy or chronic wasting disease by immunoassays and transgenic mice, *Nat. Biotechnol.* **20**, 1147–1150.

Sakaguchi, S., Katamine, S., Nishida, N., Moriuchi, R., Shigematsu, K., Sugimoto, T., Nakatani, A., Kataoka, Y., Houtani, T., Shirabe, S., Okada, H., Hasegawa, S., Miyamoto, T., Noda, T. (1996) Loss of cerebellar Purkinje cells in aged mice homozygous for a disrupted PrP gene, *Nature* **380**, 528–531.

Schätzl, H.M., Da Costa, M., Taylor, L., Cohen, F.E., Prusiner, S.B. (1995) Prion protein gene variation among primates, *J. Mol. Biol.* **245**, 362–374.

Schmitt-Ulms, G., Legname, G., Baldwin, M.A., Ball, H.L., Bradon, N., Bosque, P.J., Crossin, K.L., Edelman, G.M., DeArmond, S.J., Cohen, F.E., Prusiner, S.B. (2001) Binding of neural cell adhesion molecules (N-CAMs) to the cellular prion protein, *J. Mol. Biol.* **314**, 1209–1225.

Scott, M., Foster, D., Mirenda, C., Serban, D., Coufal, F., Wälchli, M., Torchia, M., Groth, D., Carlson, G., DeArmond, S.J., Westaway, D., Prusiner, S.B. (1989) Transgenic mice expressing hamster prion protein produce species-specific scrapie infectivity and amyloid plaques, *Cell* **59**, 847–857.

Scott, M., Groth, D., Foster, D., Torchia, M., Yang, S.-L., DeArmond, S.J., Prusiner, S.B. (1993) Propagation of prions with artificial properties in transgenic mice expressing chimeric PrP genes, *Cell* **73**, 979–988.

Scott, M.R., Köhler, R., Foster, D., Prusiner, S.B. (1992) Chimeric prion protein expression in cultured cells and transgenic mice, *Protein Sci.* **1**, 986–997.

Scott, M.R., Groth, D., Tatzelt, J., Torchia, M., Tremblay, P., DeArmond, S.J., Prusiner, S.B. (1997) Propagation of prion strains through specific conformers of the prion protein, *J. Virol.* **71**, 9032–9044.

Scott, M.R., Will, R., Ironside, J., Nguyen, H.-O.B., Tremblay, P., DeArmond, S.J., Prusiner, S.B. (1999) Compelling transgenetic evidence for transmission of bovine spongiform encephalopathy prions to humans, *Proc. Natl. Acad. Sci. U. S. A.* **96**, 15137–15142.

Scott, M.R., Safar, J., Telling, G., Nguyen, O., Groth, D., Torchia, M., Koehler, R., Tremblay, P., Walther, D., Cohen, F.E., DeArmond, S.J., Prusiner, S.B. (1997) Identification of a prion protein epitope modulating transmission of bovine spongiform encephalopathy prions to transgenic mice, *Proc. Natl. Acad. Sci. U. S. A.* **94**, 14279–14284.

Serban, D., Taraboulos, A., DeArmond, S.J., Prusiner, S.B. (1990) Rapid detection of Creutzfeldt-Jakob disease and scrapie prion proteins, *Neurology* **40**, 110–117.

Sethi, S., Lipford, G., Wagner, H., Kretzschmar, H. (2002) Postexposure prophylaxis against prion disease with a stimulator of innate immunity, *Lancet* **360**, 229–230.

Shibuya, S., Higuchi, J., Shin, R.-W., Tateishi, J., Kitamoto, T. (1998) Protective prion protein polymorphisms against sporadic Creutzfeldt-Jakob disease, *Lancet* **351**, 419.

Sigurdson, C.J., Williams, E.S., Miller, M.W., Spraker, T.R., O'Rourke, K.I., Hoover, E.A. (1999) Oral transmission and early lymphoid tropism of chronic wasting disease PrP^{res} in mule deer fawns (*Odocoileus hemionus*), *J. Gen. Virol.* **80**, 2757–2764.

Somerville, R.A., Chong, A., Mulqueen, O.U., Birkett, C.R., Wood, S.C., Hope, J. (1997) Biochemical typing of scrapie strains, *Nature* **386**, 564.

Sparkes, R.S., Simon, M., Cohn, V.H., Fournier, R.E.K., Lem, J., Klisak, I., Heinzmann, C., Blatt, C., Lucero, M., Mohandas, T., DeArmond, S.J., Westaway, D., Prusiner, S.B., Weiner, L.P. (1986) Assignment of the human and mouse prion protein genes to homologous chromosomes, *Proc. Natl. Acad. Sci. U. S. A.* **83**, 7358–7362.

Spencer, M.D., Knight, R.S., Will, R.G. (2002) First hundred cases of variant Creutzfeldt-Jakob disease: retrospective case note review of early psychiatric and neurological features, *Br. Med. J.* **324**, 1479–1482.

Spudich, S., Mastrianni, J.A., Wrensch, M., Gabizon, R., Meiner, Z., Kahana, I., Rosenmann, H., Kahana, E., Prusiner, S.B. (1995) Complete penetrance of Creutzfeldt-Jakob disease in Libyan Jews carrying the E200K mutation in the prion protein gene, *Mol. Med.* **1**, 607–613.

Stahl, N., Baldwin, M.A., Burlingame, A.L., Prusiner, S.B. (1990) Identification of glycoinositol phospholipid linked and truncated forms of the scrapie prion protein, *Biochemistry* **29**, 8879–8884.

Stahl, N., Borchelt, D.R., Hsiao, K., Prusiner, S.B. (1987) Scrapie prion protein contains a phosphatidylinositol glycolipid, *Cell* **51**, 229–240.

Stahl, N., Baldwin, M.A., Teplow, D.B., Hood, L., Gibson, B.W., Burlingame, A.L., Prusiner, S.B. (1993) Structural analysis of the scrapie prion protein using mass spectrometry and amino acid sequencing, *Biochemistry* **32**, 1991–2002.

Stekel, D.J., Nowak, M.A., Southwood, T.R.E. (1996) Prediction of future BSE spread, *Nature* **381**, 119.

Stöckel, J., Safar, J., Wallace, A.C., Cohen, F.E., Prusiner, S.B. (1998) Prion protein selectively binds copper (II) ions, *Biochemistry* **37**(20), 7185–7193.

Sulkowski, E. (1985) Purification of proteins by IMAC, *Trends Biotechnol.* **3**, 1–7.

Supattapone, S., Nguyen, H.-O.B., Cohen, F.E., Prusiner, S.B., Scott, M.R. (1999) Elimination of prions by branched polyamines and implications for therapeutics, *Proc. Natl. Acad. Sci. U. S. A.* **96**(25), 14529–14534.

Supattapone, S., Wille, H., Uyechi, L., Safar, J., Tremblay, P., Szoka, F.C., Cohen, F.E., Prusiner, S.B., Scott, M.R. (2001) Branched polyamines cure prion-infected neuroblastoma cells, *J. Virol.* **75**, 3453–3461.

Supattapone, S., Bosque, P., Muramoto, T., Wille, H., Aagaard, C., Peretz, D., Nguyen, H.-O.B., Heinrich, C., Torchia, M., Safar, J., Cohen, F.E., DeArmond, S.J., Prusiner, S.B., Scott, M. (1999) Prion protein of 106 residues creates an artificial transmission barrier for prion replication in transgenic mice, *Cell* **96**, 869–878.

Taraboulos, A., Scott, M., Semenov, A., Avrahami, D., Laszlo, L., Prusiner, S.B. (1995) Cholesterol depletion and modification of COOH-terminal targeting sequence of the prion protein inhibits formation of the scrapie isoform, *J. Cell Biol.* **129**, 121–132.

Tatzelt, J., Maeda, N., Pekny, M., Yang, S.-L., Betsholtz, C., Eliasson, C., Cayetano, J., Camerino, A.P., DeArmond, S.J., Prusiner, S.B. (1996) Scrapie in mice deficient in apolipoprotein E or glial fibrillary acidic protein, *Neurology* **47**, 449–453.

Taylor, D.M., Fernie, K., Steele, P.J., McConnell, I., Somerville, R.A. (2002) Thermostability of mouse-passaged BSE and scrapie is independent of host PrP genotype: implications for the nature of the causal agents, *J. Gen. Virol.* **83**, 3199–3204.

Telling, G.C., Haga, T., Torchia, M., Tremblay, P., DeArmond, S.J., Prusiner, S.B. (1996) Interactions between wild-type and mutant prion proteins modulate neurodegeneration in transgenic mice, *Genes Dev.* **10**, 1736–1750.

Telling, G.C., Scott, M., Mastrianni, J., Gabizon, R., Torchia, M., Cohen, F.E., DeArmond, S.J., Prusiner, S.B. (1995) Prion propagation in mice expressing human and chimeric PrP transgenes implicates the interaction of cellular PrP with another protein, *Cell* **83**, 79–90.

Telling, G.C., Parchi, P., DeArmond, S.J., Cortelli, P., Montagna, P., Gabizon, R., Mastrianni, J., Lugaresi, E., Gambetti, P., Prusiner, S.B. (1996) Evidence for the conformation of the pathologic isoform of the prion protein enciphering and propagating prion diversity, *Science* **274**, 2079–2082.

Telling, G.C., Scott, M., Hsiao, K.K., Foster, D., Yang, S.-L., Torchia, M., Sidle, K.C.L., Collinge, J., DeArmond, S.J., Prusiner, S.B. (1994) Transmission of Creutzfeldt-Jakob disease from humans to transgenic mice expressing chimeric human-mouse prion protein, *Proc. Natl. Acad. Sci. U. S. A.* **91**, 9936–9940.

Ter-Avanesyan, M.D., Dagkesamanskaya, A.R., Kushnirov, V.V., Smirnov, V.N. (1994) The *SUP35* omnipotent suppressor gene is involved in the maintenance of the non-Mendelian determinant [*psi*$^+$] in the yeast *Saccharomyces cerevisiae*, *Genetics* **137**, 671–676.

Tobler, I., Gaus, S.E., Deboer, T., Achermann, P., Fischer, M., Rülicke, T., Moser, M., Oesch, B., McBride, P.A., Manson, J.C. (1996) Altered circadian activity rhythms and sleep in mice devoid of prion protein, *Nature* **380**, 639–642.

Tremblay, P., Ball, H.L., Kaneko, K., Groth, D., Hegde, R.S., Cohen, F.E., DeArmond, S.J., Prusiner, S.B., Safar, J.G. (2004) Mutant PrP^{Sc} conformers induced by a synthetic peptide and several prion strains, *J. Virol.* **78**, 2088–2099.

Tremblay, P., Meiner, Z., Galou, M., Heinrich, C., Petromilli, C., Lisse, T., Cayetano, J., Torchia, M., Mobley, W., Bujard, H., DeArmond, S.J., Prusiner, S.B. (1998) Doxycycline control of prion protein transgene expression modulates prion disease in mice, *Proc. Natl. Acad. Sci. U. S. A.* **95**, 12580–12585.

Vey, M., Pilkuhn, S., Wille, H., Nixon, R., DeArmond, S.J., Smart, E.J., Anderson, R.G., Taraboulos, A., Prusiner, S.B. (1996) Subcellular colocalization of the cellular and scrapie prion proteins in caveolae-like membranous domains, *Proc. Natl. Acad. Sci. U. S. A.* **93**, 14945–14949.

Viles, J.H., Cohen, F.E., Prusiner, S.B., Goodin, D.B., Wright, P.E., Dyson, H.J. (1999) Copper binding to the prion protein: structural implications of four identical cooperative binding sites, *Proc. Natl. Acad. Sci. U. S. A.* **96**, 2042–2047.

Vnencak-Jones, C.L., Phillips, J.A. (1992) Identification of heterogeneous PrP gene deletions in controls by detection of allele-specific heteroduplexes (DASH), *Am. J. Hum. Genet.* **50**, 871–872.

Waggoner, D.J., Drisaldi, B., Bartnikas, T.B., Casareno, R.L., Prohaska, J.R., Gitlin, J.D. Harris, D.A. (2000) Brain copper content and cuproenzyme activity do not vary with prion protein expression level, *J. Biol. Chem.* **275**(11), 7455–7458.

Weissmann, C. (1991) Spongiform encephalopathies – the prion's progress, *Nature* **349**, 569–571.

Wells, G.A., Hawkins, S.A., Austin, A.R., Ryder, S.J., Done, S.H., Green, R.B., Dexter, I., Dawson, M., Kimberlin, R.H. (2003) Studies of the transmissibility of the agent of bovine spongiform encephalopathy to pigs, *J. Gen. Virol.* **84**, 1021–1031.

Wells, G.A.H. (2002) Report on TSE infectivity distribution in ruminant tissues (State of Knowledge, December 2001), European Commission: 10–37.

Westaway, D., Cooper, C., Turner, S., Da Costa, M., Carlson, G.A., Prusiner, S.B. (1994) Structure and polymorphism of the mouse prion protein gene, *Proc. Natl. Acad. Sci. U. S. A.* **91**, 6418–6422.

Westaway, D., Goodman, P.A., Mirenda, C.A., McKinley, M.P., Carlson, G.A. Prusiner, S.B. (1987) Distinct prion proteins in short and long scrapie incubation period mice, *Cell* **51**, 651–662.

Westaway, D., Zuliani, V., Cooper, C.M., Da Costa, M., Neuman, S., Jenny, A.L., Detwiler, L., Prusiner, S.B. (1994) Homozygosity for prion protein alleles encoding glutamine-171 renders sheep susceptible to natural scrapie, *Genes Dev.* **8**, 959–969.

Westaway, D., Mirenda, C.A., Foster, D., Zebarjadian, Y., Scott, M., Torchia, M., Yang, S.-L., Serban, H., DeArmond, S.J., Ebeling, C., Prusiner, S.B., Carlson, G.A. (1991) Paradoxical shortening of scrapie incubation times by expression of prion protein transgenes derived from long incubation period mice, *Neuron* **7**, 59–68.

White, A.R., Enever, P., Tayebi, M., Mushens, R., Linehan, J., Brandner, S., Anstee, D., Collinge, J., Hawke, S. (2003) Monoclonal antibodies inhibit prion replication and delay the development of prion disease, *Nature* **422**, 80–83.

Wickner, R.B. (1994) [URE3] as an altered URE2 protein: evidence for a prion analog in *Saccharomyces cerevisiae*, *Science* **264**, 566–569.

Wilesmith, J.W. (1991) The epidemiology of bovine spongiform encephalopathy, *Semin. Virol.* **2**, 239–245.

Wilesmith, J.W., Ryan, J.B.M. Atkinson, M.J. (1991) Bovine spongiform encephalopathy – Epidemiologic studies on the origin, *Vet. Rec.* **128**, 199–203.

Will, R.G., Ironside, J.W., Zeidler, M., Cousens, S.N., Estibeiro, K., Alperovitch, A., Poser, S., Pocchiari, M., Hofman, A., Smith, P.G. (1996) A new variant of Creutzfeldt-Jakob disease in the UK, *Lancet* **347**, 921–925.

Wille, H., Zhang, G.-F., Baldwin, M.A., Cohen, F.E. Prusiner, S.B. (1996) Separation of scrapie prion infectivity from PrP amyloid polymers, *J. Mol. Biol.* **259**, 608–621.

Wille, H., Michelitsch, M.D., Guénebaut, V., Supattapone, S., Serban, A., Cohen, F.E., Agard, D.A. Prusiner, S.B. (2002) Structural studies of the scrapie prion protein by electron crystallography, *Proc. Natl. Acad. Sci. U. S. A.* **99**, 3563–3568.

Williams, E.S., Young, S. (1980) Chronic wasting disease of captive mule deer: a spongiform encephalopathy, *J. Wildl. Dis.* **16**, 89–98.

Williamson, R.A., Peretz, D., Smorodinsky, N., Bastidas, R., Serban, H., Mehlhorn, I., DeArmond, S.J., Prusiner, S.B., Burton, D.R. (1996) Circumventing tolerance to generate autologous monoclonal antibodies to the prion protein, *Proc. Natl. Acad. Sci. U. S. A.* **93**: 7279–7282.

Wyatt, J.M., Pearson, G.R., Smerdon, T.N., Gruffydd-Jones, T.J., Wells, G.A.H., Wilesmith, J.W. (1991) Naturally occurring scrapie-like spongiform encephalopathy in five domestic cats, *Vet. Rec.* **129**, 233–236.

Yehiely, F., Bamborough, P., Costa, M.D., Perry, B.J., Thinakaran, G., Cohen, F.E., Carlson, G.A., Prusiner, S.B. (1997) Identification of candidate proteins binding to prion protein, *Neurobiol. Dis.* **3**, 339–355.

Zulianello, L., Kaneko, K., Scott, M., Erpel, S., Han, D., Cohen, F.E., Prusiner, S.B. (2000) Dominant-negative inhibition of prion formation diminished by deletion mutagenesis of the prion protein, *J. Virol.* **74**(9), 4351–4360.

13
Motor Neuron Diseases: Molecular Mechanism, Pathophysiology, and Treatments

Michael Sendtner
Institute of Clinical Neurobiology, Wuerzburg, Germany

Neurobiology. From Molecular Basis to Disease. Edited by Robert A. Meyers.

ISBN: 978-3-527-32293-0

Keywords

Amyotrophic Lateral Sclerosis (ALS)
A clinical syndrome defined by degenerative changes of lower and upper motor neurons.

Motor Neuron Disease (MND)
A variety of sporadic and inherited diseases characterized by muscle weakness due to primary degeneration of motor neurons of the spinal cord, brain stem motor nuclei, and populations of central motor neurons.

Neurotrophic Factors
Polypeptide molecules that support the survival of specific populations of neuronal cells, such as spinal motor neurons in culture and *in vivo*, via specific signal-transducing receptors.

Spinal Muscular Atrophy (SMA)
A group of disorders that are characterized by degeneration of lower (spinal and bulbar) motor neurons.

The term "motor neuron disease" refers to a variety of devastating neurological disorders characterized by a generalized degeneration of motoneurons. The common clinical manifestations of such disorders are progressive muscle weakness and often death by respiratory failure. Inherited and sporadic forms occur. Further clinical distinctions are made according to the onset of the disorder and whether the disease involves only spinal and bulbar motor neurons (spinal muscular atrophy), or both the lower and upper motor neurons. The most common forms of motor neuron disease are spinal muscular atrophy with an incidence of 1 : 6000 to 10 000 newborns, and sporadic amyotrophic lateral sclerosis with a prevalence of 1 : 100 000, which corresponds to an incidence of 1 : 10 000. About 90% of all cases of amyotrophic lateral sclerosis are sporadic. This disorder affects mostly adults, has a typical clinical appearance, is noninherited, still incurable, and invariably fatal. Recent years have seen the identification of a variety of disease mechanisms, which all lead to the degeneration of motor neurons. Thus, pharmacological approaches that can interfere with degeneration and cell death can now be developed but are complicated by the fact that such therapeutic strategies might work only in subpopulations and not in all patients with this disease.

1
Neuronal Cell Death: Implications for Motoneuron Disease

Motor neurons were among the first populations of neurons that were seen to undergo cell death to a significant extent during development. Such observations, made in the first part of the twentieth century, led to the concept of programmed cell death in the nervous system and to the proposal that target-derived neurotrophic factors play a key role in the maintenance of motor neuron survival and function during the development and in the adult. In the meantime, a variety of such neurotrophic factors have been characterized on the molecular and functional levels, and genetic analysis of patients with motor neuron disease has revealed that only very few patients show mutations or deletions in the genes for these factors. In such cases, it has been found that mutations in the genes for CNTF (ciliary neurotrophic factor) and LIF (Leukaemia inhibitory factor) might act as modifiers, but not as the cause for the disease. Both in patients and in a mouse model of familial ALS, homozygous inactivation of the CNTF gene leads to earlier disease onset. However, the cause of the disease might lie in other functional defects that are not related to the expression of neurotrophic factors or their effects on motor neurons. Studies with mouse models have shown that overexpression of mutant forms of Superoxide dismutase-I (SOD-I) leads to motor neuron disease and that mutations in genes that are relevant for axonal transport and maintenance of axonal processes seem to be more relevant than mechanisms that regulate survival and/or cell death of the motor neuron cell bodies. This indicates that the cell death of motor neuron cell bodies in motor neuron disease might be a secondary phenomenon that occurs as a consequence of axonal defects. On the other hand, a mouse model in which the gene for bcl-2, a key regulator of cellular survival, has been inactivated shows postnatal degeneration of motor neurons. Moreover, cell culture models have revealed that motor neurons derived from SOD-I transgenic mice are much more sensitive to pro-apoptotic stimuli such as Fas ligand and others. In addition, the induction of cell death after binding of excess glutamate to AMPA receptors is also discussed as a cause of motor neuron cell death in motor neuron disease (MND).

Movement of higher organisms requires nerve cells for coordination. The anatomy and physiology of the motor neurons innervating the skeletal musculature is highly conserved from lower to higher vertebrate species – much more than in other parts of the nervous system that are responsible for higher brain functions which are, for example, defective in Alzheimer's disease and other form of dementia. Thus, motor neurons appear as an ideal subject to study neurodegenerative mechanisms in cell culture models, animal models, and to compare these findings with clinical data.

2
Motor Neurons: Anatomical Aspects

The skeletal musculature is innervated by motor neurons located either in the ventral part (the ventrolateral horn) of the spinal cord or in the brain stem motor nuclei. The cell bodies of these motor neurons reside within the central nervous system, and the axons project into the nerves of the peripheral nervous system to the muscle, which they innervate. Many of these axons reach considerable length. For example, the cell bodies of motor neurons innervating the foot muscles are located in the lumbar spinal cord; thus, the total length of these cells can exceed 1 m in humans. This specific property identifies motor neurons as being among the most highly organized cell types of higher organisms.

The transmitter synthesized by motor neurons and released at the motor end plate, the specific synapses in skeletal muscle, is acetylcholine. Cholinergic neurotransmission to muscle cells is highly conserved from lower organisms to man.

Spinal motor neurons are in contact with many other cell types, including other neurons, in particular, interneurons of the spinal cord, astrocytes, oligodendrocytes in the central nervous system, and Schwann and muscle cells in the periphery. They receive synaptic inputs both from proprioceptive sensory neurons and from spinal interneurons. These interneurons receive input from upper motor neurons. The upper motor neurons can also project directly to spinal motor neurons. Interneurons in the spinal cord are responsible for spinal reflexes, and also coordinate patterns for coordinated motor neuron innervation, which are responsible for movement. Neurons of the cortical spinal tract (upper motor neurons), projecting to spinal and brain stem motor neurons either directly or via interneurons, are important for voluntary movement. The cell bodies of these upper motor neurons are located in the brain in specific regions of the precentral cortex. These cells are also affected by degenerative changes in various forms of human MND, in particular, in amyotrophic lateral sclerosis (ALS).

3
Clinical Appearance and Pathology

3.1
Inherited forms of MND: the Identification of Gene Defects, Animal Models, and Disease Mechanisms

In the case of inherited MND, more than 50 different forms have been distinguished on the basis of clinical parameters. The underlying gene defects for the major forms of MND in childhood, classical SMA, and spinal muscular atrophy with respiratory distress (SMARD) have been identified, and the gene defects for three out of five forms of familial amyotrophic lateral sclerosis are known. Furthermore, an increased size of a polymorphic triplet (CRG repeat) in the coding region of the androgen receptor gene on the X chromosome has been identified as the cause of X-linked spinal muscular atrophy. This defect leads to progressive muscle weakness and atrophy, as well as to gynecomastia and reduced fertility in affected males. Although only about 10% of adult onset ALS is familial, this group appears to be heterogeneous. The first identified gene defect that accounts for about 10 to 20% of familial ALS are point mutations in the gene for Cu^{2+}/Cn^{2+}-dependent superoxide dismutase (SOD-I). More than 50 different mutations in this gene have been identified most of them leading to autosomal dominant ALS, but there are also mutations in this gene that cause an autosomal recessive form of the disease. Clinically, there seems to be no clear correlation between disease onset or severity with specific mutations in the *SOD-I* gene, suggesting that modifiers exist.

However, some types of mutations in the *SOD-I* gene are prone to cause severe and rapid course of the disease, for example, the A4V mutation, whereas others, that is, E21G, G37R, D90A, G93C, G93V, I104F, L144S, and I151T are generally associated with survival times of more than 10 to 15 years. The D90A mutation is generally inherited as a recessive trait, and this mutation exists also as a polymorphism in the Scandinavian population. There are studies suggesting that about 100 000 heterozygous carriers for the D90A polymorphism exist in Finland alone.

The gene for another autosomal recessive form of ALS, which is coupled to mutations in the Chr2q33 region, has been identified as a mutation in *Alsin2*. This gene encodes for a GTP exchange factor, which resembles other RanGEF or RhoGEF molecules. This finding appears interesting, as such factors appear as the major signal mediators for extracellular regulators of axon growth and possibly also for axon maintenance. The gene for another rare autosomal dominant form of juvenile ALS (ALS 4) has been recently identified on chromosome 9q34 as missense mutation (T3I, L389S, and R2136H) in the senataxin gene. This gene encodes for a protein of more than 300 kD. The function of this protein is still unknown, but it contains a DNA/RNA helicase domain with homology to human IGHMBP2 that has been shown to be mutant in a specific form of spinal muscular atrophy with respiratory distress (SMARD 1). Interestingly, mutations in senataxin lead to an early form of ALS and patients normally fall ill at an age >25 years. This suggests that some forms of SMA and juvenile ALS are pathophysiologically related. Rapid progress is currently made in the identification of other forms of familial ALS.

A major breakthrough in understanding the cause of motor neuron disease in children was the identification of the underlying gene defect in classical SMA. Familial forms of lower motor neuron degeneration are among the most common genetic causes of death in children, leading to muscle weakness and respiratory failure within a few years after birth in the most severe forms. According to their severity, the autosomal inherited disorders are classified as spinal muscular atrophy type I (most severe form, also named Werdnig–Hoffmann disease), type II and type III (Kugelberg–Welander disease). All three forms map to chromosome 5 (5q13). The responsible gene defect for all three forms of SMA type is located in the survival motoneuron (*SMN*1) gene, which is positioned within a duplicated region on human chromosome 5q13. Homozygous mutation or lack of the telomeric *SMN* gene is associated with SMA, whereas mutations within the centromeric copy are not responsible for SMA. The number of centromeric copies of the *SMN* gene varies, and various lines of evidence suggest that the number of centromeric *SMN* copies (*SMN2*) and the levels of *SMN*2 gene expression correlate with the severity of the disease. In general, patients with less severe forms (SMA type III, Kugelberg–Welander) have more copies of the centromeric *SMN*2 gene than patients with type I SMA (Werdnig–Hoffmann's disease).

The *SMN* gene is highly conserved among higher organisms. Orthologous genes have been identified in mouse, *drosophila, Caenorhabditis elegans*, and the fission yeast *Schizosaccharomyces pombe*. The *SMN* gene duplication is an event that took place just after the divergence of rodents and primates. In humans, both *SMN* copies differ in that full-length protein is predominantly produced from the telomeric *SMN* gene copy. The expression of the truncated SMN2 protein is caused by a C > T exchange at position +6 in exon 7, which leads to skipping of exon 7 encoded regions in about 70% of the protein product derived from the *SMN*2 gene. This specific gene constellation has not been observed in primates, indicating that differential splicing for exon 7 with consequent development of SMA is restricted to humans. Point mutations in the telomeric *SMN* gene are only observed in 5% of

SMA patients. Hot spots of mutational events are found in exon 6 and the 5′ part of exon 7. These regions code for the *SMN* homodimerization domain.

In the past two years, many efforts have been made both in the identification of SMN-associated proteins and the analysis of cellular functions of SMN and SMN-associated proteins. The *SMN* gene is ubiquitously expressed and encodes for a protein of 294 amino acids, which is found both in the nucleus within specific structures called *gemini of coiled bodies* (gems) and in the cytoplasm, both in the cell body and in axons. A group of interacting proteins, which are colocalized in the nucleus includes gemin-2, gemin-3/gp103, a putative DEAD box RNA helicase, gemin-4, which interacts with gemin-3, but not directly with SMN, profilin, fibrillarin, and spliceosomal U snRNPs of the Sm class. The so-called SMN complex also includes other proteins called gemin-5 – 7. The SMN complex is involved in the biogenesis of spliceosomal U snRNPs. U snRNPs are involved in pre-mRNA splicing in the nucleus. Antibodies raised against gemin-2 and SMN interfere with binding of SMN to U snRNAs. This indicates that the SMN complex is essential for the assembly of U snRNPs. Indeed, gene knockout of *Smn* and *gemin-2* in mice is embryonic lethal at early stages of development before blastocysts implant in the uterus. In contrast to humans, the *SMN* gene is not duplicated in mice.

U snRNP assembly is essential for all types of cells and not only for motor neurons. Therefore, the specificity of a disease process for motor neurons remains obscure. The question of why reduced amounts of ubiquitously expressed SMN protein specifically cause degeneration of motor neurons without affecting other somatic cell types is still not fully answered. One explanation could be that SMN, in addition to its reported function, exhibits motor neuron–specific functions that are not relevant for other cells. It is not clear whether all SMN protein is bound to gemin-2 *in vivo*, and evidence from immunohistochemistry indicates that Smn – but not other proteins of the Smn complex – is enriched in axon terminals. In this specific part of motor neurons, the Smn protein seems associated with the hnRNP R protein. The interaction of SMN with hnRNP R requires the exon 7 encoded domains. A complex of SMN and hnRNP R is associated with the 3′-untranslated region of the β-actin mRNA, and mouse models of SMA, which exhibit reduced levels of Smn protein show reduced accumulation of β-actin mRNA in axon terminals. This correlates with β-actin protein levels in axon terminals and reduced axon growth in cultured motor neurons from Smn-deficient mice. Surprisingly, survival of isolated motor neurons with reduced SMN levels is not altered in comparison to wild-type controls. This indicates that the degeneration of motoneurons in the mouse model for spinal muscular atrophy might be a consequence of axonal defects. Lack of motor neuron function and degeneration of motor endplates and terminal axon also affects the capacity of these motor neurons to take up neurotrophic factors. Retrograde transport of these ligands together with their receptors to the cell bodies is reduced, and this should result in degeneration of motor neurons, which is similar to that in mouse models that lack receptors for neurotrophic factors such as gp130, LIFR-β, or CNTFR-α.

In patients with sporadic motor neuron disease, mutations were identified in the gene coding for heavy neurofilaments subunit. These mutations lead to alterations in the KSP-region, a domain with lysine-serine-proline-repeats, which is known to form the side-arm structures that cross-link the neurofilaments. This finding is particularly interesting in the light of neurofilaments pathology observed in ALS patients. However, such mutations are rare and only found in about 1% of patients with sporadic ALS.

Mutations were also found in the gene for dynactin, which points to a role by axonal transport in the pathophysiology of MND. Transgenic mice in which individual components of the axonal transport machinery are inactivated show classical signs of motor neuron disease. Also, the identification of the underlying gene defect in the mouse mutant progressive motoneuron disease (pmn) points to the relevance of axonal transport for the pathophysiology of MND. A point mutation in the tubulin-specific chaperone E gene is responsible for the motor neuron disease in this mouse model. Interestingly, disease onset and progression can significantly be modulated by pharmacological treatment with the neurotrophic factor ciliary neurotrophic factor (CNTF). This indicates that the role of neurotrophic factors is not restricted to the promotion of neuronal survival, but extends to axonal maintenance, regulation of microtubule synthesis, and regulation of axonal transport.

Interestingly, the androgen receptor, the Cu^{2+}/Cn^{2+}-dependent superoxide dismutase, and neurofilament genes are widely expressed and little is known about what makes the disease process specific to motor neurons. It can be speculated that generalized disorders might manifest themselves as MND, possibly because motor neurons are more vulnerable or sensitive to any of these defects than other cell types. Moreover, these examples demonstrate that different, unrelated gene defects have a similar clinical appearance resembling typical signs of MND indicating that MND cannot be considered to be a homogeneous disorder.

3.2 Basic Pathological Findings of ALS

As a characteristic feature of ALS (albeit with significant variation from case to case), loss of motor neurons occurs. The remaining cells appear shrunken, dark, with basophilic cytoplasm and pyknosis of the nuclei. Shrinkage of the cells, which precedes neuronal death, occurs in both upper and lower motor neurons. It has been suggested that the degeneration of lower motor neurons is a consequence of the loss of corticobulbar neurons. However, morphometric comparison of the loss of upper and lower motor neurons does not support such hypotheses. The morphological characteristics of motor neuron cell death such as disintegration of the Golgi complex, loss of Nissl structure, shrinkage, any particular pyknosis of the nuclei, very much resemble the changes occurring during apoptosis. However, experimental proof is poor that motor neuron cell death in all forms of MND is apoptotic. Several studies with postmortem ALS spinal cord have identified TUNEL-positive motor neurons. However, it is not clear whether classical apoptotic mechanisms are responsible for cell death, and when they are activated during the course of the disease. In particular, it remains to be demonstrated that apoptotic mechanisms are disturbed at early stages before first symptoms of the disease become clinically apparent.

Increasing evidence suggests that the disease process in ALS, the most common form of MND, might not be limited to motor neurons. There is also evidence of abnormal rates of degeneration in sensory neurons, but there is little doubt that the motor system is the site of predominant manifestation of the disease. Impairment of axonal transport might underlie many of the pathological characteristics. Inclusion bodies of several types have been described in the perikarya of motor neurons, such as basophilic inclusion, which are thought to consist mainly of RNA, eosinophilic Bunina bodies that resemble autophagosomes, and hyaline-inclusion, originally thought to be characteristic of familial ALS. This points to a role of abnormal protein degradation via the proteasome system, and experimental evidence mainly from cultured neurons suggests that increased proteasome activity can interfere with the pathological mechanisms in some models of MND, such as cell lines overexpressing mutant forms of SOD-I. Hyaline inclusions consist of densely packed phosphorylated neurofilaments, and it is believed that an impairment of slow anterograde axonal transport is responsible for their accumulation in the perikarya of motor neurons. Thus, defects in axonal transport and in protein degradation could play together in the pathophysiology of MND. The concept of impaired axonal maintenance is further supported by the finding that more myelinated nerve fibers are detectable in proximal than in distal parts of motor nerves of ALS patients. Failure of motor neurons to maintain distal structures such as the endplates appears as an early stage in the disease process. Evidence of the degeneration of endplates in early stages of the disease has been derived both from electrophysiological (spontaneous muscle fiber activity) and morphological studies. The disease then progresses to degeneration of the axons in a dying-back fashion, finally leading to degeneration of the cell bodies in the spinal cord and brain stem. Clinical symptoms are expected to become apparent only after it has become impossible to compensate for the loss of functional motor neurons by collateral sprouting and rearrangement of motor units. This coincides with the stage at which motor neuron nerve terminals are affected. Thus, clinical symptoms can precede the degeneration of cell bodies.

4 Sporadic MND: The Pathophysiological Mechanisms Are Still Obscure

The majority of cases (>90%) of MND occur sporadically, precluding the analysis of underlying gene defects by classical molecular genetic analysis. Candidate gene approaches have been conducted in only a few places, and it is premature to state whether the identified mutations may be responsible for the development of the disease in general or only in small subsets of patients. For example, gene defects of heavy neurofilament subunits have been observed only in a few patients (see Sect. 3.1). However, these data do not tell us whether spontaneous gene defects are pathogenic, or whether epigenetic influences (such as intoxication, infection, autoimmune disorders, or any combination of such mechanisms) might be responsible for the disorder. Moreover, defects in the genes coding for neurotrophic factors or their receptors such as CNTF or CNTFR-α have been observed in healthy individuals and in patients with neurological disorders. Targeted disruption of such genes in mice leads to slow degeneration of motor neurons and changes in motor axons in the case of ciliary neurotrophic factor gene knockout. However, symptoms of muscle weakness are only very mild. Mice with gene defects in the CNTFR-α or LIFR-β gene die around birth and thus cannot be considered as a model for human MND. The severe discrepancy between gene activation of ligands such as CNTF or LIF and for receptor components such as CNTFR-α or LIFR-β suggests that defects in the genes for neurotrophic factors are compensated for by other members of the same gene family. In this case, MND would only occur when such compensatory mechanisms fail or when several kinds of degenerative influences act together, leading to such severe cellular dysfunction of motor neurons that compensatory mechanisms are not sufficient to prevent functional deficits.

There have been suggestions that many pathological mechanisms, some of them listed in Sect. 3, may be responsible for MND, but convincing evidence that interference with any of them results in an effective treatment of MND is still lacking. The only effective drug currently used for treatment of MND, Riluzole, interferes with a variety of mechanisms including glutamate toxicity, function of ion channels, and possibly additional intracellular signaling pathways, which modulate the function and structural maintenance of motor neurons. The conclusion that can be drawn from this situation is that basic mechanisms of MND are far from being understood.

4.1 The Autoimmune Hypothesis of Motoneuron Disease

There are many reports of abnormal immunological findings in MND patients. One of the most intriguing is the toxicity of serum from ALS patients to cultures of isolated motor neurons. Moreover, abnormalities in the composition of subgroups of peripheral blood lymphocyte subsets have been described; for example, there is a decrease in mature T lymphocytes (CD3-positive cells) and a shift in the CD4/CD8-ratio in favor of CD8-positive cells, as well as an increase in natural killer cells (CD57 and CD16 positive cells). However, these findings have not been confirmed by recent studies. Another line of evidence has been derived from studies searching for autoantibodies directed at nerve antigens. Such studies have revealed antibodies against gangliosides (GM1), acetylcholine esterase, and L-type or P-type Ca^{2+}-channels. Such findings are in apparent contrast to many unsuccessful efforts to treat ALS by immunosuppressive strategies. Even total body lymph node irradiation was without effect. Some patients with high titers of anti-GM1-antibodies diagnosed as ALS exhibit peripheral nerve multifocal conduction block and are now considered to reflect a subgroup of chronic multifocal inflammatory neuropathy rather than degenerative MND. It now appears conceivable that in the course of MND epitopes on motor neurons that usually are hidden to the immune system become exposed, they allow secondary autoimmune mechanisms to come into play and add to the degenerative changes in motor neurons. Even if autoimmune mechanisms are not primary events, they could contribute to the clinical picture of MND by accelerating the functional impairment of the remaining motor neurons.

4.2 Toxic Lesions of Motor Neurons

Intoxication with ingredients of cycad plants is thought to be responsible for the ALS-Parkinsonism-dementia complex found in Chamorro populations in the

Western Pacific Islands. Since, however, the disease has been reported to occur up to several decades after exposure to the toxic agents, it is not clear whether the intoxication itself is responsible for the disease or whether it induces a long-lasting increase in the vulnerability of motor neurons to other agents. It has been suggested that β-*N*-methylamino-L-alanine (LBMAA) or the increased concentrations of glutamate and abnormal excitatory amino acids detectable in the cerebrospinal fluid of ALS patients could induce motor neuron degeneration via glutamate receptors. Glutamate toxicity via *N*-methyl-D-aspartate (NMDA) receptors is a common degenerative mechanism in other disorders of the nervous system such as stroke and CNS-trauma. However, clinical trials with NMDA-receptor antagonists such as dextromethorphan have not shown efficiency in ALS patients. However, there is evidence *in vitro* that AMPA receptors, in particular, Ca^{2+} influx through this receptor, are responsible for glutamate toxicity in motor neurons. EAAT2 variant transcripts have been found in the brain of ALS patients in combination with reduced EAAT2 protein levels. This is thought to cause elevated glutamate levels at the synapses that are located on the dendrites of bulbar and spinal motor neurons. Further evidence of glutamate toxicity in ALS is based on the efficacy of Riluzole, the only available drug for treatment of ALS up to now. This drug candidate is thought to act as a presynaptic inhibitor of glutamate release, but there is also evidence for other mechanisms of action. The role of glutamate toxicity in motor neurons has also been demonstrated after viral infection. Neuroadapted Sindbis virus mediates motor neuron death indirectly via an excitotoxic mechanism involving glutamate and calcium-permeable AMPA receptors.

5
Oxidative Stress: Is Motoneuron Degeneration in Familial ALS Due to Loss or Gain of Function of the Enzyme Superoxide Dismutase?

Mutations within the Cu/Zn-dependent superoxide dismutase (*SOD-I*) gene on chromosome 21 in about 2 to 5% of ALS patients has led to reinvestigations of all the observations and speculations that oxidative damage may play a pathogenic role in MND. SOD-I had already been the focus of interest in the context of the Alzheimer-like neurodegenerative changes associated with Down Syndrome. In the search for genes responsible for degenerative changes, patients with partial duplication of chromosome 21 translocated to other chromosomes were analyzed, and the 21q22 region has been found to be critical. Among the roughly 100 genes encoded in this region, the *SOD-I* gene was considered to be a good candidate, the duplication and consequent overexpression of which might be responsible for the degenerative changes observed in Down Syndrome: Indeed, overexpression of SOD-I in neural cell lines results in impaired neurotransmitter metabolism.

To obtain information on a possible role of increased SOD-I activity in neurodegenerative processes in more detail, transgenic mice overexpressing the human *SOD-I* gene under its homologous promoter were produced. These transgenic mice show pathological changes in motor neurons, such as withdrawal and destruction of terminal axons at the neuromuscular endplates and development of multiple small terminals in comparison to normal age-matched mice. Surprisingly, gene knockout of the *SOD-I* gene did not lead to MND, indicating that this enzyme is dispensable for maintenance of motor neurons, and its function is probably fully compensated by other enzymes such as SOD-II.

In the meantime, more than 60 mutations have been reported in patients with autosomal dominant ALS (ALS-1). These are widely scattered over the gene with hot spots in exon 4 and 5. Mutations of the conserved glycine at position 93 are quite abundant, and the A4V mutation, which is in particular found in North American families, generally leads to a severe course of ALS. The D90A mutation is quite abundant in Sweden and Finland, and in contrast to the other mutations, it is inherited as a recessive trait. 1 to 2% of the population in Scandinavia is heterozygous for the D90A allele. In Finland, about 100 000 persons are estimated to be carriers of this mutation. In general, patients with the D90A homozygous mutation show a relatively mild form of ALS with an average survival time of more than 10 years. Surprisingly, the age of onset does not differ from very severe forms such as the A4V form, indicating that other factors determine the age of onset, which act independently from those that determine survival time. Enzyme activity of SOD-I is only reduced by about 50% in most patients with familial ALS, as expected for autosomal dominant inheritance, leaving one allele of the *SOD-I* gene intact. It was suggested that the mutations lead to decreased enzyme activity either by causing an allosteric effect on the active center of the enzyme or by interfering with the formation of dimers, which is the form in which the enzyme naturally occurs in cells.

Mice with transgenic overexpression of mutant SOD-I generally show protein aggregates and other alterations in motor neurons, which become clearly detectable when the disease is clinically apparent. However, it is not fully understood whether such aggregates are responsible for the disease. Alternatively, it has been suggested that the inability of mutant SOD-I to bind copper might lead to accumulation of protein complexes and toxicity of free copper in cells and thus to neurodegeneration. Acquisition of copper by SOD-I requires a specific copper chaperone, a member of the family of metallochaperones, which regulate intracellular trafficking of metal ions. However, inactivation of this chaperone in each of the three most frequently used mouse models for ALS indicates that the disease is unaffected by elimination of copper loading. This argues against the possibility that primary toxicity arises from aberrant copper within the neurons. Morphological changes in mitochondria within motor neurons were observed as an early feature both in SOD-I^{G37R} and SOD-I^{G93A}–transgenic mice. This correlates with reduced mitochondrial function at least in SOD-I^{G93A}. Interestingly, mutant SOD-I^{G93A} has been detected within spinal cord mitochondria, and recent studies have shown that the accumulation of mutant SOD-I in mitochondria is also found in other ALS-linked mutations, and that these changes lead to damage of mitochondria-bound proteins. Therefore, it has been suggested that motor neuron–specific damage of mitochondria might be the reason for the selective toxicity of mutant SOD-I in MND.

6
Disintegration of Axonal Transport: Implications for MND

Several mouse mutants have been identified or generated in which disturbance of components of the cytoskeleton, of microtubuli, or of components of the axonal transport machinery lead to typical signs of MND. Recently, missense point mutations in the cytoplasmic dynein heavy chain have been identified as the cause for motor neuron disease in *Loa*- and *CIra1*-mutant mice. Dynein is part of the motor complex for protein and vesicle movement along microtubules in the minus-end direction and thus an important component for retrograde axonal transport. Interestingly, the defect does not only lead to reduced retrograde transport but also to progressive postnatal loss of motor neurons. Whereas motor neuron loss is not detectable at an age of 3 months, about 50% of the motor neurons are lost at 16 to 19 months. Mutations in the p150 subunit of the transporter protein dynactin have been identified in a family with autosomal dominant progressive MND. Dynactin binds both to microtubules and to dynein, and thus acts as an essential component of the axonal transport machinery, which mediates binding of transport vesicle and transport proteins to microtubules. Mice in which dynamitin, the p50 inhibitory subunit of the dynactin complex, is overexpressed also show late onset MND.

Other widely used mouse models are *Pmn* mice, which develop a rapidly progressive form of MND between the third and fifth postnatal weeks. Histologically, the mice exhibit atrophy and loss of motor axons and at a later stage loss of motor neuron cell bodies in the spinal cord and brain stem motor nuclei. In addition, neurons in the red nucleus degenerate. The axonal projections from this nucleus to the spinal motor neurons – the corticospinal tract – are functionally linked to the corticospinal tract in rodents and contribute significantly to the innervation of spinal motor neurons. It thus functionally resembles the upper motor neuron input to lower motor neurons in higher vertebrates. As the cause of the disease, a single amino acid exchange in the most C-terminal part of the tubulin-specific chaperone E (TBCE) protein was identified. The resulting protein appears less stable, but it could also be that translocation of the protein to axons, the function of the protein or interactions with other functionally relevant proteins are affected. These hypotheses need additional experimental data.

In summary, these data suggest that axonal transport plays an essential role in motor neuron maintenance, and future research on the pathophysiology of MND in humans will focus on the question whether defects in axonal transport are an early or late phenomenon in the disease.

The cytoskeleton of most animal cells contains a variety of elements also including thin actin filaments and intermediate filaments in addition to the microtubules. The β-actin mRNA is one among a small population of mRNAs, which are axonally transported. The functional relevance of axonal transport has been a point of debate for a long time. In contrast to dendritic transport, it has been argued that it is not clear whether the translocalized mRNAs are translated within the axonal compartment. Axonal transport of β-actin mRNA has been studied in much detail. A specific 3′-UTR region called *zip code* has been identified within β-actin mRNA that confers axonal localization. Proteins that bind to the zip-code region of β-actin mRNA have been named zip-code binding protein 1 (ZBP1) and ZBP2. It has been found that the formation of a particle involving β-actin mRNA and ZBP1 is necessary for β-actin translocation to growth cones. The translocation of the mRNA is increased by neurotrophic factors such as neurotrophin-3. In motoneurons, axonal terminals contain polyribosomes, which can be detected with specific antibodies against large subunit ribosomal RNA or by electron microscopy. The SMN protein, which is deficient in spinal muscular atrophy, binds to hnRNP R in the axonal compartment, and it has been shown that a complex of SMN and hnRNP R also interacts with the zip-code domain of β-actin mRNA. Motor neurons that lack Smn show reduced accumulation of β-actin mRNA in axon terminals. This correlates with highly reduced levels of β-actin protein in the axon terminals and reduced axon growth of isolated motor neurons in cell culture. In contrast, overexpression of SMN and/or hnRNP R in neuronal cell lines enhances neurite outgrowth, thus indicating that the translocation of β-actin mRNA and the local translation of this protein in axon terminals play a major role in axon growth and maintenance. Also in other animal models such as zebrafish, knockdown of SMN primarily leads to axonal disturbance and not to motor neuron cell death. Thus, spinal muscular atrophy might be considered as a disease that is caused by defective axonal translocation of β-actin mRNA. It will be interesting to know whether similar mechanisms also account for subtypes of adult onset MND. First indications have been obtained by the identification of mutations in senataxin, a protein with a DNA-helicase domain that is mutated in a rare form of inherited amyotrophic lateral sclerosis.

7
Identification of Neurotrophic Factors and Cytokines that can Regulate Survival and Function of Motor Neurons

Survival of embryonic spinal motor neurons depends on neurotrophic factors obtained from skeletal muscle and glial cells. Such factors are taken up by motor neurons by means of specific receptors and transported to motor neuron cell bodies where they act on specific parameters. In this way, these factors regulate apoptosis, transmitter synthesis, and other aspects of motor neuron function. Developing embryos and cultured embryonic spinal motor neurons have been used to identify neurotrophic factors and cytokines capable of supporting the survival of developing motor neurons. Positive effects have been observed for members of the ciliary neurotrophic factor (CNTF) family, for brain-derived neurotrophic factor (BDNF), neurotrophin (NT)-3 and -4, insulin-like growth factor (IGF) and members of the glial-derived neurotrophic factor (GDNF) family. These factors differ in their tissue distribution and temporal expression patterns, and gene knockout experiments have indicated that their modes of action might overlap with other members of the same gene family and probably of others. Interestingly, neurotrophic factors can interfere with the pathophysiology of motor neuron disease caused by different mechanisms. Positive effects are observed with IGF-I in mice that overexpress SOD-I. Pmn mutant mice show improved survival and motor performance after treatment with CNTF. This indicates that these factors can interfere with distinct pathophysiological mechanisms that have been identified for MND. The molecular mechanisms of these actions appear to be important for development of therapy and could lead the way to the identification of new small molecules that are active on the same targets.

8
Future Prospects for Treatment of MND

A large number of clinical trials with a variety of drugs and therapeutic strategies have been carried out in the past in the search for an effective treatment of MND. Among them were plasmapheresis (with the intention of removing toxic or autoaggressive molecules), drugs such as corticosteroids, corticotropin, cyclosporine A, and cyclophosphamide, which are known to be effective in autoimmune diseases. None of these trials were successful. Similar negative effects were observed in trials with growth hormone and thyrotropin-releasing factor.

A driving force behind some of the current therapeutic trials for the treatment of MND is the assumption that degeneration of motor neurons might be caused by mechanisms similar to those responsible for the death of other neurons of the central nervous system after glutamate exposure. Therefore, considerable attention has focused on clinical trials with glutamate antagonists and glutamate release inhibitors such as Riluzole and other drugs. Trials with free radical scavengers such as acetylcysteine, which are thought to reduce the generation of free radicals, a pathomechanism that has been implicated by mutations in the *SOD-I* gene in familial ALS, have not been successful so far.

Clinical trials with neurotrophic factors such as CNTF and BDNF have also failed. Among many potential reasons are, in particular, adverse pharmacokinetics. Thus, the factors are not available to the motor neuron cell bodies residing in the spinal cord. Alternatively, it could be that the receptors desensitize within a short time to neurotrophic factors, so that only short-time effects occur, which are not sufficient for a therapeutic effect of these drug candidates. Indeed, even promising trials with mouse models such as *Pmn* mice or SOD-I transgenic mice generally focus on treatment periods of only a few weeks, whereas treatment of patients would require positive drug effects over several years. Nevertheless, the positive effects of IGF-I, BDNF, and CNTF in various mouse models of motor neuron disease are a good basis for further research and the development of new therapeutic strategies that might help promote long-term effects in the degenerating motor neuron.

Another problem that needs to be solved is the apparent heterogeneity of pathomechanisms, in particular in familial and probably also sporadic ALS. It could well be that drug candidates that interfere with some pathomechanisms are without any effect in patients suffering from MND due to another defect. Therefore, the genetic and molecular basis of disease needs to be better understood, and techniques for genotyping and molecular diagnosis have to be developed so that the pathomechanism of MND can be characterized for individual patients, even in patients with sporadic ALS. Potential positive or negative effects for the various drug candidates can only be tested on this basis.

This problem seems to be easier for patients with various forms of spinal muscular atrophy. The molecular basis of classical autosomal recessive SMA, which is caused by a mutation of the *SMN* gene, is known and can be easily tested and characterized. Thus, prerequisites exist to develop drugs on a rational basis. The same is also true for other forms of childhood spinal muscular atrophy such as SMARD and for adult onset X-linked spinal muscular atrophy (Kennedy's disease), which is caused by a polyglutamine-expansion in the androgen receptor gene. Multiple lines of evidence indicate that this mutation leads to a gain of pathogenic function, which requires the translocation of the receptor after binding of testosterone. Thus, anti-androgens have been suggested as drug candidates, and first trials with candidates such as leuprorelin showed promising results by reversing the behavioral and histopathological phenotypes in a mouse model. It is to be expected that clinical trials based on such results will be initiated soon, and new drug candidates will be developed with cell culture and mouse models, which can be developed once the molecular and genetic basis of the disease is known.

See also Motor Neuron Diseases: Cellular and Animal Models.

Bibliography

Books and Reviews

Bruijn, L.I., Miller, T.M., Cleveland, D.W. (2004) Unraveling the mechanisms involved in motor neuron degeneration in ALS, *Annu. Rev. Neurosci.* **27**, 723–749.

Chou, S.M., Norris, F.H. (1993) Amyotrophic lateral sclerosis: Lower motor neuron disease spreading to upper motor neurons, *Muscle Nerve* **16**, 864–869.

Cleveland, D.W., Rothstein, J.D. (2001) From Charcot to Lou Gehrig: deciphering selective motor neuron death in ALS, *Nat. Rev. Neurosci.* **2**, 806–819.

Fischbeck, K.H., Lieberman, A., Bailey, C.K., Abel, A., Merry, D.E. (1999) Androgen receptor mutation in Kennedy's disease, *Philos. Trans. R. Soc. Lond B Biol. Sci.* **354**, 1075–1078.

Jablonka, S., Wiese, S., Sendtner, M. (2004) Axonal defects in mouse models of motoneuron disease, *J. Neurobiol.* **58**, 272–286.

Katsuno, M., Adachi, H., Tanaka, F., Sobue, G. (2004) Spinal and bulbar muscular atrophy: ligand-dependent pathogenesis and therapeutic perspectives, *J. Mol. Med.* **82**, 298–307.

Smith, R.A. (1992) *Handbook of Amyotrophic Lateral Sclerosis*, Marcel Dekker, Inc., New York.

Thoenen, H., Sendtner, M. (2002) Neurotrophins: from enthusiastic expectations through sobering experiences to rational therapeutic approaches, *Nat. Neurosci.* **5**(Suppl.), 1046–1050.

Primary Literature

Bommel, H., Xie, G., Rossoll, W., Wiese, S., Jablonka, S., Boehm, T., Sendtner, M. (2002) Missense mutation in the tubulin-specific chaperone E (Tbce) gene in the mouse mutant progressive motor neuronopathy, a model of human motoneuron disease, *J. Cell Biol.* **159**, 563–569.

Chen, Y.Z., Bennett, C.L., Huynh, H.M., Blair, I.P., Puls, I., Irobi, J., Dierick, I., Abel, A., Kennerson, M.L., Rabin, B.A., Nicholson, G.A., Auer-Grumbach, M., Wagner, K., De Jonghe, P., Griffin, J.W., Fischbeck, K.H., Timmerman, V., Cornblath, D.R., Chance, P.F. (2004) DNA/RNA helicase gene mutations in a form of juvenile amyotrophic lateral sclerosis (ALS4), *Am. J Hum. Genet.* **74**, 1128–1135.

Hafezparast, M., Klocke, R., Ruhrberg, C., Marquardt, A., Ahmad-Annuar, A., Bowen, S., Lalli, G., Witherden, A.S., Hummerich, H., Nicholson, S., Morgan, P.J., Oozageer, R., Priestley, J.V., Averill, S., King, V.R., Ball, S., Peters, J., Toda, T., Yamamoto, A., Hiraoka, Y., Augustin, M., Korthaus, D., Wattler, S., Wabnitz, P., Dickneite, C., Lampel, S., Boehme, F., Peraus, G., Popp, A., Rudelius, M., Schlegel, J., Fuchs, H., Hrabe, d.A., Schiavo, G., Shima, D.T., Russ, A.P., Stumm, G., Martin, J.E., Fisher, E.M. (2003) Mutations in dynein link motor neuron degeneration to defects in retrograde transport, *Science* **300**, 808–812.

Howland, D.S., Liu, J., She, Y., Goad, B., Maragakis, N.J., Kim, B., Erickson, J., Kulik, J., DeVito, L., Psaltis, G., DeGennaro, L.J., Cleveland, D.W., Rothstein, J.D. (2002) Focal loss of the glutamate transporter EAAT2 in a transgenic rat model of SOD1 mutant-mediated amyotrophic lateral sclerosis (ALS), *Proc. Natl. Acad. Sci. U.S.A.* **99**, 1604–1609.

Kaspar, B.K., Llado, J., Sherkat, N., Rothstein, J.D., Gage, F.H. (2003) Retrograde viral delivery of IGF-1 prolongs survival in a mouse ALS model, *Science* **301**, 839–842.

Liu, J., Lillo, C., Jonsson, P.A., Velde, C.V., Ward, C.M., Miller, T.M., Subramaniam, J.R., Rothstein, J.D., Marklund, S., Andersen, P.M., Brannstrom, T., Gredal, O., Wong, P.C.,

Williams, D.S., Cleveland, D.W. (2004) Toxicity of familial ALS-linked SOD1 mutants from selective recruitment to spinal mitochondria, *Neuron* **43**, 5–17.

Nicole, S., Diaz, C.C., Frugier, T., Melki, J. (2002) Spinal muscular atrophy: recent advances and future prospects, *Muscle Nerve* **26**, 4–13.

Puls, I., Jonnakuty, C., LaMonte, B.H., Holzbaur, E.L., Tokito, M., Mann, E., Floeter, M.K., Bidus, K., Drayna, D., Oh, S.J., Brown, R.H. Jr., Ludlow, C.L., Fischbeck, K.H. (2003) Mutant dynactin in motor neuron disease, *Nat. Genet.* **33**, 455–456.

Rossoll, W., Jablonka, S., Andreassi, C., Kroning, A.K., Karle, K., Monani, U.R., Sendtner, M. (2003) Smn, the spinal muscular atrophy-determining gene product, modulates axon growth and localization of beta-actin mRNA in growth cones of motoneurons, *J. Cell Biol.* **163**, 801–812.

Subramaniam, J.R., Lyons, W.E., Liu, J., Bartnikas, T.B., Rothstein, J., Price, D.L., Cleveland, D.W., Gitlin, J.D., Wong, P.C. (2002) Mutant SOD1 causes motor neuron disease independent of copper chaperone-mediated copper loading, *Nat. Neurosci.* **5**, 301–307.

Xu, Z., Cork, L.C., Griffin, J.W., Cleveland, D.W. (1993) Increased expression of neurofilament subunit-L produces morphological alterations that resemble the pathology of human motor neuron disease, *Cell* **73**, 23–33.

14
Motor Neuron Diseases: Cellular and Animal Models

Georg Haase
Institut de Neurobiologie de la Méditerranée, Marseille, France

Neurobiology. From Molecular Basis to Disease. Edited by Robert A. Meyers.

ISBN: 978-3-527-32293-0

Keywords

Amyotrophic Lateral Sclerosis (ALS)
A neurodegenerative disease involving the selective degeneration of motor neurons in cerebral cortex, brainstem, and spinal cord. ALS was first described in 1869 by the French neuropathologist Jean-Marie Charcot.

Chaperone
A protein that assists in the folding or assembly of another protein, without becoming part of the completed structure.

Guanine Nucleotide Exchange Factor (GEF)
A protein that facilitates the exchange of GDP for GTP in a GTP-binding protein.

Superoxide Dismutase 1 (SOD1)
An enzyme that catalyzes the conversion of superoxide anion to water and hydrogen peroxide.

Motor neuron diseases are human disorders characterized by the progressive degeneration and death of motor neurons and the denervation of skeletal muscles. Their course is always fatal and there are currently no really effective therapies. While most adult motor neuron diseases appear sporadically, others are inherited, because of genetic mutations in Superoxide Dismutase 1, Alsin, or Dynactin, or associated with risk factors such as genetic polymorphisms in Neurofilaments or Vascular Endothelial Growth Factor (VEGF). Multiple molecular mechanisms might therefore cause – or contribute to – disease. Some of the proposed disease mechanisms might operate cell-autonomously in motor neurons, for example, by perturbing energy metabolism, endosome trafficking, or axonal transport, whereas others seem to act indirectly, involving nonneuronal cells and diffusible factors such as nitric oxide and excess glutamate. Relevant cellular and animal models are now available to test these hypotheses and to investigate new therapeutic strategies.

1 Human Motor Neuron Diseases

Motor neuron diseases represent a spectrum of disorders in which neurons controlling voluntary movement progressively degenerate while other types of neurons are spared. Degeneration of upper motor neurons in cerebral cortex results in spasticity and hyperreflexia. Degeneration of lower motor neurons in brainstem and spinal cord leads to muscle paralysis and atrophy. According to these clinical and pathological criteria, motor neuron diseases can be further classified: Amyotrophic lateral sclerosis (ALS, Charcot disease or Lou Gehrig's disease) affects both upper and lower motor neurons. An ALS form with predominant brainstem symptoms is called progressive bulbar palsy. Progressive muscular atrophies represent conditions in which only lower motor neurons degenerate, whereas primary lateral sclerosis involves only upper motor neurons. ALS is the most frequent motor neuron disease in the adult, with a lifetime risk of 1 in 2000 and a mean age of onset around 55 years. ALS is inevitably fatal but its course is variable between patients. Death is most often caused by failure of respiratory muscles appearing within one to five years after diagnosis.

Histopathological studies have documented a loss of motor neurons in the cervical and lumbar spinal cord of ALS patients. The number of myelinated large caliber axons in the corticospinal tract and in the ventral roots was also found to be reduced and signs of Wallerian degeneration and atrophy have been noted. Other microscopic abnormalities such as reactive gliosis, neuronal inclusion bodies, axonal spheroids, and the loss of dendrites have also been described. In skeletal muscle, denervated fibers appear atrophic and angulated. Grouping of muscle fiber types positive for esterase, ATPase, or myosin isoforms indicates muscle reinnervation from the remaining motor neurons.

2
Genetics of Motor Neuron Diseases

While most ALS cases appear spontaneously, some (~10%) are familial (FALS). These can show autosomal dominant, recessive, or X-linked inheritance. Genetic studies in FALS (familial amyotrophic lateral sclerosis) are rendered difficult by the late onset of disease, its incomplete penetrance, and the short survival of affected family members. Nevertheless, several FALS genes have been identified (Table 1) and additional FALS loci have been mapped to chromosomes 9q21-22, 15q15-22, 16, 18, and 20.

Tab. 1 Gene mutations responsible for motor neuron diseases.

Gene	*Disease*	*Locus*	*Inheritance*	*Mutation type*
SOD1	ALS1	21q22.1	Dominant[a]	Missense
Alsin	Juvenile ALS (ALS2) progressive lateral sclerosis	2q33	Recessive	Frameshift
Tau	ALS/dementia/ Parkinson complex	17q21	Dominant	Missense or intronic
Dynactin (p150 subunit)		2p13	Dominant	Missense
Androgen receptor	Spinal and bulbar muscular atrophy (SBMA)	Xq12	Recessive	CAG triplet expansion
Survival motor neuron (SMN)	SMA	5q13	Recessive	Deletion or missense

[a]With the exception of SOD1 D90A.

2.1 SOD1 Mutations

About 10 to 20% of familial ALS cases (1–2% of total) are linked to mutations in the superoxide dismutase 1 (SOD1) gene. The *SOD1* gene is ubiquitously expressed and encodes a dimeric enzyme containing 153 amino acids, one copper, and one zinc ion per monomer. The normal function of this enzyme is to catalyze the conversion of superoxide to hydrogen peroxide. Over 100 different missense mutations scattered throughout the protein have now been shown to produce motor neuron disease. All SOD1 mutations except one (D90A) show autosomal dominant transmission. It has been proposed that SOD1 mutations cause motor neuron degeneration by catalyzing aberrant free radical chemistry, by perturbing energy homeostasis, by initiating protein aggregation, by causing excessive glutamate levels, or through a combination of these mechanisms.

2.2 Alsin Mutations

Mutations in a gene termed *ALS2 or Alsin*, located on chromosome 2q33-35, have been detected in familial ALS cases with juvenile onset and slow disease progression. The gene mutations are autosomal-recessively inherited and most of them represent small deletions. Computer algorithms have predicted that the Alsin protein contains guanine nucleotide exchange factor (GEF) domains for small GTPases. Small GTPases are molecular switches that cycle between an inactive GDP-bound and an active GTP-bound state; they are activated by GEFs, which stimulate the exchange of GDP for GTP. The 183-kDa alsin protein acts as guanine nucleotide exchange factor for the small GTPase Rab5, localizes to early endosomes, and enhances their fusion. Reported alsin mutations are predicted to result in a failure of Rab5 activation, suggesting that normal endosome fusion and trafficking are essential for motor neuron maintenance.

3 Culture Models of Motor Neuron Degeneration

In an attempt to unravel disease mechanisms and to identify or validate new therapeutic candidates, various motor neuron culture models have been developed. These include neuronal cell lines, primary motor neurons, and spinal cord slices.

3.1 Neuronal Cell Lines

Neuronal cell lines, derived from tumors or *in vitro* transformed cells, only rarely resemble motor neurons. One of these exceptions is the NSC-34 cell line, which was generated by fusion of mouse neuroblastoma cells with embryonic spinal motor neurons. Like motor neurons, NSC-34 cells display a multipolar phenotype and express choline acetyltransferase (ChAT), different ion channels, and neurofilaments. These cells can also generate action potentials and induce acetylcholine receptor clusters on cocultured myotubes. NSC-34 cells transfected with mutant SOD1 plasmids display various abnormalities including mitochondrial dysfunction, increased cytochrome c release, and decreased cell survival following oxidative stress. Gene expression profiling and proteome analysis revealed underlying molecular changes such as downregulation of neurofilaments and glutathion S-transferases and upregulation of proteins involved in NO metabolism (argininosuccinate synthase, argininosuccinate lyase, and neuronal NO synthase). Some of these changes could be confirmed *in vivo* in individual spinal cord motor neurons.

3.2
Motor Neuron Cultures

Motor neurons are notoriously difficult to purify and to maintain in primary culture. Current methods rely on three properties of these cells: their characteristic size, their expression of specific surface markers, and their dependence on trophic factors for survival and neurite outgrowth. Since motor neurons are larger in size and lower in density than most other types of neurons and glial cells, they can be enriched by differential centrifugation. As a further purification step, immuno-panning techniques for chicken and rat spinal motor neurons have been developed using monoclonal antibodies that recognize specific cell surface antigens such as SC1, BEN, or the p75 neurotrophin receptor. Spinal motor neurons can also be isolated using antibody-coated microbeads and subsequent magnetic cell sorting. Once purified, motor neurons are seeded in culture dishes coated with appropriate substrates such as polyornithine and laminin. While many motor neurons die during the first hours of culture, the remaining cells develop typical morphologies and are able to survive for several weeks when appropriate trophic factors are added. These procedures are currently limited to embryonic motor neurons from spinal cord and brainstem and not yet available for isolating cortical motor neurons.

One of the first motor neuron culture systems relevant to human ALS involved microinjection of mutant and wild-type SOD1 plasmid expression vectors into cultured motor neurons. In this system, abnormal protein aggregates appeared specifically in mutant SOD1-expressing cells. Newer studies suggest that cell death of SOD1 expressing motor neurons might also involve external triggers: In normal embryonic motor neurons, activation of the cell surface receptor Fas triggers death via two parallel pathways, the classical FADD/caspase-8 cascade and a second cascade involving Daxx, ASK1, p38, and transcriptional upregulation of neuronal NO synthase; see Fig. 1. Interestingly, motor neurons from transgenic mice for the SOD1 mutations G37R, G85R, or G93A are much more sensitive to Fas- or NO-triggered cell death than motor neurons expressing wild-type human SOD1. Cerebellar neurons, DRG neurons, or astrocytes do not display such increased sensitivity, reflecting the selective vulnerability of motor neurons in living mutant SOD1 mice. Despite their inherent technical difficulties, mutant SOD1 motor neuron cultures thus emerge as a promising tool to screen pharmacological compounds for their effects on disease-related molecular targets or cellular phenotypes.

3.3
Spinal Cord Slice Cultures

In an attempt to study motor neurons in their normal cellular environment, several groups have prepared spinal cord slice cultures from rat and mouse. These cultures can be maintained for up to 2 weeks (adult mouse) or even several months (perinatal rat). Individual motor neurons in living slice cultures are identified by their large size, the presence of action potentials, or retrograde DiI labeling prior to culturing. In fixed slices, motor neuron cell bodies

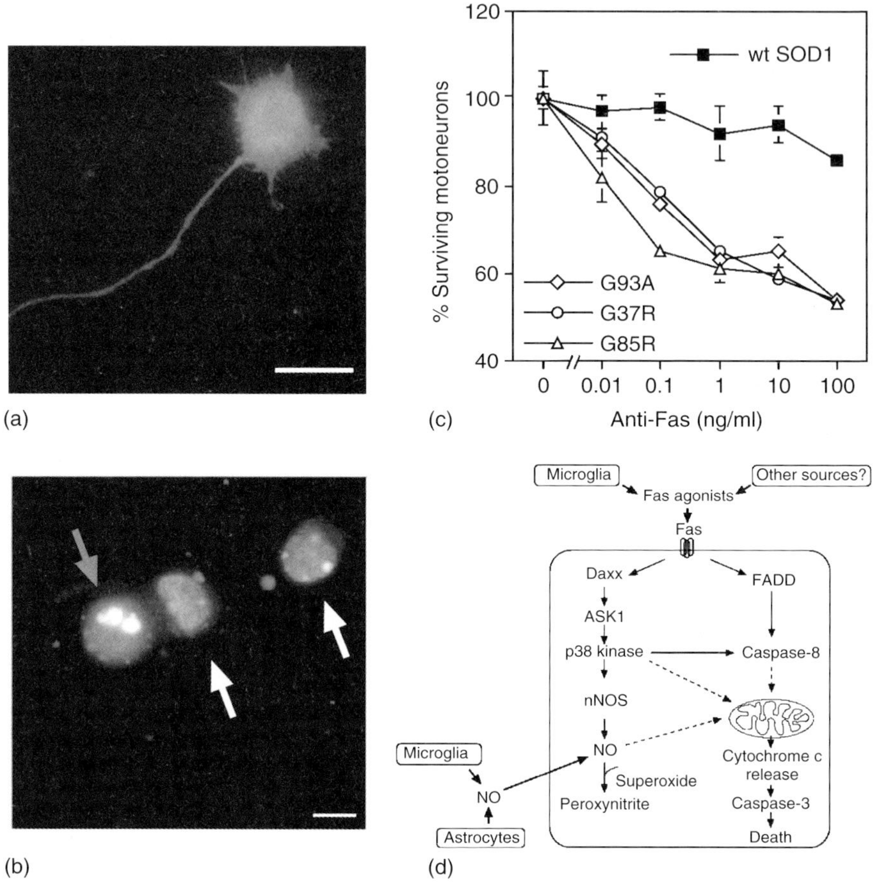

Fig. 1 Isolated motor neurons can be used to study mutant SOD1-linked motor neuron death. (a) Human superoxide SOD1 is expressed in cultured embryonic motor neurons from transgenic SOD1 G93A mice. (b) Treatment of SOD1 G85R motor neuron cultures with agonistic antibodies to the cell surface receptor Fas leads to increased apoptosis, as detected by staining with DAPI (in blue) and immunolabeling for activated caspase 3 (in red). An apoptotic motor neuron with strong caspase-3 activation and nuclear condensation c (red arrow) can be distinguished from healthy motor neurons displaying only weak caspase-3 activation and normal chromatin structure (white arrows). Scale bars: 25 μm. (c) Motor neurons from transgenic mutant SOD1 G93A, G85R, and G37R mice show higher susceptibility to Fas-triggered cell death than motor neurons from mice expressing wild-type SOD1. (d) Model of Fas-triggered motor neuron death. Cell death involves the classical FADD/caspase-8 pathway and a parallel pathway leading from Daxx, ASK1, and p38 activation to transcriptional upregulation of neuronal NO synthase (nNOS) and NO production. The presence of mutant SOD1 sensitizes motor neurons to Fas agonists and NO. Potential sources for these cell death triggers are astrocytes and microglia. (See color plate p. xxxvii); see Fig. 2.

can be identified and counted by Nissl staining or immunocytochemical labeling for the neurotransmitter-synthesizing enzyme ChAT, the transcription factors Islet 1/2, or the surface receptor Ret. Slice culture systems have also been developed for the study of cortical motor neurons.

In the context of ALS research, organotypic culture systems have been used to show that chronic inhibition of glutamate uptake can cause slow degeneration of motor neurons. In this model, antiexcitotoxic agents such as CNQX, riluzole, and topiramate, nitric oxide synthase inhibitors, and neurotrophic factors such as IGF-1 and GDNF protected against motor neuron degeneration. When organotypic slice cultures were derived from transgenic mutant SOD1 mice and exposed to excitotoxic stimuli, mutant motor neuron death could be blocked with a cyclo-oxygenase-2 inhibitor. In another spinal cord slice model, chronic administration of malonate, an inhibitor of mitochondrial electron transport, was found to induce motor neuron death, which could be prevented by a number of antiexcitotoxic agents, antioxidants, and caspase inhibitors.

4 Animal Models of Motor Neuron Degeneration

4.1 Axotomy Models

In neonate animals, motor neuron death can be experimentally induced by a peripheral nerve lesion, since developing motor neurons depend for their survival on contact with their target muscle and their axonal environment. In the most widely used models, the sciatic, facial, or hypoglossal nerve is crushed or sectioned, and the survival of the lesioned motor neurons is eventually after prior retrograde labeling. The extent of motor neuron death depends on the age of the animal and the type of lesion. When a sciatic nerve lesion is performed in 1-day old rats, 90 to 100% of the corresponding motor neurons are lost. When the same type of lesion is performed at 4 or 5 days of age, between 80 and 100% of the motor neurons survive. Lesioning the peripheral nerve by section or close to the cell body both results in slower target reinnervation and reduced motor neuron survival, as compared to lesioning the nerve by section or distally.

Axotomy-induced motor neuron death has many apoptotic features. Morphological studies on lesioned motor neurons have provided evidence for DNA fragmentation, nuclear condensation, and mitochondrial disruption. Pharmacological inhibition of caspases or mitochondrial pore proteins or overexpression of apoptosis-inhibitory proteins (IAPs) reduced the extent of axotomy-induced motor neuron death. Finally, genetic studies in transgenic mice showed that overexpression of the anti-apoptotic gene *Bcl-2* or deletion of the proapoptotic gene *Bax* protected against axotomy-induced cell death. It has also been recognized that apoptotic death of axotomized motor neurons involves activation of cell surface receptors (TNF-R and Fas) and intracellular production of nitric oxide.

4.2 Transgenic Mutant SOD1 Mice and Rats

One breakthrough in ALS research has been the generation of transgenic mice and rats expressing human ALS-linked SOD1 mutations. Several transgenic lines show clinical, histopathological, and electrophysiological signs closely resembling human motor neuron disease. Transgenic mice for the SOD1 mutants G37R, G85R, or G93A display progressive muscle atrophy and paresis, gait instability, and premature death. Similar symptoms were observed in transgenic rats expressing the SOD1 mutants G93A or H46R. In these animals, disease severity depends on the type of mutation, the number of transgenes, and the genetic background. Affected mice display a reduction in the number of spinal motor neuron cell bodies and motor axons and, to a more variable extent, astrogliosis. Other histopathological abnormalities include a fragmented Golgi apparatus, enlarged mitochondria, and various types of inclusions. Electrophysiological studies in mutant mice revealed the presence of fasciculations and fibrillations, and a decreased number of functional motor units.

Some of the molecular consequences of mutant SOD1 expression in mouse spinal cord have also been recognized: sequential activation of caspases, upregulation of apoptotic proteins such as Bad and Bax, release of cytochrome c from mitochondria, and deregulation of the cyclin-dependent kinase cdk5. How the SOD1 mutants cause these changes is still obscure. For the following reasons, altered SOD1 enzymatic activity does not seem to be causal: first, only some disease-causing SOD1 mutations are enzymatically inactive (e.g. G85R), while others retain activity (G93A or G37R). Second, no motor neuron disease is observed when SOD1 activity in mice is lowered (by knocking out the *SOD1* gene or the *CCS* gene required for SOD1 copper loading) or when SOD1 activity is elevated (by transgenic expression of wild-type SOD1). Finally, crossing transgenic mutant SOD1 mice with SOD1 knockout mice, transgenic mice overexpressing wildtype SOD1, or CCS knockout mice, has no influence on disease.

The cellular basis of mutant SOD1 toxicity also remains enigmatic: In transgenic mice, pathology is only observed when mutant SOD1 expression is ubiquitous in all cell types but not when it is restricted to astrocytes or neurons (Fig. 2). To rule out insufficient levels of transgene expression, chimeric mice have been generated that carry mixtures of wild-type and mutant

Fig. 2 Motor neuron degeneration in transgenic ALS mice is influenced by nonneuronal cells. The consequences of mutant SOD1 expression in different cell types are schematically illustrated on spinal cord cross sections. Genotypes of transgenic mice are indicated. Motor neurons are depicted as multipolar cells and glial cells as round cells. Mutant SOD1 expression in these cell types is indicated in gray or black, normal (endogenous) SOD1 expression is shown in white. (a) Normal situation. (b) In transgenic mice, ubiquitous expression of mutant (m) SOD1 causes typical motor neuron disease. (c, d) No disease is observed in transgenic mice expressing mutant SOD1, specifically in glial cells (c) or in neurons (d). (e) Crossing of Thy1: SOD1 mice with transgenic SOD1 G93A mice elevates mutant SOD1 levels (in black) in motor neurons above the ubiquitous expression levels but does not exacerbate disease. (f) Chimeric mice that carry a mixture of wild-type (wt) and mutant SOD1-expressing cells in their spinal cord show later disease onset and longer lifespan than their mutant littermates. These improvements correlate with the proportion of wild-type cells within individual spinal cords. Ubiquitin inclusions are found not only in mutant but also in wild-type motor neurons. Interestingly, mutant SOD1 motor neurons in the chimeras seem to be protected by wild-type nonneuronal cells in their environment. Taken together, these data suggest that motor neuron degeneration is the result of a complex interplay between motor neurons and surrounding nonneuronal cells.

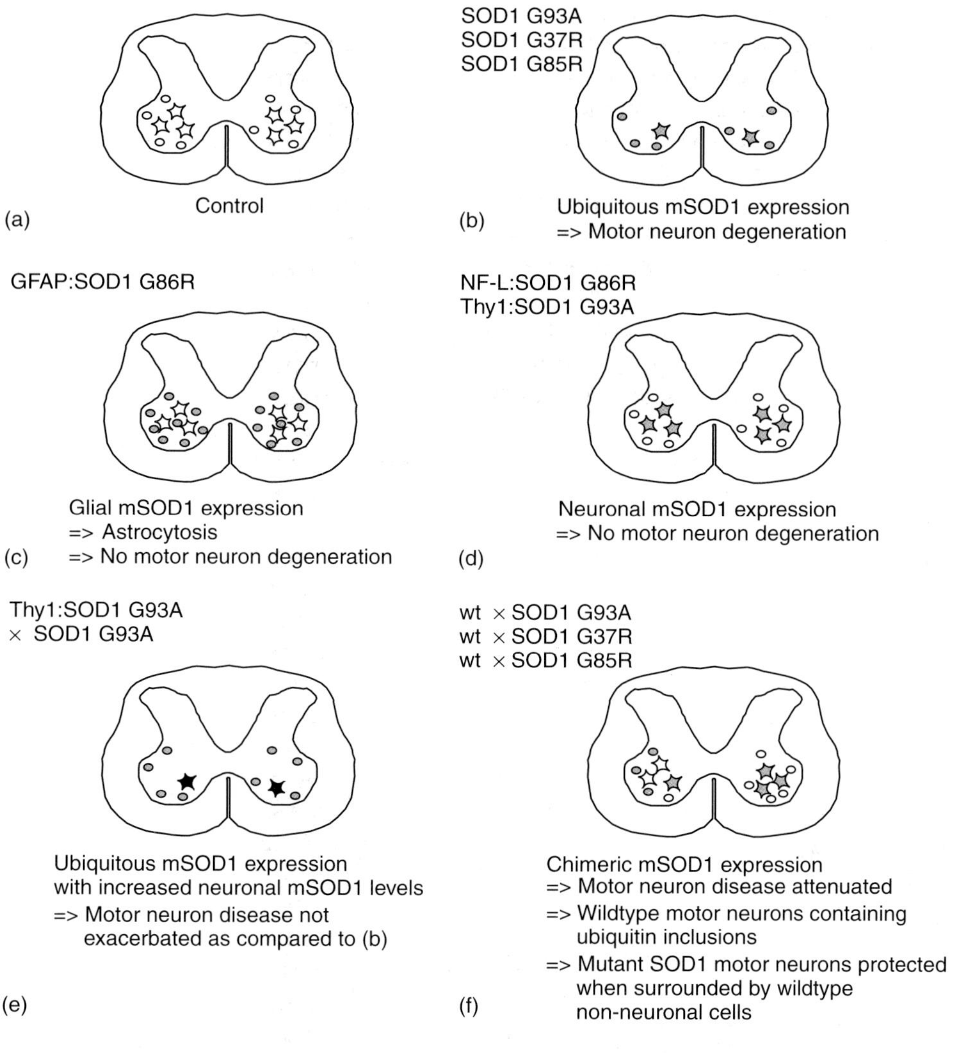

SOD1 expressing cells in the spinal cord. Interestingly, in these chimeras, degenerative signs such as ubiquitin inclusions are found not only in mutant SOD1 expressing motor neurons but also in wild-type motor neurons. Moreover, survival of mutant motor neurons seems to be enhanced by the presence of nonneuronal cells. These observations suggest that motor neuron degeneration can be triggered by non-cell-autonomous mechanisms.

4.3
pmn Mice

Mice with progressive motor neuronopathy (*pmn*) represent another popular motor neuron disease model. Homozygous *pmn* mice develop first neuromuscular symptoms at 2 weeks of age, i.e. much earlier than all reported mutant SOD1 mice. The *pmn* disease manifests with hindlimb atrophy and paresis and rapidly extends to the forelimbs and other muscle groups. Affected *pmn* mice die around 40 days of age, most probably from respiratory failure. They display fibrillations, a loss of axons in motor nerves and ventral roots and a reduced number of motor neurons in some brainstem nuclei. The genetic defect was identified as a point mutation in the *Tbce* gene, which encodes a chaperone protein for tubulin folding and microtubule assembly. The *pmn* mutation results in decreased Tbce protein stability, diminished concentrations of tubulins in peripheral nerves, and progressive loss of microtubules. Interestingly, Tbce gene mutations have also been discovered in human patients with Sanjad Sakati/Kenny Caffey disease, a rare hereditary syndrome comprising hypoparathyroidism, mental retardation, facial dysmorphism, osteosclerosis, and fulgurant infections. It is thus conceivable that the dramatic human disease course obscures motor neuron disease symptoms or that the human Tbce mutations impair another Tbce protein function than the mouse mutation.

4.4 Mutant VEGF Mice

Vascular endothelial growth factor (VEGF) is a secreted 165 amino acid protein that controls the growth and permeability of blood vessels. Under conditions of low oxygen tension, VEGF is upregulated by specific transcription factors that act on a hypoxia-responsive element located in the VEGF promoter. Mice with a targeted deletion of this hypoxia-responsive element are unable to induce VEGF in response to hypoxia. Surprisingly, these mice manifest typical signs of motor neuron degeneration at an age of 5 to 7 months, with progressive muscle atrophy, impaired motor performance, and loss of motor axons and endplates. Interestingly, three polymorphisms in the human VEGF seem associated with reduced levels of plasma VEGF and an increased risk for ALS. These studies incriminate VEGF as an ALS-modifier gene. Whether VEGF acts directly as a trophic factor for motor neurons or more indirectly, by modifying spinal cord vasculature or perfusion, is currently under investigation.

5 Therapeutic Approaches to Motor Neuron Diseases

5.1 Studies in Animal Models

In the past decade, more than 100 different therapeutic approaches have been tested in animal models of motor neuron degeneration. Lesion models have been instrumental in validating neurotrophic factors as therapeutic candidates: CNTF, BDNF, NT-4/5, GDNF, and CT-1 all enhanced survival of motor neurons after axotomy. More recently, several pharmacological compounds such as deprenyl, a monoamino-oxidase B inhibitor, MK-801, a glutamate receptor antagonist, CGP3466B, a GAPDH antagonist, and riluzole were also shown to be neuroprotective in these models.

In mutant SOD1 mice, numerous compounds have been tested but only few have resulted in robust positive effects (Table 2). Following treatment with riluzole, currently the only FDA-approved drug for ALS, the mean survival of SOD1 G93A mice was improved by 11% but histological or functional effects could not be documented. Feeding the animals with creatine also increased survival and provided some motor neuron protection. Numerous ALS

Tab. 2 Selected therapeutic trials in mouse motor neuron disease models.

Treatment	*Model*	*Delivery*		*Outcome*		
		Begin[a]	*Mode*	*Lifespan increase [%]*	*Neuromuscular function*	*Motor neuron protection*
Riluzole	SOD1 G93A	Early	Oral	11	Not reported	Not reported
	pmn	Early	Oral	<10	Yes	Not reported
Creatine	SOD1 G93A	Early	Oral	18	Yes	Yes
Minocyclin	SOD1 G37R	Early	Oral	≈6	Yes	Yes
Minocyclin+ Riluzole+ Nimodipin	SOD1 G37R	Early	Oral	13	Yes	Yes
zVAD-fmk	SOD1 G93A	Early	Intrathecal	21	Yes	Yes
NT-3 Adeno	*pmn*	Early	i.m.	50	yes	Yes
GDNF AAV	SOD1 G93A	Early	i.m.	14	yes	Yes
CT-1Adeno	SOD1 G93A	Early	i.m.	8	yes	Yes
	pmn	Early	i.m.	18	yes	Yes
IGF AAV	SOD1 G93A	Late	i.m.	18	yes	Yes

[a]Treatment begin is classified with respect to disease onset: "early" treatments are initiated in the presymptomatic period, "late" treatments are started at or after disease onset.

patients have since then taken creatine, but a recent placebo-controlled clinical trial disclosed any beneficial effect of oral creatine at 10 g per day. Encouraging results were obtained by intrathecal infusion of the broad caspase inhibitor zVAD-fmk. Recently, several studies have also reported positive effects of minocyclin, which might be related to the inhibitory actions of this tetracyclin analog on caspase activation, mitochondrial cytochrome c release, cdk5 deregulation, or microglial cell activation. When minocyclin was administered in combination with riluzole and the calcium channel blocker nimodipin, the therapeutic benefit was further enhanced.

Gene therapy strategies have been developed to overcome the side effects of systemically delivered neurotrophic factors. These strategies have been tested since 1992 in *pmn* mice and since 2001 in transgenic mutant SOD1 mice (Table 2). In *pmn* mice, the implantation of genetically engineered tumor cells or subcutaneous implantation of encapsulated fibroblasts secreting CNTF reduced degeneration of phrenic and facial motor neurons. When injected intramuscularly in newborn *pmn* mice, adenovirus vectors encoding the neurotrophic factors NT-3, CNTF, and CT-1 were shown to improve neuromuscular function, reduce motor neuron degeneration, and increase the animal's lifespan. Recently, a nonviral gene transfer method, intramuscular injection and electroporation of plasmid DNA, proved to be as efficient as the adenovirus-based strategy to attenuate disease symptoms.

In transgenic mutant SOD1 mice, adenovirus, AAV, and lentivirus vectors coding for CT-1, GDNF, and more recently IGF-1 and VEGF, administered by intramuscular injection, resulted in therapeutic benefit. Some of these therapies were also capable to slow down motor neuron degeneration when started late, after the onset of disease. In order to translate these encouraging findings into clinical therapies, a number of fundamental questions now needs to be addressed: Are the observed effects due to release of the neurotrophic factors into the bloodstream or due to retrograde vector transport from muscles to motor neuron cell bodies? What are the best factors, alone or in combination? What is the safest and most efficient vector system?

5.2 Future Concepts

Since motor axons degenerate weeks to months before cell bodies in human ALS and in corresponding mouse models, therapies aimed at axonal protection appear particularly attractive. This concept is further illustrated by studies in the mouse mutant *Wallerian degeneration slow (Wld^S, Ola)*: In normal mice, a peripheral nerve lesion leads to axonal disintegration and reactive proliferation of Schwann cells in the distal nerve stump, a process called *Wallerian degeneration*. In *Wld^S* mice, however, the distal nerve stump remains intact following a lesion and remains able to conduct action potentials during several weeks. The *Wld^S* mutation was identified as a triplication on chromosome 4 that gives rise to a fusion protein between a truncated ubiquitination factor (Ube4b) and an enzyme involved in NAD synthesis (nicotinamide mononucleotide adenylyltransferase, Nmnat). Interestingly, Wld^S protects axons not only against lesion but also in various other paradigms such as Taxol- and Vincristin-induced neurotoxicity and in the inherited motor neuron disease *pmn*. Because of this broad therapeutic potential, Wld^S and its molecular effectors have become the focus of intense research efforts.

Another innovative approach consists in knocking down the expression of pathological SOD1 alleles by using the RNA interference (RNAi) technique. RNAi has been discovered as a natural gene-silencing mechanism in which double-stranded RNA triggers the degradation of homologous messenger RNAs with extraordinary sequence specificity. The mediators of mRNA degradation are small, 21- to 23-nucleotide long, RNAs (siRNAs). Since almost most ALS-linked SOD1 mutations represent single nucleotide substitutions, it is possible to design RNA sequences that specifically silence mutant SOD1 alleles without affecting expression of the wild-type allele. These RNAs are transfected as double-stranded oligonucleotides into the target cell or expressed as small hairpin RNAs from plasmid DNA vectors or lentiviral vectors. The appropriate target cell and the level of SOD1 gene silencing necessary to achieve therapeutic effects remain to be determined.

6 Conclusions and Perspectives

In the last decade, advances in genetics, molecular and cell biology, and gene therapy have boosted research on motor neuron diseases. Several gene mutations responsible for motor neuron diseases in humans (SOD1, Alsin) and mice (Dynein, Chaperones, etc.) have been identified. These discoveries led to new pathogenic hypotheses, implicating excessive glutamate levels, toxic protein aggregates, mitochondrial damage, or axonal dysfunction. It is becoming clear that multiple disease pathways – triggered in the motor neuron or in neighboring cells – can converge and cause degeneration. Experimental therapies targeting these pathways at various levels are investigated with increasing success in relevant cellular and animal models. Translating these approaches into clinically relevant therapies represents the major challenge of the future.

See also Motor Neuron Diseases: Molecular Mechanism, Pathophysiology, and Treatments; Neuron Chemistry.

Bibliography

Books and Reviews

Beckman, J.S., Estevez, A.G., Crow, J.P., Barbeito, L. (2001) Superoxide dismutases and the death of motoneurons in ALS, *Trends Neurosci.* **24**, S15–S20.

Bruijn, L.I., Miller, T.M., Cleveland, D.W. (2004) Unraveling the mechanisms involved in motor neuron degeneration in ALS, *Annu. Rev. Neurosci.* **27**, 723–749.

Coleman, M.P., Perry, V.H. (2002) Axon pathology in neurological disease: a neglected therapeutic target, *Trends Neurosci.* **25**, 532–537.

Cleveland, D.W., Rothstein, J.D. (2001) From Charcot to Lou Gehrig: deciphering selective motor neuron death in ALS, *Nat. Rev. Neurosci.* **2**, 806–819.

Hand, C.K., Rouleau, G.A. (2002) Familial amyotrophic lateral sclerosis, *Muscle Nerve* **25**, 135–159.

Henderson, C.E., et al. (1995) in *Nerve Cell Culture: A Practical Approach*, Cohen J., Wilkin G. (Eds.) Oxford University Press, London, UK, pp. 69–81.

Hirano, A. (1982) in: Rowland L.P. (Ed.) *Human Motor Neuron Diseases*, Raven Press, New York, pp. 75–87.

Julien, J.P. (2001) Amyotrophic lateral sclerosis. unfolding the toxicity of the misfolded, *Cell* **104**, 581–591.

Nicole, S., Cifuentes-Diaz, C., Frugier, T., Melki, J. (2002) Spinal muscular atrophy: recent advances and future prospects, *Muscle Nerve* **26**, 4–13.

Orrell, R.W. (2000) Amyotrophic lateral sclerosis: copper/zinc superoxide dismutase (SOD1) gene mutations, *Neuromuscul. Disord.* **10**, 63–68.

Valentine, J.S., Hart, P.J. (2003) Misfolded CuZnSOD and amyotrophic lateral sclerosis, *Proc. Natl. Acad. Sci. U. S. A.* **100**, 3617–3622.

Primary Literature

Allen, S., et al. (2003) Analysis of the cytosolic proteome in a cell culture model of familial amyotrophic lateral sclerosis reveals alterations to the proteasome, antioxidant defenses, and nitric oxide synthetic pathways, *J. Biol. Chem.* **278**(8), 6371–6383.

Arce, V., et al. (1999) Cardiotrophin-1 requires LIFRbeta to promote survival of mouse motoneurons purified by a novel technique, *J. Neurosci. Res.* **55**(1), 119–126.

Bar, P.R. (2000) Motor neuron disease in vitro: the use of cultured motor neurons to study amyotrophic lateral sclerosis, *Eur. J. Pharmacol.* **405**(1–3), 285–295.

Beckman, J.S., et al. (2001) Superoxide dismutases and the death of motoneurons in ALS, *Trends Neurosci.* **24**(11 (Suppl.)), S15–S20.

Bommel, H., et al. (2002) Missense mutation in the tubulin-specific chaperone E (Tbce) gene in the mouse mutant progressive motor neuronopathy, a model of human motoneuron disease, *J. Cell. Biol.* **159**(4), 563–569.

Bordet, T., et al. (1999) Adenoviral cardiotrophin-1 gene transfer protects pmn mice from progressive motor neuronopathy, *J. Clin. Invest.* **104**(8), 1077–1085.

Bordet, T., et al. (2001) Protective effects of cardiotrophin-1 adenoviral gene transfer on neuromuscular degeneration in transgenic ALS mice, *Hum. Mol. Genet.* **10**(18), 1925–1933.

Bruijn, L.I., et al. (1997) ALS-linked SOD1 mutant G85R mediates damage to astrocytes and promotes rapidly progressive disease with SOD1-containing inclusions, *Neuron* **18**(2), 327–338.

Bruijn, L.I., et al. (1998) Aggregation and motor neuron toxicity of an ALS-linked SOD1 mutant independent from wild-type SOD1, *Science* **281**(5384), 1851–1854.

Carlin, K.P., et al. (2000) Dendritic L-type calcium currents in mouse spinal motoneurons: implications for bistability, *Eur. J. Neurosci.* **12**(5), 1635–1646.

Casanovas, A., et al. (1996) Prevention by lamotrigine, MK-801 and N omega-nitro-L-arginine methyl ester of motoneuron cell death after neonatal axotomy, *Neuroscience* **71**(2), 313–325.

Cashman, N.R., et al. (1992) Neuroblastoma x spinal cord (NSC) hybrid cell lines resemble developing motor neurons, *Dev. Dyn.* **194**(3), 209–221.

Chan, Y.M., et al. (2003) Inhibition of caspases promotes long-term survival and reinnervation by axotomized spinal motoneurons of denervated muscle in newborn rats, *Exp. Neurol.* **181**(2), 190–203.

Charcot, J.-M., Joffroy, A. (1869) Deux cas d'atrophie musculaire progressive avec lésion de la substance grise et des fisceaux de la moelle épinière, *Arch. Physiol.* **2**, 354, 629, 744.

Chiu, A.Y., et al. (1995) Age-dependent penetrance of disease in a transgenic mouse model of familial amyotrophic lateral sclerosis, *Mol. Cell. Neurosci.* **6**(4), 349–362.

Chou, S.M. (1992) in: Smith R.A. (Ed.) *Handbook of Amyotrophic Lateral Sclerosis*, Marcel Dekker, New York, pp. 133–181.

Cifuentes-Diaz, C., et al. (2002) Neurofilament accumulation at the motor endplate and lack of axonal sprouting in a spinal muscular atrophy mouse model, *Hum. Mol. Genet.* **11**(12), 1439–1447.

Clement, A.M., et al. (2003) Wild-type nonneuronal cells extend survival of SOD1 mutant motor neurons in ALS mice, *Science* **302**(5642), 113–117.

Coleman, M.P., et al. (1998) An 85-kb tandem triplication in the slow Wallerian degeneration (Wlds) mouse, *Proc. Natl. Acad. Sci. U. S. A.* **95**(17), 9985–9990.

Conforti, L., et al. (2000) A Ufd2/D4Cole1e chimeric protein and overexpression of Rbp7 in the slow Wallerian degeneration (WldS) mouse, *Proc. Natl. Acad. Sci. U. S. A.* **97**(21), 11377–11382.

Connelly, C.A., et al. (2000) Metabolic activity of cultured rat brainstem, hippocampal and spinal cord slices, *J. Neurosci. Methods* **99**(1–2), 1–7.

Corse, A.M., et al. (1999) Preclinical testing of neuroprotective neurotrophic factors in a model of chronic motor neuron degeneration, *Neurobiol. Dis.* **6**(5), 335–346.

Dalcanto, M.C., Gurney, M.E. (1995) Neuropathological changes in two lines of mice carrying a transgene for mutant human Cu,Zn SOD, and in mice overexpressing wild type human SOD: A model of familial amyotrophic lateral sclerosis (FALS), *Brain Res.* **676**(1), 25–40.

de Bilbao, F., Dubois-Dauphin, M. (1996) Acute application of an interleukin-1 beta-converting enzyme-specific inhibitor delays axotomy-induced motoneurone death, *Neuroreport* **7**(18), 3051–3054.

de Bilbao, F., Dubois-Dauphin, M. (1996) Time course of axotomy-induced apoptotic cell death in facial motoneurons of neonatal wild type and bcl-2 transgenic mice, *Neuroscience* **71**(4), 1111–1119.

Deckwerth, T.L., et al. (1996) BAX is required for neuronal death after trophic factor deprivation and during development, *Neuron* **17**(3), 401–411.

Drachman, D.B., Rothstein, J.D. (2000) Inhibition of cyclooxygenase-2 protects motor neurons in an organotypic model of amyotrophic lateral sclerosis, *Ann. Neurol.* **48**(5), 792–795.

Dubois-Dauphin, M., et al. (1994) Neonatal motoneurons overexpressing the bcl-2 protooncogene in transgenic mice are protected from axotomy-induced cell death, *Proc. Natl. Acad. Sci. U. S. A.* **91**(8), 3309–3313.

Durham, H.D., et al. (1993) Evaluation of the spinal cord neuron X neuroblastoma hybrid cell line NSC-34 as a model for neurotoxicity testing, *Neurotoxicology* **14**(4), 387–395.

Durham, H.D., et al. (1997) Aggregation of mutant Cu/Zn superoxide dismutase proteins in a culture model of ALS, *J. Neuropathol. Exp. Neurol.* **56**(5), 523–530.

Ferri, A., et al. (2003) Inhibiting axon degeneration and synapse loss attenuates apoptosis and disease progression in a mouse model of motoneuron disease, *Curr. Biol.* **13**(8), 669–673.

Frey, D., et al. (2000) Early and selective loss of neuromuscular synapse subtypes with low sprouting competence in motoneuron diseases, *J. Neurosci.* **20**(7), 2534–2542.

Garcia, M.L., Cleveland, D.W. (2001) Going new places using an old MAP: tau, microtubules and human neurodegenerative disease, *Curr. Opin. Cell Biol.* **13**(1), 41–48.

Gong, Y.H., et al. (2000) Restricted expression of G86R Cu/Zn superoxide dismutase in astrocytes results in astrocytosis but does not cause motoneuron degeneration, *J. Neurosci.* **20**(2), 660–665.

Greensmith, L., Vrbova, G. (1996) Motoneurone survival: A functional approach, *Trends Neurosci.* **19**(11), 450–455.

Groeneveld, G.J., et al. (2003) A randomized sequential trial of creatine in amyotrophic lateral sclerosis, *Ann. Neurol.* **53**(4), 437–445.

Guegan, C., et al. (2001) Recruitment of the mitochondrial-dependent apoptotic pathway in amyotrophic lateral sclerosis, *J. Neurosci.* **21**(17), 6569–6576.

Gurney, M.E., et al. (1994) Motor neuron degeneration in mice that express a human Cu,Zn superoxide dismutase mutation, *Science* **264**(5166), 1772–1775.

Gurney, M.E., et al. (1996) Benefit of vitamin E, riluzole, and gabapentin in a transgenic model of familiar amyotrophic lateral sclerosis, *Ann. Neurol.* **39**(2), 147–157.

Haase, G., et al. (1997) Gene therapy of murine motor neuron disease using adenoviral vectors for neurotrophic factors, *Nat. Med.* **3**(4), 429–436.

Haase, G., et al. (1999) Therapeutic benefit of CNTF in progressive motor neuronopathy depends on the route of delivery, *Ann. Neurol.* **45**(3), 296–304.

Hadano, S., et al. (2001) A gene encoding a putative GTPase regulator is mutated in familial amyotrophic lateral sclerosis 2, *Nat. Genet.* **29**(2), 166–173.

Haenggeli, C., Kato, A.C. (2002) Differential vulnerability of cranial motoneurons in mouse models with motor neuron degeneration, *Neurosci. Lett.* **335**(1), 39–43.

Hannon, G.J. (2002) RNA interference, *Nature* **418**(6894), 244–251.

Henderson, C.E., et al. (1994) GDNF: A potent survival factor for motoneurons present in peripheral nerve and muscle, *Science* **266**(5187), 1062–1064.

Hentati, A., et al. (1994) Linkage of recessive familial amyotrophic lateral sclerosis to chromosome 2q33-q35, *Nat. Genet.* **7**(3), 425–428.

Ho, T.W., et al. (2000) TGFbeta trophic factors differentially modulate motor axon outgrowth and protection from excitotoxicity, *Exp. Neurol.* **161**(2), 664–675.

Hori, N., et al. (2001) Intracellular activity of rat spinal cord motoneurons in slices, *J. Neurosci. Methods* **112**(2), 185–191.

Howland, D.S., et al. (2002) Focal loss of the glutamate transporter EAAT2 in a transgenic rat model of SOD1 mutant-mediated amyotrophic lateral sclerosis (ALS), *Proc. Natl. Acad. Sci. U. S. A.* **99**(3), 1604–1609.

Ince, P.G., et al. (1998) Amyotrophic lateral sclerosis: current issues in classification, pathogenesis and molecular pathology, *Neuropathol. Appl. Neurobiol.* **24**(2), 104–117.

Iwasaki, Y., Ikeda, K. (1999) Prevention by insulin-like growth factor-I and riluzole in motor neuron death after neonatal axotomy, *J. Neurol. Sci.* **169**(1–2), 148–155.

Iwasaki, Y., et al. (1996) Deprenyl and pergolide rescue spinal motor neurons from axotomy-induced neuronal death in the neonatal rat, *Neurol. Res.* **18**(2), 168–170.

Kaal, E.C., et al. (2000) Chronic mitochondrial inhibition induces selective motoneuron death in vitro: a new model for amyotrophic lateral sclerosis, *J. Neurochem.* **74**(3), 1158–1165.

Kashihara, Y., et al. (1987) Cell death of axotomized motoneurones in neonatal rats, and its prevention by peripheral reinnervation, *J. Physiol.* **386**, 135–148.

Kaspar, B.K., et al. (2003) Retrograde viral delivery of IGF-1 prolongs survival in a mouse ALS model, *Science* **301**(5634), 839–842.

Kennedy, W.R., et al. (1968) Progressive proximal spinal and bulbar muscular atrophy of late onset. A sex-linked recessive trait, *Neurology* **18**(7), 671–680.

Kennel, P., et al. (2000) Riluzole prolongs survival and delays muscle strength deterioration in mice with progressive motor neuronopathy (pmn), *J. Neurol. Sci.* **180**(1–2), 55–61.

Kennel, P.F., et al. (1996) Neuromuscular function impairment is not caused by motor neurone loss in FALS mice: an electromyographic study, *Neuroreport* **7**(8), 1427–1431.

Kennel, P.F., et al. (1996) Electromyographical and motor performance studies in the pmn mouse model of neurodegenerative disease, *Neurobiol. Dis.* **3**(2), 137–147.

Kirby, J., et al. (2002) Differential gene expression in a cell culture model of SOD1-related familial motor neurone disease, *Hum. Mol. Genet.* **11**(17), 2061–2075.

Klivenyi, P., et al. (1999) Neuroprotective effects of creatine in a transgenic animal model of amyotrophic lateral sclerosis, *Nat. Med.* **5**(3), 347–350.

Kong, J.M., Xu, Z.S. (1998) Massive mitochondrial degeneration in motor neurons triggers the onset of amyotrophic lateral sclerosis in mice expressing a mutant SOD1, *J. Neurosci.* **18**(9), 3241–3250.

Korinthenberg, R., et al. (1997) Congenital axonal neuropathy caused by deletions in the spinal muscular atrophy region, *Ann. Neurol.* **42**(3), 364–368.

Krassioukov, A.V., et al. (2002) An in vitro model of neurotrauma in organotypic spinal cord cultures from adult mice, *Brain Res. Brain Res. Protoc.* **10**(2), 60–68.

Kriz, J., et al. (2002) Minocycline slows disease progression in a mouse model of amyotrophic lateral sclerosis, *Neurobiol. Dis.* **10**(3), 268–278.

Kriz, J., et al. (2003) Efficient three-drug cocktail for disease induced by mutant superoxide dismutase, *Ann. Neurol.* **53**(4), 429–436.

Kunst, C.B., et al. (2000) Genetic mapping of a mouse modifier gene that can prevent ALS onset, *Genomics* **70**(2), 181–189.

La Spada, A.R., et al. (1991) Androgen receptor gene mutations in X-linked spinal and bulbar muscular atrophy, *Nature* **352**(6330), 77–79.

Lambrechts, D., et al. (2003) VEGF is a modifier of amyotrophic lateral sclerosis in mice and humans and protects motoneurons against ischemic death, *Nat. Genet.* **34**(4), 383–394.

Lefebvre, S., et al. (1995) Identification and characterization of a spinal muscular atrophy-determining gene [see comments], *Cell* **80**(1), 155–165.

Leigh, P.N., et al. (1991) Ubiquitin-immunoreactive intraneuronal inclusions in amyotrophic lateral sclerosis. Morphology, distribution, and specificity, *Brain* **114**(Pt 2), 775–788.

Lesbordes, J.C., et al. (2002) In vivo electrotransfer of the cardiotrophin-1 gene into skeletal muscle slows down progression of motor neuron degeneration in pmn mice, *Hum. Mol. Genet.* **11**(14), 1615–1625.

Li, L., et al. (1998) Characterization of spinal motoneuron degeneration following different types of peripheral nerve injury in neonatal and adult mice, *J. Comp. Neurol.* **396**(2), 158–168.

Li, M., et al. (2000) Functional role of caspase-1 and caspase-3 in an ALS transgenic mouse model, *Science* **288**(5464), 335–339.

Li, M., et al. (2000) Functional role of caspase-1 and caspase-3 in an ALS transgenic mouse model, *Science* **288**(5464), 335–339.

Lino, M.M., et al. (2002) Accumulation of SOD1 mutants in postnatal motoneurons does not cause motoneuron pathology or motoneuron disease, *J. Neurosci.* **22**(12), 4825–4832.

Liu, R., et al. (2002) Increased mitochondrial antioxidative activity or decreased oxygen free radical propagation prevent mutant SOD1-mediated motor neuron cell death and increase amyotrophic lateral sclerosis-like transgenic mouse survival, *J. Neurochem.* **80**(3), 488–500.

Lowrie, M.B., et al. (1982) Recovery of slow and fast muscles following nerve injury during early postnatal development in the rat, *J. Physiol.* **331**, 51–66.

Lyon, M.F., et al. (1993) A gene affecting Wallerian nerve degeneration maps distally on mouse chromosome 4, *Proc. Natl. Acad. Sci. U. S. A.* **90**(20), 9717–9720.

Mack, T.G., et al. (2001) Wallerian degeneration of injured axons and synapses is delayed by a Ube4b/Nmnat chimeric gene, *Nat. Neurosci.* **4**(12), 1199–1206.

Maragakis, N.J., et al. (2003) Topiramate protects against motor neuron degeneration in organotypic spinal cord cultures but not in G93A SOD1 transgenic mice, *Neurosci. Lett.* **338**(2), 107–110.

Mariotti, R., et al. (1997) Age-dependent induction of nitric oxide synthase activity in facial motoneurons after axotomy, *Exp. Neurol.* **145**(2 Pt 1), 361–370.

Martin, N., et al. (2002) A missense mutation in Tbce causes progressive motor neuronopathy in mice, *Nat. Genet.* **32**(3), 443–447.

Menzies, F.M., et al. (2002) Mitochondrial dysfunction in a cell culture model of familial amyotrophic lateral sclerosis, *Brain* **125**(Pt 7), 1522–1533.

Menzies, F.M., et al. (2002) Selective loss of neurofilament expression in Cu/Zn superoxide dismutase (SOD1) linked amyotrophic lateral sclerosis, *J. Neurochem.* **82**(5), 1118–1128.

Mourelatos, Z., et al. (1996) The Golgi apparatus of spinal cord motor neurons in transgenic mice expressing mutant Cu,Zn superoxide dismutase becomes fragmented in early, preclinical stages of the disease, *Proc. Natl. Acad. Sci. U. S. A.* **93**(11), 5472–5477.

Nagai, M., et al. (2001) Rats expressing human cytosolic copper-zinc superoxide dismutase transgenes with amyotrophic lateral sclerosis: associated mutations develop motor neuron disease, *J. Neurosci.* **21**(23), 9246–9254.

Nguyen, M.D., et al. (2001) Deregulation of Cdk5 in a mouse model of ALS: toxicity alleviated by perikaryal neurofilament inclusions, *Neuron* **30**(1), 135–147.

Oosthuyse, B., et al. (2001) Deletion of the hypoxia-response element in the vascular endothelial growth factor promoter causes motor neuron degeneration, *Nat. Genet.* **28**(2), 131–138.

Otomo, A., et al. (2003) ALS2, a novel guanine nucleotide exchange factor for the small GTPase Rab5, is implicated in endosomal dynamics, *Hum. Mol. Genet.* **12**(14), 1671–1687.

Parvari, R., et al. (2002) Mutation of TBCE causes hypoparathyroidism-retardation-dysmorphism and autosomal recessive Kenny-Caffey syndrome, *Nat. Genet.* **32**(3), 448–452.

Pasinelli, P., et al. (2000) Caspase-1 and -3 are sequentially activated in motor neuron death in Cu,Zn superoxide dismutase-mediated familial amyotrophic lateral sclerosis, *Proc. Natl. Acad. Sci. U. S. A.* **97**(25), 13901–13906.

Pennica, D., et al. (1996) Cardiotrophin-1, a cytokine present in embryonic muscle, supports long-term survival of spinal motoneurons, *Neuron* **17**(1), 63–74.

Perrelet, D., et al. (2000) IAP family proteins delay motoneuron cell death in vivo, *Eur. J. Neurosci.* **12**(6), 2059–2067.

Pramatarova, A., et al. (2001) Neuron-specific expression of mutant superoxide dismutase 1 in transgenic mice does not lead to motor impairment, *J. Neurosci.* **21**(10), 3369–3374.

Puls, I., et al. (2003) Mutant dynactin in motor neuron disease, *Nat. Genet.* **33**(4), 455–456.

Raivich, G., et al. (2002) Cytotoxic potential of proinflammatory cytokines: combined deletion of TNF receptors TNFR1 and TNFR2 prevents motoneuron cell death after facial axotomy in adult mouse, *Exp. Neurol.* **178**(2), 186–193.

Rakowicz, W.P., et al. (2002) Glial cell line-derived neurotrophic factor promotes the survival of early postnatal spinal motor neurons in the lateral and medial motor columns in slice culture, *J. Neurosci.* **22**(10), 3953–3962.

Raoul, C., et al. (1999) Programmed cell death of embryonic motoneurons triggered through the fas death receptor, *J. Cell. Biol.* **147**(5), 1049–1062.

Raoul, C., et al. (2002) Motoneuron death triggered by a specific pathway downstream of Fas. potentiation by ALS-linked SOD1 mutations, *Neuron* **35**(6), 1067–1083.

Reaume, A.G., et al. (1996) Motor neurons in Cu/Zn superoxide dismutase-deficient mice develop normally but exhibit enhanced cell death after axonal injury, *Nat. Genet.* **13**(1), 43–47.

Ripps, M.E., et al. (1995) Transgenic mice expressing an altered murine superoxide dismutase gene provide an animal model of amyotrophic lateral sclerosis, *Proc. Natl. Acad. Sci. U. S. A.* **92**(3), 689–693.

Rossiter, J.P., et al. (1996) Axotomy-induced apoptotic cell death of neonatal rat facial motoneurons: time course analysis and relation to NADPH-diaphorase activity, *Exp. Neurol.* **138**(1), 33–44.

Rothstein, J.D., et al. (1993) Chronic inhibition of glutamate uptake produces a model of slow neurotoxicity, *Proc. Natl. Acad. Sci. U. S. A.* **90**(14), 6591–6595.

Rothstein, J.D., Kuncl, R.W. (1995) Neuroprotective strategies in a model of chronic glutamate-mediated motor neuron toxicity, *J. Neurochem.* **65**(2), 643–651.

Sagot, Y., et al. (1995) Polymer encapsulated cell lines genetically engineered to release ciliary neurotrophic factor can slow down progressive motor neuronopathy in the mouse, *Eur. J. Neurosci.* **7**(6), 1313–1322.

Schmalbruch, H. (1984) Motoneuron death after sciatic nerve section in newborn rats, *J. Comp. Neurol.* **224**, 252–258.

Schmalbruch, H., et al. (1991) A new mouse mutant with progressive motor neuronopathy, *J. Neuropathol. Exp. Neurol.* **50**, 192–204.

Schmalbruch, H., Rosenthal, A. (1995) Neurotrophin-4/5 postpones the death of injured spinal motoneurons in newborn rats, *Brain Res.* **700**(1–2), 254–260.

Schnaar, R.I., Schaffner, A.E. (1981) Separation of cell types from embryonic chicken and rat spinal cord: characterization of motoneuron-enriched fractions, *J. Neurosci.* **1**(2), 204–217.

Sendtner, M., et al. (1990) Ciliary neurotrophic factor prevents the degeneration of motor neurons after axotomy, *Nature* **345**(6274), 440–441.

Sendtner, M., et al. (1992) Brain-derived neurotrophic factor prevents the death of motoneurons in newborn rats after nerve section, *Nature* **360**, 757–759.

Sendtner, M., et al. (1992) Ciliary neurotrophic factor prevents degeneration of motor neurons in mouse mutant progressive motor neuronopathy, *Nature* **358**(6386), 502–504.

Subramaniam, J.R., et al. (2002) Mutant SOD1 causes motor neuron disease independent of copper chaperone-mediated copper loading, *Nat. Neurosci.* **5**(4), 301–307.

Tandan, R., Bradley, W.G. (1985) Amyotrophic lateral sclerosis: Part 1. Clinical features, pathology, and ethical issues in management, *Ann. Neurol.* **18**(3), 271–280.

Terrado, J., et al. (2000) Soluble TNF receptors partially protect injured motoneurons in the postnatal CNS, *Eur. J. Neurosci.* **12**(9), 3443–3447.

Tian, G., et al. (1996) Pathway leading to correctly folded beta-tubulin, *Cell* **86**(2), 287–296.

Tu, P.H., et al. (1996) Transgenic mice carrying a human mutant superoxide dismutase transgene develop neuronal cytoskeletal pathology resembling human amyotrophic lateral sclerosis lesions, *Proc. Natl. Acad. Sci. U. S. A.* **93**(7), 3155–3160.

Ugolini, G., et al. (2003) Fas/tumor necrosis factor receptor death signaling is required for axotomy-induced death of motoneurons in vivo, *J. Neurosci.* **23**(24), 8526–8531.

Van Westerlaak, M.G., et al. (2001) Chronic mitochondrial inhibition induces glutamate-mediated corticomotoneuron death in an organotypic culture model, *Exp. Neurol.* **167**(2), 393–400.

Vanderluit, J.L., et al. (2003) In vivo application of mitochondrial pore inhibitors blocks the induction of apoptosis in axotomized neonatal facial motoneurons, *Cell. Death Differ.* **10**(9), 969–976.

Vukosavic, S., et al. (1999) Bax and Bcl-2 interaction in a transgenic mouse model of familial amyotrophic lateral sclerosis, *J. Neurochem.* **73**(6), 2460–2468.

Waldmeier, P.C., et al. (2000) Neurorescuing effects of the GAPDH ligand CGP 3466B, *J. Neural. Transm. Suppl.* **60**, 197–214.

Wang, L.J., et al. (2002) Neuroprotective effects of glial cell line-derived neurotrophic factor mediated by an adeno-associated virus vector in a transgenic animal model of amyotrophic lateral sclerosis, *J. Neurosci.* **22**(16), 6920–6928.

Wang, M.S., et al. (2001) The WldS protein protects against axonal degeneration: a model of gene therapy for peripheral neuropathy, *Ann. Neurol.* **50**(6), 773–779.

Wang, M.S., et al. (2002) WldS mice are resistant to paclitaxel (taxol) neuropathy, *Ann. Neurol.* **52**(4), 442–447.

Wong, P.C., et al. (1995) An adverse property of a familial ALS-linked SOD1 mutation causes motor neuron disease characterized by vacuolar degeneration of mitochondria, *Neuron* **14**(6), 1105–1116.

Xia, X.G., et al. (2003) An enhanced U6 promoter for synthesis of short hairpin RNA, *Nucleic Acids Res.* **31**(17), e100.

Yan, Q., et al. (1992) Brain-derived neurotrophic factor rescues spinal motor neurons from axotomy-induced cell death, *Nature* **360**(6406), 753–755.

Yang, Y., et al. (2001) The gene encoding alsin, a protein with three guanine-nucleotide exchange factor domains, is mutated in a form of recessive amyotrophic lateral sclerosis, *Nat. Genet.* **29**(2), 160–165.

Zhu, S., et al. (2002) Minocycline inhibits cytochrome c release and delays progression of amyotrophic lateral sclerosis in mice, *Nature* **417**(6884), 74–78.

15
Alzheimer's Disease

Jun Wang, Silva Hecimovic, and Alison Goate
Department of Psychiatry, Washington University School of Medicine, St. Louis, MO, USA

Neurobiology. From Molecular Basis to Disease. Edited by Robert A. Meyers.

ISBN: 978-3-527-32293-0

Keywords

Association Study
Tests for differences in allele or genotype frequencies between unrelated, affected, and unaffected individuals.

Autosomal Dominant Trait
A phenotype that is expressed in a heterozygote, which maps to a chromosome other than the sex chromosomes.

β-secretase
Also known as BACE1; a transmembrane aspartyl protease involved in processing of APP-generating N-terminal end of Aβ peptides.

Candidate Gene Study
An association study performed on a particular gene(s) selected according to the biology of the disease and/or their location within chromosome regions found to be implicated in disease (by linkage or association study).

Full Genome Scan
A linkage or association analysis performed between the disease and polymorphic markers that are spaced throughout the genome.

γ-secretase
A multiprotein complex involved in intramembranous cleavage of APP-generating C-terminal end of Aβ peptides. It contains four proteins: the catalytic component (presenilins) and three cofactors (nicastrin, APH-1, and PEN-2).

Neurofibrillary Tangles (NFTs)
Abnormal intraneuronal deposits abundant in the brains of patients with AD and some other neurodegenerative diseases. The main constituent of NFTs is hyperphosphorylated tau protein.

Senile Plaques
Complex extracellular deposits abundant in the brains of AD patients. The main proteinaceous component of plaques is amyloid-β peptide.

Susceptibility Locus/Gene
A region of a chromosome or a specific gene that may influence the risk to a certain condition/disease.

Familial Alzheimer's Disease
Transmitted as an autosomal dominant trait; usually has an age of onset <60 years.

Sporadic Alzheimer's Disease
Shows modest familial clustering and probably results from the synergistic action of genetic and environmental factors. Usually has an age of onset >60 years.

Abbreviations

Aβ	β-amyloid peptide
AD	Alzheimer's disease
AICD	APP Intracellular domain
APLP	amyloid precursor-like protein
APOE	apolipoprotein E
APP	β-amyloid precursor protein
BBB	blood brain barrier
CAA	cerebral amyloid angiopathy
CNS	central nervous system
CTF	C-terminal fragment
FAD	familial Alzheimer's disease
FTDP-17	frontotemporal dementia with parkinsonism linked to chromosome 17
KPI	Kunitz-type protease inhibitor
MT	microtubule
NFT	neurofibrillary tangles
NSAID	nonsteroidal anti-inflammatory drugs
NTF	N-terminal fragment
PDGF	platelet-derived growth factor
PHF	paired helical filaments
PS	presenilin
TM	transmembrane

Alzheimer's disease (AD) is the most common cause of dementia in the elderly. It is characterized by a progressive loss of memory, reasoning, judgment, and orientation, by the presence of large numbers of extracellular senile plaques and intracellular neurofibrillary tangles, as well as substantial cell loss in the brain. Genetic studies in early onset families have identified mutations in three genes that cause AD, while genetic studies in sporadic AD have identified the apolipoprotein E4 allele as a risk factor for disease. *In vitro* cell biology and transgenic mouse studies implicate all of these genes in the biology of Aβ metabolism. A consistent effect of these genetic factors is an increase in Aβ deposition. Furthermore, cell biology studies have identified targets for possible treatment of AD. Transgenic models are now being used to test the validity of these targets and therapeutic approaches to the treatment of AD.

1 Pathology and Epidemiology of Alzheimer's Disease

The German psychiatrist Alois Alzheimer first described the clinical and pathological features of Alzheimer's disease (AD) in a 55-year-old female patient; he published the discovery as a case report in 1907. It was thought that the "presenile dementia" described by Alzheimer was distinct from "senile dementia" seen in older patients, but it is now widely accepted that the two conditions are the same disease, differing primarily in the age of onset.

Alzheimer's disease is the most common cause of dementia in the western world, and the fourth leading cause of deaths in the United States. The disease is confined mainly to aged individuals and is progressive. It affects 1% of people in the developed nations and is likely to become a major problem in developing countries as the proportion of elderly persons increases in these populations. The incidence of AD rises with increasing age: 1 to 5% of people aged 65 years are affected, while 10 to 20% of those over the age of 80 have AD.

The first clinical manifestation is usually a loss of short-term memory; this is followed, over the next 6 to 20 years, by a progressive loss of memory, reasoning, judgment, and orientation (dementia). Death occurs, on average, about 10 years after the onset of symptoms. Many epidemiological surveys have been carried out to identify risk factors. There is a clear correlation between AD and a positive family history in about one-third of cases; the risk of AD is also elevated in individuals with Down syndrome. Severe head trauma with loss of consciousness increases risk for AD, particularly in those individuals who also have an apolipoprotein E4 allele. In contrast, the use of nonsteroidal anti-inflammatory agents or cholesterol-lowering agents, such as the statins, may reduce risk for AD. Another factor that may potentially decrease risk for AD is the years of education; the more the years of education, the lower the risk of disease.

Alzheimer's disease is characterized neuropathologically by the presence of large numbers of neuritic plaques and neurofibrillary tangles (NFT) within the brain cortex (Fig. 1). In addition, there is massive

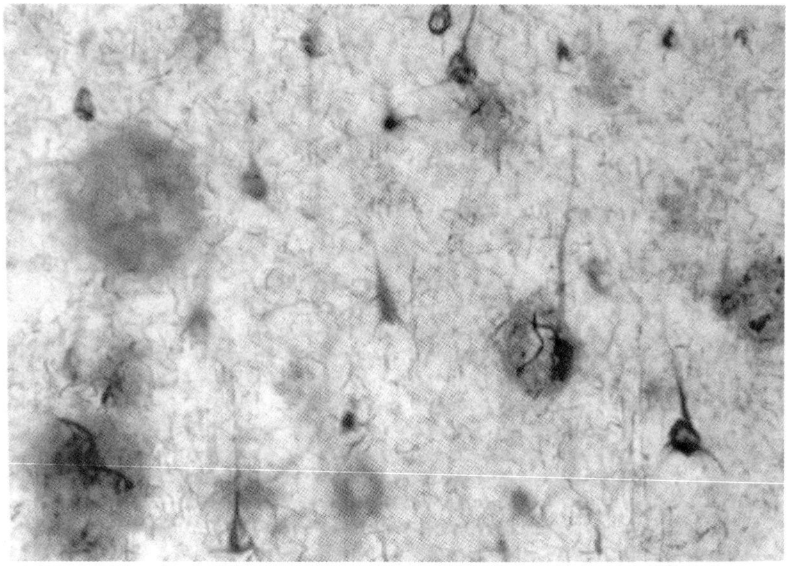

Fig. 1 Plaques and tangles in an AD brain. Shown is a brain section from an individual with mild AD. The section was double-stained with antibody PHF-1 specific to tangles and 10D5 specific to β-amyloid. Courtesy of Dr. Joel Price.

neuronal cell loss, and amyloid deposits that occur in the walls of the blood vessels of the meninges and cerebral cortex. Neither plaques nor tangles are specific to AD – they occur in intellectually normal, elderly people and in certain other diseases. However, they are more numerous and more widely distributed throughout the brains of Alzheimer's patients. Senile plaques are extracellular deposits, while tangles initially appear within the degenerating neurons. The major proteinaceous constituent of plaques is a 39 to 43 amino acid peptide called β-amyloid, or Aβ. The microtubule-associated protein, tau, is a major constituent of tangles. In the AD brain, there is extensive activation of astrocytes and microglia, the two primary mediators of inflammation in the central nervous system (CNS). A large number of inflammatory mediators and complement components are elevated in the AD brain, and colocalized with both plaques and tangles. The inflammatory response might be induced by increased Aβ levels or by plaque formation.

The deposition of Aβ has also been observed in association with several other pathological conditions. Individuals with Down syndrome, who live into their late thirties and beyond, develop a neuropathology indistinguishable from that of AD. Temporal studies in Down syndrome suggest that Aβ deposition is the earliest of the known pathological markers to appear, and that neurofibrillary tangles and neuronal cell loss occur decades after the first Aβ deposition. However, dementia correlates more closely with neurofibrillary tangles and neuron loss than with plaques. According to one hypothesis based on these observations, amyloid deposition precedes and induces tangle formation; tangles (and perhaps plaques) cause neuronal death, and the resultant neuronal damage destroys important neurotransmission pathways in the brain, producing dementia.

2
The Genetics of Alzheimer's Disease

There are two forms of AD: familial Alzheimer's disease (FAD), in which the disease is transmitted as an autosomal dominant trait; and sporadic AD, which shows modest familial clustering and probably results from the synergistic action of genetic and environmental factors. Sporadic AD can have an early (<60 yrs) or late (>60 yrs) age of onset. FAD does not appear to be clinically or neuropathologically different from the more common sporadic form of AD, except that it generally has an earlier age of onset.

2.1
The Genetics of Familial Alzheimer's Disease

The existence of families in which AD segregates as a fully penetrant, autosomal, dominant trait presents the most striking evidence for the involvement of genetic factors in the etiology of AD. Mutations in three distinct genes, β-amyloid precursor protein (APP), *PS1* (presenilin 1) and *PS2* (presenilin 2) have been identified to cause FAD.

2.1.1 β-amyloid Precursor Protein Mutations

The main component of the AD-associated senile plaques, Aβ, is derived by the proteolytic processing of APP. The gene coding for APP is located on the long arm of chromosome 21. It contains 18 exons, three of which can be alternatively spliced to produce a variety of mRNA species.

The first mutation shown to cause FAD was a valine to isoleucine substitution at residue 717 of APP (Fig. 2). To date, 22 APP mutations have been reported, of which 17 are linked to early-onset AD (www.alzforum.org). Most mutations lie within the transmembrane (TM) domain of APP in the region that is involved in the generation of the C-terminal end of Aβ. All TM domain FAD-APP mutations result in an increased production of the longer, more amyloidogenic form of Aβ (Aβ42/43). The most severe mutation (T714I) causes a 11-fold increase in the Aβ42/Aβ40 ratio; it causes both a decrease in Aβ40 and an increase in Aβ42 production. Patients carrying the T714I FAD-APP mutation die in their 40s. FAD mutations in APP can differentially affect Aβ-peptide formation by either increasing Aβ42 or decreasing Aβ40, or both. The Swedish mutation (K670N, M671L) lies in the region outside of the TM domain in which cleavage by β-secretase generates the N-terminal end of the Aβ, resulting in an increase in both Aβ40 and Aβ42 levels. Five mutations are located within the Aβ sequence but outside the TM domain: Flemish (A692G), Dutch (E693Q), Arctic (E693G), Italian (E693K), and Iowa (D694N). Patients carrying these mutations deposit Aβ in senile plaques and/or in cerebral vascular walls (cerebral amyloid angiopathy – CAA) and present clinically with either AD or hemorrhagic strokes, or both. For example, Flemish A692G patients present with both AD and strokes, while most Dutch E693Q patients suffer from strokes followed by a progressive multi-infarct dementia (HCHWA-D, hereditary cerebral hemorrhage with amyloidosis Dutch type). Pathologically, Flemish A692G patients demonstrate a classical AD pathology although with a strong CAA component, while Dutch APP patients

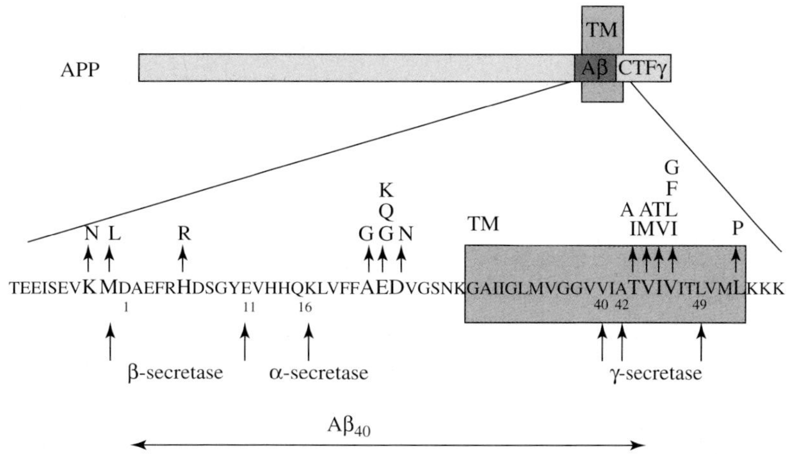

Fig. 2 FAD mutations in APP. Shown is a schematic representation of the APP protein showing the regions coding for Aβ peptide and C-terminal intracellular domain-CTFγ/AICD. The location of each FAD mutation is shown. Letters in bold are FAD-mutation sites, letters in gray are FAD substitutions. Major sites of APP processing, α-, β-, and γ-secretase sites, are also labeled. The amino acid numbering below the letters is according to the Aβ sequence.

have mainly vascular Aβ in the absence of neurofibrillary pathology. Recent analysis of pathogenic effect on Aβ levels of Arctic, Italian, Dutch, and Flemish APP mutations revealed that all, except the Flemish mutation, cause a decrease in both Aβ40 and Aβ42. In contrast, the Flemish mutation increases both Aβ40 and Aβ42 levels.

2.1.2 Presenilin Mutations

The discovery that mutations within the presenilin 1 *(PS1)* gene located on chromosome 14 (14q24.3) were responsible for most FAD cases opened up a whole new area of research into the molecular mechanisms of the pathogenesis of AD. Shortly after the discovery of *PS1*, DNA sequence database searches led to the identification of a homologous gene named presenilin 2 *(PS2)*, located on chromosome 1 (1q41.2). Mutations in *PS2* also result in FAD.

Pathogenic mutations in both of these genes have been identified; overall, mutations in *PS1*, *PS2*, and *APP* genes account for the majority of FAD cases. Although the proteins encoded by the two presenilin genes are 63% homologous at the amino acid level, intensive mutation and epidemiological analysis have revealed some interesting differences between the two proteins. The number of mutations identified in *PS1* is far greater than those in *PS2*. To date, more than 107 individual point mutations in *PS1* have been identified (Fig. 3, http://molgen-www.uia.ac.be/ADMutations) and found to be responsible for the majority of the early onset FAD. Most of these alterations are missense mutations that result in single amino acid changes in residues that are conserved between the PS1 and PS2 proteins. An intronic mutation (referred as ΔE9) results in the disruption of a splice acceptor site leading to the in-frame deletion of exon 9/10. This is the only *PS1*-FAD mutation that results in pathologically active protein that does not undergo proteolysis. Endoproteolytic

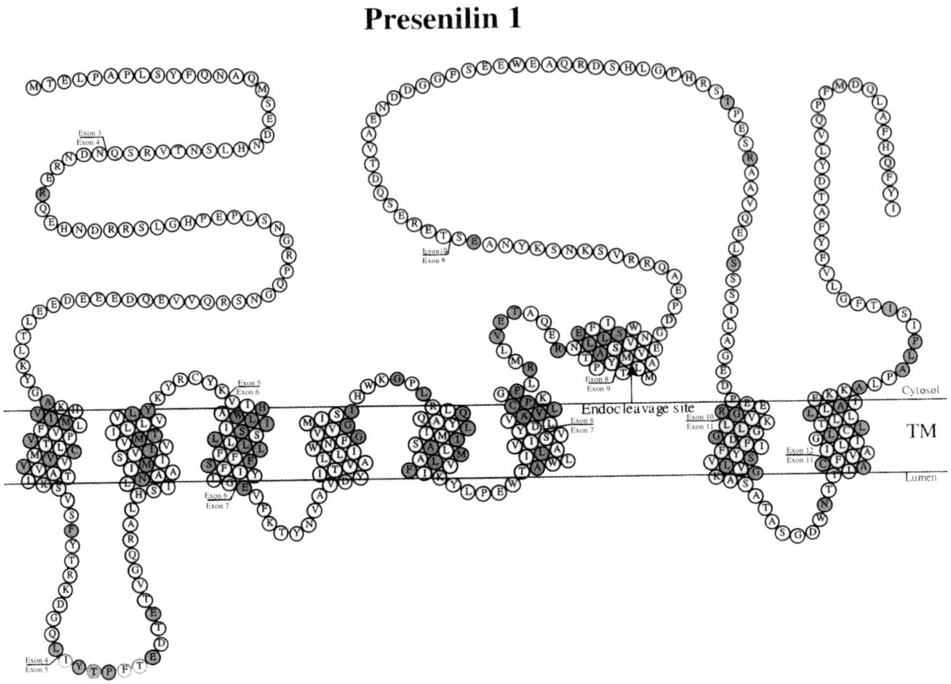

Fig. 3 FAD mutations in presenilin-1. Shown is a schematic representation of human presenilin 1 protein. The protein is predicted to have eight transmembrane domains. Arrow indicates the endocleavage site. Shaded residues are those associated with FAD mutations.

processing of the presenilins is an important aspect of the biochemistry of both PS1 and PS2 proteins, since it has been implicated in the formation of the functionally active complex (see Sect. 3). The most consistent effect mediated by the presenilin FAD mutations is increased production of Aβ42. This has been confirmed both in transfected cells and in transgenic mice. Although the mechanism by which FAD mutations in *PS1* and *PS2* cause AD is not fully understood, it is suggested that FAD mutations may alter γ-secretase activity toward cleavage of the more amyloidogenic peptide of Aβ, Aβ42/43. Although point mutations are distributed throughout the *PS1* gene, the majority of them are clustered within the TM domains and large hydrophilic loop between TM6 and TM7 (Fig. 3). The mean age of onset in all *PS1* families is about 45 years with a range of 28 to 62 years, suggesting that mutations in *PS1* result in a more aggressive form of FAD than APP mutations.

In contrast to *PS1*, only eight point mutations have been identified in *PS2* (http://molgen-www.uia.ac.be/ADMutations). The first *PS2* mutation to be identified was the N141I mutation that is responsible for disease in a large Volga German kindred. Whether the preponderance of *PS1* versus *PS2* mutations is due to the differing genomic environment of the two genes or the factors related to differing biological functions of the two molecules, remains unclear. Mutation analysis of the two proteins has also led to the identification of a phenotypic difference. Overall, the mutations in *PS1* result in relatively early and constant age of onset while those in *PS2* lead to a somewhat later and more variable age of onset. The identification of a *PS1* N135I mutation, a position and substitution that corresponds to the *PS2* Volga German mutation, and the finding that the affected members of this pedigree have a relatively constant age of onset suggests that the observed variations in age of onset are dependent on the molecule itself as opposed to the specific position or nature of the amino acid substitution.

2.2 The Genetics of Sporadic Alzheimer's Disease

While the genetics of FAD is fairly well understood, our understanding of the more common sporadic AD remains much less complete. The majority of cases have onset >60 years. Some families have stronger clustering of AD cases than others, suggesting that the genetic component of the disease can be quite variable. The estimated cumulative risk to first-degree relatives of AD-affected probands approaches 50% by age 90 compared to a disease risk of 10 to 15% in the general population. To date, only the Apolipoprotein *E*4 (*APOE4*) allele has been linked to increased risk for sporadic AD. While it is clear that *APOE* is a major risk factor for AD, epidemiological studies estimate that 42 to 68% of AD cases do not have an *APOE4* allele, indicating that additional genetic and environmental factors are involved in this form of the disease. The balance of these risk factors most likely determines both the age of onset of the disease and the number of affected first-degree relatives.

2.2.1 Apolipoprotein E

Linkage analysis in late-onset AD pedigrees led to the observation that these families inherit a genetic risk factor located on chromosome 19. The defect involved in these particular pedigrees is a susceptibility defect as opposed to a causative defect such as those present in the gene coding for APP and the presenilins. The inheritance of the susceptibility gene defect alone is not sufficient to cause the disease, and one or more secondary events, either environmental or genetic, must accompany the inheritance of the primary defect to cause the disease. Association studies have shown that the chromosome 19 susceptibility gene (localized to 19q13.2) is within the *APOE* gene locus.

APOE protein is a lipid-transport molecule and the primary apolipoprotein observed in the CNS. It is encoded by a polymorphic gene that exists as three alleles designated *APOE2*, *APOE3*, and *APOE4*. These genetic variations result in amino acid substitutions (arginine or cysteine) at positions 112 and 158 of the protein. Thus, whereas the *APOE2* polypeptide has a cysteine at both of these positions, *APOE3* has a cysteine at 112 and an arginine at 158, and *APOE4* has arginine at both positions. In most Caucasian populations, the *APOE3* is the most common allele (frequency 0.78); *APOE4* (frequency 0.14) and *APOE2* (frequency 0.08) are considered variants.

When the association of the three common *APOE* alleles with AD was examined in late-onset AD pedigrees, it was found that the *APOE4* allele frequency was 52%, compared to 16% for age-matched controls. In addition, in sporadic AD cases, the *APOE4* allele frequency was 40%. The *APOE4* association with AD was rapidly and widely confirmed in patients with familial and sporadic AD. These multiple confirmations established the *APOE4* allele as the most important genetic marker of risk for the disease identified so far, accounting for approximately 50% of the genetic component of AD. The frequency of *APOE4* allele among AD and control cases can vary in different populations, however, the association with AD is consistent.

Individuals inheriting two copies of the *APOE4* allele have an increased risk and earlier age of onset of AD than individuals with one *APOE4* allele. Likewise, individuals with one *APOE4* allele have a greater risk and earlier onset of disease than those with no *APOE4* alleles. These data demonstrate that the effect of *APOE4* on risk and age of onset is dose-related. In contrast to the effect of *APOE4* allele, the *APOE2* allele has a protective effect on AD, as evidenced by a decreased frequency of this allele in AD cases. The effect of the most commonly inherited allele, *APOE3*, on risk and age of onset falls between *APOE4* and *APOE2*. There also appears to be an interaction of *APOE* genotypes with the age of onset in FAD. When survival analysis was used in a large Colombian kindred carrying a presenilin mutation, the presence of an *APOE4* allele was observed to decrease age of onset, while presence of an *APOE2* allele increased age of onset. Although it is still not clear how an *APOE4* allele modifies risk for AD, several studies suggest that APOE may act via an Aβ-dependent mechanism.

2.2.2 The Search for New Alzheimer's Disease Susceptibility Genes

There are two genetic approaches to finding susceptibility loci: full genome scans and candidate gene studies. In candidate gene studies, genes are selected on the basis of the known biology of the disease or their location within chromosomal regions implicated in risk for late-onset AD, and are assessed individually in order to determine whether variants in each candidate are associated with disease. An alternative approach, a genome scan, does not select genes *a priori*, but instead tests for linkage or association between the disease and polymorphic markers that are spaced evenly through the genome. Association methods use unrelated case-control samples or siblings who are discordant for disease, and test for differences in allele frequencies between affected and unaffected individuals. In contrast, linkage studies use extended families or affected sibling pairs and look for regions of the genome in which there is an increased allele sharing between affected relatives.

To date, only five late-onset AD genome-wide screens have been reported. The most consistent findings among these studies are evidences for AD susceptibility loci on chromosomes 19 (*APOE*), 9, 10, and 12. Since FAD is closely associated with elevated Aβ42 levels, some investigators have hypothesized that loci, which modify plasma Aβ42, might also modify risk for AD. Using this approach, evidence for linkage was observed in the same region of chromosome 10 as was observed in the AD-linkage studies. Over the past few years, a large number of biologically interesting candidate genes as well as candidate genes that map within chromosome regions implicated in risk for late-onset AD have been evaluated in case-control populations. The genes that have received most attention are those encoding *APOE* on chromosome 19, α2-macroglobulin and the LDL-receptor related protein on chromosome 12 and insulin degrading enzyme and urokinase on chromosome 10. All of these genes have been closely linked to Aβ metabolism. However, with the exception of the *APOE* gene, whose association with

AD has been universally replicated, the relative significance of these associations is still in question.

3 Molecular and Cellular Mechanisms Underlying Alzheimer's Disease

3.1 β-amyloid Precursor Protein

β-Amyloid precursor protein (APP) is a large, type I transmembrane protein that is expressed in almost all cell types except erythrocytes. It consists of a large extracellular N-terminal region, a transmembrane domain, and a short intracellular C-terminal tail. The Aβ-peptide sequence begins close to the membrane on the extracellular part of APP and ends within the membrane-spanning domain. At least five different mRNAs arise as a result of alternative splicing of the primary transcript of the APP gene: APP770, 751, 714, 695, and 563. Transcript APP695 lacks the Kunitz-type protease inhibitor (KPI) domain and is the major form expressed in neurons, whereas APP751, which contains the KPI domain, is the major transcript in non-neuronal cell types. The APP transcripts appear to be developmentally regulated, and all cell types so far tested contain at least one of the transcripts.

In addition to differential RNA splicing, APP undergoes posttranslational modification, including phosphorylation, N- and O-linked glycosylation, sulfation and proteolytic processing. It is not clear how these different events modify APP metabolism or function. A detailed knowledge of posttranslational events is undoubtedly crucial to understanding amyloid plaque formation and to elucidating the biological functions of APP and Aβ.

The normal function of APP is still unclear, but it has been implicated in neuroprotective action. APP is a member of a multigene family that contains at least two other homologs known as amyloid precursor-like protein 1 and 2 (APLP1 and APLP2). Both $APP^{-/-}$ and $APLP2^{-/-}$ mice are viable while APP/APLP2 double knockout mice are early postnatal lethal suggesting that APP and APLPs may share important physiological functions and the APLPs may compensate for the function of APP.

3.2 Production of Aβ – Proteolytic Processing of APP

It is clear that Aβ production does occur in normal cells, suggesting that excessive Aβ production in AD is due to enhanced or altered processing of APP by normal pathways that generate Aβ, rather than by an abnormal route specific to Alzheimer's cells. APP undergoes a series of proteolytic processing events (Fig. 4) by several different proteases called secretases. It is first cleaved by either α- or β-secretase. Two disintegrin metalloproteases, ADAM 10 and TACE (TNF-α converting enzyme), are involved in α-cleavage of APP. This cleavage produces a large soluble N-terminal fragment (NTF), APPsα and an 83-amino acid membrane-bound C-terminal fragment, C83 (CTFα). Alternatively, cleavage by the β-secretase results in an N-terminal fragment APPsβ and a 99-amino acid membrane-bound fragment, C99 (CTFβ). The β-secretase has recently been identified as a transmembrane aspartyl protease called BACE1, for β-site APP-cleaving enzyme. Both C83 and C99 are substrates for γ-secretase, a mysterious enzyme that carries out an unusual proteolysis in the middle of the transmembrane domain of

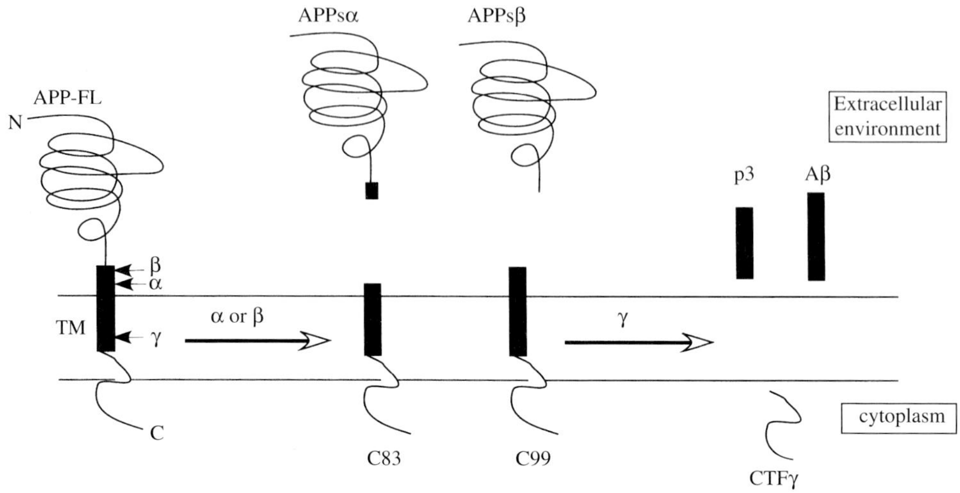

Fig. 4 Proteolytic processing of APP. APP is initially cleaved by either α- or β-secretase to generate membrane-bound C-terminal fragments, C83 or C99, which are subsequently cleaved by the γ-secretase (in the transmembrane domain) to produce p3 or Aβ.

APP, resulting in the release and secretion of the 4-kD Aβ from C99 and the 3-kD p3, a nonpathogenic peptide, from C83. γ-secretase cleavage is promiscuous and produces a spectrum of Aβ peptides varying from 39 to 43 residues in length. Over 90% of the Aβ in the brain is Aβ40 and the remaining 5 to 10% is the longer species Aβ42/43.

The C-terminal product of the γ-secretase cleavage is called CTFγ or AICD (from APP Intracellular Domain). Interestingly, CTFγ, purified from brain tissue, is not C59/57, which would be the predicted length of the C-terminal product of Aβ cleavage. Instead, sequence analysis reveals that it is C50/49, several amino acids shorter. The mechanism that accounts for such differential cleavages is currently under investigation. Several reports have suggested a role of CTFγ in APP signaling to the nucleus. Indeed, it was shown that CTFγ, when together with Fe65 protein, translocates to the nucleus and that this complex can activate transcription of a reporter gene *in vitro*.

3.3
β-secretase – BACE1

In late 1999, four groups almost simultaneously reported the discovery of BACE1, a novel aspartyl protease that exhibited all the known functional characteristics of the β-secretase. BACE1 is a type I transmembrane protein with an open reading frame of 501 amino acids. It is most closely related to the pepsin family of aspartyl proteases. BACE1 is expressed in most tissues and cell types at modest levels, but shows much higher expression levels in neurons. Structurally, it contains an N-terminal 21 amino acid signal peptide, a propeptide domain, a large lumenal domain followed by a transmembrane domain, and a short cytosolic C-terminal tail. BACE1 contains two aspartyl protease active site motifs in the lumenal domain: DTGS (residues 93–96) and DSGT (residues 289–292). Both aspartic acid residues have been shown to be essential for enzymatic activity. BACE1 contains intrinsic proteolytic activity; as a purified recombinant, BACE1 can directly cleave APP *in vitro*. Overexpression of BACE1 in cell lines dramatically increases β-secretase activity and β-cleavage products: APPsβ and C99. In addition, mice deficient in BACE1 show a normal phenotype, but Aβ generation is abolished, further supporting that BACE1 is indeed a β-secretase.

BACE1 is initially synthesized in the ER as an immature precursor protein, proBACE1, which undergoes rapid maturation involving N-glycosylation as well as removal of the propeptide domain by a furin-like convertase in the Golgi apparatus. Unlike most zymogens, proBACE1 possesses β-secretase activity. The propeptide domain does not appear to significantly inhibit protease activity, but rather acts as a chaperone to assist proper folding of the protease domain. Phosphorylation of BACE1 at Ser498 appears to be important for the intracellular trafficking of the protein between early endosomes and late endosomes/TGN compartments.

A homologous aspartyl protease, BACE2 was shortly identified by searching the EST database with the BACE1 sequence. BACE1 and BACE2 share ~64% amino acid similarity. Together they define a novel family of transmembrane aspartyl proteases. Brain expression of BACE2 is very low and most evidence suggests that it does not play a major role in β-cleavage of APP.

3.4 γ-secretase – Presenilins and Cofactors

The enzyme that is clearly central to AD pathogenesis is the γ-secretase, which cleaves C99 and produces Aβ. Evidence suggests that it is an aspartyl protease with the unusual property of cleaving substrates within the transmembrane domain. Independent from its role in processing APP, γ-secretase also mediates the intramembranous cleavage of several other type I transmembrane proteins including Notch and ErbB-4. γ-secretase cleavage of Notch releases the Notch intracellular domain, which translocates to the nucleus and acts as a cofactor to modify transcription of target genes.

The identity of the γ-secretase is rather enigmatic. It appears to be a multiprotein, high molecular weight complex that consists of presenilins (PS) and three other cofactors: nicastrin, APH-1 and PEN-2.

3.4.1 Presenilins

Presenilins are a family of polytopic transmembrane proteins that are implicated in AD pathogenesis. PS homologs have been found in species as diverse as *Drosophila melanogaster*, *Caenorhabditis elegans* and *Arabidopsis*, but not in yeast. The function of PS appears to be conserved. Deficiency in sel-12, one of the nematode PS homolog, causes egg-laying defects. This phenotype can be rescued by expressing human PS1 in the mutant worms. In mammals, there are two PS genes, *PS1* and *PS2*, that share 65% identity. *PS1* is believed to account for the majority of PS functions, while *PS2* seems to play a complementary role.

Presenilins are found primarily within intracellular membranes including the ER and Golgi as well as the plasma membrane. Structurally, they are predicted to have eight transmembrane domains with both the amino and carboxyl termini oriented toward the cytoplasm (Fig. 3). A large cytoplasmic loop is postulated between TM6 and TM7. PS are synthesized as single polypeptides that rapidly undergo endoproteolysis within the cytoplasmic loop, generating a 30-kD N-terminal fragment (NTF) and a 20-kD C-terminal fragment (CTF). The NTF and CTF form a stable complex, which appears to be the functional form of presenilins. PS endoproteolysis is tightly regulated such that overexpression of PS leads to the accumulation of a full-length protein, but does not result in an increase in the fragments. The NTF/CTF complex, but not the holoprotein, are components of a high molecular weight complex, estimated at ~250 kD or even larger. This high molecular weight complex is believed to be the γ-secretase complex.

Many lines of evidence indicate that *PS1* is the catalytic component of the γ-secretase. The two membrane-embedded aspartyl residues in *PS1* (D257 in TM6 and D385 in TM7) are essential for γ-secretase activity. The mutation of either residue abolishes γ-secretase processing of APP and Notch. The knockout of the PS genes blocks the Aβ production and Notch function. While in *PS1* single K/O cells, there is residual Aβ production, in PS1/2 double knockout cells, there is no detectable Aβ, demonstrating that there is no PS-independent γ-secretase activity. Inhibitors designed to be transition-state analogs of the γ-secretase bind to presenilin fragments. Recently, a family of signal peptide peptidases was discovered, which are apparently *bona fide* intramembranous aspartyl proteases, with similar active site motifs and similar topology to the presenilins. This finding further supports the hypothesis that presenilin is an aspartyl protease.

3.4.2 Cofactors

Nicastrin is a large type I transmembrane glycoprotein that was immunopurified from PS1 immunoprecipitates. It has no significant sequence homology to other functionally characterized proteins. Nicastrin is a 709 amino acid protein containing a putative signal peptide, a long N-terminal hydrophilic domain with numerous glycosylation sites, a TM domain and a short hydrophilic C-terminal tail of 20 residues. The mature protein migrates at ~150 kD owing to heavy glycosylation. There is evidence suggesting that the association of nicastrin with γ-secretase is tightly regulated via glycosylation. Nicastrin binds to presenilins and the active γ-secretase complex. Nicastrin deficiency in *Drosophila* abolishes Notch signaling and γ-secretase cleavage of APP. In addition, presenilin stability was compromised in the absence of nicastrin. All these data support that nicastrin is a component of the γ-secretase complex and an important regulator of the γ-secretase activity.

Genetic screens in *C. elegans* identified two proteins, APH-1 and PEN-2, which interact with worm homolog of *PS1* and nicastrin. Loss-of-function mutations of either protein produces phenotypes that resemble that of null mutants in nicastrin and presenilins. APH-1 is an integral protein containing seven predicted TM domains. There are two mammalian *APH-1* genes (*mAPH-1a* and *mAPH-1b*). *mAPH-1a* has at least two splice variants: *mAPH-1a*^{L} has seven exons and encodes a longer protein of 265 residues; mAPH-1a^S has six exons and encodes a shorter protein of 247 residues. All three variants are widely expressed, although their abundance seems to be coordinately regulated. To date, no functional difference among these variants has been documented. PEN-2 is a small protein of 101 amino acids containing two TM domains. Topology study suggests that both the N- and C-terminals of the protein face the lumen of the ER.

Human APH-1 and PEN-2 can partially rescue the *C. elegans* mutants, suggesting functional conservation among eukaryotes. The human genes must be present together to rescue the phenotypes and adding PS1 improves the rescue, indicating that these proteins act in concert. Consistent with the genetic studies, biochemical studies showed that reduction of APH-1 and PEN-2 expression in mammalian cells by RNA inhibition reduced γ-secretase activity.

While the exact functions of these cofactors remain to be elucidated, the data obtained so far suggest that all four proteins, presenilins, nicastrin, APH-1, and PEN-2, are necessary components of the γ-secretase. They appear to regulate one another for protein stability, proper maturation, and functional γ-secretase complex formation. It is also very likely that the least abundant component becomes the cellular-limiting factor that controls overall γ-secretase activity.

3.5
Neurofibrillary Tangles and Tau Protein

Neurofibrillary tangles are intracellular deposits that consist of aggregates of a fibrous substance, the paired helical filaments (PHF). These filaments are highly insoluble, even resistant to strong detergents such as SDS and guanidine hydrochloride. A major component of PHF is tau, a microtubule (MT)-associated phosphoprotein. Tau is abundantly expressed in the brain. At least six isoforms of tau are present in the brain, resulting from alternative mRNA splicing of a single gene on chromosome 17q.

Unlike APP, whose functions are not clear, tau has a biological role that is reasonably well understood. One function of tau is to promote microtubule assembly, a process controlled by phosphorylation of tau. In addition, it has been proposed that tau has a specific role in the generation of axon morphology. Increased phosphorylation of tau decreases its ability to promote tubulin polymerization to microtubules.

Tau in PHF is hyperphosphorylated, and is insoluble under conditions in which normal tau is soluble. It is believed that tangles are formed when tau becomes hyperphosphorylated and dissociates from microtubules and spontaneously forms insoluble PHF. Over 25 phosphorylation sites have been identified in PHF tau purified from AD brains. Most of the phosphorylation sites are located in the flanking region of the MT-binding domain. A variety of protein kinases, including glycogen synthase kinase-3β (GSK3β), cyclin-dependent kinase 5 (cdk5), mitogen-activated protein kinase (MAPK) and protein kinase A, have been found to be involved in hyperphosphorylation of tau. This suggests that multiple phosphorylation cascades may be activated in NFT-bearing neurons. However, it remains unclear whether these cascades are activated independently or whether one or more kinases are primarily responsible for initiating the cascade *in vivo*.

In addition to their presence in AD, filamentous tau inclusions accompanied by extensive gliosis and loss of neurons are the neuropathological hallmarks of other neurodegenerative diseases that are sometimes designated as tauopathies. There has been considerable debate in the AD literature concerning the primacy of Aβ versus NFT in the pathogenesis of AD. This was largely resolved by the identification of mutations in the tau gene in families with frontotemporal dementia with parkinsonism (FTDP-17). FTDP-17 is characterized clinically by the presence of prominent behavioral changes early in the disease. The neuropathology associated with FTDP-17 is characterized by large numbers of NFT but no senile plaques. Mutations in the tau gene in FTDP-17 can alter splicing and produce a shift from the short tau isoform with three repeats to the longer isoform, which contains four repeats. Several other point mutations have been discovered in the coding region of tau that impact the functional integrity of the tau protein itself, P301L being the most common. So far, genetic analysis has not linked tau mutation directly to AD, but tau dysfunction may still have a significant role to play in the disease, if not in its initiation, then in its progression.

3.6 Apolipoprotein E

Apolipoprotein E is a 299 amino acid glycoprotein with a molecular weight of ~34 kD. Its N-terminal domain consists of four amphipathic α-helices and the C-terminal domain contains most of the lipid-binding activity. APOE is a major apolipoprotein that regulates cholesterol uptake and release. The CNS contains high levels of APOE with abundance only second to the liver. In the CNS, APOE is synthesized predominantly by astrocytes and is secreted into the CSF, where it is a major lipid carrier.

APOE appears to have multiple functions in the brain – any or all of which could potentially influence AD pathogenesis. The molecular mechanisms by which APOE protein is involved in AD pathogenesis are unclear because reports on the allele-specific effects of APOE are controversial. Nevertheless, numerous studies have demonstrated that APOE can affect AD pathogenesis either by influencing Aβ deposition or by a direct effect on neuronal survival. APOE is a major component of senile plaques and specifically binds to soluble Aβs. Allele-specific differences in Aβ binding, aggregation, and fibrillogenesis have been reported. Most histopathologic studies demonstrated a correlation between senile plaque density and *APOE4* allele dose in AD. Animal studies also show that APOE has isoform-specific effects on Aβ deposition and plaque formation (see Sect. 4.4).

There is also evidence that different alleles of *APOE* have different effects on neurite extension, neuronal survival, and the cell's response to oxidative stress. *In vitro* studies with primary neurons and neuronal cell cultures show that lipid-bound E3 promotes neurite extension, while E4 is neutral or inhibitory. These *in vitro* findings are consistent with reported defects in neuronal remodeling in AD brains carrying an E4 allele. APOE can have allele-specific differences on both neurotoxic or neuroprotective activities. High levels of APOE or some APOE synthetic peptides were shown to cause neurite degeneration or neuronal death in cultured cells, with the E4 having the most neurotoxic effects. On the other hand, APOE appears to play a role in protection against oxidative stress, a prominent phenomenon in AD pathogenesis. Apoe -/- mice exhibit increased markers of lipid peroxidation within the brain and appear to have increased susceptibility to neuronal damage. Different *APOE* alleles are shown to have isoform-specific neuroprotective effects (E2 > E3 > E4) on oxidative cytotoxicity caused by hydrogen peroxide, or Aβ peptides. A possible mechanism for the different antioxidant activities of the *APOE* alleles is their ability to detoxify 4-hydroxynonenal, a neurotoxic lipid-peroxidation product that is believed to play a key role in neuronal death in

AD. 4-hydroxynonenal covalently binds to APOE, with more binding to E2 and E3 than E4, probably because of the cysteine residues that are present in E2 and E3. E2 and E3 may therefore protect neurons by binding to, and thereby detoxifying, 4-hydroxynonenal.

4 Transgenic Mouse Models of Alzheimer's Disease

A major hindrance to basic research into the molecular pathogenesis of AD has been the absence of a mouse or rat model. As a result of the identification of mutations that cause AD in humans, this has begun to change. Although initial efforts to make transgenic models of AD failed, more recent efforts have recapitulated at least some of the key features of the disease. Many genetically altered mice have been designed to reproduce the neuropathology of AD, however, the majority of them share only some of the neuropathological and/or cognitive impairment characteristics of AD (http://www.alzforum.org). These include mice expressing human forms of the APP, the PS, APOE and, more recently, tau. Transgenic mice expressing human APP develop amyloid plaques, but neurodegeneration and neurofibrillary tangles are not observed. The Presenilin transgenic mice produced only increased Aβ42/43 levels and did not develop signs of AD pathology, even though the same mutations caused some of the earliest forms of inherited AD in humans. Strategies emphasizing tau resulted in increased phosphorylation of tau and tangle formation, although amyloid plaques were absent. Nevertheless, crossing transgenic animals expressing mutated tau and APP has produced a mouse that closely recapitulates the neuropathology of AD. This section provides a review of the various murine models and their role in understanding the pathogenesis of AD.

4.1 APP Transgenic Mice

The most extensively studied APP transgenic mouse lines are known as PDAPP and Tg2576. The PDAPP transgenic mouse expresses a human APP770 mini gene containing the V717F FAD-APP mutation, while Tg2576 expresses human APP695 that contains the Swedish FAD-APP mutation. In PDAPP, expression of the *APP* gene is under the control of the human platelet–derived growth factor (PDGF)-β chain neuronal promoter that targets expression preferentially to neurons in the cortex, hippocampus, hypothalamus, and cerebellum of the transgenic animals. This mouse was made on a mixed-strain background (C57BL/6, DBA/2, and Swiss-Webster), while Tg2576 was made on a single C57B6/SJL background. Expression of the *APP* gene in Tg2576 mice is under the control of the prion protein promoter resulting in a less-limited expression of APP both in the CNS and the periphery. The success of both models was due to the high level of APP expression achieved (>10-fold overexpression of human APP in PDAPP mouse and >6-fold higher expression of APP in Tg2576 compared to endogenous murine APP levels). Cortical and limbic amyloid deposits begin slightly earlier in the PDAPP mouse: by 3 months of age in homozygotes and at 6 to 9 months in heterozygotes, while in Tg2576 mouse they occur between 9 to 12 months of age, elevated Aβ production is observed as early as 3 months of age. Both models develop β-amyloid deposits (in hippocampus and neocortex but limited in striatum and cerebellum), and the deposits are associated with dystrophic neurites, punctate immunoreactivity to hyperphosphorylated tau, astrocytosis, microgliosis, and vascular amyloidosis. The elevation of Aβ in the Tg2576 mouse correlates with the appearance of memory and learning deficits in the oldest group of transgenic mice as well. However, both APP transgenic models fail to show all the pathological features of human AD; no significant neuronal loss and no formation of NFTs were observed in either line.

Several other APP transgenic models were generated in order to obtain a complete or more comprehensive picture of the hallmarks of AD. The TgAPP22 transgenic mice overexpress a double mutant APP with Swedish (K670N, M671L) and London (V717I) FAD mutations in cis, while TgAPP23 only carries the Swedish mutation. In both models, the human APP751 isoform is used and its expression is under the control of the murine Thy-1 promoter. The TgAPP22 mice show twofold overexpression of the transgene over endogenous APP, and Aβ deposits were detected in the neocortex and hippocampus at 18 months of age, while the TgAPP23 has higher, sevenfold overexpression of human APP and typical Aβ plaques appear at the age of six months. Interestingly, these two murine AD models showed differences in plaque type; the majority of amyloid deposits in TgAPP22 mice brains are of the "diffuse" type, while in the TgAPP23 mice almost all extracellular amyloid deposits were fibrillar. Both substantial neurodegeneration and a reduction of neuron numbers were apparent in TgAPP23. In 14–18 months old TgAPP23 neuronal loss was 14% and reached 25% in mice with high Aβ plaque load. Additional pathological features included dystrophic neurites surrounding the plaques, hyperphosphorylated tau, but no NFTs developed.

Each of the models described above is remarkable in that the anatomical pattern of plaque formation parallels that seen in human AD. Furthermore, the morphology of amyloid plaques in aged APP-transgenic mice recapitulates amyloid pathology in human AD: the plaques span a continuum from diffuse Aβ deposits to compact core plaques with inflammation and neuritic dystrophy. However, none of these models reflects a complete picture of the neuropathology of AD.

4.2
Presenilin 1/2 Transgenic Mice

Several PS1 transgenic mice have been created to study presenilin biology *in vivo*. These mice differ in the promoter and the strains of mice used, but are similar in achieving high levels of protein production (1–3 fold over endogenous) in neuronal regions of the brain. Knockin transgenes were created expressing either the wild-type human PS1 or PS1 containing the FAD mutation A246E under the transcriptional control of the human Thy-1 promoter in PS1 null background. Both mice rescued the PS1 knockout mouse from embryonic lethality, confirming that PS1 mutation does not cause a loss of function of PS1 protein but has similar physiological properties as wild-type PS1. A second PS1 transgenic model expressing human mutant (M146L or M146V) and wild-type PS1 was generated under the control of the PDGF promoter. All three PS1 transgenics showed similar results: overexpressing mutant PS1 in the brains of transgenic mice lead to the elevation of

$A\beta42/43$, but not $A\beta40$. This effect was a direct result of the mutation and not overexpression of the human protein as the overexpression of the wild-type PS1 did not have any significant effect on $A\beta$ levels. The physiological significance of the specific elevation of $A\beta42/43$ appears to be the observation that AD patients with PS1 mutations show plaques composed primarily of $A\beta42/43$. However, all three PS1 transgenic mice, in contrast to APP transgenics, revealed no $A\beta$ deposition or other AD-associated pathology.

Recently, a PS1 mouse model carrying a PS1-P264L FAD mutation was developed using gene-targeting approach. The gene-targeted models are distinct from transgenic models because the mutant gene is expressed at normal levels, in the absence of the wild-type protein. $PS1^{P264L/P264L}$ mice had normal expression of PS1 mRNA, but levels of the N- and C-terminal protein fragments of PS1 were reduced while levels of the holoprotein were increased. $APP^{sw/sw}/PS1^{P264L/P264L}$ double gene-targeted mice had elevated levels of $A\beta42$, sufficient to cause $A\beta$ deposition beginning at six months of age. This was the first animal model that exhibited $A\beta$ deposition without overexpression of APP.

Three transgenic mice that express PS2-FAD mutation N141I under the control of PDGFβ chain promoter, chicken β-actin promoter or the neuron-specific enolase promoter were designed. The first two PS2 transgenic mice showed increased levels of $A\beta42$ in the brain of 12-month old mice. The third model utilizing neuron-specific enolase promoter did not show any difference in $A\beta42$ levels compared to the PS2 wild-type transgenic mice and both had no obvious differences in AD phenotypes (both had amyloid deposits).

4.3
Tau Transgenic Mice

One of the deficits of most AD transgenic models is the lack of tau pathology. APP and APP/PS transgenic mice, which exhibit extensive amyloid deposition also develop hyperphosphorylated tau around the amyloid deposits but it does not progress to resemble human NFTs. This suggests that some disruption of tau does occur in response to amyloid accumulation, but it does not progress to resemble human tauopathy in the mouse model. The idea that a species barrier prevents this progression has been proposed suggesting that primates and rodents differ in their response to injected amyloid not only in terms of neurodegeneration, but also in the accumulation of abnormal tau around the injected amyloid.

The most exciting development in the tau field is the creation of a transgenic mouse model with robust and reproducible tangle pathology. This transgenic mouse line, designated JNPL3, express human four repeat tau containing the most common FTDP-17 mutation, P301L, under the control of the mouse prion promoter. The mice develop motor and behavioral deficits that are associated with age- and gene-dosage-dependent development of congophilic NFT. Tau-immunoreactive pretangles are found in cortex, hippocampus, and basal ganglia, but at lower levels than commonly found in human disease. The tangles are associated with neurodegeneration, especially in the spinal cord where motor neurons were reduced by approximately 48%. This mouse has recently been used in breeding double and triple transgenic mouse models to produce a model system that more accurately recapitulates the hallmark pathology of AD (see further in the text). Recently, another

P301L tau transgenic model has been described that develops fivefold increase in the numbers of NFT soon after 18 days after the intracerebral injection of Aβ42 into the CA1 region of the hippocampus. These experiments indicate that an interaction of β-amyloid with the P301L tau mutation accelerates NFT formation.

4.4 Apolipoprotein E Transgenic Mice

A consistent consequence of carrying the *APOE4* allele is an increased number of amyloid plaques in the brain and more abundant amyloid deposition in the cerebral vasculature. The mechanism by which *APOE4* contributes to the development of neurodegeneration remains unknown, although the evidence suggests that this may be linked to the ability of APOE to interact with the amyloid β peptide and influence its concentration and structure. Recent studies with transgenic mice that overexpress the human APOE isoforms APOE2, APOE3, and APOE4 in a PDAPP hemizygous mouse (Apoe-/- background) show that isoforms that are known to increase the risk of AD enhance amyloid load and increase the neuritic pathology associated with fibrillar plaque development in aged animals (by 15 months of age APOE4 expressing mice having a 10-fold greater amyloid burden than APOE3 mice). The time course of Aβ deposition, Aβ levels, structure, and anatomic and subcellular distribution showed profound age-, species-, and isoform-dependent effect of APOE. It was also found that APOE not only impacts on the nature of aggregated Aβ, but also on the clearance of soluble Aβ from the brain across the blood–brain barrier (BBB). Aβ40 was rapidly cleared from the brain, while Aβ42 was cleared much less effectively, supporting the idea that Aβ42 production may favor amyloid deposition due to a reduced clearance across the BBB. However, Aβ clearance was not APOE isoform specific.

These recent findings are consistent with earlier work on Apoe null mice. When Tg2576 and PDAPP mice were crossed with Apoe null mice it was revealed that the absence of Apoe altered the quantity, character, and distribution of Aβ deposits in the transgenic animals, confirming that APOE affects the neuropathological phenotype in AD brains. The Aβ deposition is significantly reduced and is not thioflavine-S-positive.

4.5 Modulating and Enhancing the Phenotype: Transgenic Crosses

One of the great advantages of transgenic animals is that different mice can be mated together so that the effect of multiple transgenes on the disease phenotype can be observed. In particular, since most transgenic AD models do not exhibit all the pathological features of AD, different crosses have been made in order to design a model that will more closely describe the pathology, neurodegeneration, and memory loss seen in AD patients.

Double transgenic mice that carry *PS1-A246E* and *APPsw* FAD mutations show an elevated Aβ42/Aβ40 ratio in brain homogenates compared to the ratio observed in transgenic mice expressing *APPsw* alone or transgenic mice coexpressing wild-type human *PS1* and *APPsw*. In addition, double transgenic mice *APPswxPS1-A246E* developed numerous amyloid deposits much earlier (9 months of age) than age-matched mice expressing *APPsw* (18 months of age) and wild-type *huPS1* and *APPsw* alone. Interestingly, the majority of Aβ deposits in the double transgenic mice were not immunoreactive to Aβ42 but instead were stained with antisera to Aβ40. Similarly, crossing the Tg2576 transgenic mice, which express mutant APPsw, with mice transgenic for *PS1-M146L* (PDPS1) revealed 41% increase in Aβ42/43 levels and formation of Aβ deposits in the cortex and hippocampus as early as 12 weeks of age, compared to development of Aβ deposits in Tg2576 mice of 9–12 months of age and approximately 1.5-fold elevation of Aβ42/43 from birth in PDPS1 line. The early Aβ deposits were found to be primarily composed of fibrillar Aβ and resembled compact amyloid plaques. As the mice aged, these fibrillar deposits did not increase substantially beyond the 12 months of age. Interestingly, the diffuse deposits appeared only until later, this is opposite to the general perception that in AD the compact deposits are formed by condensation of the diffuse material.

Studies on APPxPS1 double transgenics, therefore, revealed that marginal increases in Aβ42/43 levels, evoked by PS1 mutant transgene, can accelerate the deposition process by several months. As these mice have a severe Aβ pathology for an extended time relative to singly transgenic APP mice, other features of the disease such as tau abnormalities or major cell loss may become apparent as the animals age.

Double mutants produced from crossing JNPL3 transgenic mice expressing mutant P301L tau with Tg2576 mice expressing the APPsw mutation offered proof that Aβ influences the development of NFTs. The double mutant exhibited NFT pathology that was substantially enhanced in the limbic system and the olfactory cortex. These results suggest that APP or Aβ augments the formation of NFTs in the regions of the brain vulnerable to the formation of these lesions. Recently, a triple transgenic mouse was generated expressing APPsw, PS1-M146L, and P301L tau mutant protein that exhibited both plaques and NFT. However, other characteristics of AD such as neurodegeneration and memory loss have yet to be examined.

Although most of the transgenic models for AD do not completely recreate the disease phenotype, they have shown considerable utility, both for studying the disease mechanisms and for the preliminary testing of therapeutic agents, particularly those that are designed to modulate Aβ deposition.

5 Potential Treatments

The ultimate goal of AD research is to prevent disease and/or develop treatments for this devastating disease. Unfortunately, there is still no effective treatment available. Currently, the only FDA-approved therapy for AD is cholinesterase inhibitors, which can enhance cholinergic activity and temporarily improve cognitive function in some individuals in the early stages of disease. However, these drugs only treat the symptoms but have no impact on progression of the disease.

Recent advances in dissecting the molecular and cellular mechanisms that are involved in AD pathogenesis have provided substantial knowledge for determining novel targets for drug development. Several new therapeutic approaches targeted at distinct aspects of the disease are currently being pursued. Potential therapies include: anti-inflammatory agents, cholesterol-lowering drugs, antioxidants, hormonal therapy, and approaches that inhibit Aβ production or increase Aβ degradation and clearance. Strategies targeted directly at Aβ may be most effective given the fact that Aβ deposition is the central and fundamental event that causes AD. Other approaches affecting downstream events in the neurodegenerative process are likely to be used as supplementary treatments.

5.1 Secretase Inhibitors

Two prime targets for drug development are β- and γ-secretases, the two enzymes that process APP to Aβ. Inhibitors to these two proteases are expected to have the greatest Aβ-lowering effect since inhibiting either enzyme can completely inhibit Aβ generation.

γ-secretase is considered to be central to AD pathogenesis because altering its activity invariantly favors Aβ42 production and causes AD. The major concern in developing a drug that targets the γ-secretase is its potential for side effects. It is clear that γ-secretase is important for several physiological functions through its role as the protease that mediates the intramembranous cleavage of a number of TM proteins, including Notch. One possible solution is to develop an inhibitor that selectively lowers Aβ (Aβ42 in particular) production without significantly affecting the cleavage of other γ-secretase substrates. The development of such inhibitors has been reported. Another possible approach is partial inhibition of γ-secretase activity, at inhibitor concentrations that reduce Aβ production without affecting Notch signaling. Indeed, it has been shown that compounds with no reported selectivity allow significant Aβ reduction without changing the Notch-related function. Some inhibitors, however, are reported to preferentially increase Aβ42 production when used at low concentrations.

Most evidence suggests that β-secretase (BACE1) may be a better therapeutic target than γ-secretase. BACE1 knockout mice do not generate Aβ, but are otherwise healthy and fertile without obvious deficits in the basal, neurological, and physiological functions. In addition, the X-ray structure of the BACE1 protease domain has been solved and should provide valuable knowledge for inhibitor design.

Although the knockout data are reassuring that the detrimental effects of inhibiting β-secretase may be minimal, concerns regarding the possible side effects remain. First, there are likely other β-secretase substrates since it is counterintuitive to assume that BACE1 has evolved just to generate Aβ. Indeed, one recent study suggests that β-secretase may be responsible for the cleavage and secretion of a Golgi resident sialyltransferase. It is very important to identify these substrates because they are valuable for predicting the possible side effects of BACE1 inhibitors. It will also be important to know the phenotype of BACE1/2 double knockout mice. Although BACE2 is not primarily involved in Aβ generation, it may have an important physiological role and inhibitors of BACE1 may also inhibit BACE2 because of the high degree of homology between these two proteins.

While inhibitors to both secretases may potentially be effective in AD treatment, the challenges facing drug development are significant. The inhibitors need to be highly selective for the target enzyme and target substrate to reduce possible side effects to a minimum. The inhibitors need to be able to efficiently cross the BBB. Potent peptide-based inhibitors to BACE1 have been developed. However, such large compounds are not viable drug candidates because they will not penetrate the BBB to a sufficient extent. Instead, the enzyme inhibitors with therapeutic potential are preferably small organic molecules with high specificity. The X-ray structure of BACE1 suggests that its active site is more open and less hydrophobic than that of other aspartyl proteases. It may pose challenges for the development of small-molecule inhibitors.

5.2 Vaccine Approaches

Another novel approach for Alzheimer therapy is Aβ immunization. Many studies have shown that direct immunization with Aβ peptide or anti-Aβ antibodies can greatly reduce plaque formation, neuritic dystrophy, and other AD-like pathology in APP transgenic mice. One study also shows that such an immunization strategy is also effective in reversing behavioral deficits in the transgenic mouse model. It appears that antibodies to Aβ can enhance Aβ clearance, prevent Aβ fibril formation, disrupt Aβ fibrils, as well as block Aβ toxicity.

Despite the great effects in mice, there are serious concerns about the safety of this approach when used in humans. One concern is that autoimmunity may occur in humans, although it has not been reported in mice. Another major concern is the toxicity of Aβ42, which can cross the BBB and seed fibril formation in the brain, and therefore may actually promote plaque formation. In addition, Aβ42 may cause inflammation and neurotoxicity. A phase II clinical trial using Aβ42 vaccination was eventually terminated because of cerebral inflammation observed in several patients. The first autopsy result from this trial was just published. It shows exactly the two sides of this approach. Vaccination has such a powerful effect that the patient's cerebral cortex was almost cleared of Aβ deposition, but the side effect of cerebral inflammation was so severe that the patient developed meningoencephalitis, which led to death.

Safer therapeutic approaches including using nonamyloidogenic/nontoxic Aβ derivatives as an immunogen or passive immunization with Aβ antibodies are being developed. One study shows that passive immunization with $(Fab')_2$ fragments of Aβ-specific antibody can clear Aβ in a mouse model. This is promising because $(Fab')_2$ fragments do not interact with Fc receptors and so will not activate the cellular-immune response which has been the major cause for the severe side effects.

5.3 Other Therapeutic Approaches

Several FAD-approved nonsteroidal anti-inflammatory drugs (NSAIDs), including ibuprofen, indomethacin, and sulindac have been shown to selectively reduce Aβ42 levels in cultured cells. In addition, treatment of APP transgenic mice with ibuprofen reduces Aβ42 levels in the brain and suppresses plaque pathology. It appears that NSAIDs can subtly alter γ-secretase activity so that cleavage is shifted from Aβ42 to Aβ38.

Many studies also indicate that cholesterol influences Aβ metabolism and cholesterol-reducing drugs may have a beneficial effect on AD. Individuals taking cholesterol-lowering drugs such as statins show greatly reduced risk for developing AD, whereas individuals with elevated cholesterol are at a higher risk. Statins and other cholesterol-lowering drugs have been shown to reduce Aβ levels and Aβ deposition in both cell culture systems and animal models. The role of cholesterol in Aβ metabolism appears to be quite complex and the mechanisms are not clear.

Nevertheless, treatment with statins and other cholesterol-reducing drugs may have a significant clinical benefit in the prevention of AD.

6 Summary

We have seen a dramatic increase in our knowledge of the underlying molecular mechanisms of AD pathology during the last ten years. Understanding the effects of causative mutations in *APP* and the presenilins and the genetic risk factor, *APOE4* has led to the conclusion that changes in Aβ metabolism and deposition are central to the disease process. These studies suggest that similar mechanisms underlie both FAD and sporadic AD. Transgenic models of AD now provide a new hope for the rapid development in rational treatments for AD.

See also Brain Development.

Bibliography

Books and Reviews

Duff, K. (2001) Transgenic mouse models of Alzheimer's disease: phenotype and mechanisms of pathogenesis, *Biochem. Soc. Symp.* **67**, 195–202.

Golde, T.E. (2003) Alzheimer disease therapy: Can the amyloid cascade be halted? *J. Clin. Invest.* **11**, 11–18.

Hardy, J., Selkoe, D.J. (2002) The amyloid hypothesis of Alzheimer's disease: progress and problems on the road to therapeutics, *Science* **297**, 353–356.

Myers, A., Goate, A.M. (2001) The genetics of late-onset Alzheimer's disease, *Curr. Opin. Neurol.* **14**, 433–440.

Richardson, J.A., Burns, D.K. (2002) Mouse models of Alzheimer's disease: a quest for plaques and tangles, *ILAR J.* **43**, 89–99.

Selkoe, D.J., Podlisny, M.B. (2002) Deciphering the genetic basis of Alzheimer's disease, *Annu. Rev. Genomics Hum. Genet.* **3**, 67–99.

Tandon, A., Fraser, P. (2002) The presenilins, *Genome Biol.* **3**, 3014.1–3014.9; reviews.

Vassar, R. (2002) β-Secretase (BACE) as a drug target for alzheimer's disease, *Adv. Drug Delivery Rev.* **54**, 1589–1602.

Primary Literature

Alzheimer, A. (1907) Über eine eigenartige Erkrankung der Hirnrinde, *Allg. Zeitsch. Psychiatrie Psychisch-gerichtliche Med.* **64**, 146–148.

Biernat, J., Gustke, N., Drewes, G., Mandelkow, E.M., Mandelkow, E. Phosphorylation of Ser262 strongly reduces binding of tau to microtubules: distinction between PHF-like immunoreactivity and microtubule binding, *Neuron* **11**, 153–163.

Blacker, D., Bertram, L., Saunders, A.J., Moscarillo, T.J., Albert, M.S., Wiener, H., Perry, R.T., Collins, J.S., Harrell, L.E., Go, R.C., Mahoney, A., Beaty, T., Fallin, M.D., Avramopoulos, D., Chase, G.A., Folstein, M.F., McInnis, M.G., Bassett, S.S., Doheny, K.J., Pugh, E.W., Tanzi, R.E. (2003) Results of a high-resolution genome screen of 437 Alzheimer's disease families, *Hum. Mol. Genet.* **12**, 23–32.

Borchelt, D.R., Ratovitski, T., van Lare, J., Lee, M.K., Gonzales, V., Jenkins, N.A., Copeland, N.G., Price, D.L., Sisodia, S.S. (1997) Accelerated amyloid deposition in the brains of transgenic mice coexpressing mutant presenilin 1 and amyloid precursor proteins, *Neuron* **19**, 939–945.

Borchelt, D.R., Thinakaran, G., Eckman, C.B., Lee, M.K., Davenport, F., Ratovitsky, T., Prada, C.M., Kim, G., Seekins, S., Yager, D., Slunt, H.H., Wang, R., Seeger, M., Levey, A.I., Gandy, S.E., Copeland, N.G., Jenkins, N.A., Price, D.L., Younkin, S.G., Sisodia, S.S. (1996) Familial Alzheimer's disease-linked presenilin 1 variants elevate Aβ1-42/1-40 ratio *in vitro* and *in vivo*, *Neuron* **17**, 1005–1013.

Braak, H., Braak, E. (1991) Neuropathological stageing of Alzheimer-related changes, *Acta Neuropathol. (Berl)* **82**, 239–259.

Brendza, R.P., Bales, K.R., Paul, S.M., Holtzman, D.M. (2002) Role of apoE/Aβ interactions in Alzheimer's disease: insights from transgenic mouse models, *Mol. Psychiatry* **7**, 132–135.

Cai, X.D., Golde, T.E., Younkin, S.G. (1993) Release of excess amyloid β protein from a mutant amyloid β protein precursor, *Science* **259**, 514–516.

Chung, H.M., Struhl, G. (2001) Nicastrin is required for Presenilin-mediated transmembrane cleavage in *Drosophila*, *Nat. Cell. Biol.* **3**, 1129–1132.

Citron, M., Oltersdorf, T., Haass, C., McConlogue, L., Hung, A.Y., Seubert, P., Vigo-Pelfrey, C., Lieberburg, I., Selkoe, D.J. (1992) Mutation of the β-amyloid precursor protein in familial Alzheimer's disease increases β-protein production, *Nature* **360**, 672–674.

Citron, M., Westaway, D., Xia, W., Carlson, G., Diehl, T., Levesque, G., Johnson-Wood, K., Lee, M., Seubert, P., Davis, A., Kholodenko, D., Motter, R., Sherrington, R., Perry, B., Yao, H., Strome, R., Lieberburg, I., Rommens, J., Kim, S., Schenk, D., Fraser, P., St. George Hyslop, P., Selkoe, D.J. (1997) Mutant presenilins of Alzheimer's disease increase production of 42-residue amyloid β-protein in both transfected cells and transgenic mice, *Nat. Med.* **3**, 67–72.

Corder, E.H., Saunders, A.M., Risch, N.J., Strittmatter, W.J., Schmechel, D.E., Gaskell, P.C. Jr., Rimmler, J.B., Locke, P.A., Conneally, P.M., Schmader, K.E., et al. (1994) Protective effect of apolipoprotein E type 2 allele for late onset Alzheimer disease, *Nat. Genet.* **7**, 180–184.

Corder, E.H., Saunders, A.M., Strittmatter, W.J., Schmechel, D.E., Gaskell, P.C., Small, G.W., Roses, A.D., Haines, J.L., Pericak-Vance, M.A. (1993) Gene dose of apolipoprotein E type 4 allele and the risk of Alzheimer's disease in late onset families, *Science* **261**, 921–923.

DeMattos, R.B., Bales, K.R., Cummins, D.J., Dodart, J.C., Paul, S.M., Holtzman, D.M. (2001) Peripheral anti-Aβ antibody alters CNS and plasma Aβ clearance and decreases brain Aβ burden in a mouse model of Alzheimer's disease, *Proc. Natl. Acad. Sci. U.S.A.* **98**, 8850–8855.

DeMattos, R.B., Bales, K.R., Cummins, D.J., Paul, S.M., Holtzman, D.M. (2002) Brain to plasma amyloid-β efflux: a measure of brain amyloid burden in a mouse model of Alzheimer's disease, *Science* **295**, 2264–2267.

De Strooper, B., Annaert, W., Cupers, P., Saftig, P., Craessaerts, K., Mumm, J.S., Schroeter, E.H., Schrijvers, V., Wolfe, M.S., Ray, W.J., Goate, A., Kopan, R.A. (1999) presenilin-1-dependent γ-secretase-like protease mediates release of Notch intracellular domain, *Nature* **398**, 518–522.

De Strooper, B., Saftig, P., Craessaerts, K., Vanderstichele, H., Guhde, G., Annaert, W., von Figura, K., Van Leuven, F. (1998) Deficiency of presenilin-1 inhibits the normal cleavage of amyloid precursor protein, *Nature* **391**, 387–390.

Duff, K., Eckman, C., Zehr, C., Yu, X., Prada, C.M., Perez-Tur, J., Hutton, M., Buee, L., Harigaya, Y., Yager, D., Morgan, D., Gordon, M.N., Holcomb, L., Refolo, L., Zenk, B., Hardy, J., Younkin, S. (1996) Increased amyloid-β42(43) in brains of mice expressing mutant presenilin 1, *Nature* **383**, 710–713.

Edbauer, D., Winkler, E., Haass, C., Steiner, H. (2002) Presenilin and nicastrin regulate each other and determine amyloid β-peptide production via complex formation, *Proc. Natl. Acad. Sci. U.S.A.* **99**, 8666–8671.

Ertekin-Taner, N., Graff-Radford, N., Younkin, L.H., Eckman, C., Baker, M., Adamson, J., Ronald, J., Blangero, J., Hutton, M., Younkin, S.G. (2000) Linkage of plasma Aβ42 to a quantitative locus on chromosome 10 in late-onset Alzheimer's disease pedigrees, *Science* **290**, 2303–2304.

Esch, F.S., Keim, P.S., Beattie, E.C., Blacher, R.W., Culwell, A.R., Oltersdorf, T., McClure, D., Ward, P.J. (1990) Cleavage of amyloid β peptide during constitutive processing of its precursor, *Science* **248**, 1122–1124.

Estus, S., Golde, T.E., Younkin, S.G. (1992) Normal processing of the Alzheimer's disease amyloid β protein precursor generates potentially amyloidogenic carboxyl-terminal derivatives, *Ann. N. Y. Acad. Sci.* **674**, 138–148.

Fagan, A.M., Watson, M., Parsadanian, M., Bales, K.R., Paul, S.M., Holtzman, D.M. (2002) Human and murine ApoE markedly alters A β metabolism before and after plaque formation in a mouse model of Alzheimer's disease, *Neurobiol. Dis.* **9**, 305–318.

Farrer, L.A., Bowirrat, A., Friedland, R.P., Waraska, K., Korczyn, A.D., Baldwin, C.T. (2003) Identification of multiple loci for Alzheimer disease in a consanguineous Israeli-Arab community, *Hum. Mol. Genet.* **12**, 415–422.

Flood, D.G., Reaume, A.G., Dorfman, K.S., Lin, Y.G., Lang, D.M., Trusko, S.P., Savage, M.J., Annaert, W.G., De Strooper, B., Siman, R.,

Scott, R.W. (2002) FAD mutant PS-1 gene-targeted mice: increased Aβ42 and Aβ deposition without APP overproduction, *Neurobiol. Aging* **23**, 335–348.

Francis, R., McGrath, G., Zhang, J., Ruddy, D.A., Sym, M., Apfeld, J., Nicoll, M., Maxwell, M., Hai, B., Ellis, M.C., Parks, A.L., Xu, W., Li, J., Gurney, M., Myers, R.L., Himes, C.S., Hiebsch, R., Ruble, C., Nye, J.S., Curtis, D. (2002) aph-1 and pen-2 are required for Notch pathway signaling, γ-secretase cleavage of βAPP, and presenilin protein accumulation, *Dev. Cell.* **3**, 85–97.

Games, D., Adams, D., Alessandrini, R., Barbour, R., Berthelette, P., Blackwell, C., Carr, T., Clemens, J., Donaldson, T., Gillespie, F. (1995) Alzheimer-type neuropathology in transgenic mice overexpressing V717F β-amyloid precursor protein, *Nature* **373**, 523–527.

Glenner, G.G., Wong, C.W. (1984) Alzheimer's disease: initial report of the purification and characterization of a novel cerebrovascular amyloid protein, *Biochem. Biophys. Res. Commun.* **120**, 885–890.

Goate, A., Chartier-Harlin, M.C., Mullan, M., Brown, J., Crawford, F., Fidani, L., Giuffra, L., Haynes, A., Irving, N., James, L. (1991) Segregation of a missense mutation in the amyloid precursor protein gene with familial Alzheimer's disease, *Nature* **349**, 704–706.

Goedert, M., Wischik, C.M., Crowther, R.A., Walker, J.E., Klug, A. (1988) Cloning and sequencing of the cDNA encoding a core protein of the paired helical filament of Alzheimer disease: identification as the microtubule-associated protein tau, *Proc. Natl. Acad. Sci. U.S.A.* **85**, 4051–4055.

Goldgaber, D., Lerman, M.I., McBride, O.W., Saffiotti, U., Gajdusek, D.C. (1987) Characterization and chromosomal localization of a cDNA encoding brain amyloid of Alzheimer's disease, *Science* **235**, 877–880.

Gómez-Isla, T., Price, J.L., McKeel, D.W., Morris, J.C., Growdon, J.H., Hyman, B.T. (1996) Profound loss of layer II entorhinal cortex neurons occurs in very mild Alzheimer's disease, *J. Neurosci.* **16**, 4491–4500.

Gotz, J., Chen, F., van Dorpe, J., Nitsch, R.M. (2001) Formation of neurofibrillary tangles in P301l tau transgenic mice induced by Aβ42 fibrils, *Science* **293**, 1491–1495.

Goutte, C., Tsunozaki, M., Hale, V.A., Priess, J.R. (2002) APH-1 is a multipass membrane protein essential for the Notch signaling pathway in *Caenorhabditis elegans* embryos, *Proc. Natl. Acad. Sci. U.S.A.* **99**, 775–779.

Grundke-Iqbal, I., Iqbal, K., Quinlan, M., Tung, Y.C., Zaidi, M.S., Wisniewski, H.M. (1986) Microtubule-associated protein tau. A component of Alzheimer paired helical filaments, *J. Biol. Chem.* **261**, 6084–6089.

Haass, C., Koo, E.H., Mellon, A., Hung, A.Y., Selkoe, D.J. (1992) Targeting of cell-surface β-amyloid precursor protein to lysosomes: alternative processing into amyloid-bearing fragments, *Nature* **357**, 500–503.

Haass, C., Schlossmacher, M.G., Hung, A.Y., Vigo-Pelfrey, C., Mellon, A., Ostaszewski, B.L., Lieberburg, I., Koo, E.H., Schenk, D., Teplow, D.B. (1992) Amyloid β-peptide is produced by cultured cells during normal metabolism, *Nature* **359**, 322–325.

Holcomb, L., Gordon, M.N., McGowan, E., Yu, X., Benkovic, S., Jantzen, P., Wright, K., Saad, I., Mueller, R., Morgan, D., Sanders, S., Zehr, C., O'Campo, K., Hardy, J., Prada, C.M., Eckman, C., Younkin, S., Hsiao, K., Duff, K. (1998) Accelerated Alzheimer-type phenotype in transgenic mice carrying both mutant amyloid precursor protein and presenilin 1 transgenes, *Nat. Med.* **4**, 97–100.

Holtzman, D.M. (2002) Aβ conformational change is central to Alzheimer's disease, *Neurobiol. Aging* **23**, 1085–1088.

Hong, C.S., Caromile, L., Nomata, Y., Mori, H., Bredesen, D.E., Koo, E.H. (1999) Contrasting role of presenilin-1 and presenilin-2 in neuronal differentiation *in vitro*, *J. Neurosci.* **19**, 637–643.

Hsiao, K., Chapman, P., Nilsen, S., Eckman, C., Harigaya, Y., Younkin, S., Yang, F., Cole, G. (1996) Correlative memory deficits, Aβ elevation, and amyloid plaques in transgenic mice, *Science* **274**, 99–102.

Hu, Y., Ye, Y., Fortini, M.E. (2002) Nicastrin is required for γ-secretase cleavage of the *Drosophila* Notch receptor, *Dev. Cell.* **2**, 69–78.

Hutton, M., Lendon, C.L., Rizzu, P., Baker, M., Froelich, S., Houlden, H., Pickering-Brown, S., Chakraverty, S., Isaacs, A., Grover, A., Hackett, J., Adamson, J., Lincoln, S., Dickson, D., Davies, P., Petersen, R.C., Stevens, M., de Graaff, E., Wauters, E., van Baren, J., Hillebrand, M., Joosse, M., Kwon, J.M., Nowotny, P., Heutink, P. (1998) Association of missense and 5′-splice-site mutations in tau with the inherited dementia FTDP-17, *Nature* **393**, 702–705.

Iqbal, K., Zaidi, T., Thompson, C.H., Merz, P.A., Wisniewski, H.M. (1984) Alzheimer paired helical filaments: bulk isolation, solubility, and protein composition, *Acta Neuropathol. (Berl)* **62**, 167–177.

Kang, J., Lemaire, H.G., Unterbeck, A., Salbaum, J.M., Masters, C.L., Grzeschik, K.H., Multhaup, G., Beyreuther, K., Müller-Hill, B. (1987) The precursor of Alzheimer's disease amyloid A4 protein resembles a cell-surface receptor, *Nature* **325**, 733–736.

Leissring, M.A., Murphy, M.P., Mead, T.R., Akbari, Y., Sugarman, M.C., Jannatipour, M., Anliker, B., Muller, U., Saftig, P., De Strooper, B., Wolfe, M.S., Golde, T.E., LaFerla, F.M. (2002) A physiologic signaling role for the γ-secretase-derived intracellular fragment of APP, *Proc. Natl. Acad. Sci. U.S.A.* **99**, 4697–4702.

Levy, E., Carman, M.D., Fernandez-Madrid, I.J., Power, M.D., Lieberburg, I., van Duinen, S.G., Bots, G.T., Luyendijk, W., Frangione, B. (1990) Mutation of the Alzheimer's disease amyloid gene in hereditary cerebral hemorrhage, Dutch type, *Science* **248**, 1124–1126.

Levy-Lahad, E., Wasco, W., Poorkaj, P., Romano, D.M., Oshima, J., Pettingell, W.H., Yu, C.E., Jondro, P.D., Schmidt, S.D., Wang, K. (1995) Candidate gene for the chromosome 1 familial Alzheimer's disease locus, *Science* **269**, 973–977.

Lewis, J., Dickson, D.W., Lin, W.L., Chisholm, L., Corral, A., Jones, G., Yen, S.H., Sahara, N., Skipper, L., Yager, D., Eckman, C., Hardy, J., Hutton, M., McGowan, E. (2001) Enhanced neurofibrillary degeneration in transgenic mice expressing mutant tau and APP, *Science* **293**, 1487–1491.

Lewis, J., McGowan, E., Rockwood, J., Melrose, H., Nacharaju, P., Van Slegtenhorst, M., Gwinn-Hardy, K., Paul Murphy, M., Baker, M., Yu, X., Duff, K., Hardy, J., Corral, A., Lin, W.L., Yen, S.H., Dickson, D.W., Davies, P., Hutton, M. (2000) Neurofibrillary tangles, amyotrophy and progressive motor disturbance in mice expressing mutant (P301L) tau protein, *Nat. Genet.* **25**, 402–405.

Lopez-Schier, H., St. Johnston, D. (2002) *Drosophila* nicastrin is essential for the intramembranous cleavage of notch, *Dev. Cell.* **2**, 79–89.

Masters, C.L., Simms, G., Weinman, N.A., Multhaup, G., McDonald, B.L., Beyreuther, K. (1985) Amyloid plaque core protein in Alzheimer disease and Down syndrome, *Proc. Natl. Acad. Sci. U.S.A.* **82**, 4245–4249.

Mayeux, R., Lee, J.H., Romas, S.N., Mayo, D., Santana, V., Williamson, J., Ciappa, A., Rondon, H.Z., Estevez, P., Lantigua, R., Medrano, M., Torres, M., Stern, Y., Tycko, B., Knowles, J.A. (2002) Chromosome-12 mapping of late-onset Alzheimer disease among Caribbean Hispanics, *Am. J. Hum. Genet.* **70**, 237–243.

Murrell, J., Farlow, M., Ghetti, B., Benson, M.D. (1991) A mutation in the amyloid precursor protein associated with hereditary Alzheimer's disease, *Science* **254**, 97–99.

Nicoll, J.A., Wilkinson, D., Holmes, C., Steart, P., Markham, H., Weller, R.O. (2003) Neuropathology of human Alzheimer disease after immunization with amyloid-β peptide: a case report, *Nat. Med.* **9**, 448–452.

Nilsberth, C., Westlind-Danielsson, A., Eckman, C.B., Condron, M.M., Axelman, K., Forsell, C., Stenh, C., Luthman, J., Teplow, D.B., Younkin, S.G., Naslund, J., Lannfelt, L. (2001) The 'Arctic' APP mutation (E693G) causes Alzheimer's disease by enhanced Aβ protofibril formation, *Nat. Neurosci.* **4**, 887–893.

Price, J.L., Davis, P.B., Morris, J.C., White, D.L. The distribution of tangles, plaques and related immunohistochemical markers in healthy aging and Alzheimer's disease, *Neurobiol. Aging* **12**, 295–312.

Rogaev, E.I., Sherrington, R., Rogaeva, E.A., Levesque, G., Ikeda, M., Liang, Y., Chi, H., Lin, C., Holman, K., Tsuda, T. (1995) Familial Alzheimer's disease in kindreds with missense mutations in a gene on chromosome 1 related to the Alzheimer's disease type 3 gene, *Nature* **376**, 775–778.

St. George-Hyslop, P.H., Tanzi, R.E., Polinsky, R.J., Haines, J.L., Nee, L., Watkins, P.C., Myers, R.H., Feldman, R.G., Pollen, D., Drachman, D. (1987) The genetic defect causing familial Alzheimer's disease maps on chromosome 21, *Science* **235**, 885–890.

Schenk, D., Barbour, R., Dunn, W., Gordon, G., Grajeda, H., Guido, T., Hu, K., Huang, J., Johnson-Wood, K., Khan, K., Kholodenko, D., Lee, M., Liao, Z., Lieberburg, I., Motter, R., Mutter, L., Soriano, F., Shopp, G., Vasquez, N., Vandevert, C., Walker, S., Wogulis, M., Yednock, T., Games, D., Seubert, P. (1999) Immunization with amyloid-β attenuates

Alzheimer-disease-like pathology in the PDAPP mouse, *Nature* **400**, 173–177.

Scheuner, D., Eckman, C., Jensen, M., Song, X., Citron, M., Suzuki, N., Bird, T.D., Hardy, J., Hutton, M., Kukull, W., Larson, E., Levy-Lahad, E., Viitanen, M., Peskind, E., Poorkaj, P., Schellenberg, G., Tanzi, R., Wasco, W., Lannfelt, L., Selkoe, D., Younkin, S. (1996) Secreted amyloid β-protein similar to that in the senile plaques of Alzheimer's disease is increased *in vitro* by the presenilin 1 and 2 and APP mutations linked to familial Alzheimer's disease, *Nat. Med.* **2**, 864–870.

Seubert, P., Oltersdorf, T., Lee, M.G., Barbour, R., Blomquist, C., Davis, D.L., Bryant, K., Fritz, L.C., Galasko, D., Thal, L.J. (1993) Secretion of β-amyloid precursor protein cleaved at the amino terminus of the β-amyloid peptide, *Nature* **361**, 260–263.

Seubert, P., Vigo-Pelfrey, C., Esch, F., Lee, M., Dovey, H., Davis, D., Sinha, S., Schlossmacher, M., Whaley, J., Swindlehurst, C. (1992) Isolation and quantification of soluble Alzheimer's β-peptide from biological fluids, *Nature* **359**, 325–327.

Sherrington, R., Rogaev, E.I., Liang, Y., Rogaeva, E.A., Levesque, G., Ikeda, M., Chi, H., Lin, C., Li, G., Holman, K. (1995) Cloning of a gene bearing missense mutations in early-onset familial Alzheimer's disease, *Nature* **375**, 754–760.

Shoji, M., Golde, T.E., Ghiso, J., Cheung, T.T., Estus, S., Shaffer, L.M., Cai, X.D., McKay, D.M., Tintner, R., Frangione, B. (1992) Production of the Alzheimer amyloid β protein by normal proteolytic processing, *Science* **258**, 126–129.

Sinha, S., Anderson, J.P., Barbour, R., Basi, G.S., Caccavello, R., Davis, D., Doan, M., Dovey, H.F., Frigon, N., Hong, J., Jacobson-Croak, K., Jewett, N., Keim, P., Knops, J., Lieberburg, I., Power, M., Tan, H., Tatsuno, G., Tung, J., Schenk, D., Seubert, P., Suomensaari, S.M., Wang, S., Walker, D., John, V. (1999) Purification and cloning of amyloid precursor protein β-secretase from human brain, *Nature* **402**, 537–540.

Skovronsky, D.M., Zhang, B., Kung, M.-P., Kung, H.F., Trojanowski, J.Q., Lee, V.M. (2000) *In vitro* detection of amyloid plaques in a mouse model of Alzheimer's disease, *Proc. Natl. Acad. Sci. U.S.A.* **97**, 7609–7614.

Stenh, C., Nilsberth, C., Hammarback, J., Engvall, B., Naslund, J., Lannfelt, L. (2002) The Arctic mutation interferes with processing of the amyloid precursor protein, *Neuro Report* **13**, 1857–1860.

Strittmatter, W.J., Saunders, A.M., Schmechel, D., Pericak-Vance, M., Enghild, J., Salvesen, G.S., Roses, A.D. (1993) Apolipoprotein E: high-avidity binding to β-amyloid and increased frequency of type 4 allele in late-onset familial Alzheimer disease, *Proc. Natl. Acad. Sci. U.S.A.* **90**, 1977–1981.

Sturchler-Pierrat, C., Abramowski, D., Duke, M., Wiederhold, K.H., Mistl, C., Rothacher, S., Ledermann, B., Burki, K., Frey, P., Paganetti, P.A., Waridel, C., Calhoun, M.E., Jucker, M., Probst, A., Staufenbiel, M., Sommer, B. (1997) Two amyloid precursor protein transgenic mouse models with Alzheimer disease-like pathology, *Proc. Natl. Acad. Sci. U.S.A.* **94**, 13287–13292.

Sturchler-Pierrat, C., Staufenbiel, M. (2000) Pathogenic mechanisms of Alzheimer's disease analyzed in the APP23 transgenic mouse model, *Ann. N. Y. Acad. Sci.* **920**, 134–139.

Suzuki, N., Cheung, T.T., Cai, X.D., Odaka, A., Otvos, L., Eckman, C., Golde, T.E., Younkin, S.G. (1994) An increased percentage of long amyloid β protein secreted by familial amyloid β protein precursor (β APP717) mutants, *Science* **264**, 1336–1340.

Tanzi, R.E., Gusella, J.F., Watkins, P.C., Bruns, G.A., St. George-Hyslop, P., Van Keuren, M.L., Patterson, D., Pagan, S., Kurnit, D.M., Neve, R.L. (1987) Amyloid β protein gene: cDNA, mRNA distribution, and genetic linkage near the Alzheimer locus, *Science* **235**, 880–884.

Terry, R.D., Masliah, E., Salmon, D.P., Butters, N., DeTeresa, R., Hill, R., Hansen, L.A., Katzman, R. (1991) Physical basis of cognitive alterations in Alzheimer's disease: synapse loss is the major correlate of cognitive impairment, *Ann. Neurol.* **30**, 572–580.

Thinakaran, G., Borchelt, D.R., Lee, M.K., Slunt, H.H., Spitzer, L., Kim, G., Ratovitsky, T., Davenport, F., Nordstedt, C., Seeger, M., Hardy, J., Levey, A.I., Gandy, S.E., Jenkins, N.A., Copeland, N.G., Price, D.L., Sisodia, S.S. (1996) Endoproteolysis of presenilin 1 and accumulation of processed derivatives *in vitro*, *Neuron* **17**, 181–190.

Thinakaran, G., Harris, C.L., Ratovitski, T., Davenport, F., Slunt, H.H., Price, D.L., Borchelt, D.R., Sisodia, S.S. (1997) Evidence that levels of presenilins (PS1 and PS2) are coordinately regulated by competition for limiting cellular factors, *J. Biol. Chem.* **272**, 28415–28422.

Van Nostrand, W.E., Schmaier, A.H., Farrow, J.S., Cunningham, D.D. (1990) Protease nexin-II (amyloid β-protein precursor): a platelet α-granule protein, *Science* **248**, 745–748.

Vassar, R., Bennett, B.D., Babu-Khan, S., Kahn, S., Mendiaz, E.A., Denis, P., Teplow, D.B., Ross, S., Amarante, P., Loeloff, R., Luo, Y., Fisher, S., Fuller, J., Edenson, S., Lile, J., Jarosinski, M.A., Biere, A.L., Curran, E., Burgess, T., Louis, J.C., Collins, F., Treanor, J., Rogers, G., Citron, M. (1999) B-secretase cleavage of Alzheimer's amyloid precursor protein by the transmembrane aspartic protease BAC, *Science* **286**, 735–741.

Weggen, S., Eriksen, J.L., Das, P., Sagi, S.A., Wang, R., Pietrzik, C.U., Findlay, K.A., Smith, T.E., Murphy, M.P., Bulter, T., Kang, D.E., Marquez-Sterling, N., Golde, T.E., Koo, E.H. (2001) A subset of NSAIDs lower amyloidogenic Aβ42 independently of cyclooxygenase activity, *Nature* **414**, 212–216.

Weihofen, A., Binns, K., Lemberg, M.K., Ashman, K., Martoglio, B. (2002) Identification of signal peptide peptidase, a presenilin-type aspartic protease, *Science* **296**, 2215–2218.

Wolfe, M.S., Xia, W., Ostaszewski, B.L., Diehl, T.S., Kimberly, W.T., Selkoe, D.J. (1999) Two transmembrane aspartates in presenilin-1 required for presenilin endoproteolysis and γ-secretase activity, *Nature* **398**, 513–517.

Yu, G., Nishimura, M., Arawaka, S., Levitan, D., Zhang, L., Tandon, A., Song, Y.Q., Rogaeva, E., Chen, F., Kawarai, T., Supala, A., Levesque, L., Yu, H., Yang, D.S., Holmes, E., Milman, P., Liang, Y., Zhang, D.M., Xu, D.H., Sato, C., Rogaev, E., Smith, M., Janus, C., Zhang, Y., Aebersold, R., Farrer, L.S., Sorbi, S., Bruni, A., Fraser, P., St. George-Hyslop, P. (2000) Nicastrin modulates presenilin-mediated notch/glp-1 signal transduction and βAPP processing, *Nature* **407**, 48–54.

16
Molecular Genetics of Down Syndrome

Charles J. Epstein
University of California, San Francisco, CA, USA

Keywords

Acrocentric
A type of chromosome in which the centromere is very close to one end.

Autosomes
All of the chromosomes in the genome except for the sex chromosomes (X and Y).

Neurobiology. From Molecular Basis to Disease. Edited by Robert A. Meyers.

ISBN: 978-3-527-32293-0

Centromere
The structure within each chromosome at which the fibers required to move the chromosomes during meiosis or cell division (mitosis) attach.

Meiosis
The process within the germ cells during which genetic recombination occurs, and in which the number of chromosomes within the egg or sperm is reduced from the 46 found in somatic cells to 23.

Trisomy
The presence in the genome of three rather than two copies of a specific chromosome.

Down syndrome (DS) is the commonest of the genetically caused forms of mental retardation. It occurs with a frequency of approximately 1 per 800 to 1 per 1000 live births, and its incidence increases with increasing maternal age. DS is caused by the presence of an extra chromosome 21 within the genome, which, in turn, results in a 50% increase in the expression of the genes contained on the chromosome. By mechanisms currently undefined, the increased expression of several genes on human chromosome 21 results in a syndrome characterized by mental and growth retardation, a distinctive set of major and minor congenital malformations, a variety of cellular abnormalities, and, later in life, by the development of Alzheimer disease.

1 Phenotype of Down Syndrome

The most immediately apparent, if not the most serious, manifestations of Down syndrome (DS) are the many minor dysmorphic features that collectively constitute its distinctive physical phenotype. Salient among these are upslanting palpebral fissures, epicanthic folds, flat nasal bridge, brachycephaly, short broad hands, incurved fifth fingers, loose skin of the nape of the neck, open mouth with protruding tongue, and transverse palmar creases. Although any individual with DS will have many of the characteristic features and can be easily recognized as having the disorder, none of these features is present in all persons with DS. DS affects the nervous system in three principal ways: hypotonia, which occurs in virtually all newborns and infants; delayed psychomotor development in infancy and mental retardation throughout life; and neuronal degeneration during the adult years. The latter process, which is pathologically identical with Alzheimer disease (presenile and/or senile dementia), results in significant pathologic changes in the brain and may further compromise the already impaired mental functioning. DS is associated with two types of major congenital malformations. Most frequent (about 40%) is congenital heart disease, usually of the endocardial cushion type or one of its variants. Gastrointestinal tract abnormalities occur in about 4.5% of individuals, more than half of whom have duodenal stenosis or atresia. Structural abnormalities of the thymus and functional defects in T-cell function leading to an increased susceptibility to infection are present, and there is a 10- to 18-times normal incidence of childhood leukemia with a frequent occurrence of acute megakaryoblastic leukemia.

2 Cytogenetics

Down syndrome is the phenotypic manifestation of trisomy 21. As such, it occurs when a third copy of chromosome 21 is present in the genome, either as a free chromosome or as part of a Robertsonian fusion chromosome (in which the long arms of two acrocentric chromosomes are joined at the centromeres). Except in 2 to 4% of cases in which mosaicism, with two populations of cells, one diploid and one trisomic, exists, all cells of the body are trisomic. Although these cytogenetic abnormalities involve most or all of chromosome 21, there are rare cases involving translocations in which only part of the long arm of the chromosome is triplicated. Depending on the region of the chromosome that is involved, such cases may or may not express the classical DS phenotype.

It has long been recognized that the risk of having a child with DS increases with maternal age and that the distribution of maternal age in the population of women having children is the primary determinant of the overall incidence of DS. A variety of estimates of the incidence of DS in the newborn population have been made, and most recent figures are in the vicinity of 1 per 1000. The figures can be broken down to provide maternal age–specific rates per 1000 live births: 0.6 at 20 years, 1.0 at 30 years, 2.6 at 35 years, 9.1 at 40 years, 24.9 at 44 years, and 41.2 at 46 years. Analyses using DNA markers have shown that maternal nondisjunction, the failure of paired chromosomes to separate properly, accounts for 92% of all cases of trisomy 21, 65% occurring at meiosis I, 23% at meiosis II, and 3% postzygotically (at mitosis). Eight percent of cases are paternal in origin, 3% at meiosis I and 5% at meiosis II. A reduced rate of recombination is associated with meiosis I errors and an increased rate with meiosis II errors.

3 Structure of Chromosome 21

Chromosome 21 is an acrocentric chromosome with a genetic length estimated to be 46 cM. It is the smallest of the human autosomes constituting approximately 1.7% of the length of the haploid genome. In physical terms, chromosome 21 is an acrocentric chromosome with the centromere very close to one end and with a very small short arm. The short arm (21p) terminates in a satellite region that may vary in size. Proximal to the satellite is the stalk (secondary constriction) that, as the nucleolar organizer region (NOR), contains multiple copies of the ribosomal RNA genes (*RNR4*) and stains characteristically with silver. The degree of silver staining appears to be a representation of the molecular activity of the ribosomal RNA genes that the chromosome contains. The region of 21p adjacent to the centromere contains highly repeated DNA sequences that consist of the satellite (including alphoid) and the "724" families of sequences. None of these gene families is unique to chromosome 21. It is believed that these families of repeated gene sequences may be involved in the juxtaposition or association of the satellite regions (satellite association) of the acrocentric chromosomes during mitotic metaphase and with the formation of the nucleolus during interphase.

The major part of chromosome 21 is the long arm (21q) that has a characteristic banding pattern consisting of 3 or 4 bands at low resolution and as many as 11 dark and light bands resolvable by prometaphase banding. With one possible exception, all genes of known function (other than those for ribosomal RNA) are located on this arm of chromosome 21, and only this arm is essential for normal development and function. The presence of a Robertsonian fusion chromosome in which the short arms of two acrocentric chromosomes (sometimes both chromosomes 21) are deleted does not cause detectable abnormalities if the genome is otherwise balanced.

The sequence of the long arm of chromosome 21 was completed in 2000 with an estimated 99.7% coverage. This arm has a length of 33.7 Mbp and is relatively gene-poor, with less than half the number of identified genes found on the similarly sized chromosome 22. In the original annotation, it was judged to contain approximately 127 known genes, 98 predicted genes (of which 69% have no similarity to known proteins), and 59 pseudogenes. These numbers have been repeatedly revised as annotation of the sequence has proceeded, and the most recent figures are 178 confirmed and 36 predicted genes. There is evidence that not all expressed genes have been identified and that some predicted genes do not exist. Among the known genes are at least 10 kinases, 5 genes in the ubiquitination pathway, 5 cell adhesion molecules, 7 ion channels, 5 members of the interferon receptor family, and several transcription factors. About 22.4% of the chromosome consists of interspersed Alu sequences and LINE1 elements.

4
Phenotypic Mapping

Cases in which only part of chromosome 21 is triplicated have been intensively studied to arrive at a phenotypic map that will permit a correlation between particular phenotypic features of DS and specific regions or loci on the chromosome. The consensus of studies carried out before molecular markers became available is that the full DS phenotype, as manifested by mental retardation, congenital heart disease, characteristic facial appearance, hand anomalies, and dermatoglyphic changes, appears when band 21q22 is duplicated, and of this, subbands 21q22.1 and probably 21q22.2 are required. Molecular analysis has been used to define the extent of the triplication of regions of chromosome 21, and the results have been used to generate phenotypic maps. Although interpretations vary, it seems clear that many of the phenotypic features appear to be associated with imbalance in the region surrounding and distal to *D21S55*. However, there is also evidence for contributions of genes outside the *D21S55* region, especially in the proximal part of 21q, although the region responsible for congenital heart disease seems to be confined to the distal part of 21q. The genes that, when present in extra copies, contribute to impaired cognition appear to be located in several regions of the chromosome.

5 Pathogenesis

The immediate consequence of an aneuploid state is a gene dosage effect for each of the loci present on the unbalanced chromosome or chromosome segment. Such gene dosage effects have been reported for several chromosome 21 loci. With the exception of the amyloid precursor protein (APP) in the brain from fetuses with DS, and possibly a few other chromosome 21–encoded genes, the measured increase in activities or concentrations in trisomic cells is close to the theoretically expected value of 1.5. Taken in the aggregate, these results confirm the existence of quite precise dosage effects in cells aneuploid for chromosome 21. However, changes in the synthesis of proteins coded for by genes elsewhere in the genome have been observed, presumably as the result of genomic dysregulation resulting from the increased expression of chromosome w31 genes. However, this dysregulation is limited and certainly does not affect the vast majority of expressed genes.

On the basis of studies in animal models (see the following section), decreased platelet serotonin uptake and prostaglandin synthesis, and abnormal myoneural junctions, all of which occur in DS, have been ascribed to increased activity of CuZn-superoxide dismutase (SOD1). With these exceptions, it has not been possible to relate any component of the DS phenotype to overexpression of specific loci. However, a great amount of attention has been devoted to the pathogenesis of the cognitive deficits and Alzheimer disease in DS. The effects of increased expression of several of the loci on chromosome 21 have been considered, and it has been possible to construct plausible explanations for how overexpression of each locus might affect the development, function, and integrity of the central nervous system. A recent case in point is the gene for cystathionine synthase, overexpression of which in DS results in depressed levels of plasma homocysteine. It has been suggested that the altered homocysteine metabolism could compromise the folate-dependent resynthesis of methionine and thereby create a functional folate deficiency, the consequences of which on the central nervous system function could be many. However, in considering such possibilities, it is important to keep in mind that whereas imbalance of an individual locus could well have significant effects on the central nervous system, it is unlikely that it will provide a unitary explanation for the entire range of deficits found in DS. Nevertheless, it is possible that the development of Alzheimer disease in DS is related to the overproduction of APP, perhaps superimposed on an intrinsically defective nervous system.

6 Animal Models

The fact that many of the consequences of aneuploidy in humans arise during the period of morphogenesis and affect the development and function of the central nervous system places special stumbling blocks in the way of their investigation. Research on events occurring during gestation, especially early gestation, is both technically impractical and, currently and for the foreseeable future, ethically and legally impossible. Similarly, there are severe limitations to the study of the central nervous system function in living individuals, and experimental genetic alterations are impossible. It is for this reason that interest has turned to the development of animal models for DS.

Since ultimate concern is with humans and with human disease, we require models that will duplicate the human condition – in developmental and functional terms – as closely as possible. No model can be an exact one, since no other organism duplicates the human with respect to all of his or her biologic and genetic attributes. Nevertheless, models based on other mammals, which share numerous biologic similarities with humans, seem most appropriate. Primates with the genetic and clinical equivalent of trisomy 21 and DS have been observed but do not offer significant advantages over humans for investigational purposes. Therefore, the mouse has been the model animal of choice for several reasons, including ease of manipulation and genetic control and the similarities to humans in the processes of morphogenesis and probably of central nervous system function, in neurobiological if not psychological terms. Furthermore, despite considerable rearrangement of the mammalian genome, sizable chromosomal regions carrying many genes remain intact and structurally similar in both humans and mice.

Models based on the mouse are not without their problems. It may not always be easy or even possible to obtain postnatally viable animals when sizable regions of the genome are unbalanced. Furthermore, even if it were possible, these animals will never be able to duplicate the higher central nervous system functions, such as cognition, which are so vulnerable to the effects of aneuploidy in humans. As difficult as it is to define mental retardation in an aneuploid human, it is even more difficult to specify the proper functional homology in an aneuploid mouse. Nevertheless, elementary forms of learning are common to all animals with an evolved nervous system, and there appears to be a conservation of the relevant biochemical mechanisms. Therefore, the mouse can be considered to be a legitimate model for learning and memory in humans.

6.1 Trisomic Mice

The modeling of human trisomy 21 in the mouse started at the two ends of the possible spectrum of such models. On one end is the trisomic mouse with an extra copy of mouse chromosome 16, the mouse chromosome that most closely resembles human chromosome 21. At the other extreme is the transgenic mouse with an extra copy or copies of a single human chromosome 21 gene. More recently, however, several intermediate possibilities, with duplications of only parts of mouse chromosome 16 or the insertion of mouse or human yeast artificial chromosome (YAC) transgenes, have been developed.

With perhaps one exception, all the loci in human 21q from the centromere to the *MX1* locus in the distal part of band 21q22.3, approximately 130 to 150 in number, map to the distal part of mouse chromosome 16 (see http://www.informatics.jax.org). The remaining loci, not on mouse chromosome 16, have been mapped to mouse chromosomes 10 and 17. These loci are outside the region shown by phenotypic mapping to be responsible for the facial features of DS. In addition to coding regions, there are also a large number of sequences conserved between the human and mouse genomes that appear to have a functional significance but cannot be identified as protein-coding or RNA genes.

Because of the genetic discrepancies between mouse chromosome 16 and human chromosome 21, trisomy for all of mouse chromosome16, while reproducing much of the genetic imbalance associated with human trisomy 21, results in an overall degree of imbalance that is more extensive than in human trisomy and in an overall phenotype that is more severe. For this reason, the results of studies with mouse trisomy 16 (Ts16) must be viewed with concern insofar as modeling DS is concerned. To correct this situation, segmentally trisomic mice (designated Ts65Dn) in which only the human chromosome 21 orthologous region is triplicated were produced. In these animals, approximately 115 of the original number of 225 genes on human chromosome 21q are represented. The trisomic animals do not have the major anomalies present in Ts16. Rather, their salient phenotypic abnormalities are confined to the craniofacial structures and the central nervous system, although they do exhibit male sterility. As in persons with DS, there is a reduction in the size and granule cell density of the cerebellum, a reduced sensitivity to pain, elevation of brain *myo*-inositol, and abnormalities of the dendritic spines. Although motor learning is normal, Ts65Dn mice have major deficits in learning and behavior, and show abnormalities in long-term potentiation (LTP) and depression and in hippocampal cAMP generation. The pattern of behavioral and learning deficits suggests impairment of both hippocampal spatial learning and of prefrontal functions, and the LTP and cAMP abnormalities are also indicative of hippocampal dysfunction. Further evidence for abnormality of the hippocampus of the Ts65Dn mouse is provided by the presence of giant dendritic spines with aberrant connections in hippocampal neurons and by suppression of LTP by inhibitory inputs. The finding in Ts65Dn mice of an age-dependent loss of function and atrophy of cholinergic neurons in the medial septum of the basal forebrain, which correlates with the loss of cognitive function and appears to be related to an abnormality of nerve growth factor transport, is also consistent with hippocampal dysfunction. It is of note, in this regard, that a reduction in Trk-A immunoreactive cholinergic neurons has been observed in the nucleus basalis of persons with DS.

A second segmentally trisomic mouse, Ts1Cje, is functionally trisomic for the region of mouse chromosome 16 distal to *Sod1*. This trisomy involves about 87 chromosome 21–orthologous genes and still includes the region responsible for the major phenotypic abnormalities in DS. Their learning deficits are similar to but somewhat less severe than those of the Ts65Dn animals. Similarly, there is a reduction in the volume of the cerebellum, but granule cell density is much less affected. However, Ts1Cje animals do not appear to exhibit degeneration of the cholinergic neurons, suggesting that the gene(s) responsible for this phenotype are located in the region of difference between the two segmental trisomies, a region containing about 30 chromosome 21–orthologous genes. A third segmentally trisomic mouse, designated Ms1Ts65, carries a triplication of this region of difference. These animals have only minimal learning deficits, but they are still not normal.

Taken together with the comparison of Ts65Dn and Ts1Cje, the results with Ms1Ts65 indicate that the major genes responsible for the behavioral and learning abnormalities of Ts65Dn mice are located distal to *Sod1*. These studies constitute the first step in phenotypic mapping of the segmental-trisomy DS models by a subtractive approach in which a known phenotype (as in Ts65Dn) is compared with the phenotype obtained after reduction of the region of trisomy (as in Ts1Cje). The same approach has been used to compare the alterations in craniofacial morphogenesis produced by Ts65Dn and Ts1Cje. Both trisomics demonstrate alterations in the formation of the craniofacial skeleton that are similar to what is observed in DS, with the changes in Ts1Cje being largely the same as those in Ts65Dn. This again demonstrates that the major genes responsible for these skeletal abnormalities in the Ts65Dn mice are located distal to *Sod1*. The ability to generate these various segmental trisomics and to compare their phenotypes directly provides a precedent for the systematic creation of smaller and smaller segmental trisomics and the dissection of the trisomic phenotype.

6.2 Transgenic Mice

In contrast with the formation of trisomic animals, an approach that permits the analysis of the effects of increased dosage of individual human chromosome 21 genes is the construction of transgenic mice. Several strains of such transgenic mice carrying *SOD1*, the gene for human CuZnSOD, have been made and shown to have from 1.6- to 6-fold increased activities of CuZnSOD in the brain and other tissues. Several effects of this increased activity of CuZnSOD that resemble changes found in persons with DS have been found. Transgenic mouse platelets have diminished levels of serotonin and decreased rates of serotonin accumulation that are attributed to a diminished pH gradient across the granule membrane. There is a reduction in prostaglandin E_2 and prostaglandin D_2 biosynthesis in transgenic primary fetal cells and in the cerebellum and hippocampus. The neuromuscular junctions of the transgenic tongue muscles are abnormal, with atrophy, degeneration, withdrawal and destruction of terminal axons, development of multiple small terminals, decreased ratio of terminal axon area to postsynaptic membrane, and hyperplastic secondary folds. Transgenic mice also have anatomical abnormalities of the thymus and premature thymic involution. Impairment of LTP has also been reported.

Many of the abnormalities in transgenic tissues just discussed are presumed to be mediated by alterations in the metabolism of oxygen free radicals resulting from the increased CuZnSOD activity (this enzyme mediates the first step in the ultimate conversion of superoxide anions to water).

Limited work has been done with other types of transgenic mice carrying individual human chromosome 21 genes or their murine homologues. Although they do not develop the full pathology of AD, mice expressing the human APP-751 isoform gene do have diffuse and amorphous extracellular deposits of β-amyloid that resemble those of early AD and manifest deficits in spatial learning. Transgenic mice overexpressing the Ets2 transcription factor develop craniofacial and other skeletal anomalies, with brachycephaly and short necks, thought to resemble anomalies found in DS and mouse Ts16. However, similar anomalies are not found in any of the segmental trisomics of chromosome 16, even though *Ets2* is present in three copies. Overexpression of *HMG14* results in abnormalities of the thymus and epithelial cysts, and mice transgenic for the mouse single-minded gene, *mSim2*, have impaired spatial learning.

The transgenes just described involve loci present on mouse chromosome 16 as well as human chromosome 21. Therefore, the absence of similar findings in trisomic mice involving the same regions is of concern. The following two loci, however, are represented on mouse chromosome 10, and the trisomy 16 models are, therefore, not relevant. Transgenic mice overexpressing the gene for the protein subunit S100β were reported to be hyperactive, to have learning impairments, and to manifest astrocytosis, axonal proliferation, and a significant loss of dendrites in the hippocampus. Mice overexpressing liver-type phosphofructokinase had elevated activity during fetal life but not during adult life.

Another approach to the generation of transgenic models for DS has been the insertion of genomic segments larger than individual genes in the form of yeast artificial chromosomes. A mouse produced in this manner had abnormalities of learning and behavior too, but less severe than that observed with the segmental trisomies, Ts1Cje and Ts65Dn. Since the major and possibly only gene in the integrated segment of human DNA was the human minibrain gene (*DYRK*), it was suggested that imbalance of this locus is responsible for the observed abnormalities. Support for this inference is provided by mice transgenic for the *Dyrk1A* gene. These animals display delayed neuromotor development and impairment in spatial learning with defective reference memory, findings considered indicative of hippocampal and prefrontal cortex dysfunction.

6.3 Gene Expression Analyses

The cloning and sequencing of both the mouse and human genomes has made it possible to undertake *in situ* hybridization studies of the expression of chromosome 21–orthologous genes in diploid mouse embryos, fetuses, and postnatal animals. Broad expression of the human chromosome 21 orthologues is present in early embryos and becomes more restricted as development proceeds. The number of genes expressed and their patterns of expression differ among tissues, but it is possible to identify several clusters of coexpressing genes. By correlating expression patterns with specific features of the DS phenotype, candidate genes have been proposed for congenital heart and intestinal defects.

See also Brain Development.

Bibliography

Books and Reviews

Antonarakis, S.E., Lyle, R., Chrast, R., Scott, H.S. (2001) Differential gene expression studies to explore the molecular pathophysiology of Down syndrome, *Brain Res. Rev.* **36**, 265–274.

Capone, G.T. (2001) Down syndrome: advances in molecular biology and the neurosciences, *Dev. Behav. Pediatr.* **22**, 40–59.

Dierssen, M., Fillat, C., Crnic, L., Arbonés, M., Flórez, J., Estivill, X. (2001) Murine models for Down syndrome, *Physiol. Behav.* **73**, 859–871.

Epstein, C.J. (1986) *The Consequences of Chromosome Imbalance. Principles, Mechanisms, and Models*, Cambridge University Press, New York.

Epstein, C.J. (Ed.) (1993) *The Phenotypic Mapping of Down Syndrome and Other Aneuploid Conditions*, Wiley-Liss, New York.

Epstein, C.J. (2001) Down Syndrome (trisomy 21), in: Scriver, C.R., Beaudet, A.L., Sly, W.S., Valle, D. (Eds.) *The Metabolic and Molecular Bases of Inherited Disease*, 8th edition, McGraw-Hill, New York, pp. 1123–1256.

Epstein, C.J., Hassold, T., Lott, I.T., Nadel, L., Patterson, D. (Eds.) (1995) *Etiology and Pathogenesis of Down Syndrome*, Wiley-Liss, New York.

Galdzicki, Z., Siarey, R., Pearce, R., Stoll, J., Rapoport, S.I. (2001) On the cause of mental retardation in Down syndrome: extrapolation from full and segmental trisomy 16 models, *Brain Res. Rev.* **35**, 115–145.

Hassold, T., Hunt, P. (2001) To err (meiotically) is human: the genesis of human aneuploidy, *Nat. Rev. Genet.* **2**, 280–291.

Pueschel, S.M., Rynders, J.E. (Eds.) (1982) *Down Syndrome. Advances in Biomedicine and the Behavioral Sciences*, Ware, Cambridge, MA.

Reeves, R.H., Baxter L.L., Richtsmeier, J.T. (2001) Too much of a good thing: mechanisms of gene action in Down syndrome, *Trends Genet.* **17**, 83–88.

Primary Literature

Altafaj, X., Dierssen, M., Baamonde, C., Marti, E., Visa, J., Guimerà, J., Oset, M., González, J.R., Flórez, J., Fillat, C., Estivill, X. (2001) Neurodevelopmental delay, motor abnormalities and cognitive deficits in transgenic mice overexpressing *Dyrk1A* (*minibrain*), a murine model of Down's syndrome, *Hum. Mol. Genet.* **10**, 1915–1923.

Chrast, R., Scott, H.S., Papasavvas, M.P., Rossier, C., Antonarakis, E.S., Barras, C., Davisson, M.T., Schmidt, C., Estivill, X., Dierssen, M., Pritchard, M., Antonarakis, S.E. (2000) The mouse brain transcriptome by SAGE: differences in gene expression between P30 brains of the partial trisomy 16 mouse model of Down syndrome (Ts65Dn) and normals, *Genome Res.* **10**, 2006–2021.

Cooper, J., Salehi, A., Delcroix, J.-D., Howe, C.L., Belichenko, P.V., Chua-Couzens, J., Kilbridge, J.K., Carlson, E.J., Epstein, C.J., Mobley, W.C. (2001) Failed retrograde transport of NGF in a mouse model of Down's syndrome: reversal of cholinergic neurodegenerative phenotypes following NGF infusion, *Proc. Natl. Acad. Sci. U.S.A.* **98**, 10439–10444.

Dermitzakis, E.T., Reymond, A., Lyle, R., Scamuffa, N., Ucla, C., Deutsch, S., Stevenson, B.J., Flegel, V., Bucher, P., Jongeneel, C.V., Antonarakis, S.E. (2002) Human chromosome 21 gene expression atlas in the mouse, *Nature* **420**, 578–582.

Ema, M., Ikegami, S., Hosoya, T., Mimura, J., Ohtani, H., Nakao, K., Inokuchi, K., Katsuki. M., Fujii-Kuriyama, Y. (1999) Mild impairment of learning and memory in mice overexpressing the mSim2 gene located on chromosome 16: an animal model of Down's syndrome, *Hum. Mol. Genet.* **8**, 1409–1415.

Engidawork, E., Lubec, G. (2001) Protein expression in Down syndrome brain, *Amino Acids* **21**, 331–361.

Gitton, Y., Dahmanane, N., Balk, S., Ruiz, I., Altaba, A., Neidhardt, L., Scholtze, M., Herrmann, B.G., Kahlem, P., Benkahla, A., Schrinner, S., Yildirimman, R., Herwig, R., Lehrach, H., Yaspo, M.-L. (2002) A gene expression map of human chromosome 21 orthologues in the mouse, *Nature* **420**, 586–600.

Hattori, M., Fujiyama, A., Taylor, T.D., Watanabe, H., Yada, T., Park, H.-S., Toyoda, A. et al. (2000) The DNA sequence of human chromosome 21, *Nature* **405**, 311–319.

Holtzman, D.M., Bayney, R.M., Li, Y.W., Khosrovi, H., Berger, C.N., Epstein, C.J., Mobley, W.C. (1992) Dysregulation of gene expression in mouse trisomy 16, an animal model of Down syndrome, *EMBO J.* **11**, 619–627.

Hyde, L.A., Frisone, D.F., Crnic, L.S. (2001) Ts65Dn mice, a model for Down syndrome, have deficits in context discrimination learning suggesting impaired hippocampal function, *Behav. Brain Res.* **118**, 53–60.

Korenberg, J.R., Chen, X., Schipper, R., Sun, Z., Gonsky, R., Gerwehr, S., Carpenter, N., Daumer, C., Dignan, P., Disteche, C., Graha, J.M. Jr., Hudgins, L., McGillivray, B., Miyazaki, K., Ogasawara, N., Park, J.P., Pagon, R., Pueschel, S., Sack, G., Say, B., Schuffenhauer, S., Soukup, S., Yamanaka, T. (1994) Down syndrome phenotypes: the consequences of chromosome imbalance, *Proc. Natl. Acad. Sci. U.S.A.* **91**, 4997–5001.

Kapranov, P., Cawley, S.E., Drenkow, J., Bekiranov, S., Strausberg, R.L., Fodor, S.P.A., Gingeras, T.R. (2002) Large-scale transcriptional activity in chromosomes 21 and 22, *Science* **296**, 916–919.

Lamb, N.E., Feingold, E., Savage, A., Avramopoulos, D., Freeman, S., Gu, Y., Hallberg, A., Hersey, J., Karadima, G., Pettay, D., Saker, D., Shen, J., Taft, L., Mikkelsen, M., Petersen, M.B., Hassold, T., Sherman, S.L. (1997) Characterization of susceptible chiasma configurations that increase the risk for maternal nondisjunction of chromosome 21, *Hum. Mol. Genet.* **6**, 1391–1399.

Mouse Genome Sequencing Consortium (2002) Initial sequencing and comparative analysis of the mouse genome, *Nature* **420**, 520–562.

Mural, R.J., Adams, M.D., Myers, E.W., Smith, H.O., Miklos, G.L.G., Wides, R. et al. (2002) A comparison of whole-genome shotgun-derived mouse chromosome 116 and the human genome, *Science* **296**, 1661–1671.

Pinter, J.D., Eliez, S., Schmitt, J.E., Capone, G.T., Reiss, A.L. (2001) Neuroanatomy of Down's syndrome: a high resolution MRI study, *Am. J. Psychiatry* **158**, 1659–1665.

Pletcher, M.T., Wiltshire, T., Cabin, D.E., Villanueva, M., Reeves, R.H. (2001) Use of comparative physical and sequence mapping to annotate mouse chromosome 16 and human chromosome 21, *Genomics* **74**, 45–54.

Pogribna, M., Melnyk, S., Pogribny, I., Chango, A., Yi, P., James, S.J. (2001) Homocysteine metabolism in children with Down syndrome: in vitro modulation, *Am. J. Hum. Genet.* **69**, 88–95.

Reeves, R.H., Irving, N.G., Moran, T.H., Wohn, A., Kitt, C., Sisodia, S.S., Schmidt, C., Bronson, R.T., Davisson, M.T. (1995) A mouse model for Down syndrome exhibits learning and behaviour deficits, *Nat. Genet.* **11**, 177–184.

Reymond, A., Camargo, A.A., Deutsch, S., Stevenson, B.J., Parmigiani, R.B., Ucla, C., Bettoni, F., Rossier, C., Lyle, R., Guipponi, M., de Souza, S., Iseli, C., Jongeneel, C.V., Bucher, P., Simpson, A.J., Antonarakis, S.E. (2002) Nineteen additional unpredicted transcripts from human chromosome 21, *Genomics* **79**, 824–832.

Reymond, A., Friedli, M., Henrichsen, C.N., Chapot, F., Deutsch, S., Ucla, C., Rossier, C., Lyle, R., Guipponi, M., Antonarakis, S.E. (2001) From PREDs and open reading frames to cDNA isolation: revisiting the human chromosome 21 transcription map, *Genomics* **78**, 46–54.

Reymond, A., Marigo, V., Yaylaoglu, M.B., Leoni, S., Ucla, C., Scamuffa, N., Caccioppoli, C., Dermitzakis, E.T., Lyle, R., Banfi, S., Eichele, G., Antonarakis, S.E., Ballabio, A. (2002) Human chromosome 21 gene expression atlas in the mouse, *Nature* **420**, 582–586.

Richtsmeier, J.T., Zumwalt, A., Carlson, E.J., Epstein, C.J., Reeves, R.H. (2002) Craniofacial anomalies in segmentally trisomic mouse models for Down syndrome, *Am. J. Med. Genet.* **107**, 317–324.

Sago, H., Carlson, E.J., Smith, D.J., Rubin, E.M., Crnic, L.S., Huang, T.-T., Epstein, C.J. (2000) Genetic dissection of the region associated with behavioral abnormalities in mouse models of Down syndrome, *Pediatr. Res.* **48**, 606–613.

Sinet, P.M., Théophile, D., Rahmani, Z., Chettouh, Z., Blouin, J.L., Prieur, M., Noel, B., Delabar, J.M. (1994) Mapping of the Down syndrome phenotype on chromosome 21 at the molecular level, *Biomed. Pharmacother.* **48**, 247–252.

Weitzdoerfer, R., Fountoulakis, M. (2002) Reduction of actin-related protein complex 2/3 in fetal Down syndrome brain, *Biochem. Biophys. Res. Commun.* **293**, 836–841.

17
Fragile X-linked Mental Retardation

Ilse Gantois[1], *R. Frank Kooy*[1], *and Ben A. Oostra*[2]
[1] *University of Antwerp, Antwerp, Belgium*
[2] *Department of Clinical Genetics, Erasmus MC, Rotterdam, The Netherlands*

Neurobiology. From Molecular Basis to Disease. Edited by Robert A. Meyers.

ISBN: 978-3-527-32293-0

Keywords

Allele
Alternative form of a gene found at the same locus on homologous chromosomes.

CGG-repeat Expansion
Repeated-sequence CGG at the DNA level. When the number of repeats increases, this is referred to as an expansion.

FMR1
Gene that is shut down and causes fragile X mental retardation.

FMRP
The protein product of the *FMR1* gene.

Fragile Site
Can be induced in chromosomes by culturing cells in a folate-deficient medium.

Full Mutation
Is defined as a mutation in the *FMR1* gene with phenotypic symptoms.

Genotype
Genetic constitution of an individual.

Isoforms
Alternative splicing products of a gene.

Normal Transmitting Male
A phenotypically normal male who is a carrier of a premutation in the *FMR1* gene and may show cytogenetic expression of the fragile site.

Phenotypic
Observable characteristics.

Premutation
Is defined as a mutation in the *FMR1* gene without phenotypic symptoms.

Ribosome
A subcellular structure involved in protein production.

Fragile X syndrome is the most common form of inherited mental retardation. The disease is caused by methylation-induced transcriptional silencing of the fragile X mental retardation 1 gene (*FMR1*) as a result of the expansion of a CGG repeat in the 5′ UTR of the gene. This leads to the loss of fragile X mental retardation protein product (FMRP). FMRP is an RNA-binding protein that associates with translating polyribosomes as part of a large messenger ribonucleoprotein (mRNP) and modulates the translation of its RNA ligands. The *in vivo* targets of the FMRP are beginning to be elucidated and have been shown to include a subset of RNAs with a planar structure called *G-quartet*.

To study the physiological role of the FMR1 protein, animal models have been developed, including genetically modified mouse and *Drosophila*. The loss-of-function mouse model shows some features comparable to human patients including enlarged testes, a subtle behavioral phenotype, and discrete anomalies of dendritic spines. Studies in *Drosophila* indicate that dFmr1 in flies plays an important role in synaptogenesis and axonal arborization, which may underlie the observed deficits in flight ability and circadian rhythm in the dFmr1-mutant flies.

1 Introduction

Mental retardation can be defined as a failure to develop cognitive abilities and achieve a level of intelligence that would be appropriate for the age group. A percentage of the general population, variably estimated between 0.5 and 3%, is reported to be functioning two standard deviations below the average (i.e. to have an IQ of less than 70). Genetic determinants underlie approximately half of these conditions and an excess of affected males has often been reported, especially in the mild-to-moderate mental retardation range with an IQ between 70 and 35. Excess of males with mental retardation compared to females was eventually interpreted as being due to X-linked factors. One of the first described X-linked heritable mental retardation syndromes is the fragile X or Martin–Bell syndrome. After trisomy 21 or Down syndrome, it is the most frequent cause of mental retardation.

2 Description of Fragile X Syndrome

2.1 Some Historical Aspects

In 1943, Martin and Bell reported a large family in which 11 mentally retarded boys were born to mothers of normal intelligence and hypothesized that the mental defect of the boys was due to a sex-linked recessive gene, which was suppressed in the two grandfathers. More than a quarter of a century later, Lubs reported a unique cytogenetic marker at the end of the long arm of the X chromosome (Xq27.3) in cultured lymphocyte cells from four retarded male patients in a family with X-linked mental retardation. As more pedigrees were identified in which the marker on the X chromosome segregated with mental retardation, it became clear that the fragile site could be used as a diagnostic tool. In 1978, Turner observed macroorchidism (enlarged testicles) in a large proportion of the patients with the fragile site. Reinvestigation of the pedigree described by Martin and Bell revealed a fragile site and the typical features of the syndrome in the affected patients. In recognition of the first published pedigree, the mental retardation syndrome associated with the fragile site on the X chromosome and macroorchidism was named *Martin–Bell syndrome*, though these days it is more usually called *fragile X syndrome*.

2.2 Fragile Site

Before the cloning of the responsible gene in 1991, the laboratory confirmation of the diagnosis of fragile X syndrome was dependent on cytogenetic detection of the fragile site at Xq27.3 in cultured lymphocytes, fibroblasts, or amniocytes (Fig. 1). This fragile site was called FRAXA (fragile site, X chromosome, A site), as it was the first fragile site identified on the X chromosome. On metaphase spreads, fragile sites are microscopically visible as a gap or a break. The fragile site is induced by culturing cells under folic acid deprivation before chromosome spreads are made. The percentage of cells that show fragile site expression may vary from individual to individual, a phenomenon for which there is no explanation. In fragile X males, the fragile site is usually seen in 2 to 60% of the cells, but the site can only be detected in 50% of obligate carrier females, limiting the utility of cytogenetic carrier detection in pedigrees.

2.3 Unusual Pattern of Inheritance

From the initial observations onward, it was clear that fragile X syndrome has an unusual pattern of inheritance (Fig. 2). In X-linked recessive diseases, female carriers are usually unaffected and transmit their carrier status to half of their daughters, while half of their sons may develop the disease. However, in the case of the fragile X syndrome, 30% of carrier females are mentally retarded, although often less severely than males. In addition, 20% of males with cytogenetic expression of the fragile site are not affected. Such males, called *normal transmitting males* (NTMs), pass the mutation on to their unaffected daughters. However, these carrier daughters have a 40% risk of transmitting the disorder to their sons. In comparison, sons of affected females have a recurrence risk of 50%. Only transmission through the female line can give rise to a full phenotype in the next generation. Other peculiarities include the

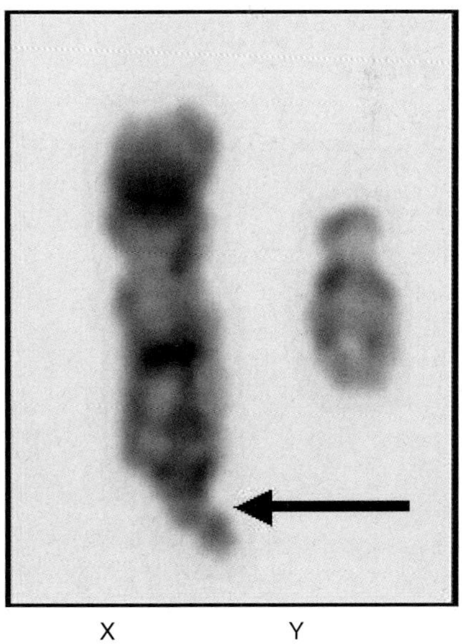

Fig. 1 Cytogenetic expression of the fragile site at Xq27.3 in a patient is clearly visible and indicated with an arrow.

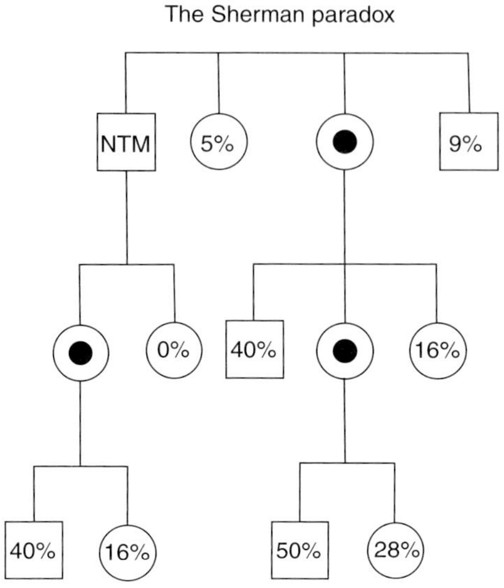

Fig. 2 A schematic representation of a fragile X family illustrates the Sherman paradox. The position within the pedigree affects the chance (indicated as a percentage) of being affected with the fragile X syndrome. NTM, normal transmitting male.

observation that more patients occur in the youngest generation of the pedigree. This increase in the penetrance of the disease with generations is called the *Sherman paradox*. The abnormal inheritance pattern could only be explained after the cloning of the gene and elucidation of the peculiar mutational mechanism.

Fig. 3 Male fragile X patient displaying the typical facial characteristics of the syndrome with a long face and prominent ears.

2.4 Phenotype

Fragile X syndrome has an incidence of about 1 in 4000 males and 1 in 6000 females. Males suffer from mild to severe mental retardation with an IQ between 20 and 60, while females are usually less severely affected. Macroorchidism develops during or after puberty and is seen in more than 80% of all adult patients. Fragile X patients show physical characteristics such as a typical long face and prominent ears (Fig. 3), but these symptoms can be absent in younger children. Hyperextensible joints, hand calluses, and flat feet can also be observed and may point to connective tissue dysplasia. Associated behavioral problems may include hyperactivity, decreased attention span, repetitive behavior, poor eye contact, tactile defensiveness, and shyness. Speech and language skills are usually impaired in older boys. In 13 to 23% of the male patients, autistiform features have been observed, occurring mostly before adulthood. Notwithstanding this observation in a proportion of fragile X patients, there seems to be no general

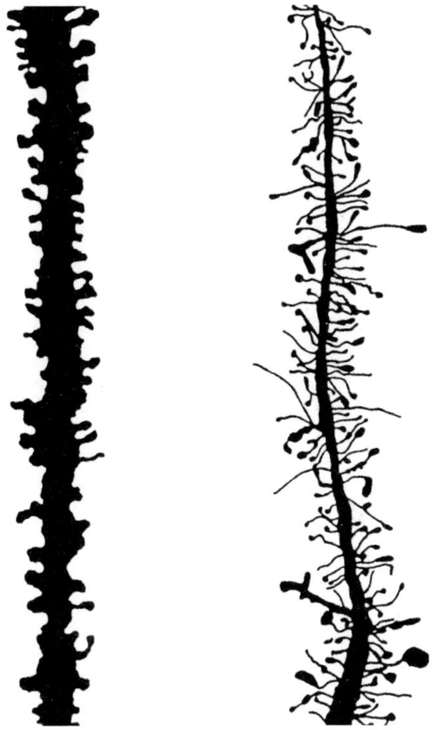

Fig. 4 Example of a typical spine morphology on dendrites from (a) a fragile X syndrome patient and (b) an unaffected control. Dendrites of the fragile X patient display more immature spines and a higher spine density. (Reprinted with permission from Fiala et al. (2002).)

association between fragile X syndrome and autism. In addition, fragile X boys suffer from a greater variability in total sleep time and sleep-rhythm maintenance when compared to controls. Fragile X patients have a normal life span.

2.5
Neuroanatomy

Brain abnormalities in fragile X patients have been determined using magnetic resonance imaging (NMR). A reduction in the size of the posterior cerebellar vermis and enlargement of the fourth ventricle compared to controls was reported. An increased volume of the hippocampus and the lateral ventricles in fragile X males was observed together with a significantly larger volume of the caudate nucleus. This is an important observation since the caudate nucleus, together with the lenticular and thalamic nucleus, forms the subcortical gray that allows the cerebral cortex to have an influence on behavior. In 3D reconstructions, fragile X brains appear to be larger than control brains. Volumetric changes were not confirmed by physical measurement of the brains of two deceased fragile X patients in a separate study.

An impaired neuronal development has been observed in fragile X patients. The spines on apical dendrites of pyramidal cells in the Golgi-impregnated visual and temporal cortex of fragile X patients are long, tortuous, immature, and thin, and fragile X patients exhibit an increased spine density on distal segments of apical dendrite shafts (Fig. 4). The higher spine density contrasts with other known mental retardation syndromes such as Down syndrome, in which a lower spine density is observed.

3
Identification of the Causative Gene

3.1
The Cloning of the Gene

Initially, the genetic defect underlying the fragile X syndrome was unknown and diagnosis was only based upon the observation of the fragile site at Xq27.3 in cultured cells. Positional cloning enabled the isolation of the gene and helped understand the molecular mechanism leading to the disorder. Linkage analysis with polymorphic markers placed the fragile X locus between factor IX at Xq27.1 and factor VIII at Xq28, conforming the locus for the gene at or nearby the fragile site. Hybrid cell lines with translocation break points at the fragile site were very useful in developing and mapping new probes near the gene. Two cell lines, termed Micro21D and QIX, each containing a reciprocal part of the human X chromosome, were especially helpful. Micro21D contained the Xpter–Xq27.3 part of the human X chromosome and Q1X contains the Xq27.3–Xqter part of the human X chromosome.

Using markers linked to the fragile site, several YACs ranging from 200 to 950 kb could be isolated. Among these, YAC 209G4 contained 475 kb human DNA and was shown to cross the fragile site by *in situ* hybridization. A cDNA clone was isolated from a human fetal brain library using cosmids from this particular YAC and this gene was assigned the name *FMR1* (Fragile X Mental Retardation 1). Further analysis of this gene sequence showed that the translation breakpoints of Micro21D and Q1X lie in exon 1 of the gene. Surprisingly, exon 1 contained a trinucleotide repeat that shows length variation between control and fragile X patients. Expansion of the CGG repeat was shown to be involved in both the inactivation of the *FMR1* gene and the cytogenetic expression of the fragile site.

3.2 Dynamic Mutation

Studying the dynamic inheritance pattern of the CGG repeat resolved the complex inheritance of the fragile X syndrome. In more than 99% of all fragile X patients, the syndrome is caused by a dynamic expansion of a CGG repeat in the 5′ untranslated region (5′UTR) of the *FMR1* gene (Fig. 5). In the normal population, the CGG repeat is polymorphic with allele sizes between 5 and 50 repeat units, 30 units being the most frequent allele size. Alleles of this size range are inherited stably from generation to generation and the gene is transcribed and translated into protein. In patients, a trinucleotide CGG expansion of more than 200 to several thousands of CGG repeats, referred to as a *full mutation*, is observed. Because of the repeat expansion, the CGG repeat and the surrounding promoter region of the *FMR1* gene is methylated, inhibiting mRNA translation and causing absence of the protein product. Absence of the FMR1 protein product, FMRP, causes the characteristic features of the fragile X syndrome. The alleles with a repeat size between 50 and 200 CGG units are called *premutation alleles*. Premutations are unmethylated and produce FMR1 mRNA and FMRP. Premutation carriers are not affected with fragile X syndrome. However, repeats in the premutation size range may be instable upon transmission to the next generation. Expansion to a full mutation occurs only through maternal transmission, explaining why only progeny of female premutation carriers is at risk for the fragile X syndrome. The observation that the larger the premutation, the more likely it will expand to a full mutation in the next generation, resolved the Sherman paradox.

The number of CGG repeats itself does not always allow discrimination between the three categories of repeat sizes. Repeats of 200 to 230 CGG units can be either large premutations or small full mutations. However, discrimination between the two types of mutations is possible on

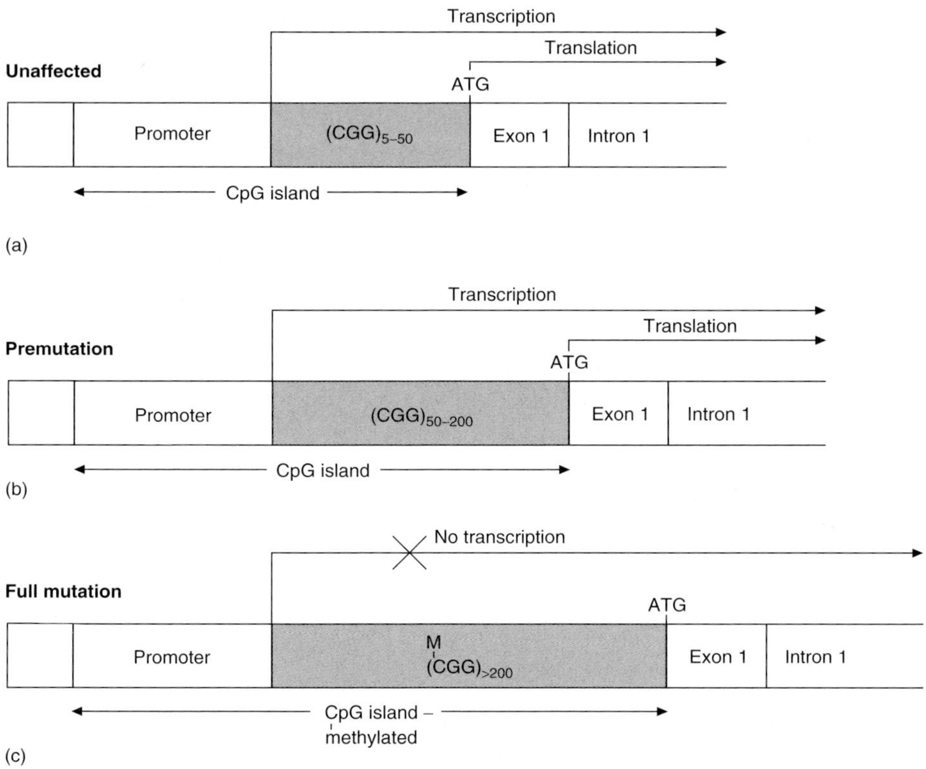

Fig. 5 Schematic representation of the transcription of (a) the *FMR1* gene in normal controls, (b) premutation carriers, and (c) full-mutation patients. The normal and the premutation size alleles (a and b) are transcribed and translated into FMRP. The full-mutation allele (c) is hypermethylated and therefore not transcribed.

the basis of the methylation status. Alleles comprising 45 to 60 CGG units are referred to as being of intermediate size. These may inherit stably, but may also increase or decrease by a few CGG units in length during transmission and become premutation alleles at risk for expansion to a full mutation in future generations. Factors influencing repeat stability include, but are limited to, the amount of AGG interspersions in the CGG repeat. In the *FMR1* gene, on average every 10th CGG repeat unit is interspersed with an AGG triplet. AGG repeats stabilize the repeat; fewer the AGG interspersions, the more prone it would be to intergenerational expansion. The instability threshold for FMR1 alleles is estimated to be about 34 to 38 uninterrupted CGG repeats, and the expansion of an interrupted CGG repeat of 59 units to a full mutation in the next generation has been observed.

4 *FMR1* Gene and Protein

4.1 Gene Structure and Expression Pattern

The *FMR1* gene has a genomic size of approximately 38 kb and consists of

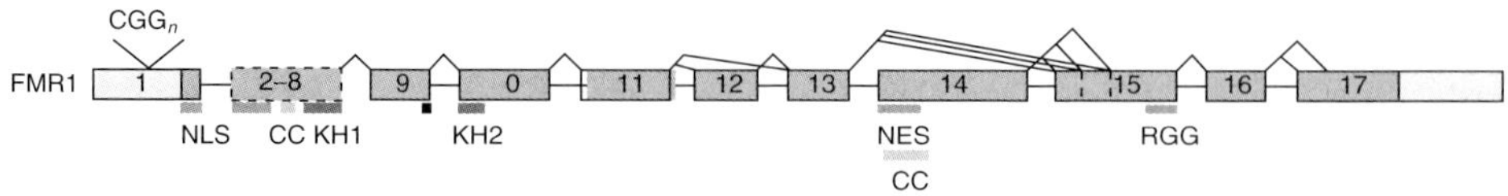

Fig. 6 The *FMR1* gene containing 17 exons and a CGG repeat in the 5′ UTR. The splice sites as well as the positions of the functional domains are indicated. NLS, nuclear localization signal; NES, nuclear export signal; CC, coiled coils; KH/RGG, KH, and RGG domains are important for RNA binding.

17 exons (Fig. 6). The transcript length is 4.4 kb, and all elements necessary for proper FMR1 expression *in vivo* are present in a 2.8 kb fragment immediately upstream of the ATG. The promoter sequence is GC-rich, as found in many housekeeping genes. Footprinting analyses of the promoter site revealed consensus-binding sites for ubiquitously expressed transcription factors such as Sp1, H4TF1/Sp1-like, and c-myc. In addition, an AP-2 binding site, specific for neuronal expression, was identified. Upstream of the CGG triplet repeat, a 20-nucleotide-long internal ribosome entry site (IRES) was identified. The function of this element in FMR1 is unclear, but IRES elements of other mRNAs play a role in neuronal regulation of translation in response to synaptic stimuli.

By Western blotting using antibodies directed against FMRP, different isoforms were identified with a molecular mass around 70 to 80 kDa and a maximal protein length of 631 amino acids. Excessive alternative splicing occurs especially in the 3′ terminal half of the gene, in exons 12, 14, 15, and 17 (Fig. 6). This can potentially give rise to 12 different mRNAs and protein isoforms. Skipping of exon 14 introduces a frameshift in the open reading frame, resulting in a different amino acid order in the carboxyl terminal half of the protein. The function of the different isoforms is still unknown and the different splice products are not tissue-specific. FMRP shows near-ubiquitous expression. It is highly expressed in ovary, thymus, eye, spleen, and esophageal epithelium with an abundant expression in brain and testis and a moderate expression in colon, uterus, thyroid, and liver, and little or no expression in the heart, aorta. or muscle. In the brain, the FMRP expression is restricted to differentiated neurons and Purkinje cells and is absent in nonneuronal cells. Neuronal FMRP is concentrated in the perikaryon and proximal dendrites and is also expressed in the synapses but is not detected in the axons. Especially high expression was found in Purkinje cells of the cerebellum and in neurons of the granular layer, the cortex, and the hippocampus. In the testis, the FMRP is expressed in the cytoplasm of primordial germ cells and in the early spermatogonia adjacent to the basal membrane of the tubuli seminiferi. Immunocytochemical studies revealed that FMRP is predominantly localized in the cytoplasm with high levels found near the rough endoplasmic reticulum, but some nuclear expression was also observed by both light and electron microscopy.

4.2 Structural Domains of FMRP

Two types of RNA binding domains have been identified in the protein, including two KH domains that show homology

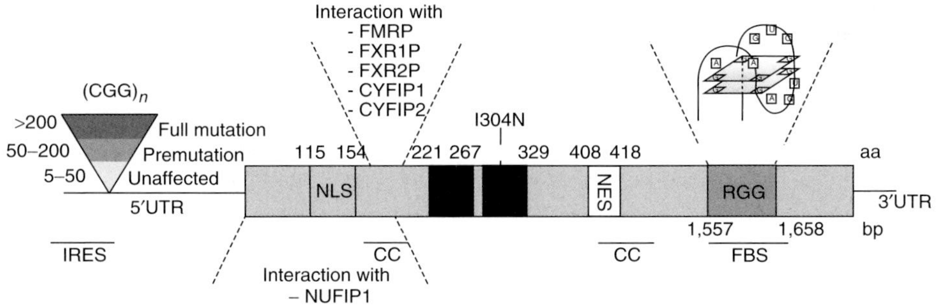

Fig. 7 Schematic representation of the FMR1 mRNA and protein. The known domains are indicated: IRES, internal ribosome entry site; NLS, nuclear localization signal; CC, coiled coils; KH, K-protein homology domain; NES, nuclear export signal; RGG box, Arg-Gly-Gly triplet; FBS, FMRP binding site. In addition, the Ile304Asn missense mutation found in a single, most severely affected patient is indicated. In the 5′ UTR, the different repeat classes are represented. The G-quartet structure present in the mRNA is depicted above FBS that overlaps with the RGG box. The interacting regions of FMRP with its paralog genes, *FXR1* and *FXR2*, and with other identified proteins, CYFIP proteins and NUFIP1, are indicated.

to hnRNP K domains and an RGG box containing a conserved Arg-Gly-Gly triplet. In addition, a nuclear localization signal (NLS), a nuclear export signal (NES), two coiled coils, and a G-quartet-binding structure have been identified (Fig. 7).

The KH domain consists of the consensus sequence, I/L/V-I-G-X_2-G-X_2-I. It is usually present in more copies in a protein and is able to bind ssRNA, mRNA, and rRNA. Using NMR (nuclear magnetic resonance), the KH domain was shown to consist of a stable three-stranded antiparallel β-sheet, orientated against three α-helices. The $\beta\alpha\alpha\beta\beta\alpha$ folds to a surface for RNA binding and a conserved Gly-Lys-X-Gly is centered on the loop between the first two helices. The importance of the KH domain for proper FMRP function is illustrated by a rare, unique patient with an Ile304Asn mutation located in the second KH domain. The mutated isoleucine is highly conserved in the KH domain sequence. The patient has a phenotype much more severe than any other known fragile X patient, including profound mental retardation and excessive macroorchidism. This patient still generates FMRP. The mutant FMRP has impaired RNA-binding capacity under elevated salt conditions. The mutation disrupts the normal folding of the KH domain. The FMRP–RNA binding can be disrupted by deletion of the C-terminus of the protein, containing the RGG box, suggesting that the RGG box plays an important role in FMRP–RNA binding. The RGG coding region of the FMR1 protein shows a high affinity for its own mRNA and other G-quartet-containing mRNAs, further described in Chapter 3.5. All these findings let us assume that FMRP has multiple binding regions to associate with RNA. *In vitro* translated FMRP binds directly to RNA homopolymers. It shows highest affinity for poly(G), a lower affinity for poly(U), and no affinity for poly(A) or poly(C). FMRP is able to bind selectively 4% of all brain mRNAs including its own mRNA, confirming that FMRP is an RNA binding protein. The creation of deletion mutants of FMRP, missing the N-terminal, the C-terminal, or one of the KH domains, revealed that only the first KH domain interacts with RNA. This result agrees with the findings that the Ile304Asn mutant is still able to bind some RNAs since the mutation is situated in the second KH domain, although the capacity to bind poly(U) homopolymers is destroyed. Both the N-terminal and the C-terminal show RNA-binding abilities and the N-terminal binds poly(G) with a similar preference as the first KH domain.

FMRP contains both an NES and an NLS signal. The NES is approximately 10 amino acids long and resembles the NES motif of HIV1 Rev. It consists of a small stretch of hydrophilic amino acids containing four critically spaced, large, hydrophobic amino acids. The NES is located in exon 14 and absence of this exon results in trapping of the FMRP in the nucleus. The nuclear export mediated by the NES of FMRP is exportin1-dependent, since Leptomycin B treatment, an antibiotic acting specifically on exportin1, causes the accumulation of FMRP in the nucleoplasm. The NLS contains clusters of basic amino acids, but the sequence does not resemble other known nuclear localization signals. It is located between residues 115 and 154. The nuclear localization of FMRP mediated by the NLS is an energy-dependent process. The presence of both an NES and an NLS may suggest that the fragile X protein shuttles through the cell into and out of the nucleus. In accordance with this hypothesis, the protein has been observed in the nuclear pore during transfer between the nucleus and the cytoplasm.

FMRP contains two coiled coils involved in protein–protein interaction and located in exon 7 and in exons 11 to 14 respectively. The first coiled coil is potentially involved in binding of FMRP interacting proteins, while the second was suggested to play a role in FMRP–ribosome binding.

4.3
FMRP Associates with Ribosomes as Part of an mRNP

In human cell lines, FMRP colocalizes primarily with polyribosomes and rough endoplasmic reticulum. High levels of FMRP were found in perikarya, where it is concentrated in regions rich in ribosomes, particularly near or between rough endoplasmic reticulum cisternae. FMRP has also been detected in the granular component of nucleoli in COS cells. The nucleolus is a cellular component formed in response to the requirement for new ribosome synthesis. The FMRP–ribosome association is sensitive to 500 mM salt concentration and to low concentrations of RNase digestion that does not disturb the ribosome assembly *per se*. This suggests that FMRP is bound to the ribosomes via RNA. These findings are in contradiction with earlier experiments that suggested that FMRP binds to the 60S ribosomal subunit via part of exon 13 and 14, containing the second coiled coil domain. Whatever the role of the FMRP–60S ribosome binding, the RNA–protein interaction seems the most important as nearly all cytoplasmic FMRP can be cocaptured with poly(A) mRNA. Moreover, after EDTA treatment, the FMRP complex is released from the ribosomes.

In the cell, FMRP forms a messenger ribonucleoparticle, mRNP, with a sedimentation value of 60S and a size $>$660 kDa. The composition of the mRNP complex is not known, but some proteins involved in the mRNP particle include FXR1 and FXR2, paralogs of FMR1 (Chapter 4), NUFIP and CYFIP proteins (Chapter 5), nucleolin, a known protein in other mRNPs, and an ortholog of the mouse protein 1/p50. It is not known whether all mRNP particles containing FMRP have the same composition. If this is not the case, the absence of FMRP may influence the functioning of the different types of mRNPs in a different way.

4.4
FMRP as a Possible Inhibitor of Translation

The association of FMRP with the polyribosomes suggested a potential controlling role in mRNA translation and stability. *In vitro*, FMRP is able to suppress translation of mRNA. This inhibition is dosage-dependent and is observed for all mRNAs tested at low FMRP concentrations in an *in vitro* synthesis system and in microinjected *Xenopus laevis* oocytes. This may suggest that cells with absent FMRP translate certain mRNAs in excess. Nevertheless, a study by Brown et al. (2001) (Chapter 3.6) showed that increased and decreased amounts of specific polysome-bounded mRNAs occur in fragile X patients. It was suggested that elevated levels of FMRP are sequestering mRNAs into incompetent granules causing the repression of reporter genes. Recently, it was reported that only a minor fraction of newly synthesized FMRP is associated with polyribosomes, while the majority of it is found in cytoplasmic granules that are resistant to nonionic detergents. Increasing the amount of FMRP resulted in trapping of the mRNAs in these granules together with several other mRNA-binding proteins. This suggested a potential dual role for FMRP both as a transporter of specific mRNAs and also as a repressor of translation in specific conditions.

FMRP containing the Ile304Asn mutation fails to repress translation despite the fact that the RNA-binding capacity of mutated FMRP is only mildly affected. The mutated protein appears as a part of a smaller, abnormal mRNP that is not associated with translating polyribosomes, perhaps because FMRP is also able to form dimers but the Ile304Asn protein fails to do so. This suggests that the loss of association to ribosomes and to form normal mRNP particles causes the severe phenotype of this patient.

4.5
mRNA Targets of FMRP

As FMRP has RNA-binding capacities, identifying the mRNAs bound by FMRP in the cell is a prerequisite for understanding its function. An interaction of FMRP with the 3′ UTR of the myelin basic protein mRNA was found using purified recombinant FMRP. Using a biotinylated-FMRP affinity resign, Sung et al. (2000) identified nine mRNAs from adult brain that are able to bind FMRP, including a neuronal NT2 EST and Tip60a, a Tat interactive protein. Also the *Xenopus* elongation factor 1A, xEF-1A, binds strongly to human FMRP. FMRP was demonstrated to inhibit mRNA translation of this gene, and in the absence of FMRP, the translation of human EF-1A is derepressed.

Recently, the C-terminal region of FMRP was identified as major site of specific mRNA interaction. A specific, high-affinity binding site for its own mRNA in the RGG box region of FMRP was identified. A 100-nucleotide mRNA fragment, situated in the region encoding the RGG box, was shown to contain a purine-quartet motif predicted to form a planar structure. This so-called G-quartet structure was shown to be involved in the binding of FMR1-mRNA by FMRP (Fig. 7). At the same time, Darnell et al. (2001) identified

FMRP-bound mRNAs, by screening a pool of 96-mer RNAs containing 52 bases of random sequence. A common motif preferentially bound by FMRP was identified as a G-quartet.

In vivo studies also revealed G-quartet-containing mRNAs as targets of FMRP. mRNAs associated with FMRP-mRNPs that interact with active translating ribosomes in mouse brain were isolated and hybridized on microarrays containing a significant proportion of murine genes. By subtracting the expression patterns of knockout and control mice, RNAs specifically bound to FMRP were identified. In a similar way, mRNA populations in the polyribosomal fractions of lymphoblasts of fragile X patients were compared with controls. The studies revealed a long list of FMRP-associated mRNAs. A total of 426 mRNAs were identified in both the human and the mouse studies. Moreover, 14 of the mRNAs were identified in both the experiments of Darnell et al. (2001) and Brown et al. (2001). Eight of them revealed a G-quartet structure including, for example, NAP22, which is present in axon terminals and dendritic spines and plays a role in maturation or maintenance of synapses. In addition, in both studies MAP1B was identified, which is involved in transport within neurons and is the human ortholog of the *Drosophila* futsch gene (Chapter 6).

Using the Fmr1 knockout mouse model, Zalfa et al. (2003) revealed that FMRP regulates translation of specific dendritic mRNAs. The translational efficiency, the percent of a given mRNA that is associated with polysomes (PMP = Percent of Messenger on Polysomes), was quantified and shown to be different for specific RNAs, including those that encode for cytoskeletal proteins such as Arc and MAP1B and signal transduction molecules such as α-CaMKII. Interestingly, interaction occurred through a ribonucleoprotein complex that includes the small dendritic noncoding RNA BC1. It was shown that BC1 directly binds to FMRP and is able to base-pair with regions of complementarity between BC1 and the regulated mRNAs. BC1 may be required for FMRP association with the targeted mRNAs and for determining the binding specificity.

In vivo binding of mRNAs to FMRP in hippocampal neurons was determined using antibody-positioned RNA amplification, a novel technique that allows identification of RNAs that are directly associated with FMRP in the cell. Approximately 83 genes believed to interact directly with FMRP were identified. The FMRP interacting mRNAs were mostly involved in diverse aspects of synaptic plasticity and neuronal maturation. G-quartets are present in a small percentage of the mRNAs indicating that other factors in addition to the G-quartet are important in FMRP–RNA binding. Interestingly, some of the proteins encoded by the RNA showed a discrete change in abundance and subcellular distribution. The identified mRNAs include the glucocorticoid receptor α (GRα). It is a low-affinity receptor for corticosteroid hormones, involved in the electrical activity in the hippocampus in response to mineralocorticoid receptor stimulation. The receptor showed a change in dendritical distribution in the hippocampus of the knockout mouse. Diminished responsiveness of the receptor is compatible with learning problems observed in fragile X patients.

4.6 Function of FMRP in Neurons

The ability of FMRP to bind RNA, its association with ribosomes, and the observation that the protein shuttles through

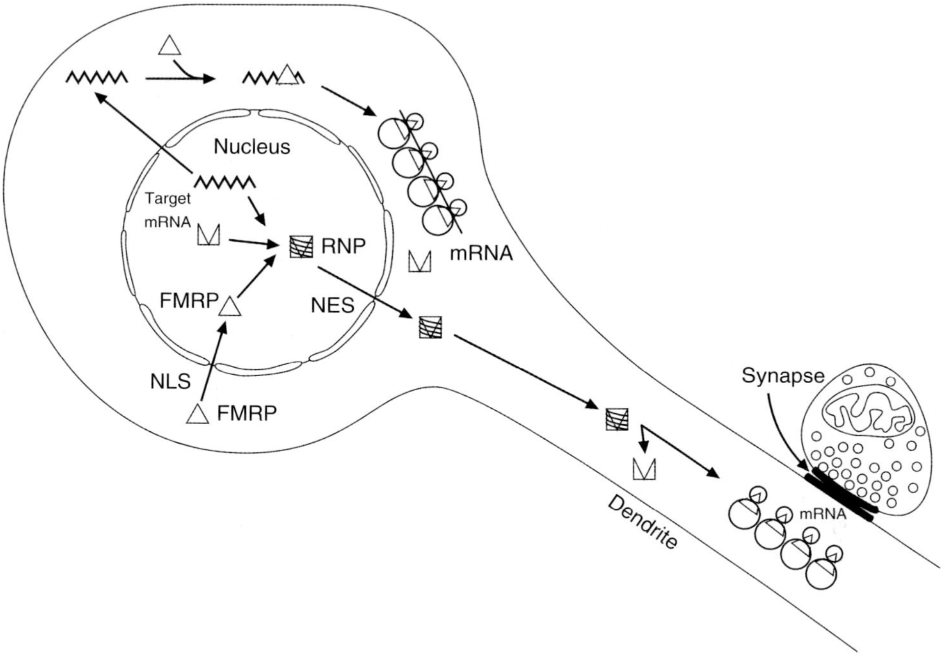

Fig. 8 A hypothetical model of the FMRP function in the neuron. FMRP is transported from the cytoplasm to the nucleus, where it assembles into an mRNP particle, containing specific mRNAs and other proteins. The mRNP particle is transported to the cytoplasm and mRNAs are presented to the polyribosomes in the perikaryon and the dendrites nearby the synapses.

the cell lead to the hypothesis that FMRP may play a role in mRNA transport from the nucleus to specific destinations in the cell (Fig. 8). This hypothesis is supported by the observation that FMRP associates in the nucleus with mRNAs and proteins to form an mRNP particle that is transported through the cytoplasm and is presented to the ribosomes for translation. It is not clear what happens with FMRP after translation. FMRP may be released from the ribosomes as a free protein after translation termination or may be trapped in cytoplasmic granules.

Because of the putative role of FMRP in cellular mRNA transport, dendritic mRNA transport was investigated in an FMR1–EGFP stable transfected PC12 cell line with an inducible expression system (Tet-On) for regulated expression of the FMRP–GFP fusion protein. After Dox-induction, FMRP–GFP was localized in granules in the neurites of PC12 cells. Using time-lapse microscopy, the trafficking of FMRP–GFP granules into the neurites of living PC12 cells was demonstrated. Motile FMRP–GFP granules displayed two types of movements: oscillatory (bidirectional) and unidirectional anterograde. The average velocity of the granules was $0.19\ \mu m\ s^{-1}$ with a maximum speed of $0.71\ \mu m\ s^{-1}$. The movement of FMRP-labeled granules was microtubule-dependent. Colocalization studies showed that the FMRP–GFP granules also contained RNA, ribosomal subunits, kinesin heavy chain, and FXR1P molecules as constituents. This observation stresses the putative role for FMRP–mRNP particles in dendritic mRNA transport to the synapse.

In the synapse, FMRP might regulate the translation of certain mRNAs. Interestingly, local protein synthesis plays an important role in neuronal processes, including learning and memory. FMRP can be detected in polysomes of synaptoneurosomes, neuronal preparations highly enriched in synapses. Synaptoneurosomes respond to stimulation by metabotropic glutamate agonists with fast-increasing polyribosome formation and accelerated protein synthesis. The synthesis of FMRP in the synaptoneurosomes increases after stimulation with the neurotransmitter glutamate, suggesting that synthesis of this protein in synaptoneurosomes is triggered by a class I glutamate receptor. On the other hand, different experiments reported that Fmr1 knockout mice have lowered expression levels of the glutamate receptor mGluR1 in the cerebral cortex and in the granule cell layer of the cerebellum, but not in the hippocampus. Thus, it has been suggested that this receptor alteration disturbs local protein synthesis and leads to impaired cortical synaptic plasticity. These findings suggest a specific function in the modulation of FMRP synthesis at postsynaptic sites and, in addition, that FMRP has a role in neurotransmitter-actuated polyribosome assembly. FMRP synthesis can also be affected by the brain-derived neurotrophic factor, BDNF, a known regulator of synaptic plasticity. BDNF downregulates the FMRP expression in cultured hippocampal neurons as well as in mouse brains. This downregulation by BDNF is a response to increased tyrosine kinase receptor signaling. The decreased FMR1-mRNA amounts could be correlated to a decreased amount of FMRP in the cell.

The *in vivo* trigger of FMRP synthesis is not known, but FMRP increases in the barrel cortex after whisker stimulation in rats, a model of experience-dependent plasticity. This increase was notably observed in subcellular fractions enriched for synaptoneurosomes and polyribosomes, suggesting a site-specific production of the protein. The altered level of FMRP most likely influences the translation of specific mRNAs in the synapse. Inhibition of the synthesis of its own mRNA, MAP1B, Arc, and α-CaMKII has been demonstrated, but it remains unclear whether all cellular RNAs that are bound by FMRP are translationally repressed.

Thus, FMRP plays an important role in translation of proteins in the dendrites after specific triggering. Abnormal synaptic protein synthesis in the absence of FMRP could underlie variable symptoms of the fragile X syndrome, including the presence of immature spines and impaired synaptic maturation.

5 Paralog Genes FXR1 and FXR2

FMR1 has two paralog genes called *FXR1* and *FXR2* (fragile X–related protein 1 and protein 2). FXR1 was identified by cross-hybridization with an FMR1 probe, while FXR2 was identified by yeast two-hybrid assays using whole FMRP as bait. The *FXR1* and *FXR2* genes have been localized to 3q28 and to 17p13.1 respectively. Overall similarity with FMR1 is high, especially in the amino-terminal half of the gene, and both FXR1 and FXR2 proteins contain identical functional domains as FMRP, including two KH domains, an RGG box, an NES, and an NLS (Fig. 9). Sequences

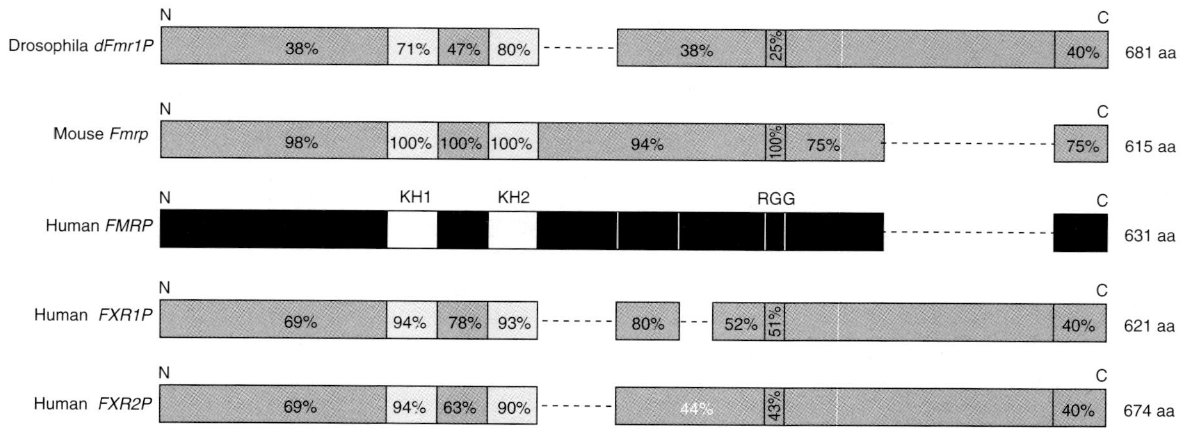

Fig. 9 Protein identity between human FMRP and human FXR1P and FXR2P, mouse Fmrp and Drosophila dFmr1P based on the alignment of Wan et al. (2000). The three RNA binding domains in the mouse Fmrp show 100% identity with the FMRP domains, whereas in the other proteins high identity is present in the KH domains and moderate identity in the RGG box region. Identity in the carboxyl end is not indicated because the low homology of the protein in this region does not allow an unambiguous alignment.

corresponding to exons 11 and 12 in FMRP are not identified in the two paralog genes. FMRP shows 86% and 70% homology to FXR1P and FXR2P respectively. Specific differences include the carboxyl end of FXR1P that is very different from that of FMRP and the FXR2P amino-terminus that contains an extra 10 amino acids compared to FXR1P and FMRP. The 5′ and 3′ UTR of FXR1 and FXR2 are up to five times shorter than that of FMR1. The 3′ UTR of FXR1 shows a high similarity with FMR1 and is highly conserved during evolution, while that of FXR2 shows no similarity with FMR1. No CGG repeat is present in the FXR1 sequence, while the 5′ UTR of FXR2 contains a short $(CGG)_4$ unit.

FXR1 has seven isoforms with the two major ones having a predominant mRNA, 2.2 to 2.4 kb in size. The protein FXR1P has a molecular mass of respectively 70 and 78 kDa. There is a tissue-specific alternative splicing for FXR1 pre-mRNA in the 3′ carboxyl part. The *FXR2* gene generates a 3 kb-sized mRNA and the protein FXR2P has a mass of 95 kDa.

The expression pattern of FXR1P and FXR2P shows some similarities but also important differences with that of FMRP. In the testis, FMRP is only expressed in the spermatogonia, whereas FXR1P is expressed in postmeiotic spermatides at high levels, especially in the microtubules of sperm tails, and FXR2P is expressed in the tubuli seminiferi. A specific isoform of FXR1P is highly expressed in the muscles and the heart, where FMRP expression is nearly absent. In the brain, both homolog proteins are expressed in neuronal cytoplasm. In fragile X patients, similar expression levels of FXR1 and FXR2 are noticed in all tissues compared to controls.

On the basis of the sequence similarity between FMR1, FXR1, and FXR2, it has been suggested that FXR1P and FXR2P may have a similar function in the cell as FMRP and can partially compensate for its loss in fragile X patients, explaining the relatively mild phenotype of patients. Similarities include the localization in the cytoplasm, the association with polyribosomes, and the capability to bind RNA. In addition, both the paralog proteins can form homodimers and heterodimers with each other and with FMRP (Fig. 7), though, like FMRP, FXR1P, and FXR2P are predominantly cytoplasmic proteins, with some expression observed in the nucleus. FXR1P is able to shuttle between the nucleus and the cytoplasm, while FXR2P and some of the isoforms of FXR1P shuttle between the nucleolus and the cytoplasm. However, though some functional redundancy seems likely, differences between the FXR proteins and FMRP are also apparent. Unlike FMRP, FXR1P and FXR2P do not inhibit translation and have a different localization in some tissues. Moreover, it is unclear whether FXR1P and FXR2P can bind G-quartet-containing mRNAs to the same extent as FMRP.

6 Amino-terminal Interacting Proteins

Using the N-terminal domain of FMR1 containing the NLS, Bardoni et al. (1999) identified a novel protein that interacts with FMRP by a yeast two-hybrid assay (Fig. 7). The interacting protein was called NUFIP (nuclear FMRP interacting protein). It shows no homology with known proteins in the databases. NUFIP maps to chromosome 13q14, while a pseudogene maps to chromosome 6q12.

NUFIP contains two C_2H_2 zinc finger motifs and an NLS. The protein is mainly localized in the nucleus in a dotlike distribution but is also observed in the cytoplasm, where it associates with ribosomes. NUFIP interacts with FMRP both via the N-terminal and the C-terminal half, suggesting the presence of two protein-interacting regions. Despite the fact that the 5′ region of the *FMRP* gene, used as bait to pick up NUFIP, is highly similar between FMRP, FXR1P, and FXR2P, no interaction between NUFIP and the other FXR proteins could be detected. The mouse NUFIP shows a 95% homology with the human NUFIP and has an expression pattern comparable with that of FMRP, with high expression in the neurons of the hippocampus and the cortex, and in the Purkinje cells of the cerebellum, but lower expression in the granular cell layer. No expression was detected in glia cells. NUFIP is also expressed in fibroblasts and lymphoblasts. Not only is the expression pattern of NUFIP similar to that of FMRP but, in addition, the RNA-binding capacities of NUFIP do resemble those of FMRP as NUFIP binds with high efficiency to poly(G), with less efficieny to poly(U), and shows no affinity to poly(A) or poly(C) homopolymers. The RNA binding domain of NUFIP has not been identified yet.

More recently, two additional FMRP interacting proteins, named CYFIP1 and CYFIP2 (cytoplasmatic FMRP interacting protein 1 and 2), were isolated. CYFIP1 and CYFIP2 are 88% identical on protein levels and both are conserved during evolution but do not show similarities with genes in the database. The CYFIP proteins are expressed in the synaptic terminals of neurons in the brains, where it colocalizes with FMRP to ribosomes in the cytoplasm. The first coiled coil region of FMRP, encoded by exon 7 and mediating FMRP–protein interaction, is involved in the interaction of FMRP with CYFIP proteins (Fig. 7). Despite the high sequence similarity between the CYFIP proteins, CYFIP1 interacts only with FMRP, whereas CYFIP2 interacts with FMRP, FXR1P, and FXR2P. Not much is known about the function of the CYFIP proteins, but CYFIP1 appeared to be an interactor of the Rac1 protein, a member of the Rho–GTPase family, which plays a role in maturation and maintenance of dendritic spines by controlling the actin cytoskeleton.

7 Animal Models

7.1 Comparison between Species

Orthologs of the *FMR1* gene have been identified in different species such as mouse, rat, chicken, *Xenopus laevis*, *Danio rerio*, monkey, and *Drosophila*. The protein is well conserved between species. The murine, frog, chicken, and fly proteins are 97, 86, 92, and 31% identical to the human FMRP respectively. Homology is nearly 100% within the KH domain, while a lower similarity in the RGG box region is observed. Outside the coding region, the 3′ UTR is well conserved among species. The 5′ UTR region is highly GC-rich in all species analyzed, and a CGG repeat has been detected in most mammals. Chicken contains a 99-unit-long CCT repeat in place of the CGG repeat, flanked on both sides by dinucleotide repeats. No instability upon generational transmission was observed in these repeats in species other than man and monkey.

7.2 What Have We Learned from the Mouse Model?

As in humans, the murine *FMR1* gene is located on the X chromosome. All functional domains present in the human protein, 2 KH domains, an RGG box, two coiled coils, an NLS, and an NES are well preserved (Fig. 9). In the 5′ UTR, a triplet repeat containing 9 to 11 CGG repeats has been identified, depending on the mouse strain, interspersed by 1 CGA triplet. No fragile site and no spontaneous CGG-repeat instability have been observed in the mouse. The expression pattern of the murine *FMR1* gene is comparable with that of the human gene with abundant expression in the testis and the hippocampus.

Approximately 10 years ago, a knockout mouse model was constructed. Using homologous recombination, the Fmr1 sequence was interrupted by a neomycin cassette in exon 5 of the gene. Although the knockout mutation in the mouse *FMR1* gene differs from the dynamic mutation in most fragile X patients, both mutations lead to the same result: absence of protein product. The fragile X knockout mouse is fertile, and visual inspection revealed no obvious abnormalities. It has been subjected to a wide range of pathological, physiological, and behavioral tests.

7.2.1 Macroorchidism

Macroorchidism is present in the great majority of fragile X patients after puberty. In the knockout mouse, macroorchidism can be observed from day 15 after birth, increasing to a difference of 30% in weight at half a year of age. Similar to observations in human patients, pathological observation revealed no structural abnormalities in the testis of the knockout mouse. It was concluded that increased Sertoli cell proliferation in the fragile X mice in a well-defined period shortly before and after birth leads to higher spermatogenic cell number and testicular weight.

7.2.2 Cognitive Testing

Spatial learning in mice can be measured using the Morris water-maze test. In this test, mice have to find a hidden platform in a basin filled with opaque water during 12 learning trials. In the reversal phase, the platform is moved to the opposite position in the basin, and mice have to find the platform in its new position during four additional trials. It is particularly in this reversal phase that mild learning deficit is observed in the knockout mice. Similar observations are reported when mice were tested using the plus-shaped water maze and the terrestrial radial maze. The nature of the spatial deficit is not clear. The deficit may point to hippocampal defects, but other brain areas can also play a role in the observed differences in this reversal test.

7.2.3 Motor Activity and Anxiety

Fragile X patients are reported to be hyperactive and anxious. Fragile X knockout mice are more active than their control littermates. However, as the difference in activity between groups is mild and variance within groups is high, hyperactivity may be too small to measure when registered during a single time period as illustrated in Fig. 10.

A seemingly contradictory observation is that Fmr1 knockout mice displayed increased anxiety-related responses in the exploratory behavior test, whereas lesser anxiety is reported when measured in the open field test. However, no difference in anxiety was registered in the elevated plus

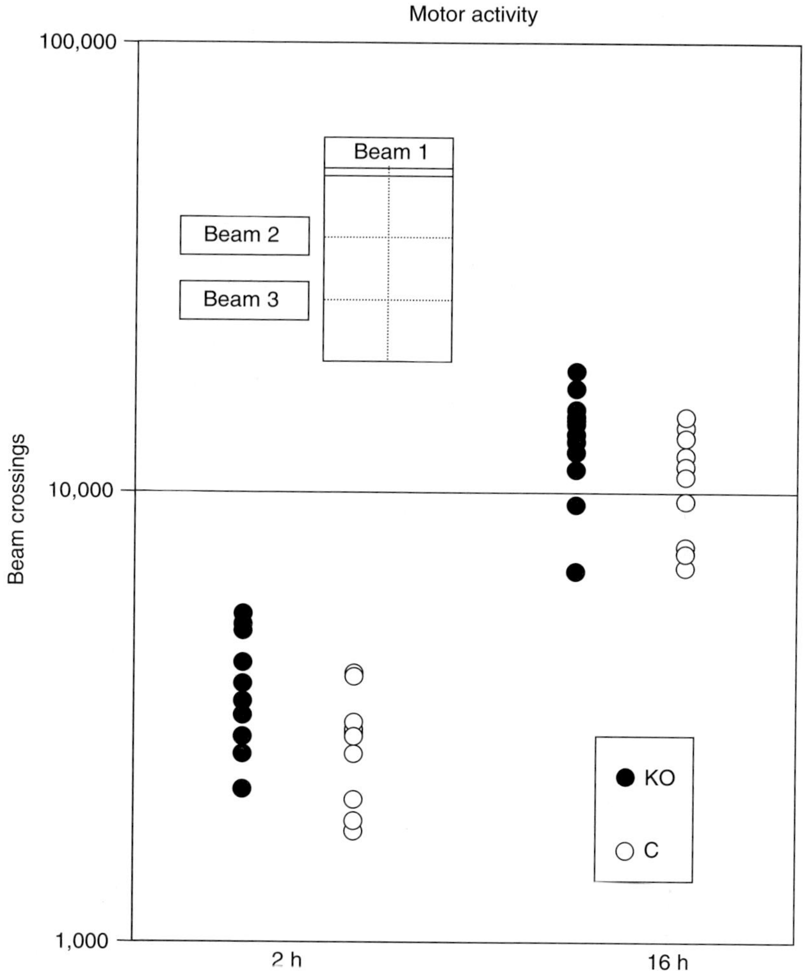

Fig. 10 Motor activity testing of the fragile X knockout mouse in comparison with control littermates. In one experiment, the total number of beam crossings was significantly elevated after 2 h of registration ($P < 0.005$), while after 16 h no statistical difference (N.S.) between both groups was observed. When the individual data were considered (black circles, knockout; white circle, control), it is noticed that a single outlier in the group of individual mice influenced the test result.

maze, considered to be the most suitable test to measure anxiety.

7.2.4 Acoustic Startle Reflex

The acoustic startle response is a behavioral tool to assess brain mechanisms of sensorimotor integration and is mediated by neurons in the lower brainstem. The knockout mice exhibited increased auditory startle-response amplitudes to low-intensity stimuli and decreased responses to high-intensity stimuli. The increased response to low-amplitude stimuli is compatible with the hyperarousal of fragile

X patients, while the decreased reactivity of the fragile X knockout mouse to high-amplitude stimuli is compatible with the decreased functioning of the neuronal connections in fragile X patients.

Prepulse inhibition was also affected in fragile X mice. In this variant of the acoustic startle response, a prepulse preceding the stimulus itself suppresses the response. Prepulse inhibition is dependent on a more complex forebrain circuitry. Inhibition was stronger in fragile X knockout mice compared to controls, suggesting involvement of the forebrain pedunculopontine nucleus pathway in fragile X syndrome.

7.2.5 Audiogenic Seizures

Spontaneous seizures are observed in more than 20% of the fragile X patients. These spontaneous seizures are never observed in the knockout mouse, but an increased sensitivity to epileptic seizures in response to auditory stimuli is detected in the knockout mouse. This may indicate that absence of Fmrp results in increased cortical excitability and is consistent with the sensory hypersensitivity of fragile X patients.

7.2.6 Neuroanatomy and Physiology of the Knockout Brain

Magnetic resonance studies of mouse brains did not reveal size differences in brain areas reported to be of abnormal size in human patients. As in human fragile X patients, impaired neuronal maturation is observed in the knockout mice. In cortical layer V of knockout mouse brains, pyramidal neurons have more elongated, thin, immature, and tortuous spines along the dendrites. In addition, a trend toward a higher spine density, characteristic for human fragile X patients, was observed. Since the knockout mice have an excess of immature spines, synaptic connectivity may be impaired. This may suggest that the fragile X protein is required in synaptic pruning and maturation processes in neurons, although at present it has not been established whether the neural abnormalities are the cause or the consequence of the mental retardation.

A detailed study of the hippocampus did reveal possible subtle neuroanatomic abnormalities. Sections through the hippocampus revealed excessive sprouting of the intra- and infra-pyramidal mossy fibers in subfield CA3 in knockout mice. This suggests increased axonal branching and synaptogenesis in the knockout mouse, perhaps related to the increased sensitivity to seizures. However, a decrease in size of the mossy fibers in the same hippocampal area was reported in an independent study. The reason for the discrepancy between both studies is not clear but could be related to strain differences. FVB mice with an unusually small CA3 subfield were used in the first-named study, whereas C57BL/6 mice with a relatively large CA3 subfield were used in the second study.

Detailed studies of dendritic processes of spiny stellate neurons in barrels, the cylindrical arrangements of neuronal cell bodies that define processes in the somatosensory cortex for the large facial whiskers, revealed additional subtle abnormalities. During normal development, dendritic processes extend in all directions from the cell body. Branches directed toward the interior of the barrel continue proliferating during development and those directed away from the interior are withdrawn. In knockout mice, no differences were observed for branches directed toward the barrel interior, while the branches directed away from the interior remained more extensive compared to the controls.

Metabolic rates are also altered in the knockout mouse. Qin et al. (2002) measured the cerebral metabolic rates for glucose, CMR_{glc}, as an indicator of the level of functional activity in 38 brain regions of fragile X knockout mice and wild-type littermates. In 26 of the 38 regions, the CMR_{glc} level was significantly increased in the Fmr1 knockout mice, particularly in the regions of the limbic system and primary sensory and posterior parietal cortical areas. Interestingly, these areas are involved in motor activity and exploration, processes that are affected in the knockout mice. Thus, the increased level of the energy metabolism in cortical structures correlates with abnormal dendritic spine morphology and an increased excitatory state.

7.2.7 FMRP is Involved in Long-term Plasticity

Long-term potentiation (LTP), a long-lasting increase in synaptic efficiency, is believed to be involved in learning and memory. No altered LTP was measured in the hippocampus of the knockout mouse, but Li et al. observed severely reduced LTP when measured in the cortex. Reduced LTP could be correlated with a decreased quantity of GluR1 expression in the central synapses.

Long-term depression (LTD), a second form of hippocampal synaptic plasticity, is the opposite of LTP. It is triggered by the activation of group 1 metabotropic glutamate receptors. LTD appeared significantly enhanced in the hippocampus of the knockout mouse. While at first an unexpected finding, increase in LTD is consistent with the finding that Fmrp is a negative regulator of translation. As LTD is protein synthesis–dependent, absence of the translational inhibitor, FMRP, may inhibit the protein synthesis necessary for LTD (Fig. 11).

Since LTD and LTP are involved in learning and memory, it is tempting to speculate that abnormalities in synaptic connections in the knockout mouse may be related to the cognitive deficits in the fragile X syndrome.

7.2.8 FXR1/FXR2 Knockout Mice

It could be speculated that the abnormalities observed in the fragile X knockout mouse are relatively mild because the paralog genes *Fxr1* and *Fxr2* partially take over *Fmrp* function. Knockout mice for both paralog genes have been generated. Mice homozygous for the *Fxr1* knockout die shortly after birth because of respiratory problems (Siomi, personal communication). Mice heterozygous for the *Fxr1* mutation seem to be lighter and smaller than control littermates but not much is known about their phenotype. The *Fxr2* knockout mouse is viable and fertile and shows no obvious abnormalities. It has been subjected to a similar range of behavioral and pathological tests as the fragile X knockout mouse. Like the *Fmr1* knockout mouse, the *Fxr2* knockout mouse shows no obvious pathological abnormalities, is hyperactive, and performs poorly in the Morris water-maze test. Double or triple mutants of *Fmr1*/*Fxr* mice have not been described as yet, but such multiple knockouts may show us more of the possible redundancy in function of the Fxr proteins.

7.3 What Have We Learned from Drosophila?

7.3.1 The Ortholog, dFmr1

dFmr1, the *Drosophila melanogaster* ortholog of the *FMR1* gene, appears to

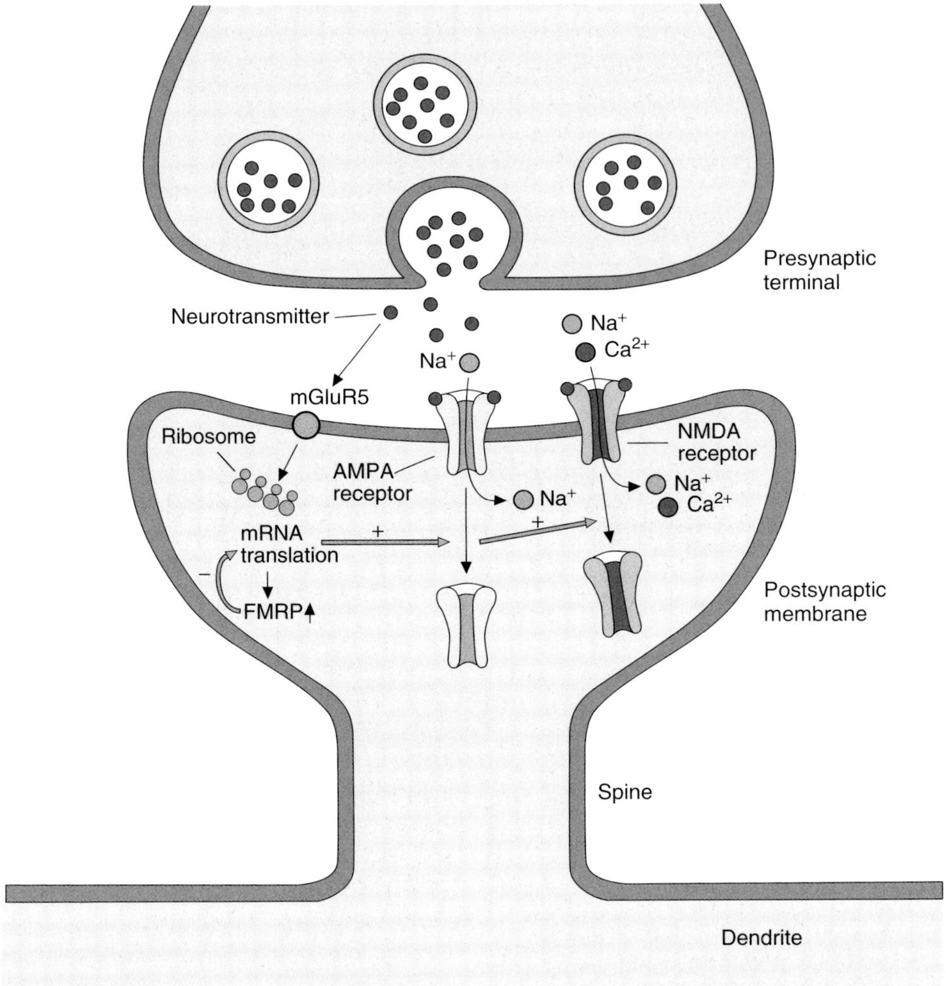

Fig. 11 Hypothetical model of the regulation of FMRP translation in a hippocampal synapse adapted from Huber et al. (2002). Stimulation of mGluR5 induces local mRNA synthesis. This results in novel protein synthesis that, in turn, stimulates the internalization of the AMPA and NMDA glutamate receptors, both essential for long-term plasticity. One of the proteins stimulated by mGluR5 activity is FMRP. As FMRP is a negative regulator of transcription, an increase in FMRP may result in a local decrease in mRNA translation. The resulting decrease in protein synthesis may slow down the internalization of the glutamate receptors, resulting in an increase in LTD.

be the only gene of this protein family in *Drosophila* with equal similarity to FMR1, FXR1, and FXR2 (Fig. 9). No CGG repeat is present in the 422-bp-long 5′ UTR region. Alternative splicing in dFmr1 occurs across the gene giving rise to different splice forms, but nothing is known about the expression of these splice products yet.

Like the human FMRP, the dFmr1 protein contains 2 KH domains, an RGG box with 50% identity to the human FMR1 RGG box, an NES, and an NLS. The dFmr1 NES shows little sequence conservation when compared to human FMRP and it remains to be determined whether it is a functional nuclear export signal. The region involved in the binding of FMRP to the 60S ribosomal subunit is also present in dFmr1.

The protein can be observed in the cytoplasm, but a nuclear localization has not been detected. dFmr1 is expressed in early embryo starting from gastrulation. It is highly expressed in neurons but not in glia. It is not enriched in synapses. dFmr1 is moderately expressed in muscles, in presynaptic motor neurons, and in postsynaptic muscles during embryonic development and in the larvae.

Using heteropolymer binding essays, dFmr1 shows high affinity for poly(G), weaker affinity for poly(U), and no affinity for poly(C) and poly(A), characteristics identical to the human protein. Point mutations have been introduced in the highly conserved isoleucine residues, corresponding to the human Ile304Asn mutation, within each of the KH domains (I244N and I307N). Like mutated human FMRP, mutated dFmr1 loses the capacity to bind poly(U); dFmr1 binds with high capacity to human FMR1, but shows only moderate binding capacities with FXR1 and FXR2 and with itself.

7.3.2 dFmr1 is Involved in Neuronal Maturation

Different mutant flies, including dFmr1 null mutants, dFmr1 overexpression mutants, and I244N and I307N KH point mutants, have been constructed to study the role of dFmr1 in *Drosophila*. The introduction of dFmr1 null mutants is not lethal, but some mutants fail to eclose properly, and only a few "escapers" survive. Overexpression of dFmr1 leads to severe rough-eye phenotype, whereas null mutants have no detectable effect in the structure of the eye. Photoreceptor development and function are not effected by loss of dFmr1, but overexpression of dFmr1 in the antennal disc of the eye causes apoptosis.

As observed in humans and mice, absence of dFmr1 and even different concentrations of dFmr1 compared to controls causes aberrant synaptic growth and plasticity. While overexpression of dFmr1 gives undergrowth of synapses and causes presynaptic enlargements of single synaptic boutons of twice the size, the null mutant of dFmr1 shows synaptic overgrowth and overelaboration of synaptic terminals but has no visible effect on the size of the synaptic boutons. Null mutants of dFmr1 exhibit aberrant small ectopic neurite branches, probably causing a misregulation of the intrinsic branching program. There is also a mistargeting of axons and looping fibers, potentially because of guidance defects. In dFmr1 null mutants, fiber extension abnormal morphology in the dorsal-cluster (DC) neurons is observed, as shown in Fig. 12. The level of dFmr1 determines the severity of the phenotype. The phenotype of

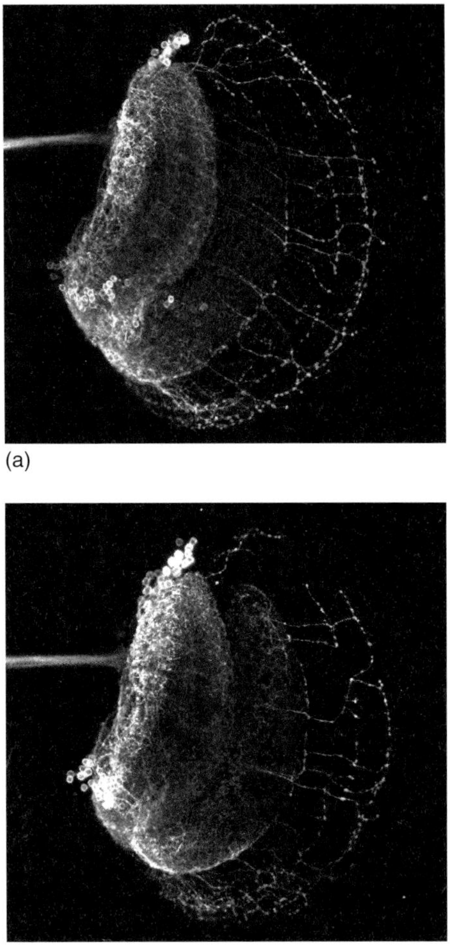

(a)

(b)

Fig. 12 Illustration of abnormal neuronal migration in *Drosophila* dFmr1 null mutant. Comparison of (a) heterozygous with (b) homozygous mutants shows a reduction in the number of dorsal-cluster neuron axons crossing from the lobula into the medulla. This finding suggests that in dorsal-cluster neurons, dFMR1 is required for proper extension and pathfinding of axons. (Photograph kindly provided by Katsu Okamura and Haru Siomi.)

the overexpression mutant is more severe than that of the loss-of-function mutant. dFmr1 overexpression caused a complete failure of axon extension from the lobula to the medulla. Overexpression of an I307N mutant dFmr1 causes defects similar to those observed in loss-of-function mutants but are less severe. These findings suggest that dFmr1 acts as an inhibitor of axon extension and as a regulator of axon growth and synapse formation in the cell.

7.3.3 MAP1B may be Involved in the Fragile X Syndrome

MAP1B is a protein that regulates the growth of dendrites, axons, and synapses. The *Drosophila* homolog for MAP1B has been identified as futsch. Using dFmr1-specific antibodies, futsch mRNA could be coprecipitated, suggesting that it may be part of the dFmr1–mRNP complex. MAP1B has also been identified as an *in vivo* target of FMRP and its mRNA contains a G-quartet. The level of futsch mRNA in *Drosophila* is inversely related to the level of dFmr1. Null mutants of dFmr1 display an increase in futsch mRNA, while overexpression mutants show a decrease in futsch-mRNA level. Overexpression of futsch in the fly causes neuromuscular junction overgrowth and increased bouton number, synaptic area, and branching. Introducing double futsch/dFmr1 null mutations rescued the structural and functional defects compared to dFmr1 mutant flies. This may suggest that dFmr1 regulates synaptic maturation and plasticity by functioning as a translational repressor of futsch mRNA.

7.3.4 dFmr1 is Involved in Circadian Rhythm Regulation

Most dFmr1 loss-of-function mutant flies develop until the pharate adult stage but fail to eclose in the early morning. Moreover, while wild-type flies eclose during the early morning, dFmr1 null mutants may eclose at all times of the day. This suggests a defect in the circadian rhythm. This could be demonstrated by activity measurements. dFmr1 mutants show severely decreased activity levels without any day–night periodicity when the frequency with which flies cross an infrared beam was measured in the locomotor activity test, whereas wild-type flies exhibit a rhythmic pattern of rest and activity during a 23.5 h circadian cycle. The circadian defect was rescued by introducing the wild-type *dFmr1* gene as a transgene into the null mutant, while overexpression of dFmr1 lengthened the circadian pattern. Futsch mutants and dFmr1/futsch double mutants have no circadian defect and no change in locomotor activity or strength of rhythmicity.

Independently, Dockendorff et al. (2002) and Morales et al. (2002) concluded that clock genes are not affected in the dFmr1 mutant flies, suggesting that the dFmr1 mutations may have a strong effect on the pathway downstream from the endogenous clock components. Consistent with this hypothesis, the amplitude of the oscillations of the cAMP responsive-element binding protein, CREB, was found to be reduced in the mutant dFmr1 flies, suggesting that dFmr1 affects this known molecular output of the clock.

7.3.5 **dFmr1 may be Involved in RNA Interference**

dFmr1 was found to interact with L5 and L11, two large ribosomal subunit proteins. This demonstrates that, like the human fragile X protein, the dFmr1 protein associates with the 60S ribosomal subunit in *Drosophila*. The finding that not the dFmr1 region equivalent to the ribosomal binding domain in human FMRP but the 285–amino acid–long C-terminal end of the dFmr1 protein binds both ribosomal proteins was peculiar. The L5 protein is associated with 5S rRNA and forms a ternary complex by binding dFmr1 protein, suggesting that dFmr1 interaction with ribosomes takes place via the L11 protein and the L5/5S rRNA complex.

Two other proteins found associated with the dFmr1 protein in *Drosophila* are AGO2 and Dmp68. AGO2, a member of the Argonaute gene family, is an important component of the RNA-induced silencing complex (RISC), which recognizes and cleaves homologous mRNAs and mediates RNAi in Drosophila. Dicer, an enzyme that processes both miRNA and siRNA was also part of the complex containing AGO2 and dFmr1. p68, the human ortholog of Dmp68, unwinds short dsRNA in an ATP-dependent manner. Dmp68 is also required for efficient RNAi. Binding to both proteins suggests that dFmr1 may be involved in RNAi.

7.3.6 **dFmr1 is Modified After Translation**

CKII is a protein that phosphorylates nuclear and nonnuclear proteins and is highly conserved among species. Recent studies identified *Drosophila* casein kinase II as a protein phosphorylating dFmr1. The phosphorylated Ser406, located upstream of the RGG box in dFmr1, is a conserved amino acid between the fly and human FMR1 and FXR proteins. *In vitro*, replacement of the serine with alanine residues in dFmr1 protein and human FMRP abolishes CKII phosphorylation. Phosphorylation affects the efficiency of dFmr1 to form homomers and to bind other RNAs. Though phosphorylation was observed *in vivo*, the significance of this posttranslational modification remains unclear.

8 Phenotypic Effect of the Premutation

With the advent of a DNA-based diagnostic test, the study of premutation carriers has increased. Individuals with a premutation were initially thought to have no phenotypic manifestations. However, a number of studies have reported mild learning disabilities and emotional problems in a small subgroup of premutation carriers. Subsequent study demonstrated that these findings challenge the view that premutation carriers of FraX are unaffected. Premutation carriers have an increased risk for premature ovarian failure (POF) in females. Hundscheid et al. (2000) found that paternally inherited premutations were more likely to give rise to POF than maternally inherited premutations. This imprinting was not seen by the other groups, perhaps as a result of differences in the age of the study population or differences in inclusion criteria.

Recent studies in premutation carriers showed that the length of the CGG repeat has an inverse correlation with the FMRP levels, while it has a positive correlation with the FMR1 mRNA levels. Tassone et al. (2000) described premutation carriers with more than 100 CGG repeats having a reduced number of FMRP-positive cells but with at least fivefold elevated mRNA levels. These findings lead to the hypothesis of a compensatory mechanism. If the translation of FMR1 mRNA is less efficient or defective, low FMRP levels initiate via feedback induction FMR1 transcription. The resulting high FMR1 mRNA levels can only partially compensate the defect since normal FMRP levels are not achieved in cells with large premutation repeats. It is yet unknown whether FMRP levels itself are the crucial factor at work in this compensatory mechanism or whether it is coupled with the length of the CGG repeat. The expanded CGG repeat may lead to proportionally more open promoter conformation and consequently enhanced transcriptional activity. The normal FMR1 message levels in the mutant I304N cell line argues against a role for the lack of functional FMRP as a modulator of transcriptional activity of the *FMR1* gene. Another candidate for the upregulation of the FMR1 expression might be CGGBP1, a CGG-binding protein, that has shown the ability to regulate expression from the *FMR1* gene with a small CGG repeat. It can be hypothesized that in the case of an elongated CGG repeat, the CGGBP1 might be sequestered owing to extensive binding to the elongated repeat in the mRNA, resulting in a lowered active concentration of CGGBP1. A lower CGGBP1 concentration results in an increased FMR1 transcription.

Elevated FMR1 mRNA levels and/or reduced FMRP levels result in a phenotype recently described in older males carrying a premutation. These males may exhibit a unique neurodegenerative syndrome characterized by progressive intention tremor and ataxia. More advanced cases are accompanied by memory and executive function deficits, anxiety, and eventual dementia. MR imaging studies (T_2 signal) of the brain of symptomatic adult male premutation carriers showed a characteristic imaging, including hyperintensities of the middle cerebellar peduncle, cerebellar white matter lateral, superior, and inferior to the dentate nuclei, and volume loss involving the pons, mesencephalon, cerebellar cortex, cerebral cortex, white matter of the cerebral hemispheres, and corpus callosum. Neurohistological studies on the brains of four symptomatic elderly premutation carriers demonstrated neuronal degeneration in the cerebellum and the presence of eosinophilic intranuclear inclusions in both neurons and astroglia. Furthermore, the inclusions showed a positive reaction with antiubiquitin antibodies, which suggests a link with the proteasome degradation pathway. The origin and constitution of the inclusions is poorly understood; however, elevated FMR1 mRNA levels have been proposed to be important for the formation of the inclusions.

A "knock-in" CGG triplet mouse has been generated that showed moderate CGG-repeat instability upon both maternal and paternal transmission. These mice showed an increased expression of FMR1 mRNA and a decreased expression of FMRP. The presence of ubiquitin-positive intranuclear inclusions was seen in neurons of the CGG expanded-repeat mouse. An increase in both the number and the size of the inclusions was observed during the course of life, which correlates with the progressive character of the cerebellar tremor/ataxia syndrome in humans. The observations in expanded-repeat mice suggest a correlation between the presence of intranuclear inclusions in distinct regions of the brain and the clinical features in symptomatic premutation carriers. The presence of inclusions in expanded-repeat mice supports a direct role of the *FMR1* gene, by either CGG expansion *per se* or mRNA level, in the formation of the inclusions, which in humans could have also be explained by a synergistic effect with another genetic locus.

These two clinical features are remarkable in that they are unique to premutation carriers, while full-mutation individuals are not affected. As the number of male premutation carriers is 10 times the number of males with a full mutation, the number of individuals who are at risk for the ataxia/tremor phenotype is large. It is not known whether females with a premutation suffer from this asimilar phenotype. If so, the age of onset can be expected to be later than in males.

9 The FMR1 Mutation

The FMR1 CGG repeat containing less than 50 CGGs is very stable. However, the CGG repeat above 50 repeat units is remarkably unstable. In the lower premutation range, the repeat can increase and decrease in size while staying in the premutation range. In the upper premutation range, the repeat can undergo massive expansions to more than several hundred repeats in full-mutation alleles. Two reports have suggested that full-mutation expansions are more likely to occur in male than in female offspring. However, a recent multicenter study found no significant differences in the proportion of full-mutation alleles in male or female fetuses.

9.1 Mechanism of CGG-repeat Expansion

Repeat instability is influenced by several factors. First, only transmission through the female germ line can give rise to a full mutation in the next generation. In contrast, transmission through the male germ line, irrespective of whether the male has either a full mutation or a premutation in his somatic cells, always reveals a premutation in the daughters. Second, the instability of premutation alleles in females is positively correlated with the repeat size. They may expand to a different-sized premutation or a full mutation, larger repeats being more prone to expansion than smaller ones. Third, not only the length of the repeat is a risk factor for repeat instability but also the composition of the repeat. Cryptic AGG interruptions in the CGG repeat might provide stability to the repeat and instability might result from loss of AGG interruptions. In the normal population, the CGG repeat is interrupted by AGG trinucleotides, most often at positions 10 and 20. Loss of AGG repeats leads to alleles with longer perfect CGG-repeat tracts, which are prone to expansion, especially when the number of pure CGG units exceeds 30. Premutation alleles have either no or only one AGG repeat unit and are therefore unstable.

Loss of these AGG interruptions always occurs at the 3′ end of the repeat tract, and as a consequence, the longest pure CGG tract found within a premutation or full mutation is always located at the most 3′ end. The reason for this polar variation might involve the mechanisms underlying the repeat instability. The polarity of expansion of the CGG repeats in humans mimics the orientation-dependent instability of CGG repeats cloned into *Escherichia coli* and *Saccharomyces cerevisiae*. This finding suggests that a difference in leading- versus lagging-strand synthesis may be involved in repeat expansion.

Nonstructural, cis-acting factors may be related to a specific haplotype background as certain haplotypes of polymorphic markers in and around the *FMR1* gene are overabundant in fragile X chromosome. Dombrowski et al. (2002) found, in a population-based study, that gray zone alleles or protomutations (40–55 repeats) also are often associated with a specific fragile X syndrome-related haplotype. However, others, in a much smaller number of unstable events, did not find evidence for a specific haplotype as a risk factor for instability.

Studies in yeast and *E. coli* suggest that direction of replication, genetic background (including repair systems), transcription, and growth conditions have an influence on repeat instability too. Repeat instability is thought to be instigated by the formation of d(CGG) secondary structures. Several structures of the CGG repeat, such as hairpins and tetrahelical structures, have been suggested. The stability of the different structures seems to differ between both templates. Pausing of DNA polymerases at both hairpin and tetrahelical structures was observed *in vitro* as well as *in vivo*. This stalling of DNA polymerase might increase the probability of DNA polymerase slippage and this can result in expansion of the CGG repeat.

White et al. (1999) proposed the influence of RAD27, the yeast homolog of human FEN1, as a trans-acting factor on instability. RAD27 is necessary for correct processing of Okazaki fragments and its absence strongly increased CGG instability. Interestingly, White et al. (1999) did not observe any increase in the frequency of large expansion events when studying various CGG arrays in *mlh1*, *msh2*, *msh3*, and *msh6* mutant backgrounds. In fact, it has been surprisingly concluded that repeat instability requires mismatch repair proteins.

It is likely that a multistep process is involved in repeat instability since it has never been observed that a normal repeat expands to a full-mutation repeat upon one transmission. The first step assumed to be involved in the initial expansion of trinucleotide repeats is slipped-strand

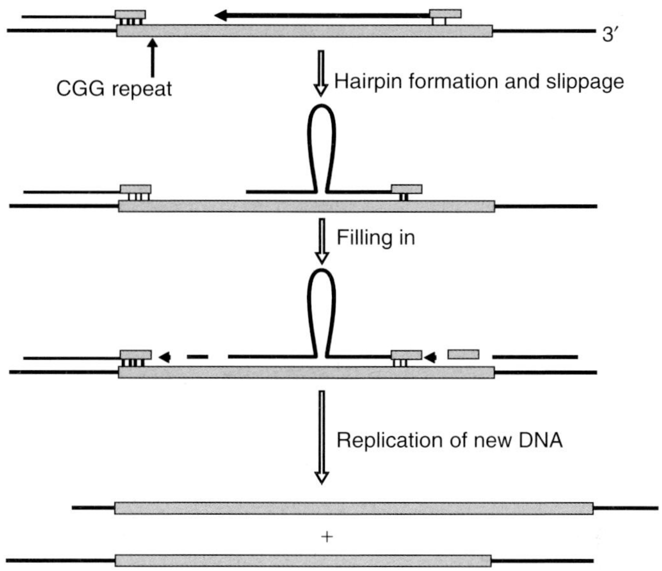

Fig. 13 Model for instability and expansion of the CGG repeat. During replication of the CGG repeat, a hairpin is formed followed by filling in of the gap, resulting in expansion of the repeat during multiple rounds of replication.

mispairing between DNA strands during DNA replication (Fig. 13) or possibly repair. Pausing of DNA polymerases allows the formation of secondary structures on the nascent lagging strand, which, if not correctly excised, should result in a small (up to 10 CGG-repeat units) increase. This slippage model can explain relatively small changes that occur within the normal and premutation alleles. The slippage model is the favored model, although other models, such as unequal meiotic sister chromatid exchange and/or gene conversion, are difficult to rule out.

For transition from premutation to full mutation, a different mechanism must be considered and it is thought that extremely unstable structures are formed in the nascent lagging strand owing to slippage on both ends of Okazaki fragments when these fragments contain solely CGG repeats. Incorrect repair will then lead to large expansions but might also result in contractions or deletions.

9.2 Timing of CGG-repeat Expansion

The timing of the repeat expansion and the difference observed in male and female transmission of the repeat is an intriguing feature of the mode of inheritance of the fragile X syndrome. How and when expansions from premutation to full mutation occur is not known.

Oocytes from full-mutation female fetuses have shown to contain a full mutation without a detectable premutation, suggesting a prezygotic model for expansion. In male fetuses with a full mutation in their somatic cells, the developing testes also show absence of FMRP in most cells. Testis tissue from a 13-week-old full-mutation fetus showed only full mutations in the germ cells, while testis tissue from an older fetus showed evidence of both full and premutations in the germ cells. Adult full-mutation males show only a premutation in their sperm cells. The most likely model explaining these findings is the prezygotic model in which the full mutation is transmitted in the oocyte and the mutation in sperm-producing cells is regressed to a premutation size. While data by Malter et al. (1997) point to a prezygotic expansion, expansion occurring very early in embryogenesis (before day 3–5) cannot be ruled out.

The somatic mosaicism, often seen in patients, can be explained by instability that occurs postzygotically. Instability in the full-mutation range can occur, but also regression of the full mutation to a premutation can occur in some cells in somatic tissue. To study the exact timing of repeat expansion, an animal model will be required. The availability of an animal model to study the CGG-repeat instability will allow studying the repeat length at different time points during development.

10 Diagnostics

The identification of the *FMR1* gene and the mutational mechanism responsible for the fragile X syndrome introduced reliable DNA-based diagnostic tests, focused on the detection of the CGG-repeat amplification in patients and carriers, with or without associated hypermethylation. The two DNA methods to determine the CGG-repeat size are Southern blot hybridization and the polymerase chain reaction (PCR).

The availability of antibodies directed against FMRP allowed a diagnostic test to identify fragile X patients on the basis of the presence of FMRP in cells from unaffected individuals and its absence in cells from fragile X patients. Table 1 gives an overview of the different mutations in fragile X patients and carriers.

Tab. 1 Overview of the different mutations in fragile X patients and carriers.

Individual	Mutation	MR-associated CpG island	Number of repeats	Methylation of FMR-1	mRNA transcription	Risk of affected offspring
Normal	–	–	6–50	–	+	–
NTM	Pre-	–	50–200	–	+	–
Daughter of NTM/female carrier of a premutation	Pre-	–	50–200	–	+	+
Female carrier of a full mutation	Full[a]	70%[b]	>200	+[c]	+[d]	+
Male patient	Full[a]	+	>200	+	–	–

Note: MR: mental retardation; NTM: normal transmitting male.
[a]Somatic mosaicism is seen in about 40% of male patients (Nolin, 1994 #180). It is also observed in female carriers.
[b]In about 70% of females carrying a full mutation, mild-to-moderate mental retardation is found.
[c]Methylation is found on the inactive X and on the X chromosome carrying the full mutation.
[d]mRNA is only transcribed from the normal active X.

10.1 DNA Diagnostics

Very few fragile X cases have been reported without amplification of the CGG repeat and with either point mutations or deletions in other parts of the *FMR1* gene and DNA diagnostics is concentrating on detecting the repeat size.

The ability to detect both premutations and full mutations at the DNA level has made the diagnosis of the widespread disease available and has made genetic counseling possible through reliable detection of female and male carriers and patients, and can be used for prenatal diagnosis. Fragile X syndrome testing is based on the detection of CGG-repeat elongation in patients. This is based either on a Southern blot of leukocyte DNA digested with specific restriction endonucleases or on direct amplification of the CGG repeat using flanking primers.

10.1.1 Southern Blotting

The length of the CGG repeat can be determined by Southern blot analysis and mostly *Eco*RI or *Hind*III are used. Normal alleles will give a fragment of 5.2 kb. Longer fragments indicate the presence of a pre- or full mutation. Premutations are visible as fragments somewhat larger than 5.2 kb (up to 5.7 kb). Full mutations are visible as diffuse broad bands or "smears" owing to the presence of different repeat lengths (above 200 repeats) in different cells. Digestion with *Pst*I gives smaller fragments and, therefore, allows a better detection and sizing of premutations. By using an additional methylation-sensitive enzyme such as *Eag*I or *Bss*HII, the length and the methylation status can be determined. This is applied to distinguish large premutations from small full mutations. For prenatal Southern blot diagnosis, digestion with a single enzyme (*Hind*III) is sufficient, as methylation has not occurred yet at the time (11 weeks) when chorionic samples were taken. DNA analysis is usually carried out using radioactive-labeled DNA probes; however, a good alternative is the use of nonradioactive (digoxigenin-labeled) probes followed by chemiluminescent detection.

10.1.2 Example of DNA Diagnostics in a Fragile X Family

Postnatal and prenatal diagnosis of fragile X syndrome can now be exercised with a high degree of accuracy by detecting the CGG-repeat amplification at the DNA level. Figure 14 shows a fragile X family in which premutations and full mutations are found. In the DNA of the grandmother (1) one normal 5.2 kb band (indicated by an arrow) is seen representing her 2 normal X chromosomes. The grandfather (2) is a normal transmitting male with a premutation of 70 repeats in the *FMR-1* gene; his DNA shows a band slightly larger than 5.2 kb. He passed his premutation on to his three daughters (4, 8, and 14) who are phenotypically normal carriers. Besides the 5.2 kb band of their normal chromosome, they have a premutation band of increased size (75 and 80 repeats). Daughter 4 has 1 affected daughter (6) who is mosaic for a premutation and a full mutation. Daughter 8 has 2 affected sons (12 and 13) with full mutations. Daughter 14 has one affected son (16) who is mosaic for a premutation and a full mutation.

10.1.3 PCR Amplification Analysis

Detection and accurate sizing of premutations can be done by PCR. The high CG content requires a specific PCR protocol. Initially, radioactivity was used to detect

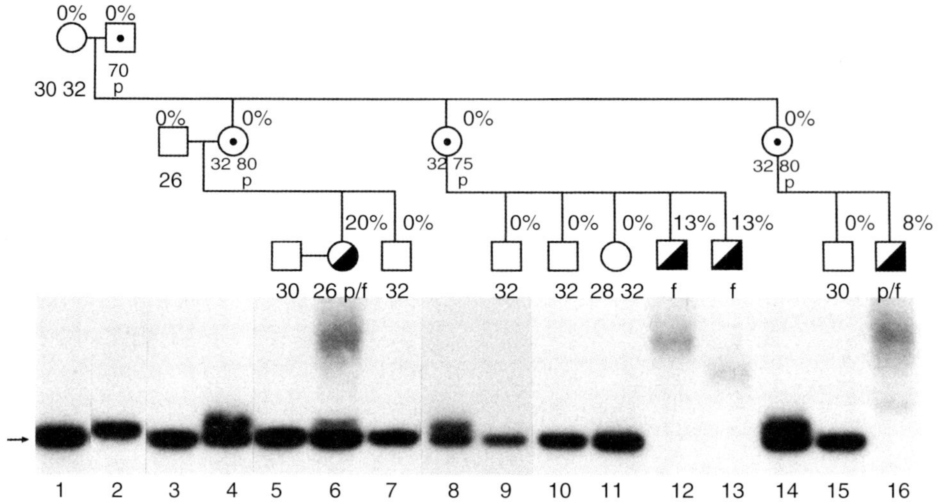

Fig. 14 Detection of the *FMR1* gene in a fragile X family by means of Southern blot analysis. {symbol individual 2} = normal transmitting male; {symbol individual 4} = normal female carrier; {symbol individual 6} = female carrier with cytogenetic expression; {symbol individual 12} = mentally retarded male; percentage of cytogenetic expression (if determined); number of CGG repeats for the different X chromosomes (determined by PCR) is indicated; p, premutation; f, full mutation. (Example of fragile X DNA diagnosis provided by Dr D. J. J. Halley and Dr A. M. W. van der Ouweland.)

the size of the PCR products. Alternatively, a nonradioactive method was developed using fluorescent-labeled precursors during the PCR amplification, and the size of the PCR product can be determined on an automated ABI sequencer. This last method gives a very accurate sizing of the PCR product in both males and females, starting from a very small amount of DNA.

Amplification of full mutations is very inefficient because it is difficult to amplify stretches of more than 100 to 200 CGG repeats, and it is therefore not a reliable method. Recently, an alternative PCR approach was described for the diagnosis of fragile X syndrome based on the methylation-sensitive conversion of C residues to U by bisulfite on single-strand DNA and subsequent amplification of the antisense strand with specific primers. A PCR with primers for methylated C residues will amplify the CpG dinucleotide region upstream of the CGG repeats exclusively in affected males. Moreover, the bisulfite treatment dramatically reduces the C + G component of the region; thus, the high T_m and the strong secondary structures are no longer obstacles for PCR amplification. This method can be used to identify males with a full mutation but cannot be applied for females with a full mutation owing to the presence of a nonmutated, methylated, inactive X chromosome.

10.2 Protein Test

Recently, alternative immunocytochemical tests have been described that are based on the direct detection of FMRP by a monoclonal antibody and the indirect visualization of antibody–antigen complexes by alkaline phosphatase enzyme activity. The antibody test has the advantage of having the test result within 24 h, whereas routine DNA analysis by Southern blot generally takes more than 7 days.

A rapid method based on antibody detection of the FMR1 protein in cells of blood smears has been described and validated by Willemsen et al. FMRP in the cytoplasm of cells from unaffected individuals could be detected, whereas the cytoplasm of cells from affected males were not labeled for FMRP. For males, the diagnostic power of the test is high, because there is no overlap between the values of unaffected individuals and affected males with the full mutation. The diagnostic power of the test in identifying females with the full mutation is less specific compared to male patients. The X-inactivation process (lyonization) in cells from females with the full mutation may result in FMRP expression in a high percentage of their lymphocytes (up to 80%).

More recently, the protein test has been adapted for hair roots (Willemsen 1999), avoiding the need for blood drawing and allowing the analysis of the hair bulb cells that are of ectodermal origin and possibly less subject to selection effects in mosaic individuals than blood cells (of mesodermal origin). The diagnostic power of this new noninvasive test is high and it can be readily used to identify male patients. In addition, it was shown that the hair root test has the potential to serve as a prognostic indicator of the mental status in females with the full mutation.

11 Therapeutic Options

To study whether the fragile X syndrome is a potentially treatable disorder, rescue mice have been generated. Rescue mice are knockout mice in which an *FMR1* gene is introduced in the germ line by transgenesis. Introduction of FMRP as a cDNA under a CMV promoter in the Fmr1 knockout mouse resulted in expression of FMRP in the brain at roughly 50% of the control level. However, neither the phenotype, as measured by testicular weight, nor the cognitive capacities, as measured in the Morris water-maze, was rescued. Immunohistochemistry of the brain of the rescue mice showed FMRP expression in neurons, but also in nonneuronal cells, perhaps as a result of the CMV promoter used to drive FMR1 transcription.

Subsequently, Peier et al. (2000) created YAC transgenics, containing the whole human FMR1 with a large amount of flanking sequences, as a rescue model. Since the *FMR1* gene in the YAC is under the control of its own promoter, the gene is expressed in tissues similar to control mice. However, the overall level of FMRP seemed higher than that in control littermates. The testicular weight of the rescue mouse was restored, but the results of the behavioral test were controversial. In some tests, the rescue mice seemed "overcorrected"; for example, in the open field test, knockouts were significantly more active than controls, but the rescue mice were significantly less active than controls.

The generation of these two rescue mice suggests that concentration of FMRP as well as cell specificity and timing of expression are very important.

11.1 CGG Methylation and Reactivation

Rare individuals with expanded but unmethylated repeats have been described. Such individuals are generally unaffected with the fragile X syndrome, demonstrating that methylation and not repeat elongation *per se* causes the typical features of the syndrome. Addition of the demethylating agent 5-azadeoxycytidine to cultured lymphoblasts of fragile X patients did enable the production of FMRP by these cells. However, the percentage of reactivated cells was disappointingly low. As methylation in the cell is followed by deacetylation, it was hypothesized that a combination of demethylating and acetylating drugs might enhance demethylation of the CGG repeat. Indeed, a synergistic effect of the two types of drugs was demonstrated to enhance the reactivation of the FMRP production approximately fivefold.

Though reactivation of protein expression *in vivo* is a promising result, it should be realized that demethylating agents are highly toxic, and as long as no drugs are developed that specifically target the FMR1-CGG repeat and that are less toxic, the use of demethylating and acetylating agents in patients remains elusive.

11.2 AMPAkine Trial in Relationship with LTD Abnormality

Synaptic plasticity is believed to be involved in learning and memory, and it is therefore tempting to speculate that the enhanced LTD observed in fragile X mice (see Chapter 6.2.7) is potentially correlated to the cognitive deficits observed in the fragile X syndrome. LTD is directly dependent on the AMPA receptor. The availability of a group of drugs called AMPAkines that directly targets the AMPA receptor prompted a trial with one of these agents called CX516 in a small group of fragile X patients. So far, no negative side effects have been reported, but the results on ameliorating the symptoms of the fragile X syndrome are awaited.

12 Epilogue

More than a decade of research following the discovery of the causative gene has revealed a wealth of information regarding the possible role of the lack of the *FMR1* gene in the abnormalities of the fragile X syndrome. Functional domains identified in the fragile X protein suggested a role in RNA transport through the cell. Subsequent experimental evidence revealed that FMRP transports specific mRNAs as an mRNP particle along the dendrites to the synapses in neurons and plays a role in translation after binding to polyribosomes. A large amount of possible mRNAs bound by FMRP whose translation may be regulated by FMRP has been determined, including a specific group of mRNAs that contain a planar G-quartet structure. A subset of the bound mRNAs identified by different research groups plays an important role in synapse maturation and plasticity. Though certain neurotransmitters, hormones or stress, are able to influence the FMRP levels, the relevant *in vivo* stimuli, if any, are not known at present. In addition, the effect of the variable levels of FMRP on individual bound mRNAs is not known. However, in general FMRP seems to inhibit translation. Further functional studies have to determine the importance of each individual gene whose translation is influenced by FMRP levels in fragile X syndrome. Thus, misregulation or mistrafficking of a subset of mRNAs as a consequence of the absence of FMRP could lead to impaired synaptic maturation and plasticity and lie at the basis of the mental retardation and associated abnormalities in fragile X syndrome. Further elucidation of the role of FMRP in patients and animal models may help us understand why absence of the fragile X protein leads to the pathogenesis of the syndrome.

See also Brain Development.

Bibliography

Books and Reviews

Bardoni, B., Mandel, J.L. (2002) Advances in understanding of fragile X pathogenesis and FMRP function, and in identification of X linked mental retardation genes, *Curr. Opin. Genet. Dev.* **12**, 284–293.

Chiurazzi, P., Oostra, B.A. (2000) Genetics of mental retardation, *Curr. Opin. Pediatr.* **12**, 529–535.

Gantois, I., Kooy, R.F. (2002) Targeting fragile X, *Genome Biol.* **3**, 1014.1–1014.5.

Hagerman, R.J., Hagerman, P.J. (Eds.) *Fragile X Syndrome: Diagnosis, Treatment, and Research*, Johns Hopkins University Press, Baltimore, MA, pp. 3–109.

Hagerman, R.J., Hagerman, P.J. (2001) Fragile X syndrome: a model of gene-brain-behavior relationships, *Mol. Genet. Metab.* **74**, 89–97.

Jin, P., Warren, S.T. (2003) New insights into fragile X syndrome: from molecules to neurobehaviors, *Trends Biochem. Sci.* **28**, 152–158.

Kooy, R.F. (2003) Of mice and the fragile X syndrome, *Trends Genet.* **19**, 148–154.

Kooy, R.F., Willemsen, R., Oostra, B.A. (2000) Fragile X syndrome at the turn of the century, *Mol. Med. Today* **6**, 194–199.

O'Donnell, W.T., Warren, S.T. (2002) A decade of molecular studies of fragile X syndrome, *Annu. Rev. Neurosci.* **25**, 315–338.

Oostra, B.A., Chiurazzi, P. (2001) The fragile X gene and its function, *Clin. Genet.* **60**, 399–408.

Willemsen, R., Oostra, B.A. (2000) FMRP detection assay for the diagnosis of the fragile X syndrome, *Am. J. Med. Genet.* **97**, 183–188.

Primary Literature

Abitbol, M., Menini, C., Delezoide, A.-L., Rhyner, T., Vekemans, M., Mallet, J. (1993) Nucleus basalis magnocellularis and hippocampus are the major sites of FMR-1 expression in the human fetal brain, *Nat. Genet.* **4**, 147–153.

Adinolfi, S., Bagni, C., Musco, G., Gibson, T., Mazzarella, L., Pastore, A. (1999) Dissecting FMR1, the protein responsible for fragile X syndrome, in its structural and functional domains, *RNA* **5**, 1248–1258.

Ashley, C.T., Sutcliffe, J.S., Kunst, C.B., Leiner, H.A., Eichler, E.E., Nelson, D.L., Warren, S.T. (1993a) Human and murine FMR1: alternative splicing and translational initiation downstream of the GGG repeat, *Nat. Genet.* **4**, 244–251.

Ashley, C.T., Wilkinson, K.D., Reines, D., Warren, S.T. (1993b) FMR-1 protein: conserved RNP family domains and selective RNA binding, *Science* **262**, 563–566.

Bächner, D., Mance, A., Steinbach, P., Wöhrle, D., Just, W., Vogel, W., Hameister, H., Poustka, A. (1993) Enhanced expression of the murine FMR1 gene during germ cell proliferation suggests a special function in both the male and the female gonad, *Hum. Mol. Genet.* **2**, 2043–2050.

Bakker, C.E., Kooy, R.F., D'Hooge, R., Tamanini, F., Willemsen, R., Nieuwenhuizen, I., de Vries, B.B.A., Reyniers, E., Hoogeveen, A.T., Willems, P.J., De Deyn, P.P., Oostra, B.A. (2000) Introduction of a *FMR1* transgene in the fragile X knockout mouse, *Neurosci. Res. Commun.* **26**, 265–277.

Bakker, C.E., Verheij, C., Willemsen, R., van der Helm, R., Oerlemans, F., Vermey, M., Bygrave, A., Hoogeveen, A.T., Oostra, B.A., Reyniers, E., De Boulle, K., D'Hooge, R., Cras, P., van Velzen, D., Nagels, G., Martin, J.-J., De Deyn, P.P., Darby, J.K., Willems, P.J. (1994) Fmr1 knockout mice: a model to study fragile X mental retardation, *Cell* **78**, 23–33.

Bardoni, B., Giglio, S., Schenck, A., Rocchi, M., Mandel, J.L. (2000) Assignment of NUFIP1 (nuclear FMRP interacting protein 1) gene to chromosome 13q14 and assignment of a pseudogene to chromosome 6q12, *Cytogenet. Cell. Genet.* **89**, 11–13.

Bardoni, B., Schenck, A., Mandel, J.L. (1999) A novel RNA-binding nuclear protein that interacts with the fragile X mental retardation (FMR1) protein, *Hum. Mol. Genet.* **8**, 2557–2566.

Bardoni, B., Sittler, A., Shen, Y., Mandel, J.L. (1997) Analysis of domains affecting intracellular localization of the FMRP protein, *Neurobiol. Dis.* **4**, 329–336.

Bat, O., Kimmel, M., Axelrod, D.E. (1997) Computer simulation of expansions of DNA triplet repeats in the fragile X syndrome and Huntington's disease, *J. Theor. Biol.* **188**, 53–67.

Bliss, T.V.P., Collingridge, G.L. (1993) A synaptic model of memory: long-term potentiation in the hippocampus, *Nature* **361**, 31–39.

Bontekoe, C.J.M., Bakker, C.E., Nieuwenhuizen, I.M., van der Lans, H., de Lange, D., Hirst, M.C., Oostra, B.A. (2001) Instability of a $(CGG)_{98}$ repeat in the *Fmr1* promoter, *Hum. Mol. Genet.* **10**, 1693–1699.

Bontekoe, C.J.M., McIlwain, K.L., Nieuwenhuizen, I.M., Yuva-Paylor, L.A., Nellis, A., Willemsen, R., Fang, Z., Kirkpatrick, L., Bakker, C.E., McAninch, R., Cheng, N.C., Merriweather, M., Hoogeveen, A.T., Nelson, D., Paylor, R., Oostra, B.A. (2002) Knockout mouse model for Fxr2: a model for mental retardation, *Hum. Mol. Genet.* **11**, 487–498.

Brown, W.T., Houck, G.E., Jeziorowska, A., Levinson, F.N., Ding, X., Dobkin, C., Zhong, N., Henderson, J., Brooks, S.S., Jenkins, E.C. (1993) Rapid fragile X carrier screening and prenatal diagnosis using a nonradioactive PCR test, *JAMA* **270**, 1569–1575.

Brown, V., Jin, P., Ceman, S., Darnell, J., O'Donnell, W.T., Tenenbaum, S.A., Jin, X., Feng, Y., Wilkinson, K.D., Keene, J.D., Darnell, R.B., Warren, S.T. (2001) Microarray identification of FMRP-associated brain mRNAs and altered mRNA translational profiles in fragile X syndrome, *Cell* **107**, 477–487.

Brown, V., Small, K., Lakkis, L., Feng, Y., Gunter, C., Wilkinson, K.D., Warren, S.T. (1998) Purified recombinant Fmrp exhibits selective RNA binding as an intrinsic property of the fragile X mental retardation protein, *J. Biol. Chem.* **273**, 15521–15527.

Brunberg, J.A., Jacquemont, S., Hagerman, R.J., Berry, K., Grigsby, J., Leehey, M.A., Tassone, F., Brown, W.T., Greco, C.M., Hagerman, P.J. (2002) Fragile X premutation carriers: characteristic MR imaging findings of adult male patients with progressive cerebellar and cognitive dysfunction, *AJNR Am. J. Neuroradiol.* **23**, 1757–1766.

Castren, M., Lampinen, K.E., Miettinen, R., Koponen, E., Sipola, I., Bakker, C.E., Oostra, B.A., Castren, E. (2002) BDNF regulates the expression of fragile X mental retardation protein mRNA in the hippocampus, *Neurobiol. Dis.* **11**, 221–229.

Ceman, S., Brown, V., Warren, S.T. (1999) Isolation of an FMRP-associated messenger ribonucleoprotein particle and identification of nucleolin and the fragile X-related proteins as components of the complex, *Mol. Cell. Biol.* **19**, 7925–7932.

Ceman, S., Nelson, R., Warren, S.T. (2000) Identification of mouse YB1/p50 as a component of the FMRP-associated mRNP particle, *Biochem. Biophys. Res. Commun.* **279**, 904–908.

Chen, L., Toth, M. (2001) Fragile X mice develop sensory hyperreactivity to auditory stimuli, *Neuroscience* **103**, 1043–1050.

Chiang, P.W., Carpenter, L.E., Hagerman, P.J. (2001) The 5′-untranslated region of the FMR1 message facilitates translation by internal ribosome entry, *J. Biol. Chem.* **276**, 37916–37921.

Chiurazzi, P., MacPherson, J., Sherman, S., Neri, G. (1996) Significance of linkage disequilibrium between the fragile X locus and its flanking markers, *Am. J. Med. Genet.* **64**, 203–208.

Chiurazzi, P., Pomponi, M.G., Pietrobono, R., Bakker, C.E., Neri, G., Oostra, B.A. (1999) Synergistic effect of histone hyperacetylation and DNA demethylation in the reactivation of the *FMR1* gene, *Hum. Mol. Genet.* **8**, 2317–2323.

Chiurazzi, P., Pomponi, M.G., Willemsen, R., Oostra, B.A., Neri, G. (1998) In vitro reactivation of the FMR1 gene involved in fragile X syndrome, *Hum. Mol. Genet.* **7**, 109–113.

Comery, T.A., Harris, J.B., Willems, P.J., Oostra, B.A., Irwin, S.A., Weiler, I.J., Greenough, W.T. (1997) Abnormal dendritic spines in fragile X knockout mice: maturation and pruning deficits, *Proc. Natl. Acad. Sci. USA* **94**, 5401–5404.

Corbin, F., Bouillon, M., Fortin, A., Morin, S., Rousseau, F., Khandjian, E.W. (1997) The fragile X mental retardation protein is associated with poly(A)+ mRNA in actively translating polyribosomes, *Hum. Mol. Genet.* **6**, 1465–1472.

Coy, J.F., Sedlacek, Z., Bächner, D., Hameister, H., Joos, S., Lichter, P., Delius, H., Poustka, A. (1995) Highly conserved 3′ UTR and expression pattern of FXR1 points to a divergent gene regulation of FXR1 and FMR1, *Hum. Mol. Genet.* **4**, 2209–2218.

Crawford, D.C., Zhang, F., Wilson, B., Warren, S.T., Sherman, S.L. (2000) Fragile X CGG repeat structures among African-Americans: identification of a novel factor responsible for repeat instability, *Hum. Mol. Genet.* **9**, 1759–1769.

D'Hooge, R., Nagels, G., Franck, F., Bakker, C.E., Reyniers, E., Storm, K., Kooy, R.F., Oostra, B.A., Willems, P.J., De Deyn, P.P. (1997) Mildly impaired water maze performance in male *Fmr1* knockout mice, *Neuroscience* **76**, 367–376.

Darlow, J.M., Leach, D.R. (1998) Evidence for two preferred hairpin folding patterns in d(CGG).d(CCG) repeat tracts in vivo, *J. Mol. Biol.* **275**, 17–23.

Darnell, J., Jensen, K., Jin, P., Brown, V., Warren, S.T., Darnell, R.B. (2001) Fragile X mental retardation protein targets G quartet mRNAs important for neuronal function, *Cell* **107**, 489–499.

Das, S., Kubota, T., Song, M., Daniel, R., Berry, K., Prior, T.W., Popovich, B., Rosser, L., Arinami, T., Ledbetter, D.H. (1997) Methylation analysis of the fragile X syndrome by PCR, *Genet. Test.* **1**, 151–155.

De Boulle, K., Verkerk, A.J.M.H., Reyniers, E., Vits, L., Hendrickx, J., van Roy, B., van den Bos, F., de Graaff, E., Oostra, B.A., Willems, P.J. (1993) A point mutation in the FMR1 gene associated with fragile X mental retardation, *Nat. Genet.* **3**, 31–35.

De Diego Otero, Y., Severijnen, L.A., van Cappellen, G., Schrier, M., Oostra, B., Willemsen, R. (2002) Transport of fragile X mental retardation protein via granules in neurites of PC12 cells, *Mol. Cell. Biol.* **22**, 8332–8341.

de Graaff, E., Rouillard, P., Willems, P.J., Smits, A.P.T., Rousseau, F., Oostra, B.A. (1995) Hotspot for deletions in the CGG repeat region of *FMR1* in fragile X patients, *Hum. Mol. Genet.* **4**, 45–49.

de Vries, B.B., Fryns, J.-P., Butler, M.G., Canziani, F., Wesby-van Swaay, E., van Hemel, J.O., Oostra, B.A., Halley, D.J.J., Niermeijer, M.F. (1993) Clinical and molecular studies in fragile X patients with a Prader-Willi-like phenotype, *J. Med. Genet.* **30**, 761–766.

de Vries, B.B.A., van den Ouweland, A.M.W., Mohkamsing, S., Duivenvoorden, H.J., Mol, E., van Gelsema, K.R., Halley, D.J.J., Sandkuijl, L.A., Oostra, B.A., Tibben, A., Niermeijer, M.F. (1997) Screening and diagnosis for the fragile X syndrome among the mentally retarded: an epidemiological and psychological survey, *Am. J. Hum. Genet.* **61**, 660–667.

de Vries, B.B.A., Wiegers, A.M., Smits, A.P.T., Mohkamsing, S., Duivenvoorden, H.J., Fryns, J.-P., Curfs, L.M.G., Halley, D.J.J., Oostra, B.A., van den Ouweland, A.M.W., Niermeijer, M.F. (1996) Mental status of females with an FMR1 gene full mutation, *Am. J. Hum. Genet.* **58**, 1025–1032.

Deelen, W., Bakker, C., Halley, D.J., Oostra, B.A. (1994) Conservation of CGG region in FMR1 gene in mammals, *Am. J. Med. Genet.* **51**, 513–516.

Devys, D., Lutz, Y., Rouyer, N., Bellocq, J.-P., Mandel, J.-L. (1993) The FMR-1 protein is cytoplasmic, most abundant in neurons and appears normal in carriers of a fragile X premutation, *Nat. Genet.* **4**, 335–340.

Dobkin, C., Rabe, A., Dumas, R., El Idrissi, A., Haubenstock, H., Brown, W.T. (2000) Fmr1 knockout mouse has a distinctive strain-specific learning impairment, *Neuroscience* **100**, 423–429.

Dockendorff, T.C., Su, H.S., McBride, S.M., Yang, Z., Choi, C.H., Siwicki, K.K., Sehgal, A., Jongens, T.A. (2002) Drosophila lacking dfmr1 activity show defects in circadian output and fail to maintain courtship interest, *Neuron* **34**, 973–984.

Dombrowski, C., Levesque, S., Morel, M.L., Rouillard, P., Morgan, K., Rousseau, F. (2002) Premutation and intermediate-size FMR1 alleles in 10572 males from the general population: loss of an AGG interruption is a late event in the generation of fragile X syndrome alleles, *Hum. Mol. Genet.* **11**, 371–378.

Drouin, R., Angers, M., Dallaire, N., Rose, T.M., Khandjian, W., Rousseau, F. (1997) Structural and functional characterization of the human FMR1 promoter reveals similarities with the hnRNP-A2 promoter region, *Hum. Mol. Genet.* **6**, 2051–2060.

Eberhart, D.E., Malter, H.E., Feng, Y., Warren, S.T. (1996) The fragile X mental retardation protein is a ribonucleoprotein containing both nuclear localization and nuclear export signals, *Hum. Mol. Genet.* **5**, 1083–1091.

Eichler, E.E., Holden, J.J.A., Popovich, B.W., Reiss, A.L., Snow, K., Thibodeau, S.N., Richards, C.S., Ward, P.A., Nelson, D.L. (1995a) Length of uninterrupted CGG repeats determines instability in the *FMR1* gene, *Nat. Genet.* **8**, 88–94.

Eichler, E.E., Kunst, C.B., Lugenbeel, K.A., Ryder, O.A., Davison, D., Warren, S.T., Nelson, D.L. (1995b) Evolution of the cryptic *FMR1* CGG repeat, *Nat. Genet.* **11**, 301–307.

Eichler, E.E., Nelson, D.L. (1996) Genetic variation and evolutionary stability of the FMR1 CGG repeat in six closed human populations, *Am. J. Med. Genet.* **64**, 220–225.

Eichler, E.E., Richards, S., Gibbs, R.A., Nelson, D.L. (1993) Fine structure of the human FMR1 gene, *Hum. Mol. Genet.* **2**, 1147–1153.

Faust, C.J., Verkerk, A.J., Wilson, P.J., Morris, C.P., Hopwood, J.J., Oostra, B.A., Herman, G.E. (1992) Genetic mapping on the mouse X chromosome of human cDNA clones for the fragile X and Hunter syndromes, *Genomics* **12**, 814–817.

Feng, Y., Absher, D., Eberhart, D.E., Brown, V., Malter, H.E., Warren, S.T. (1997) FMRP associates with polyribosomes as an mRNP, and the I304N mutation of severe fragile X syndrome abolishes this association, *Mol. Cell* **1**, 109–118.

Feng, Y., Gutekunst, C.-A., Eberhart, D.E., Yi, H., Warren, S.T., Hersch, S.M. (1997) Fragile X mental retardation protein: nucleocytoplasmic shuttling and association with somatodendritic ribosomes, *J. Neurosci.* **17**, 1539–1547.

Feng, Y., Lakkis, L., Devys, D., Warren, S.T. (1995a) Quantitative comparison of FMR1 gene expression in normal and premutation alleles, *Am. J. Hum. Genet.* **56**, 106–113.

Feng, Y., Zhang, F., Lokey, L.K., Chastain, J.L., Lakkis, L., Eberhart, D., Warren, S.T. (1995b) Translational suppression by trinucleotide repeat expansion at FMR1, *Science* **268**, 731–734.

Fiala, J.C., Spacek, J., Harris, K.M. (2002) Dendritic spine pathology: cause or consequence of neurological disorders? *Brain. Res. Rev.* **39**, 29–54.

Fridell, R.A., Benson, R.E., Hua, J., Bogerd, H.P., Cullen, B.R. (1996) A nuclear role for the fragile X mental retardation protein, *EMBO J.* **15**, 5408–5414.

Fu, Y.-H., Kuhl, D.P.A., Pizzuti, A., Pieretti, M., Sutcliffe, J.S., Richards, S., Verkerk, A.J.M.H., Holden, J.J.A., Fenwick, R.G., Warren, S.T., Oostra, B.A., Nelson, D.L., Caskey, C.T. (1991) Variation of the CGG repeat at the fragile X site results in genetic instability: resolution of the Sherman paradox, *Cell* **67**, 1047–1058.

Gantois, I., Bakker, C.E., Reyniers, E., Willemsen, R., D'Hooge, R., De Deyn, P.P., Oostra, B.A., Kooy, R.F. (2001) Restoring the phenotype of fragile X syndrome: insight from the mouse model, *Curr. Mol. Med.* **1**, 447–455.

Genc, B., Muller-Hartmann, H., Zeschnigk, M., Deissler, H., Schmitz, B., Majewski, F., Gontard, A., Doerfler, W. (2000) Methylation mosaicism of 5′-(CGG)(n)-3′ repeats in fragile X, premutation and normal individuals, *Nucleic Acids Res.* **28**, 2141–2152.

Godfraind, J.-M., Reyniers, E., De Boulle, K., D'Hooge, R., De Deyn, P.P., Bakker, C.E., Oostra, B.A., Kooy, R.F., Willems, P.J. (1996) Long-term potentiation in the hippocampus of fragile X knockout mice, *Am. J. Med. Genet.* **64**, 246–251.

Gold, B., Radu, D., Balanko, A., Chiang, C.S. (2000) Diagnosis of fragile X syndrome by Southern blot hybridization using a chemiluminescent probe: a laboratory protocol, *Mol. Diagn.* **5**, 169–178.

Gould, E.L., Loesch, D.Z., Martin, M.J., Hagerman, R.J., Armstrong, S.M., Huggins, R.M. (2000) Melatonin profiles and sleep characteristics in boys with fragile X syndrome: a preliminary study, *Am. J. Med. Genet.* **95**, 307–315.

Greco, C.M., Hagerman, R.J., Tassone, F., Chudley, A.E., Del, B., Jacquemont, S., Leehey, M., Hagerman, P.J. (2002) Neuronal intranuclear inclusions in a new cerebellar tremor/ataxia syndrome among fragile X carriers, *Brain* **125**, 1760–1771.

Greenough, W.T., Klintsova, A.Y., Irwin, S.A., Galvez, R., Bates, K.E., Weiler, I.J. (2001) Synaptic regulation of protein synthesis and the fragile X protein, *Proc. Natl. Acad. Sci. USA* **98**, 7101–7106.

Gronskov, K., Hallberg, A., Brondum, N. (1998) Mutational analysis of the FMR1 gene in 118 mentally retarded males suspected of fragile

X syndrome: absence of prevalent mutations, *Hum. Genet.* **102**, 440–445.

Gunter, C., Paradee, W., Crawford, D.C., Meadows, K.A., Newman, J., Kunst, C.B., Nelson, D.L., Schwartz, C., Murray, A., Macpherson, J.N., Sherman, S.L., Warren, S.T. (1998) Re-examination of factors associated with expansion of CGG repeats using a single nucleotide polymorphism in FMR1, *Hum. Mol. Genet.* **7**, 1935–1946.

Hamel, B.C.J., Smits, A.P.T., de Graaff, E., Smeets, D.F.C.M., Schoute, F., Eussen, B.H.J., Knight, S.J.L., Davies, K.E., Assman-Hulsmans, C.F.C.H., Oostra, B.A. (1994) Segregation of FRAXE in a large family: clinical, psychometric, cytogenetic, and molecular data, *Am. J. Hum. Genet.* **55**, 923–931.

Hansen, R.S., Gartler, S.M., Scott, C.R., Chen, S.-H., Laird, C.D. (1992) Methylation analysis of CGG sites in the CpG island of the human FMR1 gene, *Hum. Mol. Genet.* **1**, 571–578.

Heitz, D., Devys, D., Imbert, G., Kretz, C., Mandel, J.L. (1992) Inheritance of the fragile X syndrome: size of the fragile X premutation is a major determinant of the transition to full mutation, *J. Med. Genet.* **29**, 794–801.

Hinds, H.L., Ashley, C.T., Sutcliffe, J.S., Nelson, D.L., Warren, S.T., Housman, D.E., Schalling, M. (1993) Tissue specific expression of FMR-1 provides evidence for a functional role in fragile X syndrome, *Nat. Genet.* **3**, 36–43.

Hinton, V.J., Brown, W.T., Wisniewski, K., Rudelli, R.D. (1991) Analysis of neocortex in three males with the fragile X syndrome, *Am. J. Med. Genet.* **41**, 289–294.

Hirst, M.C., White, P.J. (1998) Cloned human FMR1 trinucleotide repeats exhibit a length- and orientation-dependent instability suggestive of in vivo lagging strand secondary structure, *Nucleic Acids Res.* **26**, 2353–2358.

Holden, J.J.A., Percy, M., Allingham-Hawkins, D., Brown, W.T., Chiurazzi, P., Fisch, G., Gane, L., Gunter, C., Hagerman, R., Jenkins, E.C., Kooy, R.F., Lubs, H.A., Murray, A., Neri, G., Schwartz, C., Tranebjaerg, L., Villard, L., Willems, P.J. (1999) Eighth international workshop on the fragile X syndrome and X-linked mental retardation, August 16–22, 1997, *Am. J. Med. Genet.* **83**, 221–236.

Huber, K.M., Gallagher, S.M., Warren, S.T., Bear, M.F. (2002) Altered synaptic plasticity in a mouse model of fragile X mental retardation, *Proc. Natl. Acad. Sci. USA* **99**, 7746–7750.

Hull, C., Hagerman, R.J. (1993) A study of the physical, behavioral, and medical phenotype, including anthropometric measures, of females with fragile X syndrome, *Am. J. Dis. Child.* **147**, 1236–1241.

Hundscheid, R.D., Sistermans, E.A., Thomas, C.M., Braat, D.D., Straatman, H., Kiemeney, L.A., Oostra, B.A., Smits, A.P. (2000) Imprinting effect in premature ovarian failure confined to paternally inherited fragile X premutations, *Am. J. Hum. Genet.* **66**, 413–418.

Huot, M.E., Mazroui, R., Leclerc, P., Khandjian, E.W. (2001) Developmental expression of the fragile X-related 1 proteins in mouse testis: association with microtubule elements, *Hum. Mol. Genet.* **10**, 2803–2811.

Inoue, S., Shimoda, M., Nishinokubi, I., Siomi, M.C., Okamura, M., Nakamura, A., Kobayashi, S., Ishida, N., Siomi, H. (2002) A role for the Drosophila fragile X-related gene in circadian output, *Curr. Biol.* **12**, 1331–1335.

Irwin, S.A., Galvez, R., Greenough, W.T. (2000) Dendritic spine structural anomalies in fragile X mental retardation syndrome, *Cereb. Cortex* **10**, 1038–1044.

Irwin, S.A., Patel, B., Idupulapati, M., Harris, J.B., Crisostomo, R.A., Larsen, B.P., Kooy, F., Willems, P.J., Cras, P., Kozlowski, P.B., Swain, R.A., Weiler, I.J., Greenough, W.T. (2001) Abnormal dendritic spine characteristics in the temporal and visual cortices of patients with fragile X syndrome: a quantitative examination, *Am. J. Med. Genet.* **98**, 161–167.

Irwin, S.A., Swain, R.A., Christmon, C.A., Chakravarti, A., Weiler, I.J., Greenough, W.T. (2000) Evidence for altered Fragile-X mental retardation protein expression in response to behavioral stimulation, *Neurobiol. Learn. Mem.* **73**, 87–93.

Ishizuka, A., Siomi, M.C., Siomi, H. (2002) A Drosophila fragile X protein interacts with components of RNAi and ribosomal proteins, *Genes. Dev.* **16**, 2497–2508.

Ivanco, T.L., Greenough, W.T. (2002) Altered mossy fiber distributions in adult Fmr1 (FVB) knockout mice, *Hippocampus* **12**, 47–54.

Kang, S., Ohshima, K., Shimizu, M., Amirhaeri, S., Wells, R.D. (1995) Pausing of DNA synthesis in vitro at specific loci in CTG and CGG triplet repeats from human hereditary disease genes, *J. Biol. Chem.* **270**, 27014–27021.

Kates, W.R., Abrams, M.T., Kaufmann, W.E., Breiter, S.N., Reiss, A.L. (1997) Reliability and validity of MRI measurement of the amygdala and hippocampus in children with fragile X syndrome, *Psychiatry. Res.* **75**, 31–48.

Kenneson, A., Zhang, F., Hagedorn, C.H., Warren, S.T. (2001) Reduced FMRP and increased FMR1 transcription is proportionally associated with CGG repeat number in intermediate-length and premutation carriers, *Hum. Mol. Genet.* **10**, 1449–1454.

Khandjian, E.W., Bardoni, B., Corbin, F., Sittler, A., Giroux, S., Heitz, D., Tremblay, S., Pinset, C., Montarras, D., Rousseau, F., Mandel, J. (1998) Novel isoforms of the fragile X related protein FXR1P are expressed during myogenesis, *Hum. Mol. Genet.* **7**, 2121–2128.

Khandjian, E.W., Corbin, F., Woerly, S., Rousseau, F. (1996) The fragile X mental retardation protein is associated with ribosomes, *Nat. Genet.* **12**, 91–93.

Kiledjian, M., Dreyfuss, G. (1992) Primary structure and binding activity of the hnRNP U protein: binding RNA through RGG box, *EMBO. J.* **11**, 2655–2664.

Koch, M. (1999) The neurobiology of startle, *Prog. Neurobiol.* **59**, 107–128.

Kolehmainen, K. (1994) Population genetics of fragile X: a multiple allele model with variable risk of CGG repeat expansion, *Am. J. Med. Genet.* **51**, 428–435.

Kooy, R.F., D'Hooge, R., Reyniers, E., Bakker, C.E., Nagels, G., De Boulle, K., Storm, K., Clincke, G., De Deyn, P.P., Oostra, B.A., Willems, P.J. (1996) Transgenic mouse model for the fragile X syndrome, *Am. J. Med. Genet.* **64**, 241–245.

Kooy, R.F., Oostra, B.A., Willems, P.J. (1997) Molecular Detection of Dynamic Mutations, in: Adolph, K.W. (Ed.) *Human Genome Methods*, CRC Press, Boca Raton, FL, pp. 23–53.

Kooy, R.F., Reyniers, E., Verhoye, M., Sijbers, J., Bakker, C.E., Oostra, B.A., Willems, P.J., Van der Linden, A. (1999) Neuroanatomy of the fragile X knockout mouse brain studied using *in vivo* high resolution magnetic resonance imaging (MRI), *Eur. J. Hum. Genet.* **7**, 526–532.

Kovtun, I.V., Goellner, G., McMurray, C.T. (2001) Structural features of trinucleotide repeats associated with DNA expansion, *Biochem. Cell. Biol.* **79**, 325–336.

Kunst, C.B., Warren, S.T. (1994) Cryptic and polar variation of the fragile X repeat could result in predisposing normal alleles, *Cell* **77**, 853–861.

Laggerbauer, B., Ostareck, D., Keidel, E.M., Ostareck-Lederer, A., Fischer, U. (2001) Evidence that fragile X mental retardation protein is a negative regulator of translation, *Hum. Mol. Genet.* **10**, 329–338.

Li, J., Pelletier, M.R., Perez Velazquez, J.-L., Carlen, P.L. (2002) Reduced cortical synaptic plasticity and GluR1 expression associated with fragile X mental retardation protein deficiency, *Mol. Cell. Neurosci.* **19**, 138–151.

Li, Z., Zhang, Y., Ku, L., Wilkinson, K.D., Warren, S.T., Feng, Y. (2001) The fragile X mental retardation protein inhibits translation via interacting with mRNA, *Nucleic Acids Res.* **29**, 2276–2283.

Loesch, D.Z., Hay, D.A., Mulley, J. (1994) Transmitting males and carrier females in fragile X–revisited, *Am. J. Med. Genet.* **51**, 392–399.

Loesch, D.Z., Huggins, R., Petrovic, V., Slater, H. (1995) Expansion of the CGG repeat in fragile X in the FMR1 gene depends on the sex of the offspring, *Am. J. Hum. Genet.* **57**, 1408–1413.

Lubs, H.A. (1969) A marker X chromosome, *Am. J. Hum. Genet.* **21**, 231–244.

Lugenbeel, K.A., Peier, A.M., Carson, N.L., Chudley, A.E., Nelson, D.L. (1995) Intragenic loss of function mutations demonstrate the primary role of FMR1 in fragile X syndrome, *Nat. Genet.* **10**, 483–485.

Malter, H.E., Iber, J.C., Willemsen, R., de Graaff, E., Tarleton, J.C., Leisti, J., Warren, S.T., Oostra, B.A. (1997) Characterization of the full fragile X syndrome mutation in fetal gametes, *Nat. Genet.* **15**, 165–169.

Manley, K., Pugh, J., Messer, A. (1999) Instability of the CAG repeat in immortalized fibroblast cell cultures from Huntington's disease transgenic mice, *Brain. Res.* **835**, 74–79.

Martin, J.P., Bell, J. (1943) A pedigree of mental defect showing sex-linkage, *J. Neurol. Psychol.* **6**, 154–157.

Mazroui, R., Huot, M.E., Tremblay, S., Filion, C., Labelle, Y., Khandjian, E.W. (2002) Trapping of messenger RNA by fragile X mental retardation protein into cytoplasmic granules induces translation repression, *Hum. Mol. Genet.* **11**, 3007–3017.

Mineur, Y.S., Crusio, W.E. (2002) Behavioral and neuroanatomical characterization of FVB/N inbred mice, *Brain. Res. Bull.* **57**, 41–47.

Miyashiro, K.Y., Beckel, M., Purk, T.P., Becker, K.G., Barret, T., Liu, L., Carbonetto, S., Weiler, I.J., Greenough, W.T., Eberwine, J. (2003) RNA cargoes associating with FMRP reveal deficits in cellular functioning in Fmr1 null mice, *Neuron* **37**, 417–431.

Morales, J., Hiesinger, P.R., Schroeder, A.J., Kume, K., Verstreken, P., Jackson, F.R., Nelson, D.L., Hassan, B.A. (2002) Drosophila fragile X protein, DFXR, regulates neuronal morphology and function in the brain, *Neuron* **34**, 961–972.

Morris, R.G.M., Garrud, P., Rawlins, J.N.P., O'Keefe, J. (1982) Place navigation impaired in rats with hippocampal lesions, *Nature* **297**, 681–683.

Morton, N.E., MacPherson, J.N. (1992) Population genetics of the fragile-X syndrome: multiallelic model for the FMR1 locus, *Proc. Natl. Acad. Sci. USA* **89**, 4215–4217.

Mostofsky, S., Mazzocco, M.M.M., Aakalu, G., Warsofsky, L.S., Denckla, M.B., Reiss, A.L. (1998) Decreased cerebellar posterior vermis size in fragile X syndrome. Correlation with neurocognitive performance, *Neurology* **50**, 121–130.

Moutou, C., Vincent, M.-C., Biancalana, V., Mandel, J.-L. (1997) Transition from premutation to full mutation in fragile X syndrome is likely to be prezygotic, *Hum. Mol. Genet.* **6**, 971–979.

Muller-Hartmann, H., Deissler, H., Naumann, F., Schmitz, B., Schroer, J., Doerfler, W. (2000) The human 20-kDa 5′-(CGG)(n)-3′-binding protein is targeted to the nucleus and affects the activity of the FMR1 promoter, *J. Biol. Chem.* **275**, 6447–6452.

Murray, A., Webb, J., Grimley, S., Conway, G., Jacobs, P. (1998) Studies of FRAXA and FRAXE in women with premature ovarian failure, *J. Med. Genet.* **35**, 637–640.

Musco, G., Stier, G., Joseph, C., Castiglione Morelli, M.A., Nilges, M., Gibson, T.J., Pastore, A. (1996) Three-dimensional structure and stability of the KH domain: molecular insights into the fragile X syndrome, *Cell* **85**, 237–245.

Musumeci, S.A., Bosco, P., Calabrese, G., Bakker, C., De Sarro, G.B., Elia, M., Ferri, R., Oostra, B.A. (2000) Audiogenic seizures susceptibility in transgenic mice with fragile X syndrome, *Epilepsia* **41**, 19–23.

Musumeci, S.A., Hagerman, R.J., Ferri, R., Bosco, P., Dalla, B., Tassinari, C.A., De, S., Elia, M. (1999) Epilepsy and EEG findings in males with fragile X syndrome, *Epilepsia* **40**, 1092–1099.

Nielsen, D.M., Derber, W.J., McClellan, D.A., Crnic, L.S. (2002) Alterations in the auditory startle response in Fmr1 targeted mutant mouse models of fragile X syndrome, *Brain. Res.* **927**, 8–17.

Nimchinski, E.A., Oberlander, A.M., Svoboda, K. (2001) Abnormal development of dendritic spines in *FMR1* knock-out mice, *J. Neurosci.* **21**, 5139–5146.

Nolin, S.L., Brown, W.T., Glicksman, A., Houck, G.E. Jr., Gargano, A.D., Sullivan, A., Biancalana, V., Bröndum-Nielsen, K., Hjalgrim, H., Holinski-Feder, E., Kooy, F., Longshore, J., MacPherson, J., Mandel, J.-L., Matthijs, G., Rousseau, F., Steinbach, P., Väisänen, M.-L., von Koskull, H., Sherman, S.L. (2003) Expansion of the fragile X CGG repeat in females with premutation or intermediate alleles, *Am. J. Hum. Genet.* **72**, 454–464.

Nolin, S.L., Houck, G.E., Gargano, A.D., Blumstein, H., Dobkin, C.S., Brown, W.T. (1999) FMR1 CGG-repeat instability in single sperm and lymphocytes of fragile-X premutation males, *Am. J. Hum. Genet.* **65**, 680–688.

Oberlé, I., Boué, J., Croquette, M.F., Voelckel, M.A., Mattei, M.G., Mandel, J.L. (1992) Three families with high expression of a fragile site at Xq27.3, lack of anomalies at the FMR-1 CpG island, and no clear phenotypic association, *Am. J. Med. Genet.* **43**, 224–231.

Oberlé, I., Rousseau, F., Heitz, D., Kretz, C., Devys, D., Hanauer, A., Boué, J., Bertheas, M.F., Mandel, J.-L. (1991) Instability of a 550-base pair DNA segment and abnormal methylation in fragile X syndrome, *Science* **252**, 1097–1102.

Oostra, B.A., Jacky, P.B., Brown, W.T., Rousseau, F. (1993) Guidelines for the diagnosis of fragile X syndrome, *J. Med. Genet.* **30**, 410–413.

Panagopoulos, I., Lassen, C., Kristoffersson, U., Aman, P. (1999) A methylation PCR approach for detection of fragile X syndrome, *Hum. Mutat.* **14**, 71–79.

Paradee, W., Melikian, H.E., Rasmussen, D.E., Kenneson, A., Conn, P.J., Warren, S.T. (1999) Fragile X mouse: strain effects of knockout phenotype and evidence suggesting deficient amygdala function, *Neuroscience* **94**, 185–192.

Partington, M.W. (1984) The fragile X syndrome II: preliminary data on growth and development in males, *Am. J. Med. Genet.* **17**, 175–194.

Peier, A.M., McIlwain, K.L., Kenneson, A., Warren, S.T., Paylor, R., Nelson, D.L. (2000) (Over) correction of FMR1 deficiency with YAC transgenics: behavioral and physical features, *Hum. Mol. Genet.* **9**, 1145–1159.

Pieretti, M., Zhang, F., Fu, Y.-H., Warren, S.T., Oostra, B.A., Caskey, C.T., Nelson, D.L. (1991) Absence of expression of the FMR-1 gene in fragile X syndrome, *Cell* **66**, 817–822.

Price, D.K., Zhang, F., Ashley, C.T. Jr., Warren, S.T. (1996) The chicken FMR1 gene is highly conserved with a CCT 5′-untranslated repeat and encodes an RNA-binding protein, *Genomics* **31**, 3–12.

Qin, M., Kang, J., Smith, C.B. (2002) Increased rates of cerebral glucose metabolism in a mouse model of fragile X mental retardation, *Proc. Natl. Acad. Sci. USA* **99**, 15758–15763.

Reiss, A.L., Abrams, M.T., Greenlaw, R., Freund, L., Denckla, M.B. (1995) Neurodevelopmental effects of the FMR1 full mutation in humans, *Nat. Med.* **1**, 159–167.

Reiss, A.L., Freund, L., Tseng, J.E., Joshi, P.K. (1991) Neuroanatomy in fragile X females: the posterior fossa, *Am. J. Hum. Genet.* **49**, 279–288.

Reiss, A.L., Lee, J., Freund, L. (1994) Neuroanatomy of fragile X syndrome: the temporal lobe, *Neurology* **44**, 1317–1324.

Reyniers, E., Martin, J.-J., Cras, P., Van Marck, E., Handig, I., Jorens, H.Z.J., Oostra, B.A., Kooy, R.F., Willems, P.J. (1999) Post-mortem examination of two fragile X brothers with an FMR1 full mutation, *Am. J. Med. Genet.* **84**, 245–249.

Reyniers, E., Vits, L., De Boulle, K., van Roy, B., van Velzen, D., de Graaff, E., Verkerk, A.J.M.H., Jorens, H.Z.J., Darby, J.K., Oostra, B., Willems, P.J. (1993) The full mutation in the FMR-1 gene of male fragile X patients is absent in their sperm, *Nat. Genet.* **4**, 143–146.

Richards, R.I., Holman, K., Friend, K., Staples, A., Sutherland, G.R., Oudet, C., Biancalana, V., Mandel, J.L. (1994) FRAXAC2 instability, *Nat. Genet.* **7**, 123.

Richards, B.W., Sylvester, P.E., Brooker, C. (1981) Fragile X-linked mental retardation: the Martin-Bell syndrome, *J. Ment. Defic. Res.* **25**Pt 4, 253–256.

Riddle, J.E., Cheema, A., Sobesky, W.E., Gardner, S.C., Taylor, A.K., Pennington, B.F., Hagerman, R.J. (1998) Phenotypic involvement in females with the FMR1 gene mutation, *Am. J. Ment. Retard.* **102**, 590–601.

Rousseau, F., Heitz, D., Biancalana, V., Blumenfeld, S., Kretz, C., Boué, J., Tommerup, N., Van Der Hagen, C., DeLozier-Blanchet, C., Croquette, M.-F., Gilgenkrantz, S., Jalbert, P., Voelckel, M.-A., Oberlé, I., Mandel, J.-L. (1991) Direct diagnosis by DNA analysis of the fragile X syndrome of mental retardation, *N. Engl. J. Med.* **325**, 1674–1681.

Rousseau, F., Heitz, D., Tarleton, J., MacPherson, J., Malmgren, H., Dahl, N., Barnicoat, A., Mathew, C., Mornet, E., Tejada, I., Maddalena, A., Spiegel, R., Schinzel, A., Marcos, J.A.G., Schorderet, D.F., Schaap, T., MacCioni, L., Russo, S., Jacobs, P.A., Schwartz, C., Mandel, J.L. (1994) A multicenter study on genotype-phenotype correlations in the fragile X syndrome, using direct diagnosis with probe StB12.3: the first 2,253 cases, *Am. J. Hum. Genet.* **55**, 225–237.

Rousseau, F., Rouillard, P., Morel, M.L., Khandjian, E.W., Morgan, K. (1995) Prevalence of carriers of premutation-size alleles of the FMRI gene–and implications for the population genetics of the fragile X syndrome, *Am. J. Hum. Genet.* **57**, 1006–1018.

Samadashwily, G.M., Raca, G., Mirkin, S.M. (1997) Trinucleotide repeats affect DNA replication in vivo, *Nat. Genet.* **17**, 298–304.

Schaeffer, C., Bardoni, B., Mandel, J.L., Ehresmann, B., Ehresmann, C., Moine, H. (2001) The fragile X mental retardation protein binds specifically to its mRNA via a purine quartet motif, *EMBO J.* **20**, 4803–4813.

Schenck, A., Bardoni, B., Moro, A., Bagni, C., Mandel, J.L. (2001) A highly conserved protein family interacting with the fragile X mental retardation protein (FMRP) and displaying selective interactions with FMRP-related proteins FXR1P and FXR2P, *Proc. Natl. Acad. Sci. USA* **98**, 8844–8849.

Schlotterer, C., Tautz, D. (1992) Slippage synthesis of simple sequence DNA, *Nucleic Acids Res.* **20**, 211–215.

Schwemmle, S., de Graaff, E., Deissler, H., Gläser, D., Wöhrle, D., Kennerknecht, I.,

Just, W., Oostra, B.A., Drabkin, H., Vogel, W., Steinbach, P. (1997) Characterization of FMR1 promoter elements by in vivo-footprinting analysis, *Am. J. Hum. Genet.* **60**, 1354–1362.

Sherman, S.L., Jacobs, P.A., Morton, N.E., Froster-Iskenius, U., Howard-Peebles, P.N., Nielsen, K.B., Partington, N.W., Sutherland, G.R., Turner, G., Watson, M. (1985) Further segregation analysis of the fragile X syndrome with special reference to transmitting males, *Hum. Genet.* **69**, 3289–3299.

Sherman, S.L., Morton, N.E., Jacobs, P.A., Turner, G. (1984) The marker (X) chromosome: a cytogenetic and genetic analysis, *Ann. Hum. Genet.* **48**, 21–37.

Shimizu, M., Gellibolian, R., Oostra, B.A., Wells, R.D. (1996) Cloning, characterization and properties of plasmids containing CGG triplet repeats from the FMR-1 gene, *J. Mol. Biol.* **258**, 614–626.

Sinden, R.R. (2001) Neurodegenerative diseases. Origins of instability, *Nature* **411**, 757, 758.

Siomi, H., Choi, M., Siomi, M.C., Nussbaum, R.L., Dreyfuss, G. (1994) Essential role for KH domains in RNA binding: impaired RNA binding by a mutation in the KH domain of FMR1 that causes fragile X syndrome, *Cell* **77**, 33–39.

Siomi, H., Dreyfuss, G. (1995) A nuclear localization domain in the hnRNP A1 protein, *J. Cell. Biol.* **118**, 551–560.

Siomi, M.C., Higashijima, K., Ishizuka, A., Siomi, H. (2002) Casein kinase II phosphorylates the fragile X mental retardation protein and modulates its biological properties, *Mol. Cell. Biol.* **22**, 8438–8447.

Siomi, H., Matunis, M.J., Michael, W.M., Dreyfuss, G. (1993a) The pre-mRNA binding K protein contains a novel evolutionarily conserved motif, *Nucleic Acids Res.* **21**, 1193–1198.

Siomi, H., Siomi, M.C., Nussbaum, R.L., Dreyfuss, G. (1993b) The protein product of the fragile X gene, FMR1, has characteristics of an RNA binding protein, *Cell* **74**, 291–298.

Siomi, M.C., Siomi, H., Sauer, W.H., Srinivasan, S., Nussbaum, R.L., Dreyfuss, G. (1995) FXR1, an autosomal homolog of the fragile X mental retardation gene, *EMBO J.* **14**, 2401–2408.

Siomi, M.C., Zhang, Y., Siomi, H., Dreyfuss, G. (1996) Specific sequences in the fragile X syndrome protein FMR1 and the FXR proteins mediate their binding to 60S ribosomal subunits and the interactions among them, *Mol. Cell. Biol.* **16**, 3825–3832.

Sittler, A., Devys, D., Weber, C., Mandel, J.-L. (1996) Alternative splicing of exon 14 determines nuclear or cytoplasmatic localisation of fmr1 protein isoforms, *Hum. Mol. Genet.* **5**, 95–102.

Slegtenhorst-Eegdeman, K.E., de Rooij, D.G., Verhoef-Post, M., Ruiz, A., Uilenbroek, T.J., van de Kant, H.J.G., Bakker, C.E., Oostra, B.A., Grootegoed, J.A., Themmen, A.P.N. (1998) Macro-orchidism in *Fmr1* knockout mice is caused by increased sertoli cell proliferation during normal testes development, *Endocrinology* **139**, 156–162.

Smeets, H.J.M., Smits, A.P.T., Verheij, C.E., Theelen, J.P.G., Willemsen, R., van de Burgt, I., Hoogeveen, A.T., Oosterwijk, J.C., Oostra, B.A. (1995) Normal phenotype in two brothers with a full FMR1 mutation, *Hum. Mol. Genet.* **4**, 2103–2108.

Sullivan, A.K., Crawford, D.C., Scott, E.H., Leslie, M.L., Sherman, S.L. (2002) Paternally transmitted FMR1 alleles are less stable than maternally transmitted alleles in the common and intermediate size range, *Am. J. Hum. Genet.* **70**, 1532–1544.

Sung, Y.J., Conti, J., Currie, J.R., Brown, W.T., Denman, R.B. (2000) RNAs that interact with the fragile X syndrome RNA binding protein FMRP, *Biochem. Biophys. Res. Commun.* **275**, 973–980.

Sung, Y.J., Dolzhanskaya, N., Nolin, S.L., Brown, T., Currie, J.R., Denman, R.B. (2003) The fragile X mental retardation protein FMRP binds elongation factor 1A mRNA and negatively regulates its translation in vivo, *J. Biol. Chem.* **278**, 15669–15678.

Sutcliffe, J.S., Nelson, D.L., Zhang, F., Pieretti, M., Caskey, C.T., Saxe, D., Warren, S.T. (1992) DNA methylation represses FMR-1 transcription in fragile x syndrome, *Hum. Mol. Genet.* **1**, 397–400.

Sutherland, G.R. (1977) Fragile sites on human chromosomes: demonstration of their dependence on the type of tissue culture medium, *Science* **197**, 265, 266.

Tamanini, F., Meijer, N., Verheij, C., Willems, P.J., Galjaard, H., Oostra, B.A., Hoogeveen, A.T. (1996) FMRP is associated to the ribosomes via RNA, *Hum. Mol. Genet.* **5**, 809–813.

Tamanini, F., van Unen, L., Bakker, C.E., Sacchi, N., Galjaard, H., Oostra, B.A.,

Hoogeveen, A.T. (1999) The FXR proteins exist as homo-multimers within messenger RNP particles, *Biochem. J.* 517–523.

Tamanini, F., Willemsen, R., van Unen, L., Bontekoe, C., Galjaard, H., Oostra, B.A., Hoogeveen, A.T. (1997) Differential expression of FMR1, FXR1 and FXR2 proteins in human brain and testis, *Hum. Mol. Genet.* **6**, 1315–1322.

Tassone, F., Hagerman, R.J., Chamberlain, W.D., Hagerman, P.J. (2000) Transcription of the FMR1 gene in individuals with fragile X syndrome, *Am. J. Med. Genet.* **97**, 195–203.

Tassone, F., Hagerman, R.J., Ikle, D.N., Dyer, P.N., Lampe, M., Willemsen, R., Oostra, B.A., Taylor, A.K. (1999) FMRP expression as a potential prognostic indicator in fragile X syndrome, *Am. J. Med. Genet.* **84**, 250–261.

Tassone, F., Hagerman, R.J., Loesch, D.Z., Lachiewicz, A., Taylor, A.K., Hagerman, P.J. (2000a) Fragile X males with unmethylated, full mutation trinucleotide repeat expansions have elevated levels of FMR1 messenger RNA, *Am. J. Med. Genet.* **94**, 232–236.

Tassone, F., Hagerman, R.J., Taylor, A.K., Gane, L.W., Godfrey, T.E., Hagerman, P.J. (2000b) Elevated levels of FMR1 mRNA in carrier males: a new mechanism of involvement in the fragile-X syndrome, *Am. J. Hum. Genet.* **66**, 6–15.

Tassone, F., Hagerman, R.J., Taylor, A.K., Mills, J.B., Harris, S.W., Gane, L.W., Hagerman, P.J. (2000c) Clinical involvement and protein expression in individuals with the FMR1 premutation, *Am. J. Med. Genet.* **91**, 144–152.

Taylor, A.K., Safanda, J.F., Fall, M.Z., Quince, C., Lang, K.A., Hull, C.E., Carpenter, I., Staley, L.W., Hagerman, R.J. (1994) Molecular predictors of cognitive involvement in female carriers of fragile X syndrome, *JAMA* **271**, 507–514.

Todd, P.K., Mack, K.J. (2000) Sensory stimulation increases cortical expression of the fragile X mental retardation protein in vivo, *Mol. Brain. Res.* **80**, 17–25.

Todd, P.K., Malter, J.S., Mack, K.J. (2003) Whisker stimulation-dependent translation of FMRP in the barrel cortex requires activation of type I metabotropic glutamate receptors, *Brain. Res. Mol. Brain. Res.* **110**, 267–278.

Turner, G., Till, R., Daniel, A. (1978) Marker X chromosomes, mental retardation and macroorchidism, *N. Engl. J. Med.* **299**, 1472.

Turner, G., Turner, B. (1974) X-linked mental retardation, *J. Med. Genet.* **11**, 109–113.

Turner, G., Webb, T., Wake, S., Robinson, H. (1996) Prevalence of fragile X syndrome, *Am. J. Med. Genet.* **64**, 196, 197.

Usdin, K., Woodford, K.J. (1995) CGG repeats associated with DNA instability and chromosome fragility form structures that block DNA synthesis in vitro, *Nucleic Acids Res.* **23**, 4202–4209.

Van Dam, D., D'Hooge, R., Hauben, U., Reyniers, E., Gantois, I., Bakker, C.E., Oostra, B.A., Kooy, R.F., De Deyn, P.P. (2000) Spatial learning, contextual fear conditioning and conditioned emotional response in *Fmr1* knockout mice, *Behav. Brain. Res.* **117**, 127–136.

van den Ouweland, A.M.W., Deelen, W.H., Kunst, C.B., Uzielli, M.L., Nelson, D.L., Warren, S.T., Oostra, B.A., Halley, D.J. (1994) Loss of mutation at the FMR1 locus through multiple exchanges between maternal X chromosomes, *Hum. Mol. Genet.* **3**, 1823–1827.

Verheij, C., Bakker, C.E., de Graaff, E., Keulemans, J., Willemsen, R., Verkerk, A.J.M.H., Galjaard, H., Reuser, A.J.J., Hoogeveen, A.T., Oostra, B.A. (1993) Characterization and localization of the FMR-1 gene product associated with fragile X syndrome, *Nature* **363**, 722–724.

Verheij, C., de Graaff, E., Bakker, C.E., Willemsen, R., Willems, P.J., Meijer, N., Galjaard, H., Reuser, A.J.J., Oostra, B.A., Hoogeveen, A.T. (1995) Characterization of FMR1 proteins isolated from different tissues, *Hum. Mol. Genet.* **4**, 895–901.

Verkerk, A.J.M.H., de Graaff, E., De Boulle, K., Eichler, E.E., Konecki, D.S., Reyniers, E., Manca, A., Poustka, A., Willems, P.J., Nelson, D.L., Oostra, B.A. (1993) Alternative splicing in the fragile X gene FMR1, *Hum. Mol. Genet.* **2**, 399–404.

Verkerk, A.J.M.H., Pieretti, M., Sutcliffe, J.S., Fu, Y.-H., Kuhl, D.P.A., Pizzutti, A., Reiner, O., Richards, S., Victoria, M.F., Zhang, F., Eussen, B.E., van Ommen, G.-J.B., Blonden, L.A.J., Riggins, G.J., Chastain, J.L., Kunst, C.B., Galjaard, H., Caskey, C.T., Nelson, D.L., Oostra, B.A., Warren, S.T. (1991) Identification of a gene (FMR-1) containing a CGG repeat coincident with a breakpoint cluster region exhibiting length variation in fragile X syndrome, *Cell* **65**, 905–914.

Wan, L., Dockendorff, T.C., Jongens, T.A., Dreyfuss, G. (2000) Characterization of dFMR1, a Drosophila melanogaster homolog of the fragile X mental retardation protein, *Mol. Cell. Biol.* **20**, 8536–8547.

Warren, S.T., Zhang, F., Licameli, G.R., Peters, J.F. (1987) The fragile X site in somatic cell hybrids: an approach for molecular cloning of fragile sites, *Science* **237**, 420–423.

Weiler, I.J., Greenough, W.T. (1993) Metabotropic glutamate receptors trigger postsynaptic protein synthesis, *Proc. Natl. Acad. Sci. USA* **90**, 7168–7171.

Weiler, I.J., Irwin, S.A., Klintsova, A.Y., Spencer, C.M., Brazelton, A.D., Miyashiro, K., Comery, T.A., Patel, B., Eberwine, J., Greenough, W.T. (1997) Fragile X mental retardation protein is translated near synapses in response to neurotransmitter activation, *Proc. Natl. Acad. Sci. USA* **94**, 5395–5400.

Wells, R.D. (1996) Molecular basis of genetic instability of triplet repeats, *J. Biol. Chem.* **271**, 2875–2878.

White, P.J., Borts, R.H., Hirst, M.C. (1999) Stability of the human fragile X (CGG)(n) triplet repeat array in Saccharomyces cerevisiae deficient in aspects of DNA metabolism, *Mol. Cell. Biol.* **19**, 5675–5684.

Willemsen, R., Bontekoe, C., Tamanini, F., Galjaard, H., Hoogeveen, A., Oostra, B.A. (1996) Association of FMRP with ribosomal precursor particles in the nucleolus, *Biochem. Biophys. Res. Commun.* **225**, 27–33.

Willemsen, R., Hoogeveen-Westerveld, M., Reis, S., Holstege, J., Severijnen, L., Nieuwenhuizen, I., Schrier, M., VanUnen, L., Tassone, F., Hoogeveen, A., Hagerman, P., Mientjes, E., Oostra, B. (2003) The FMR1 CGG repeat mouse displays ubiquitin-positive intranuclear neuronal inclusions; implications for the cerebellar tremor/ataxia syndrome, *Hum. Mol. Genet.* **12**, 949–959.

Willemsen, R., Los, F., Mohkamsing, S., van den Ouweland, A., Deelen, W., Galjaard, H., Oostra, B. (1997) Rapid antibody test for prenatal diagnosis of fragile X syndrome on amniotic fluid cells: a new appraisal, *J. Med. Genet.* **34**, 250, 251.

Willemsen, R., Mohkamsing, S., De Vries, B., Devys, D., van den Ouweland, A., Mandel, J.L., Galjaard, H., Oostra, B. (1995) Rapid antibody test for fragile X syndrome, *Lancet* **345**, 1147, 1148.

Willemsen, R., Smits, A., Mohkamsing, S., van Beerendonk, H., de Haan, A., De Vries, B., van den Ouweland, A., Sistermans, E., Galjaard, H., Oostra, B.A. (1997) Rapid antibody test for diagnosing fragile X syndrome: a validation of the technique, *Hum. Genet.* **99**, 308–311.

Winnepenninckx, B., Rooms, L., Kooy, R.F. (2003) Mental retardation: a review of the genetic causes, *Br. J. Dev. Disabil.* **49**, 29–44.

Wisniewski, K.E., Segan, S.M., Miezejeski, C.M., Sersen, E.A., Rudelli, R.D. (1991) The Fra(X) syndrome: neurological, electrophysiological, and neuropathological abnormalities, *Am. J. Med. Genet.* **38**, 476–480.

Yu, S., Pritchard, M., Kremer, E., Lynch, M., Nancarrow, J., Baker, E., Holman, K., Mulley, J.C., Warren, S.T., Schlessinger, D., Sutherland, G.R., Richards, R.I. (1991) Fragile X genotype characterized by an unstable region of DNA, *Science* **252**, 1179–1181.

Zalfa, F., Giorgi, M., Primerano, B., Moro, A., Di, P., Reis, S., Oostra, B., Bagni, C. (2003) The Fragile fragile X Syndrome syndrome Protein protein FMRP Associates associates with BC1 RNA and Regulates regulates the Translation translation of Specific specific mRNAs at synapses., *Cell* **112**, 317–327.

Zhang, Y.Q., Bailey, A.M., Matthies, H.-J.G., Renden, R.B., Smith, M.A., Speese, S.D., Rubin, G.M., Broadie, K. (2001) Drosophila fragile X-related gene regulates the MAP1B homolog futsch to control synaptic structure and function, *Cell* **107**, 591–603.

Zhang, Y., O'Connor, J.P., Siomi, M.C., Srinivasan, S., Dutra, A., Nussbaum, R.L., Dreyfuss, G. (1995) The fragile X mental retardation syndrome protein interacts with novel homologs FXR1 and FXR2, *EMBO J.* **14**, 5358–5366.

18
Triplet Repeat Diseases

Stephan J. Guyenet and Albert R. La Spada
University of Washington, Seattle, WA, USA

Neurobiology. From Molecular Basis to Disease. Edited by Robert A. Meyers.

ISBN: 978-3-527-32293-0

Keywords

Aggregate
Accumulation of proteinaceous and/or ribonucleic acid material into a structure that is visible in a cell at the light microscope level.

Anticipation
Worsening severity of a disease phenotype as the causal (typically dominant) genetic mutation is transmitted from one generation to the next in a family segregating the disease of interest.

Gain of function
Refers to a type of mutation that imparts a novel activity or action to the gene product containing the mutation.

Loss of function
Refers to a type of mutation that eliminates the action of the gene product encoded by the gene within which the mutation resides.

Repeat Expansion
An elongation of a repeat to a larger size that no longer falls within the size distribution range typically seen in the normal population; this process is now recognized as a mechanism of human genetic mutation.

Trinucleotide
Three DNA base pairs of specific sequence composition (e.g. cytosine-adenine-guanine).

The repeat expansion disorders are a group of human diseases that are caused by the elongation of a DNA repeat sequence. In this chapter, we provide an overview of the discovery of repeat expansion as an important cause of human disease, and we summarize the molecular genetics and mechanistic basis of 27 microsatellite repeat disorders. Comparison of the many repeat expansion disorders reveals distinct categories of repeat diseases, allowing us to propose a classification of the repeat expansion disorders based upon mutation sequence and pathogenic mechanism. The four types of repeat expansion disorders defined by this approach are the CAG/polyglutamine repeat diseases; the loss-of-function repeat diseases; the RNA gain-of-function repeat diseases; and the polyalanine diseases. Although the genetic basis for most of these diseases was determined less than a decade or so ago, considerable advances have been made in our understanding of how "dynamic mutations" produce molecular pathology and human disease.

1 A Novel Mechanism of Genetic Mutation Emerges

1.1 Repeat Sequences of All Types and Sizes

Long before the sequencing of the human genome was undertaken, the discovery of bacterial enzymes that recognize specific DNA sequences ("recognition sites") and cleaved them, yielded a new methodology for differentiating individuals ("molecular fingerprinting"). This methodology also found application in the mapping of inherited genetic diseases, taking advantage of human variation in the form of so-called restriction fragment length polymorphisms (RFLP's). In the search for even more informative genetic markers, investigators uncovered a variety of very short repeat sequences ("short tandem repeats" – STR's; "simple sequence length polymorphisms" – SSLP's) that were highly dispersed throughout the human genome. These latter repeat sequences came to be known as "microsatellites" to differentiate them from the "minisatellites" that were being used for molecular fingerprinting. Minisatellites were originally defined as tandem arrays of 14–100 bp repeating sequences. In the case of minisatellites, the repeat sequence was in essence a "consensus" as deviations from the exact repeat sequence were common. Microsatellites, however, were typically pure perfect repeats of less than 13 bp, the most common microsatellites being dinucleotide repeats, trinucleotide repeats, or tetranucleotide repeats. The discovery of minisatellites and microsatellites yielded a virtual bonanza of reagents for molecular fingerprinting and genetic linkage mapping, while also providing evolutionary biologists and population geneticists with intriguing material to attempt to reconstruct evolutionary relationships and inter- or intraspecies relationships. Although human geneticists and evolutionary biologists were applying microsatellites in different ways for their own studies, they shared the commonly held belief that such repeats were neutral and therefore unlikely to be of much functional consequence, let alone play a role in causing human disease. Of course, all that would soon change in the last decade of the twentieth century.

1.2
Trinucleotide Repeat Expansion as a Cause of Disease: Unique Features Explain Unusual Genetics

In 1991, two groups working on seemingly unrelated inherited genetic diseases independently made paradigm-shifting discoveries. In one case, a CAG trinucleotide repeat expansion within the first coding exon of the androgen receptor (*AR*) gene was found to be the cause of an X-linked neuromuscular disorder known as spinal and bulbar muscular atrophy (SBMA or Kennedy's disease). CAG encodes the amino acid glutamine; thus, elongation of a polyglutamine tract within the AR protein was hypothesized to be the molecular basis for the motor neuron degeneration in SBMA. In the other case, a disorder known as the fragile X syndrome of mental retardation (FRAXA), also X-linked but much more common, was reported to result from expansion of a CGG repeat. In the latter case, although originally envisioned to encode a polyarginine tract, the CGG repeat turned out to be in the 5′ untranslated region of a novel gene, the so-called *FMR-1* gene (for "fragile X mental retardation-1").

The identification of triplet repeat expansions as the cause of two inherited diseases was an exciting turning point in the field of human molecular genetics not only because of the novel nature of these findings, but also because of the unusual genetic characteristics of this new type of mutation. Analysis of families segregating FRAXA revealed the existence of three distinct allele categories defined by the length of the pure CGG repeat: a normal size range; a disease size range; and an intermediate, "premutation" size range. As discussed below, individuals carrying premutation-sized CGG repeats never develop the mental retardation phenotype, but instead are at risk for passing on even larger CGG repeats to their children or grandchildren who then display the mental retardation phenotype. An important tenet that emerged from these early studies was that expanded CGG repeats (larger than the normal size range) displayed an exceptionally high rate of further mutation. This observation reversed a commonly held view of the genetic material – the notion that any single nucleotide in the human genome displayed a mutation rate of $\sim 1 \times 10^{-5}$. In the case of expanded CGG repeats in premutation carriers, the rate approached unity (10^0)! Besides displaying this high mutation rate, there were other unusual features: (1) the CGG repeats showed a marked tendency to further expansion, suggesting that repeat mutation was a polar process; and (2) the sex of the individual transmitting the premutation-sized CGG allele determined whether a large expansion into the disease range would be possible. For FRAXA, expansion into the disease range could only occur if the premutation allele was transmitted by a female carrier. All of these unusual aspects of FRAXA CGG repeat genetics thoroughly accounted for the bizarre non-Mendelian inheritance patterns described in FRAXA families as the "Sherman paradox."

The recognition of repeat expansion as the cause of the neurodegenerative disorder SBMA and as an explanation for the puzzling genetics of FRAXA set the stage for further discoveries in the field of neurogenetics (the study of inherited neurological disorders). One disorder known as *myotonic dystrophy* (DM), the most common of all muscular dystrophies, had been the center of a genetic controversy that had gone on for nearly a century. The controversy involved the debate over whether the clinical phenomenon of anticipation truly existed or was simply an artifact of clinical study (Anticipation may be defined as a progressively earlier age of disease onset with increasing disease severity in successive generations of a family segregating an inherited disorder). Although anticipation was initially proposed as a defining feature of DM in 1918, a number of leading geneticists, among them Penrose, L.S., dismissed its authenticity, claiming that it was a product of ascertainment bias due to better clinically defining the profound variable expressitivity in this disorder. This view persisted from its promulgation in the 1950s, reinforced by the concept that the genetic material is seldom subject to alteration or modification that could be heritable. However, with the discovery that expanded trinucleotide repeats could further expand, and that indeed the expansion process was a prerequisite for the FRAXA disease phenotype – accounting for maternal inheritance and greater percentages of affected individuals in more recent generations – a role for repeats in diseases displaying anticipation was entertained. Consequently, the third repeat mutation disorder to be identified – just one year after the SBMA and FRAXA discoveries – was DM. Studies of the causal CTG repeat expansion in DM demonstrated a strong correlation between disease severity (i.e. age of onset and rate of progression) and the length of the CTG repeat. In this way, it became clear that anticipation was a genuine phenomenon and that expanded repeat mutational instability was its long awaited molecular explanation. The mutational instability characteristics of expanded repeats have led to their designation as so-called "dynamic mutations."

2
Repeat Diseases and Their Classification

2.1
Summary of Repeat Diseases

The list of diseases caused by microsatellite repeat expansions (involving repeats of

3–12 bp) now includes more than 25 disorders (Table 1). Certain aspects of the list of repeat diseases deserve emphasis. Twenty of the repeat diseases either principally or exclusively affect the neuraxis – that is, anywhere from the brain to the cerebellum/brainstem, to the spinal cord, to the peripheral nerve or muscle – and are degenerative disorders. Almost all of these inherited neurological disorders are caused by large expanded repeats that display the property of pronounced genetic instability (dynamic mutation). On the other hand, the developmental malformation syndromes for which repeat expansion mutations have been implicated all involve modestly sized disease repeats by comparison, and these repeats do not exhibit such dynamic mutation genetic instability.

2.2
Differences in Repeat Sequence Composition and Location within Gene

When faced with the task of categorizing the various repeat expansion diseases into different classes, a number of approaches are possible. We have found that consideration of the sequence of the repeat and its location within the gene are the most useful characteristics to apply for grouping the different repeat diseases. As shown in Table 1, there are many different types of repeats varying in length and sequence composition. However, among the recurrent sequence types are CAG trinucleotide repeats, CTG trinucleotide repeats, and GCG trinucleotide repeats. The rest of the repeat sequences are unique in composition and differ widely in size, ranging from 3 to 12 bp as noted above. Comparison of the location of a repeat within the gene it is affecting also yields different types of repeats. The largest single-repeat location category is within the coding region of a gene, which applies to both CAG (glutamine) and GCG (alanine) repeats, and has been proposed but not yet demonstrated for CTG (leucine) repeats. The rest of the locations defined for repeats vary widely, ranging from the gene's promoter to its 5′ untranslated region to an intron to the 3′ untranslated region, and no more than two to three repeat expansions can be placed in each of these categories at this time.

2.3
Classification Based upon Mechanism of Pathogenesis and Nature of Mutation

To allow us to reconstruct how the different repeat expansion mutations cause molecular pathology in the various disorders that they cause, we have chosen to categorize the 25 repeat disorders into four classes (Table 2). The first class of disorders, the Type 1 repeat diseases, are the "CAG-polyglutamine disorders." This class of repeat diseases includes nine inherited neurodegenerative disorders (SBMA, Huntington's disease, dentatorubral pallidoluysian atrophy, and six forms of spinocerebellar ataxia) that all share the common feature of being caused by a CAG repeat located within the coding region of a gene. Upon CAG repeat expansion, a mutant protein with an extended polyglutamine tract is produced, making the protein then adopt an abnormal conformation and misfold to initiate the pathogenic cascade. The resultant pathology is believed to primarily stem from a gain of function of the mutant protein imparted by the expanded polyglutamine tract. As discussed below, much effort has gone into trying to define what the gain-of-function effect is for each polyglutamine disease protein and into determining if

Tab. 1 Compilation of repeat expansion diseases in humans.

Disease	*Gene*	*Repeat*	*Normal*	*Premutation*	*Disease*
Blepharophimosis-ptosis-epicanthus inversus	*FOXL2*	GCG	14	–	24
Cleidocranial dysplasia	*RUNX2*	GCG	11–17	–	27
Congenital central hypoventilation	*PHOX-2B*	GCG	20	–	25–29
Dentatorubralpallidoluysian atrophy (Haw River syndrome)	*Atrophin-1*	CAG	7–35	–	49–88
Fragile X syndrome of mental retardation	*Fmr-1*	CGG	6–53	45–200[a]	200–>2000
FRAXE mental retardation	*Fmr-2*	GCC	6–40	61–200	200–1000
Friedreich's ataxia	*Frataxin*	GAA	7–38	33–60	66–1700
Hand-foot-genital	*HOX-A13*	GCG	18	–	24–27
Huntington's disease	*Htt/IT15*	CAG	6–35	27–36	39–250
Huntington's disease like 2	*JPH3*	CAG	7–27	–	50–60
Holoprosencephaly	*ZIC2*	GCG	15	–	25
Myoclonus epilepsy type 1	*Cystatin B*	CCCCGCCCCGCG	2–3	12–17	30–150
Myotonic dystrophy type 1	*DMPK*	CTG	5–37	50–80	50–4000
Myotonic dystrophy type 2	*ZNF9*	CCTG	104–176	?	175–11 000
Oculopharyngeal muscular dystrophy	*PABP2*	GCG	6–7	–	8–17
Spinocerebellar ataxia type 1	*Ataxin-1*	CAG	6–44	–	39–>10
Spinocerebellar ataxia type 2	*Ataxin-2*	CAG	13–33	–	31–>10
Spinocerebellar ataxia type 3 (Machado-Joseph disease)	*Ataxin-3*	CAG	12–40	–	55–84
Spinocerebellar ataxia type 6	*CACNA1A*	CAG	4–18	–	19–33
Spinocerebellar ataxia type 7	*Ataxin-7*	CAG	4–35	–	37–306
Spinocerebellar ataxia type 8	?	CTG	16–>800	?	71–>1000
Spinocerebellar ataxia type 10	*Ataxin-10*	ATTCT	10–22	?	800–4500
Spinocerebellar ataxia type 12	*PPP2R2B*	CAG	7–32	–	55–78
Spinocerebellar ataxia type 17	*TBP*	CAG	25–48	–	43–66
Syndromic and nonsyndromic X-linked mental retardation	*ARX*	GCG	10–16	–	17–23
Synpolydactyly	*HOX-D13*	GCG	9–15	–	16–29
X-linked spinal & bulbar muscular atrophy	*AR*	CAG	5–34	–	37–70

[a] Male premutation carriers of the FMR-1 CGG repeat develop the Fragile X tremor-ataxia syndrome.
–, no premutation alleles identified, so they may not exist.
?, existence of premutation alleles is unknown.

Tab. 2 Classification of the repeat expansion diseases.

CAG/polyglutamine disorders	*Loss-of-function disorders*	*RNA gain-of-function disorders*	*Polyalanine disorders*
Dentatorubral pallidoluysian atrophy	Fragile X syndrome of MR	Myotonic dystrophy type 1	Blepharophimosis-ptosis-epicanthus inversus
Huntington's disease	FRAXE MR	Myotonic dystrophy type 2	
Spinal and bulbar muscular atrophy	Friedreich's ataxia	Spinocerebellar ataxia type 8	Cleidocranial dysplasia
Spinocerebellar ataxia type 1	Myoclonus epilepsy type 1	Fragile X tremor-ataxia syndrome	Congenital central hypoventilation
Spinocerebellar ataxia type 2			Hand-foot-genital syndrome
Spinocerebellar ataxia type 3			Holoprosencephaly
Spinocerebellar ataxia type 6			Oculopharyngeal muscular dystrophy
Spinocerebellar ataxia type 7			Synpolydactyly
Spinocerebellar ataxia type 17			X-linked MR (syndromic & nonsyndromic)

MR = mental retardation.

shared pathways of toxicity are initiated in the different diseases. At least one, and perhaps a greater number of these disorders may principally involve a simultaneous dominant-negative partial loss of the normal function of the disease protein. The next class of disorders, the Type 2 repeat diseases, are a much more disparate group of repeat disorders. The Type 2 repeat diseases are the "Loss-of-function repeat disorders." These disorders include different repeats that vary in sequence composition and gene location, but share a final common pathway of disease pathogenesis – a loss of function of the disease gene within which they occur. This group includes various classic trinucleotide repeat disorders such as the two fragile X syndromes of mental retardation (FRAXA and FRAXE) and Friedreich's ataxia – but also encompasses the dodecamer repeat expansion in progressive myoclonic epilepsy type 1, and possibly the CAG repeat expansion in Huntington's disease like-2 (*HDL2*) gene. Strong evidence for a loss-of-function pathway in the form of nonrepeat loss-of-function mutations supports many of these classifications. The third group of repeat diseases, the Type 3 disorders, comprise a shared class because all of them have been proposed to involve the production of a toxic RNA species. This category of repeat diseases is thus called the *RNA gain-of-function disorders.* Included among these disorders are two closely related forms of DM, the common and classic myotonic dystrophy type 1 (DM1) and its uncommon phenocopy, myotonic dystrophy type 2 (DM2). Another member of this group is the recently described fragile X tremor-ataxia syndrome (FXTAS) in male premutation carriers – a fascinating example of two different disease pathways operating upon the same expanded repeat mutation based upon size range differences. One form of spinocerebellar ataxia (SCA8) with an unclear mechanism of pathogenesis has also been provisionally placed into this category, based upon current working models of how its repeat causes disease. The last class of repeat disease, the Type 4 disorders, are the "GCG-polyalanine disorders" that are grouped together because all involve short GCG repeat tracts falling within the coding regions of unrelated genes that become expanded to moderately sized GCG repeats. With the exception of oculopharyngeal muscular dystrophy, all are developmental malformation syndromes, and while gain-of-function polyalanine toxicity has been proposed for a number of these disorders, loss of function due to the polyalanine expansion seems more likely for others. Finally, a number of repeat disorders, spinocerebellar ataxia type 10 (SCA10), spinocerebellar ataxia type 12 (SCA12), and Huntington's disease like 2 (HDL2), currently defy classification because very little is known about their molecular basis. These diseases will be considered in the final section of this chapter.

3
Type 1: The CAG/Polyglutamine Repeat Diseases

3.1
Spinal and Bulbar Muscular Atrophy

Spinal and bulbar muscular atrophy (SBMA; Kennedy's disease) is a late-onset neurodegenerative disease with an inheritance pattern resembling X-linked recessive. It has a prevalence of about 1 in 50 000 males. Patients suffer a late-onset, progressive degeneration of primarily lower motor neurons in the anterior horn of the spinal cord and in the bulbar region of the brainstem; however, variable involvement of sensory neurons in the dorsal root ganglia also occurs. SBMA typically presents with cramps, followed by proximal muscle weakness and atrophy. Patients often exhibit dysarthria, dysphagia, and fasciculations of the tongue and lips. Affected individuals have symptoms of mild androgen insensitivity, such as gynecomastia, reduced fertility, and testicular atrophy.

SBMA is caused by a polymorphic (CAG)*n* repeat in the first exon of the *AR* gene, which is expressed as a glutamine tract. Unaffected individuals carry 5 to 34 triplet repeats, while affected individuals carry 37 to 70 repeats. SBMA exhibits a paternal expansion bias. The disease does not appear to involve a simple loss-of-function mechanism, as complete loss of AR does not result in motor neuron degeneration.

AR is widely expressed in males and females, and is a member of the steroid receptor–thyroid receptor superfamily with a highly conserved DNA binding domain, ligand binding domain and transactivation domain (Fig. 1). In its inactive state, it forms an apo-receptor complex with heat-shock proteins (HSPs) 70 and 90, and resides in the cytoplasm. Upon binding androgen, it dissociates from these HSP chaperone proteins and translocates to the nucleus. Once in the nucleus, AR dimers transactivate certain genes, many of which are responsible for generating and maintaining male characteristics. Although the glutamine expansion does not affect the binding of its ligand, androgen (testosterone), the glutamine tract is in the major transactivation domain, and may affect transactivation competence. However, the effect of the polyglutamine expansion upon AR transactivation competence remains controversial.

Many lines of evidence suggest that AR must translocate to the nucleus to exert its toxicity. Nuclear inclusions (NIs) are present in motor neurons of the spinal cord and brainstem in SBMA patients. In a Drosophila model of SBMA, retinal expression of mutant AR only yielded a degenerative phenotype in the presence of ligand. As in humans, androgen binding causes the nuclear translocation of AR in mice. Transgenic male SBMA mice produce testosterone and will only

Fig. 1 Diagram of the androgen receptor. The androgen receptor is a member of the steroid receptor-thyroid hormone receptor superfamily, and consequently displays a stereotypical architecture. The CAG repeat – polyglutamine tract (striped box) resides within the amino-terminal domain, which mediates transcription activation through an "activation function" domain (charcoal gray box). Additional conserved domains include the DNA binding domain (light gray box), nuclear localization signal (black box), and the ligand binding domain (checkered box). Expansion of the CAG repeat to alleles of $\geq$37 triplets is the cause of spinal and bulbar muscular atrophy (SBMA).

develop motor neuron disease if expressing polyglutamine-expanded full-length AR protein, yet when castrated, such SBMA transgenic mice do not develop a phenotype. In the same study, female transgenic mice hemizygous for expanded AR develop motor neuron degeneration when exposed to exogenous androgen. Importantly, even rare human females who are homozygous for an AR CAG repeat expansion mutation do not develop SBMA, despite widespread expression of mutant AR throughout the CNS (central nervous system). Thus, SBMA is not a true X-linked recessive disorder, but rather is classified as a sex-limited disorder, since expression of the disease phenotype is dependent upon male levels of androgen.

A number of studies on SBMA patients and mouse models have characterized the NIs seen in this disease. While NIs are widely distributed in lower motor neurons of the spinal cord and brainstem, NIs also occur in a variety of nonneuronal tissues that appear to function normally. NIs colocalize with components of the proteasome and with molecular chaperones. How this contributes to the phenotype is unknown, although HSP70 overexpression attenuates toxicity in cell culture and in transgenic mouse. The presence of proteasome components and HSPs in NIs may thus be revealing a protective intervention by the cell, or alternatively may be deleterious due to depletion of these important cellular proteins. Interestingly, only antibodies raised to amino-terminal fragments of AR detect NIs, indicating that proteolysis may play a role in the disorder. Caspase-3 has been shown to cleave AR in a polyglutamine tract length-dependent manner *in vitro*, and this may have pathogenic significance, as truncated AR is more toxic than full-length protein in cell culture studies and transgenic mouse models. The phosphorylation of AR is modulated by androgen, and this posttranslational modification appears to enhance caspase-3 cleavage.

One possible mechanism of expanded AR toxicity is through transcription interference. Polyglutamine-expanded AR interacts with the transcription coactivator CREB-binding protein (CBP) in a polyglutamine tract length-dependent manner, colocalizes with CBP in spinal cord NIs from patients, and can interfere with CBP-dependent transcription. CBP is a transcription cofactor that regulates the expression of vascular endothelial growth factor (VEGF), among other genes. VEGF is important in motor neuron health, as deletion of a portion of its promoter called the *hypoxia response element* (HRE) causes motor neuron degeneration even in normoxic mice. Pathologically expanded AR reduces VEGF transcript expression in males, with VEGF165 isoform expression reduced at both the RNA and protein levels. Adding VEGF165 to a motor neuron-like cell line (MN-1) expressing polyglutamine-expanded AR significantly rescues its cell death, again supporting the role of transcription interference in SBMA and suggesting that VEGF165 may serve as a neurotrophic factor for motor neurons.

While no effective treatment for SBMA has yet been validated in human patients, the role of testosterone in SBMA disease progression has received considerable attention. Interestingly, testosterone supplementation was initially used as a treatment for SBMA. Rather than aggravating the disease as one might fear, it was reported to attenuate the phenotype slightly, but did not significantly retard disease progression. Such a beneficial effect of testosterone may be due to anabolic effects on muscle strength or to a downregulation of AR in response to elevated androgen levels. Still another possibility is that AR-mediated transcription in motor neurons somehow performs a trophic function in the face of damage or injury. Very recent work, however, strongly indicates that elimination of ligand by surgical or pharmacological castration is a very effective treatment in SBMA transgenic mouse models, even showing efficacy when SBMA mice display an advanced phenotype. As ligand binding is associated with nuclear translocation of mutant AR and aberrant effects upon nuclear transcription appear crucial for SBMA disease progression, abrogation of nuclear localization may account for the success of castration. Consistent with this hypothesis are results of studies with flutamide, a drug that appears to block AR-dependent transactivation without preventing its nuclear translocation. While pharmacological castration with leuprolide is highly effective therapeutically, flutamide does not ameliorate symptoms or disease progression in an SBMA mouse model. Attempted translation of these preclinical trial results to human SBMA patients is currently underway.

3.2 Huntington's Disease

Huntington's disease (HD) is an autosomal dominant disorder with a prevalence of 1 per 15 000 persons worldwide. It is a debilitating disease that often presents clinically in the fourth or fifth decade of life with chorea (i.e. spontaneous, involuntary dancelike movements). Personality change and cognitive impairment may precede the clinical onset by years, and ultimately culminate in dementia after onset of the movement disorder. Chorea gives way to bradykinesia and rigidity late in the disease. CNS atrophy occurs most prominently in the striatum, which is reduced to a fraction of its original size (Fig. 2). However, significant neurodegeneration and neuron loss in the cortex is also typical, while cerebellum, brainstem, and spinal cord are relatively spared – except in juvenile-onset cases.

HD is caused by a (CAG)*n* repeat expansion in the *huntingtin* gene (*htt*), which encodes a 350-kDa protein containing 67 exons. Unaffected individuals carry 6 to 35 repeats, while affected individuals carry 39 to 250 repeats. The largest repeats cause juvenile-onset HD and display a paternal transmission bias. The htt protein is ubiquitously expressed in brain tissue,

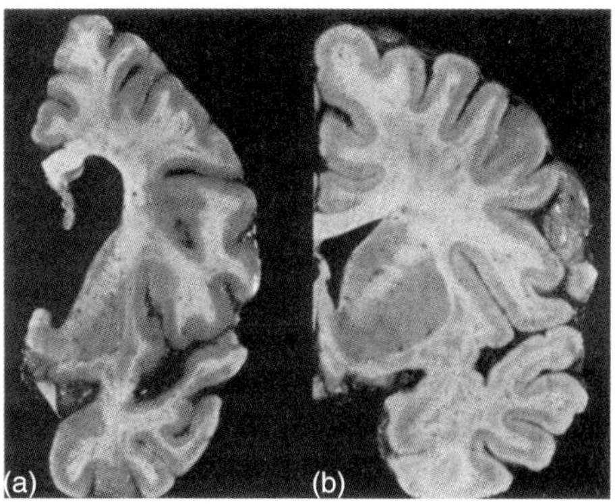

Fig. 2 Huntington's disease (HD) neuropathology. Hemi-coronal sections of postmortem brains from (a) a classic, adult-onset HD patient and (b) a normal control reveal marked degeneration of the striatum (midmedial region) and considerable atrophy of cortical regions. (From Robataille et al. (1997) *Brain pathol.* **7**, 901. Used with permission).

with highest levels in striatal interneurons and cortical pyramidal cells. Two htt mRNA transcripts have been detected: one 10 kb and the other 13 kb in length, expressed most highly in CNS neurons. Ultrastructurally, the wild-type full-length protein is predominantly cytoplasmic and is located in the pre- and postsynaptic regions of dendrites and axons. Htt protein associates with microtubules, vesicles, and organelles.

HD appears to have a predominantly dominant mechanism due to its inheritance pattern, evidence that homozygotes are no more severely affected than heterozygotes, and the observation that heterozygous deletion of huntingtin does not cause HD. The fact that no HD-causing loss-of-function mutations have been documented in the huge *htt* gene further supports a gain-of-function mechanism. However, postnatal elimination of htt expression in regions of the cortex can cause striatal degeneration in mice, so the notion that gain of function fully accounts for the HD phenotype is being reexamined. As htt is a regulator of transcription activation and/or mediator of vesicular brain-derived neurotrophic factor (BDNF) transport up and down axons, many investigators now envision the effects of expanded htt as twofold: simultaneously causing protein misfolding leading to gain-of-function toxicity, and inducing partial loss of an ill-defined normal function.

In 1997, it was first reported that mutant htt forms dense amyloid-type aggregates in the nucleus, perikarya, and neuropil of neurons from a transgenic mouse model and in human patients. The role of aggregation in the polyglutamine diseases and in a wide range of neurodegenerative disorders including Alzheimer's disease, Parkinson's disease, amyotrophic lateral sclerosis, and the prion diseases, thus emerged as an important theme at the end of the last decade. While initially it was thought that the formation of large aggregates is the basis of polyglutamine neurotoxicity, the weight of evidence now suggests little correlation between visible aggregate formation and disease pathology. However, the occurrence of aggregates along with neuropathology suggests that aggregation is inextricably linked to the pathogenicity of polyglutamine disease proteins. We will address the role of polyglutamine aggregation in neurotoxicity in detail in a separate section.

Many theories have been proposed to account for how polyglutamine-expanded htt causes neurotoxicity. A thorough discussion of this literature goes beyond the scope of this chapter, so the reader is referred to the relevant books and reviews in the Bibliography for a more intensive treatment of this topic. Major theories of htt neurotoxicity that will be considered herein include transcription dysregulation, proteasome inhibition, mitochondrial dysfunction/excitotoxicity, and proteolytic cleavage. As multiple independent toxicity events may be occurring concomitantly, these disease pathways are not mutually exclusive.

Expanded htt protein may interfere with transcriptional processes. Many transcription factors, such as CBP, contain glutamine tracts that mediate important protein–protein interactions. CBP interacts directly with htt, and mutant htt toxicity is ameliorated in striatal neurons *in vitro* when CBP lacking the htt interaction domain is overexpressed. CBP mediates the transcription of a number of neuronal survival factors, such as BDNF, and functions by acetylating histones, one aspect of chromatin remodeling that permits transcription to occur. Studies of HD fruit fly and mouse models have highlighted the potential importance of histone acetylation by demonstrating that histone deacetylase inhibitors (HDAC Is) can successfully rescue degenerative phenotypes in these model organisms. Htt can affect transcription mediated by p53, Sp1, and REST interaction, thereby potentially altering the expression of a large number of genes. Differential expression of survival factors, neurotransmitter receptors, and a number of other genes may thus contribute to HD pathogenesis.

The ubiquitin-proteasome protein degradation pathway has also been implicated in HD. Nuclear inclusions of htt colocalize with ubiquitin, which indicates that the expanded protein has been identified as misfolded and thus targeted for proteasomal degradation. However, polyQ sequences longer than 9 glutamines are impossible for eukaryotic proteasomes to cleave. The proteasomes of htt-transfected cells are consequently less capable of degrading proteins other than htt, as demonstrated by the reduced degradation of GFP-tagged proteins in culture. If proteasome components are clogged with mutant protein and/or sequestered into inclusions, they may be unable to degrade other misfolded or damaged proteins that carry out important functions. Alternatively, accumulation of improperly degraded proteins may interfere with other normal cellular processes, such as autophagy or mitochondrial oxidative phosphorylation.

Perhaps the longest and most thoroughly studied htt toxicity pathway involves metabolic disturbances and excitotoxicity. Glucose metabolism and oxygen consumption are reduced in HD brains as measured by PET. Severe deficits in the activity of complexes II/III and IV of the mitochondrial electron transport chain (ETC) are evident in HD brains. Inhibitors of complex II of the mitochondrial electron transport chain, such as 3-nitropropionic acid (3-NPA), can cause a selective degeneration of the striatum in rat and primate models when injected systemically, since the striatum has among the highest energy demands of all neuronal regions. 3-NPA reduces levels of ATP, resulting in mitochondrial and cellular depolarization with activation of voltage-dependent NMDA receptors. Excitotoxic damage may act in concert with increased production of free radicals to cause selective striatal degeneration in HD. Some of the earliest animal models of HD were thus generated by exposing the striatum or entire brain of rats or primates to metabolic or excitotoxic insults. In 1976, the first such model of HD was created by injection of the glutamate analog and excitotoxin kainic acid into the striatum of rats. Such lesioned rats exhibited a selective degeneration of neurons in the striatum reminiscent of HD. Recent studies have suggested that impaired Ca++ flux due to aberrant interaction of htt with the inositol phosphate-3 receptor (IP3-R) may underlie excitotoxic pathology. Studies of mitochondria from HD patients reveal abnormal mitochondrial Ca++ handling and decreased mitochondrial depolarization thresholds, in support of this hypothesis. One very recent study has found that deletion of peroxisome proliferator-activated receptor (PPAR) gamma coactivator 1 alpha (PGC-1α), a key mediator of mitochondrial biogenesis, yields HD-like striatal degeneration in mice. Thus, metabolic insults and excitotoxicity may be key steps in HD disease pathogenesis.

Another important theory of htt neurotoxicity posits that proteolytic cleavage of htt is a required step in neuronal dysfunction and the degeneration process. As noted above, the htt protein is enormous, with full-length product consisting of >3100 amino acids. Analysis of human

HD material has indicated an absence of midprotein and C-terminal epitopes in htt aggregates. Careful biochemical studies of htt have characterized a variety of putative caspase and calpain cleavage sites. Various studies suggest that the more truncated the polyglutamine-expanded htt protein, the more toxic it is in cell culture and in animal models, and the more likely it is to enter the nucleus and produce toxicity there. In the case of htt, a series of proteolytic cleavage steps culminating with cleavage to an ~100 amino acid peptide fragment by an aspartyl protease has been proposed to yield a final "toxic fragment." As it turns out, the "toxic fragment hypothesis" (as it has also been called) may be applicable to a number of polyQ diseases, which will be reviewed later in this section.

3.3
Dentatorubral Pallidoluysian Atrophy

Dentatorubral pallidoluysian atrophy (DRPLA) is a rare, autosomal dominant neurodegenerative disorder most prevalent in Japan. Adult-onset DRPLA typically involves progressive cerebellar ataxia, choreoatheosis, epilepsy, and dementia, while juvenile-onset cases also display myoclonus, epilepsy, and mental retardation. Neuropathological abnormalities include degeneration of the dentate nucleus of the cerebellum, rubral nucleus, and globus pallidus, as well as a more generalized degeneration and gliosis involving the brainstem, cerebellum, cortex, and pons. DRPLA is caused by a polymorphic (CAG)*n* repeat in the carboxy-terminal coding region of the *atrophin-1* gene on chromosome 12. Normal individuals carry 6–35 repeats, while affected individuals carry 49–88 repeats. Anticipation is prominent in DRPLA, as very large repeat expansions can occur in a single generation, typically via paternal transmission.

The DRPLA disease protein, atrophin-1, is widely expressed and appears to be predominantly cytoplasmic. Atrophin-1 contains both a putative nuclear localization signal (NLS) and nuclear export signal (NES), and nuclear localization of normal atrophin-1 is observed. Nuclear localization of polyglutamine-expanded atrophin-1 has been linked to increased toxicity in cell culture models. Ubiquitinated NIs are present in neurons and glia from patient brains, and are also immunoreactive for small ubiquitinlike modifier (SUMO) protein. Atrophin-1 is a substrate for c-Jun N-terminal kinase (JNK), with JNK's affinity for atrophin-1 inversely proportional to the size of the polyglutamine tract expansion.

Although the relevance of JNK phosphorylation to atrophin-1 function is unknown, other insights into the function of atrophin-1 have been reported. Using *Drosophila melanogaster* as a model system, Zhang et al. took advantage of the existence of a fly ortholog of atrophin-1 (*Atro*) and created lines of flies carrying mutations in the *Atro* gene. These flies demonstrated severe developmental abnormalities due to complex patterning defects, and subsequent experiments revealed that *Atro* is a transcription corepressor whose activity diminishes with increasing polyglutamine tract length. Independent studies of human atrophin-1 in cell culture studies found evidence for transcription interference of polyglutamine-expanded atrophin-1 with CREB-mediated gene activation. Transcription dysregulation may thus play a prominent role in DRPLA. Another study suggests that disturbed carbohydrate metabolism may contribute to the DRPLA neurodegenerative phenotype.

3.4 Spinocerebellar Ataxia Type 1

Spinocerebellar ataxia type 1 (SCA1) is an autosomal dominant disorder characterized primarily by a progressive cerebellar ataxia. It accounts for about 6% of all autosomal dominant cerebellar ataxias (ADCAs) worldwide. SCA1 patients suffer from coordination difficulties including dysarthria, dysphagia, and ophthalmoplegia. The cerebellum undergoes atrophy, gliosis, and severe loss of Purkinje cells, accompanied by degeneration of the dentate nucleus, inferior olive, and some brainstem nuclei. Disease onset typically occurs in the third or fourth decade of life, but presentation in childhood or adolescence to late life may be seen. SCA1 is caused by the expansion of a coding (CAG)*n* repeat in the amino-terminal coding region of the *ataxin-1* gene. The ataxin-1 CAG repeat is highly polymorphic, ranging in size from 6 to 44 triplets in unaffected individuals. The repeat length associated with disease ranges from 39 to more than 100 CAGs. Unaffected individuals with more than 20 repeats have CAT triplet interruptions within their (CAG)*n* repeat tracts. Such interruptions stabilize the repeat expansion, while absence of the CAT repeat interspersion is noted in SCA1 patient alleles. As in many other polyglutamine disorders, there is strong evidence supporting a gain-of-function mechanism in SCA1.

Ataxin-1 is widely expressed in the CNS and throughout the periphery, although expression levels are several-fold higher in nervous system tissues. The protein is predominantly nuclear in the CNS, however, some cytoplasmic staining is apparent in Purkinje cells of the cerebellum and in brainstem nuclei. Ataxin-1 knockout mice do not develop SCA1, although they do exhibit impairments in motor and spatial learning. These mice also have decreased paired-pulse facilitation in the CA1 region of the hippocampus, suggesting that ataxin-1 may normally function in synaptic plasticity and learning.

Large ataxin-1 containing NI's occur in the brainstem of affected individuals, and are immunoreactive for ubiquitin, the 20 S proteasome subunit, and HSPs HDJ2 and Hsc70. Work done on SCA1 in transgenic mice by the Orr and Zoghbi laboratories has been crucial in formulating models of not only SCA1 disease pathogenesis but also for influencing views of the molecular basis of all polyglutamine diseases. Indeed, the first mouse model for a polyglutamine disease was generated by transgenic overexpression of polyglutamine-expanded ataxin-1 in Purkinje cells (Fig. 3). This SCA1 transgenic mouse model has laid the foundation for

Fig. 3 The original spinocerebellar ataxia type 1 (SCA1) mouse model. (a) Diagram of the *Pcp2-SCA1* transgene construct. A Purkinje cell-specific expression cassette based upon inclusion of the promoter (straight line), first two noncoding exons (black boxes), and first intron (bent line) of the Purkinje cell protein 2 (*Pcp2*) gene and a SV40 polyadenylation sequence (open box) was the basis for this landmark work. Ataxin-1 cDNAs (gray box) containing either 30 CAGs (control) or 82 CAGs (expanded) were inserted into the *Pcp2* expression cassette. Sites of various PCR primer sets are also shown. (b) *Pcp2-SCA1* CAG-82 mice display ataxia. Still photographs of a 30-week-old *Pcp2-SCA1* CAG-82 mouse from a transgenic line that greatly overexpresses the mutant ataxin-1 transgene illustrates the inability of this mouse to maintain its balance when ambulating. Loss of balance when walking is consistent with the gait ataxia seen in human SCA1 patients. (From Burright et al. (1995) *Cell* **82**, 937, used with permission).

numerous follow-up studies of polyglutamine disease pathogenesis. For example, expression of mutant ataxin-1 lacking an intact self-association domain precluded aggregate formation, but permitted neurotoxicity, demonstrating that NIs are not required for SCA1 in mice. In a later study, crossing of SCA1 transgenic mice with mice lacking a ubiquitin ligase enzyme yielded SCA1 mice incapable of aggregate formation. These mice were more severely affected than their transgenic counterparts

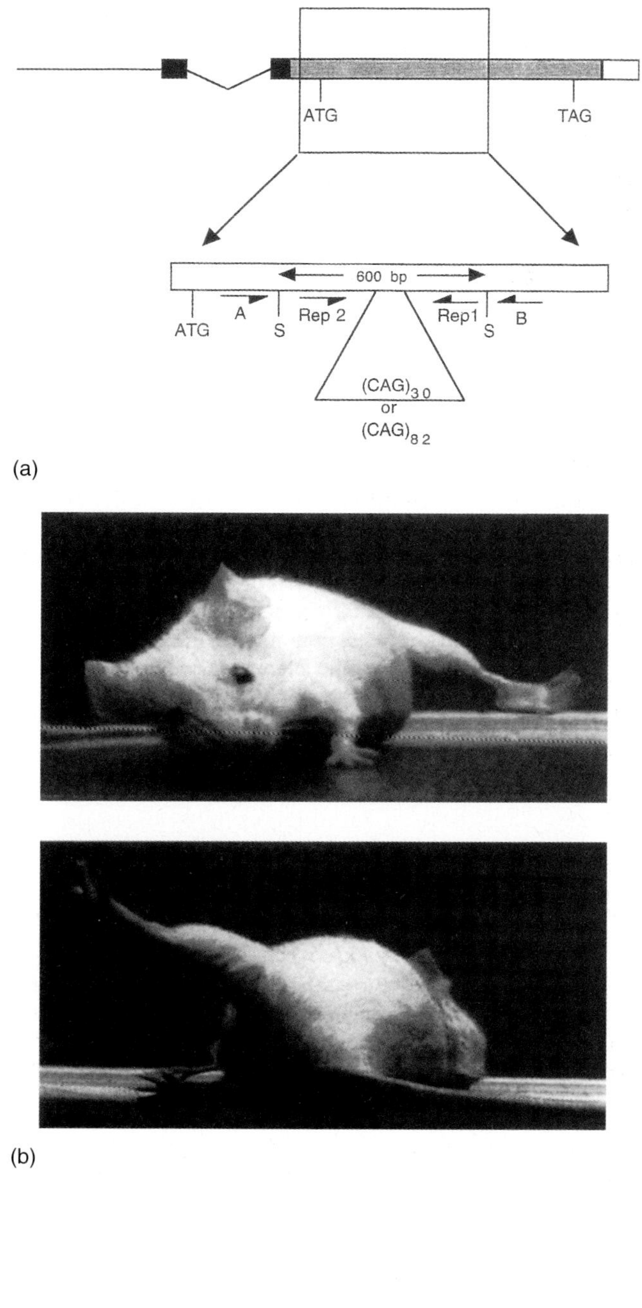

due to absence of the ubiquitin ligase. This work supported the view that visible aggregate formation may represent a protective cellular response for neutralizing misfolded polyglutamine-containing peptides. In another study, mice expressing ataxin-1 with a mutated NLS were found not to develop SCA1, suggesting that nuclear localization is absolutely required for SCA1 molecular pathology. Although targeting ataxin-1 to Purkinje cells appears sufficient to recapitulate a dramatic SCA1-like disease in mice, a subsequent knockin mouse model of SCA1 indicated that expression in other regions of the CNS yields a more representative disease phenotype.

Over the past few years, numerous leads have emerged in the search for the pathogenic basis of SCA1. In one line of investigation, phosphorylation of serine 776 of the ataxin-1 protein was shown to affect pathogenesis, highlighting the importance of this posttranslational modification. SCA1 transgenic mice in which serine 776 had been mutated to an alanine, exhibited a dramatically attenuated phenotype and a complete lack of NI's. In an accompanying study, interaction of polyglutamine-expanded ataxin-1, but not wild-type ataxin-1, with several isoforms of the phosphoserine/threonine binding protein 14-3-3 was reported. This extremely abundant peptide is thought to serve a regulatory function by binding proteins and determining their subcellular localization, among other things. The interaction between ataxin-1 and 14-3-3 was shown to be dependent upon the Akt-mediated phosphorylation of serine 776. A novel mechanism for ataxin-1 toxicity was thus proposed: upon Akt phosphorylation of polyglutamine-expanded ataxin-1, ataxin-1 binds 14-3-3, is stabilized, and ultimately accumulates in the nucleus. The downstream effects of the nuclear accumulation of mutant ataxin-1 on neuronal function, however, remain undefined.

Transcriptional dysregulation is also a reasonable hypothesis for SCA1 pathogenesis. One study has demonstrated that ataxin-1 interacts with polyglutamine binding protein 1 (PQBP-1), and that this interaction results in interference with RNA polymerase-dependent transcription. Independent studies have supported a role for ataxin-1 as a transcription corepressor. In pull-down assays and in Drosophila, ataxin-1 interacts with the proteins SMRT (silencing mediator of retinoid and thyroid hormone receptors) and HDAC3 (histone deacetylase 3), both transcriptional repressors. Aggregates of polyglutamine-expanded ataxin-1 sequester SMRTER, the Drosophila ortholog of SMRT. Transcription repressors and histone deacetylases also appeared in an earlier screen for modulators of the Ataxin-1 phenotype in the fly. In addition, two separate studies of gene expression alterations in presymptomatic SCA1 transgenic mice have uncovered changes in the levels of transcripts encoding proteins involved in Ca++ flux and metabolism. It thus appears that transcription dysregulation may be a key step in SCA1 disease pathogenesis.

3.5
Spinocerebellar Ataxia Type 2

Spinocerebellar ataxia type 2 (SCA2) is an autosomal dominant, progressive cerebellar ataxia that accounts for about 13% of all ADCA cases. Although patients display problems with voluntary coordinated movements as in the rest of the SCAs, its main distinguishing clinical feature is extremely slow saccadic eye movements. Other symptoms may include hyporeflexia, myoclonus, and action tremor. Patients suffer a gradual degeneration of the cerebellum, inferior olive, pons, and spinal cord. Both Purkinje and granule cells degenerate in the cerebellum, and CNS atrophy in some patients can be widespread. Interestingly, in certain cases, there is involvement of the substantia nigra, with such patients displaying a prominent degree of parkinsonism. The disorder is caused by a coding (CAG)*n* expansion in a novel gene of unknown function, named *ataxin-2*. The SCA2 CAG repeat displays little polymorphism in the normal population, with 95% of the population possessing 22 or 23 repeats in each allele. The remaining 5% of alleles in unaffected individuals do nonetheless range from 15–31 CAG repeats. Such unaffected individuals typically have two CAA interruptions in their CAG repeat tract. Affected SCA2 patients typically display CAG repeats numbering from 32–63 triplets, and such disease alleles are always uninterrupted CAG tracts. There is a nonlinear inverse correlation between expansion size and age of onset in SCA2, with some studies documenting a paternal transmission bias for larger expansions.

Both wild-type and expanded ataxin-2 mRNA is widely expressed, with highest levels in the substantia nigra and Purkinje cells of the cerebellum. The 140-kDa protein is cytoplasmic and its function remains uncertain. Since it contains Sm1 and Sm2 motifs common in proteins involved in RNA splicing and protein–protein interactions, involvement in RNA processing has been proposed. A yeast two-hybrid screen indicated that ataxin-2 has a binding partner, ataxin-2 binding protein-1 (A2BP1). A2BP1 is highly conserved throughout the animal kingdom and its expression pattern corresponds well with that of ataxin-2. A2BP1 contains a domain that is also conserved in RNA-binding proteins, the RNP motif. Thus, a complex including ataxin-2 and A2BP1 may be involved in RNA processing or metabolism.

The *ataxin-2* gene is evolutionarily conserved. The murine ortholog of ataxin-2 does not contain a polyglutamine repeat tract, however, but instead possesses a single glutamine residue at the analogous location. (Absence of a substantial glutamine repeat tract is typical for mouse orthologs of polyglutamine disease proteins.) In the case of ataxin-2, study of the Drosophila ortholog (*Datx2*), which does show two regions of marked amino acid similarity and does contain polyglutamine repeat regions, has yielded some potentially important insights into ataxin-2s normal function. Modulation of *Datx2* dosage resulted in mutant phenotypes whose cause could be traced to aberrant actin filament formation (Fig. 4). This work suggests that alteration of ataxin-2-mediated regulation of cytoskeletal structure could affect dendrite formation or other aspects of neuronal function in SCA2. As SCA2 is one of the few polyglutamine disorders to display prominent cytosolic aggregates, such a model of SCA2 pathogenesis seems plausible. In other experiments, eliminating the *C. elegans* ortholog of ataxin-2 (ATX-2) yielded a lethal phenotype, indicating that ATX-2 is required for early embryonic development of this nematode worm. It remains to be seen how these observations in worms and flies will apply to ataxin-2 function in mammals, especially since no simple or conditional knockout of the mouse *ataxin-2* gene has been performed.

3.6 Spinocerebellar Ataxia Type 3/Machado–Joseph Disease

Spinocerebellar ataxia type 3 (SCA3) is the most common inherited dominant

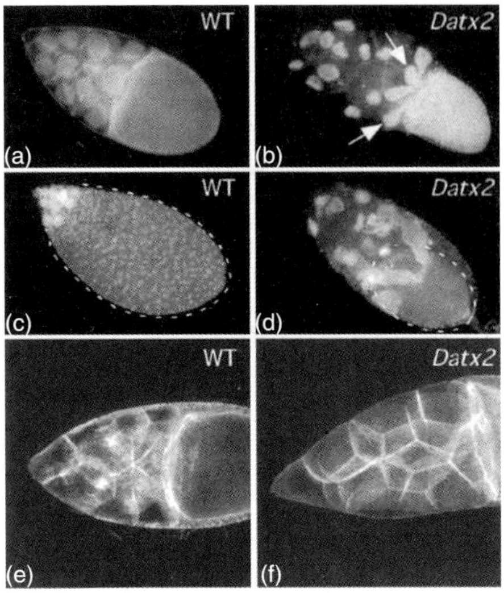

Fig. 4 Studies of the ataxin-2 ortholog in *Drosophila melanogaster* reveals a role for Drosophila ataxin-2 (*Datx2*) in actin filament formation during oogenesis. Egg chambers from normal (Wild-Type (WT)) and mutant flies with reduced expression (*Datx2*) of Drosophila ataxin-2 were stained with DAPI (blue) and phalloidin (red) to indicate nuclei and filamentous actin respectively (a–d). The egg chambers of WT flies prior to cytoplasmic transport (a) display well demarcated, separated but interconnected cells (blue) as expected, while the egg chambers of *Datx2* flies (b) contain irregularly arranged cells. After cytoplasmic transport, the egg chambers of WT flies (c) show one greatly enlarged oocyte (dashed line) with only a small section of compressed cells, while the egg chambers of *Datx2* flies (d) have failed to yield an enlarged oocyte, instead retaining dispersed and large adjacent cells. Confocal images of egg chambers prior to cytoplasmic transport stage reveal a prominent actin filament network in WT flies (e), but a remarkably transparent actin filament network in *Datx2* flies (f). The decreased density of the actin filament network underlies the cytoplasmic "dumping" defect in the *Datx2* flies. (From Satterfield, T.F., Jackson, S.M., Pallanck, L.J. (2002) A Drosophila homolog of the polyglutamine disease gene *SCA2* is a dosage-sensitive regulator of actin filament formation, *Genetics* **162**, 1687–1702, used with permission of *Genetics*.) (See color plate p. xxxviii).

spinocerebellar ataxia worldwide. SCA3 is also known as Machado–Joseph disease (MJD) because of its initial description in a group of Portuguese residents of the Azores islands. It is a progressive, autosomal dominant cerebellar ataxia whose clinical features include ophthalmoplegia, dystonia, dysarthria, and signs of lower motor neuron disease, such as tongue and facial fasciculations. Degeneration occurs in the spinocerebellar tracts, dentate nuclei, red nuclei, substantia nigra, and spinal cord. This disease is unique among the SCAs because the cerebellar cortex and inferior olive are largely spared. SCA3 is caused by a (CAG)*n* repeat near the 3′ end of the coding region of a novel gene (*ataxin-3*). Normal alleles range from 12–40 CAG repeats while affected individuals carry 55–84 repeats. Unlike many other triplet repeat disorders, there is a substantial gap between the largest normal repeat allele and the smallest disease-causing repeat allele. Some researchers have proposed that this could be due to a SCA3 founder effect. The presence of SCA3 in every major racial population worldwide, however, would require the founder to be truly ancient. There is the typical inverse correlation between repeat number and age of onset in SCA3, and in this disease, paternal expansion bias is characteristic, as documented by repeat sizing of sperm.

Ataxin-3 is a ubiquitously expressed 42-kDa protein, making it the smallest of the polyglutamine proteins. Ataxin-3 is highly conserved in eukaryotes, with homology to ENTH and VHS domain proteins involved in regulatory adaptor functions and membrane trafficking. Ataxin-3 has several splice variants which reside in the nucleus and the cytoplasm, and it appears developmentally regulated. The existence of NIs was first documented in the brains of patients with SCA3, and this feature of polyglutamine-expanded ataxin-3 remains evident in SCA3 cell culture and mouse models. Ataxin-3 NIs are ubiquitinated and contain numerous transcription factors.

A number of recent discoveries regarding the domain structure and normal function of ataxin-3 suggest pathways by which polyglutamine repeat expansion could result in disease pathogenesis. Comparison of ataxin-3 amino acid sequences across a wide range of eukaryotic species revealed the presence of an extremely highly conserved amino-terminal sequence that was named the *josephin domain*. Study of this region indicates that it may play a role in aggregate formation in concert with the expanded polyglutamine tract. A rather intriguing feature of ataxin-3 that was discovered independently is the presence of multiple ubiquitin interaction motifs (UIMs), and the demonstration that ataxin-3 is a polyubiquitin binding protein. This suggests a role for ataxin-3 in mediating protein refolding and degradation.

In addition to a possible normal role in protein surveillance, other studies have found ataxin-3 directly interacts with the histone acetyltransferases CBP and p300, and can block histone acetyltransferase activity by inhibiting access of such coactivators to their histone substrates. This appears to be mediated by the interaction of ataxin-3 with histones. *In vitro* and *in vivo* studies of ataxin-3 further revealed an interaction with histones and the chromatin remodeling machinery that led to a repression of transcription activation. Thus, polyglutamine-expanded ataxin-3 may also affect transcriptional processes once it begins to accumulate in the nuclear compartments of the cell types where it is expressed.

3.7 Spinocerebellar Ataxia Type 6

Spinocerebellar ataxia type 6 (SCA6) is an autosomal dominant, slowly progressing cerebellar ataxia that accounts for ~10–20% of ADCA worldwide. It is characterized predominantly by cerebellar dysfunction that may have an episodic component. Other common features may include dysarthria, nystagmus, loss of vibration sense and proprioception, and imbalance. Histopathological changes include loss of Purkinje cells, cerebellar granule neurons, and neurons in the dentate nucleus and inferior olive. SCA6 is caused by a coding (CAG)*n* expansion in exon 47 of the gene *CACNA1A*, which encodes the α_{1A} subunit of the P/Q-type voltage-gated calcium channel. This gene is located on chromosome 19 at band p13. The SCA6 CAG repeat is small compared to other polyglutamine disorders, with a pathogenic range of only 19–33 triplets. Unaffected individuals carry alleles of 4–18 repeats. There is an inverse correlation between repeat length and age of disease onset, and minimal intergenerational and somatic instability has been reported for the SCA6 repeat expansion in affected patients.

The 9.8 kb *CACNA1A* transcript implicated in SCA6 is expressed throughout the CNS, and most highly in cerebellar Purkinje cells and granule neurons. The α_{1A} subunit is the pore-forming component of the channel, which is important in neurotransmitter release at the synapse. Polyglutamine expansions in this subunit cause variable changes in calcium transmission rates, depending on the system and the β subunit coexpressed. Expansions do not cause a reduction in membrane channel density in HEK293 cells, suggesting that aggregation does not occur at pathogenic repeat lengths. Ubiquitin-negative neuronal inclusions are visible in the cytoplasm of SCA6 patient Purkinje cells, however.

In addition to displaying a disease allele range that does not overlap with the other polyQ diseases, there are several other reasons to suspect that SCA6 is different from the rest of the polyglutamine repeat group. The first difference is the existence of two disorders, episodic ataxia type 2 (EA2) and familial hemiplegic migraine (FMH), that are both allelic to SCA6, as both are caused by point mutations in the *CACNA1A* gene. EA2 resembles SCA6 as affected EA2 individuals suffer from a slowly progressive form of episodic ataxia, accompanied by nystagmus, dysarthria, loss of balance and sometimes cerebellar atrophy. EA2 is usually caused by truncation mutations in *CACNA1A*, resulting in loss-of-function of the calcium channel. CACNA1A knockout mice similarly develop ataxia and late-onset cerebellar atrophy, characteristic of SCA6 and EA2 patient phenotypes. Another reason SCA6 is different from other polyQ disorders is that the CACNA1A protein probably does not undergo a structural change that converts it into an aggregate-prone, beta-sheet adopting, amyloid-like conformer. Indeed, even the largest SCA6 polyglutamine tract is below the threshold required for stable β-pleated sheet formation in other polyglutamine disorders. Consistent with this prediction, channel localization of mutant polyglutamine-expanded CACNA1A protein is not affected in cell culture models, and its channel function is not completely abolished. Cytoplasmic aggregates in patient material are ubiquitin-negative, suggesting that the protein may not be grossly misfolded. This evidence supports a model whereby a dominant-negative loss of function of the P/Q-type voltage-gated calcium channel due to association of the mutant α_{1A} subunit with other subunits causes SCA6. The coincidence that this disorder is caused by a polyQ tract expansion and causes a progressive cerebellar ataxia cannot be ignored, however, more data will be required to soundly refute a possible concomitant toxic gain-of-function effect.

3.8 Spinocerebellar Ataxia Type 7

Spinocerebellar ataxia type 7 (SCA7) is an autosomal dominant, progressive cerebellar ataxia. It is unique among autosomal dominant SCAs, as patients typically develop visual impairment in addition to their cerebellar ataxia. The visual impairment is due to a cone-rod dystrophy that results in retinal degeneration. Patients first develop problems distinguishing colors, but ultimately go blind in this type of retinal degeneration. SCA7 patients may present either with cerebellar ataxia or visual impairment. The likelihood of their presentation is dictated by the size of their CAG repeat disease allele, with larger repeats typically favoring presentation with visual impairment. Affected individuals display prominent dysarthria, and can develop increased reflexes, decreased vibration sense, and oculomotor disturbances. Neuronal degeneration and reactive gliosis occur in the cerebellar cortex, dentate nucleus, inferior olive, pontine nuclei, and occasionally in the basal ganglia. NIs are widespread. Infantile-onset SCA7 has been documented, and in this fatal form of the disease, nonneuronal tissues such as the heart and the kidney are severely affected. SCA7 is caused by a highly polymorphic (CAG)n repeat in the 5′ coding region of the *ataxin-7* gene. Unaffected individuals carry 4–35 repeats, while affected individuals carry 37–306 repeats. The SCA7 trinucleotide repeat is one of the most unstable of all polyglutamine disease genes, sometimes expanding by as much as 250 repeats in a single generation. There is a pronounced paternal expansion bias, with large expansions occurring in male germ cells, frequently causing embryonic lethality that results in reduced transmission. Marked repeat instability can occur in the brain, and produce large somatic expansions on occasion.

Ataxin-7 is a ubiquitously expressed protein of 892 amino acids. It is expressed most highly in heart, skeletal muscle, and pancreas. Expression levels within the CNS are highest in the cerebellum and brainstem. One splice variant, ataxin-7b, is expressed predominantly in the CNS (32). Ataxin-7 contains a functional arrestin domain, a protein interaction domain that is highly selective for phosphorylated forms of its interacting protein(s). This suggests that it interacts with specific phospho-proteins, although none have been discovered to date. It also contains two SH3 domains, which are protein interaction domains that bind proline-rich sequences and mediate a number of cell signaling processes. Ataxin-7 also contains three putative NLSs and one NES.

Ataxin-7 is conserved throughout eukaryotes, and its yeast ortholog SGF73 is part of a multisubunit histone acetyltransferase complex called *SAGA* (*Spt/Ada/Gcn5 acetyltransferase*). The human orthologs of SAGA comprise the so-called STAGA (SPT3/TAF9/ADA2/GCN5 acetyltransferase) complex, and are essential transcription coactivators required for the transcription of certain genes. STAGA components immunoprecipitate with ataxin-7. Although pathogenic expansion of ataxin-7 does not alter its ability to be integrated into the STAGA complex, the presence of the polyglutamine-expanded ataxin-7 has a dominant-negative effect upon the GCN5 histone acetyltransferase activity of the STAGA complex, resulting in transcription dysregulation. The transcription dysregulation caused by polyglutamine-expanded ataxin-7 likely causes a disease phenotype by altering the ability of certain transcription factors to activate expression of their target genes.

The best characterized example of transcription dysregulation by polyglutamine-expanded ataxin-7 is its interference with the cone-rod homeobox protein (CRX), a glutamine domain containing transcription factor expressed only in the retina and the pineal gland. Ataxin-7 interacts directly and functionally with CRX, according to studies performed *in vitro* and in a mouse model of SCA7. Importantly, the interaction between ataxin-7 and CRX appears to involve the glutamine tract regions found in both proteins. Autosomal dominant mutations in CRX can cause a cone-rod dystrophy in humans, further supporting a model in which CRX's diminished transactivation competence is central to the SCA7 retinal degeneration phenotype. Several transcription factors, including CBP, can be found in SCA7. CRX may be but one of a number of transcription factors whose function is diminished by polyQ-expanded ataxin-7 interaction and dysregulation of STAGA complex-mediated gene expression.

One intriguing feature of SCA7 was discovered upon generation of transgenic mice expressing ataxin-7 with the mouse prion protein promoter. This promoter drives expression in every tissue, with the occasional notable exception of the Purkinje cells of the cerebellum. Despite lack of Purkinje cell expression of ataxin-7, mice

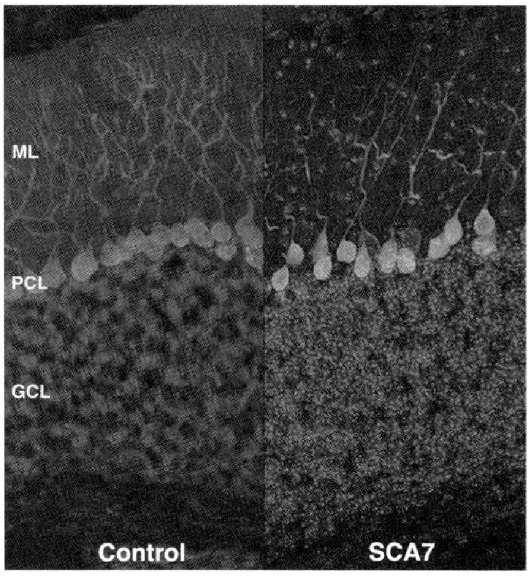

Fig. 5 Noncell autonomous Purkinje cell degeneration in a mouse model of spinocerebellar ataxia type 7 (SCA7). Confocal microscopy analysis of cerebellar sections from a SCA7 transgenic mouse (SCA7) created with an ataxin-7 CAG-92 containing murine prion protein expression vector and from an age- and sex-matched nontransgenic littermate (Control). Staining with an anti-ataxin-7 antibody (magenta), a calbindin antibody (green), and DAPI (blue) reveals a healthy, normal-appearing cerebellum characterized by properly oriented Purkinje cells with extensive dendritic arborization in the "Control" mice. However, SCA7 transgenic mice display pronounced Purkinje cell degeneration as evidenced by decreased dendritic arborization and displacement of Purkinje cell bodies. Interestingly, although numerous neurons in the granule cell layer (GCL) and the molecular layer (ML) display aggregates of ataxin-7, there is no accumulation of mutant ataxin-7 in the degenerating Purkinje cells due to lack of appreciable expression there. As the Purkinje cells degenerate without expressing the mutant protein, the degeneration is described as noncell autonomous. (Adapted from Garden, G.A., Libby, R.T., Fu, Y.H., Kinoshita, Y., Huang, J., Possin, D.E., Smith, A.C., Martinez, R.A., Fine, G.C., Grote, S.K., et al. (2002) Polyglutamine-expanded ataxin-7 promotes noncell-autonomous Purkinje cell degeneration and displays proteolytic cleavage in ataxic transgenic mice, *J. Neurosci.* **22**, 4897–4905, used with permission of the *Journal of Neuroscience*.) (See color plate p. xxxix).

developed a cerebellar ataxia accompanied by degeneration of the Purkinje cells (Fig. 5). This noncell-autonomous degeneration may point to a disease mechanism involving withdrawal of trophic support by communicating neurons (olivary, deep cerebellar, brainstem, or granule neurons) or dysfunction of glutamate transporters expressed by surrounding glia. Damage to inferior olivary neurons or Bergmann glia can indeed cause the degeneration of Purkinje cells, lending credence to this theory.

Another interesting feature of SCA7 that is common to other neurodegenerative disorders is the prominence of morphological and functional degeneration without pronounced apoptosis. Some neurons in SCA7 humans and mouse models exhibit indentations in the nuclear envelope, reduced arborization, ectopy, and increased autophagy. (Autophagy is the bulk degradation of cellular components by a membrane-bound autophagosome and is accelerated by a number of cellular insults.) Caspase-3, a proteolytic enzyme classically associated with apoptosis, is activated at abnormally high levels in SCA7 patient brains. This may suggest that caspase activation, rather than causing apoptosis, is contributing to a nonapoptotic degenerative process.

In normal human neurons, ataxin-7 is variably located in the nucleus or the cytoplasm. In SCA7 patients, ataxin-7 gradually undergoes a shift in localization into NIs. This process has been replicated in SCA7 transgenic mice, in which ataxin-7 immunoreactivity shifts from the cytoplasm to the nucleus, and ultimately forms foci there. In patient's brains, these NIs colocalize with promyelocytic leukemia (PML) protein, which is an integral part of nuclear bodies (NBs). NBs are associated with transcription regulation and the ubiquitin-proteasome pathway; thus, accumulation of ataxin-7 in NBs may represent an attempt by the cell to degrade misfolded protein.

3.9 Spinocerebellar Ataxia Type 17

Spinocerebellar ataxia type 17 (SCA17) is the most recently discovered polyglutamine repeat disease. It is an autosomal dominant, progressive disorder characterized by dementia as well as cerebellar ataxia and involuntary movement abnormalities. Its symptoms are diverse and heterogeneous, but typically begin with behavioral disturbances and cognitive impairment. This is followed by ataxia, rigidity, dystonia, hyperreflexia, and rarely parkinsonism. Neuropathologically, patients may suffer degeneration of the cortex, cerebellum (including Purkinje cells), inferior olive, caudate nucleus, and medial thalamic nuclei. SCA17 is caused by the expansion of a coding (CAG)n repeat in the TATA-binding protein (TBP) gene on chromosome 6q27. Unaffected individuals carry 25–48 repeats, while affected individuals carry 43–66 repeats. The area of overlap between affected and unaffected alleles indicates incomplete penetrance at intermediate repeat lengths. The pathogenic threshold is higher than for most other polyglutamine disorders, yet there is an inverse correlation between repeat length and age of onset.

TBP is a ubiquitously expressed transcription initiation factor that is a core component of the RNA polymerase II transcription factor D (TFIID) complex. TBP possesses DNA binding activity in the TFIID complex, and is therefore required for the transcription of numerous genes. SCA17 patients have NIs immunoreactive for TBP, polyglutamine, and ubiquitin. Given TBP's central role in transcription regulation, the basis of SCA17 disease pathogenesis is proposed to involve transcription dysregulation, but this is yet to be proven.

3.10 Role of Aggregation in Polyglutamine Disease Pathogenesis

In most polyglutamine disorders and in many neurodegenerative diseases in general, protein aggregation is a prominent feature. It occurs in almost all polyglutamine diseases, despite the lack of domain or structural similarity between the different disease proteins. Aggregates have long been considered reliable indicators of disease, although their pattern and onset often do not correspond well with the cell-type specificity of disease pathology. As the role of aggregation in pathogenesis has been one of the most hotly debated issues in the field of neurodegeneration, it may have implications for Alzheimer's disease, Parkinson's disease, and amyotrophic lateral sclerosis. Why? There are several characteristics of polyglutamine-mediated aggregation that are presumably common to all of the disorders that show aggregation. First of all, aggregates are rich in β-pleated sheets and have many of the properties of amyloid. This is supported by Congo red birefringence and immunoreactivity with antiamyloid antibodies. Various studies have shown that the conformational change is associated with the production of visible aggregates rather than in the soluble nonaggregated phase, where polyglutamine tracts are thought to remain in a random coil conformation.

In vitro, the kinetics of aggregation are consistent with nucleated-growth polymerization, in which the rate-limiting step is the formation of misfolded peptide assemblies, often referred to as *oligomers*. A mechanism involving nucleated-growth polymerization would predict that only one (or a few) visible aggregates would appear in each affected cell, and this appears to be the case for most polyglutamine disorders. In the cellular milieu however, where membranes and intermolecular interactions compartmentalize proteins, this prediction may not be valid. The discovery of polyglutamine microaggregates confirms this suspicion. The threshold for aggregation of polyglutamine tracts is similar to the disease threshold for most polyglutamine diseases. *In vitro*, longer polyglutamine tracts have a lower concentration threshold for aggregation and nucleate more quickly than shorter tracts. This is one possible explanation for the correlation between tract length and disease onset and progression.

Some data support a role for aggregation in polyglutamine disease pathogenesis. Some proponents of this theory invoke the fact that aggregates sequester many proteins besides the disease protein. Intermolecular interactions between transcription factors often involve glutamine tracts or glutamine-rich regions. Some studies have shown that normal proteins with glutamine tracts or glutamine-rich regions are enriched in polyglutamine aggregates in cell culture and animal models. Several transcription factors colocalize with such aggregates, including CBP, TBP, and numerous TBP-associated factors (TAFs). CBP is responsible for the prosurvival effects of BDNF, and its soluble concentration is lowered in HD patient's brains. Postnatal mice lacking CREB and its homolog CREM develop a progressive degeneration of the hippocampus and striatum.

Titration of enzymes and factors required for protein refolding and degradation away from the soluble phase and into aggregates has been proposed as a potential cause of cell toxicity in neurons with aggregates. Caspase activation is another way in which aggregation could be linked to pathology. Caspase recruitment into aggregates can lead to their activation, which could result in dysfunction or cell death in neurons. Further supporting the role of aggregates in polyglutamine protein toxicity, injection of preformed polyglutamine aggregates into the nuclei of cultured cells causes cell death, while the injection of nonpolyglutamine aggregates does not. At the same time, there is strong evidence that soluble polyglutamine protein, rather than aggregates, is the primary source of toxicity. Importantly, the pattern of aggregates observed in human patients often does not correlate with the pattern of neuronal dysfunction. For example, in the striatum of HD patients, large interneurons contain aggregates more frequently than medium spiny neurons, yet the former neurons are largely spared, while the latter are most vulnerable. HD transgenic mice expressing full-length mutant htt will develop inclusions in many brain regions many of which are neuropathologically unaffected, while few inclusions are detected in the striatum, the region of the brain displaying the most prominent pathology.

Aggregation and toxicity have been directly dissociated in other polyglutamine disease model systems. Studies of the SCA1 knockin mouse model revealed that neurons lacking aggregates were more susceptible to dysfunction and demise, while those neurons displaying prominent aggregates were protected. Similarly, SCA7 transgenic mice exhibit retinal pathology before the occurrence of visible aggregates. Finally, in a very provocative study, SCA1 transgenic mice lacking the ubiquitin ligase E6-AP were significantly less capable of forming large and numerous aggregates in neurons in comparison to SCA1 transgenic mice on a wild-type background, but displayed earlier onset of an ataxic phenotype and accelerated neurodegeneration. This suggests that the aggregates may be protective, although the toxicity of compromising the proteasome may have contributed to the accelerated phenotype. Thus ensued a contentious debate over the role of aggregate formation in polyglutamine disease – with some workers espousing the view that aggregates were responsible for disease pathology, others suggesting that the aggregation process was a protective coping mechanism of the cell and thereby beneficial, and still others insisting that aggregates were incidental and irrelevant. This debate was complicated by the existence of "microaggregates," small clumps of aggregated protein visible only at the electron microscope level.

Over the last few years, studies deconstructing polyglutamine tract aggregation into a multistep process suggest a parsimonious explanation for the divergent views. Using a variety of biophysical approaches, including transmission electron microscopy (TEM), Fourier transform infrared spectroscopy (FTIR), and atomic force microscopy (AFM), one study dissected the process of huntingtin (htt) exon 1 peptide aggregation and found evidence for a number of sequential morphological and structural intermediates (Fig. 6). By showing that misfolded htt exon 1–44Q adopted intermediate structures, such work opened up the possibility that intermediates (not visible at the light microscope level) are the toxic species and that the ultimate visible aggregated forms of htt exon 1–44Q are neutralized versions of mutant protein. To examine the role of aggregate formation in polyglutamine toxicity, another group then tracked survival time versus diffuse, soluble htt protein expression levels in htt-transfected striatal neurons undergoing aggregate formation. Comparison of neurons that developed aggregates over time versus those that did not confirmed that the level of soluble nonaggregated mutant polyglutamine protein was the more reliable predictor of cellular toxicity. Such studies have shifted our attention to the role of the intermediates (oligomers, protofibrils, etc.) as the toxic species instead of the final, visible aggregates.

A reasonable model for polyglutamine toxicity predicts that the process starts with a protein that misfolds because of the presence of an expanded polyglutamine tract. The misfolded protein is initially detectable in the soluble phase due to the cell's ability to maintain the protein in a properly folded state and direct it to the degradation machinery. However, the refolding capacity of the cell is ultimately exceeded, and since the degradation machinery cannot turnover the misfolded mutant polyglutamine protein, accumulation occurs. Misfolded polyglutamine proteins can spontaneously change their structural properties and adopt abnormal conformations. These abnormally folded proteins can then nucleate, forming oligomers. Oligomers then form protofibrils that grow into fibrils. The transition to the fibril stage is characterized by the attainment of a β-sheet structure, so at this point the structures are amyloid-like. Fibrils then grow into fibers (also known as microaggregates), which then assemble into aggregates visible under a light microscope. According to such a model, blocking an intermediate step could be therapeutically effective. Consistent with this, one

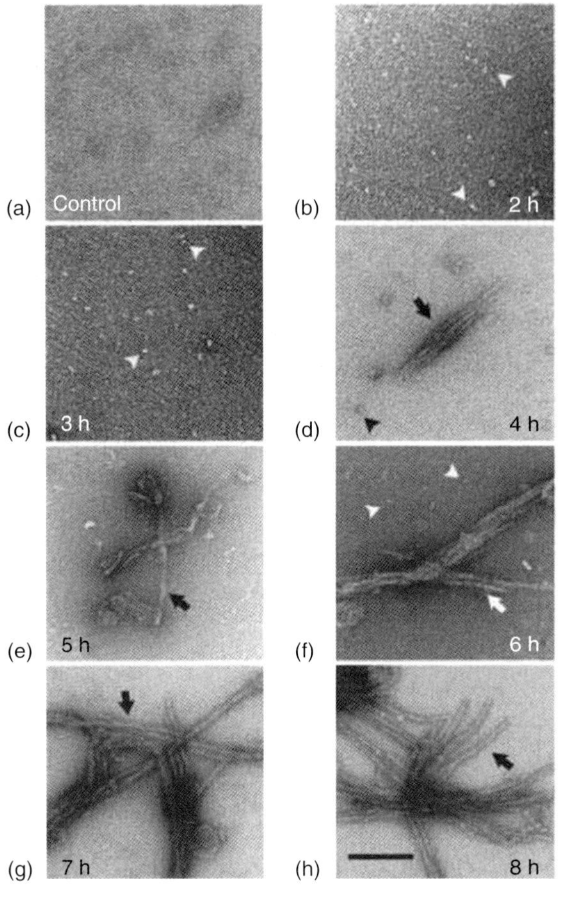

Fig. 6 Polyglutamine-expanded huntingtin protein undergoes a variety of structural alterations on its way to becoming a visible aggregate. In this experiment, transmission electron microscopy tracked the structure of huntingtin exon 1 peptide with 44 glutamines after release from a linked affinity tag. (a) Prior to affinity tag release, nothing is observed. (b-c) After affinity tag release, globular structures become apparent. (d) This is followed by formation of short fibers (4–5 nm in diameter). (e-f) Fiber formation becomes more prominent. However, during this time, large globular assemblies remain (arrowheads). (g-h) Eventually, only fibers are present, and the fibers begin to adopt a uniform appearance, suggesting consolidation into protofibrils. (From Poirier, M.A., Li, H., Macosko, J., Cai, S., Amzel, M., Ross, C.A. (2002) Huntingtin spheroids and protofibrils as precursors in polyglutamine fibrilization, *J. Biol. Chem.* **277**, 41032–41037, used with permission of the *Journal of Biological Chemistry*.).

group has nicely shown that Congo red can bind polyglutamine peptides once they are amyloid-like and prevent their conversion into fibers, while an independent group has shown that Congo red delivery to an HD mouse model is a highly effective treatment intervention.

One satisfying aspect of this model is that it allows us to simultaneously view aggregates as harmful, protective, and innocuous. How? First, aggregates clearly must be toxic as their creation is predicated upon the production of earlier intermediate toxic forms. At the same time, the final visible aggregates are less toxic than the earlier intermediates, so anything that enhances their sequestration into visible aggregates is beneficial. Finally, since oligomers are difficult to detect, it is often not possible to know whether cells are successfully sequestering the toxic precursors into aggregates or if high levels of toxic intermediates are building up. So, aggregates are thus also incidental, since their presence does not provide us with insight into the steady state levels of the toxic precursors. In conclusion, advances in our understanding of the role of aggregate formation in polyglutamine disease processes suggest that aggregates may be simultaneously viewed as harmful, beneficial, and incidental.

3.11 Protein Context

The importance of protein context in the pathogenesis of polyglutamine disorders is a subject of great interest to many researchers, because it is directly related to the mechanisms of toxicity in each disease. It is evident in the polyglutamine disorders that protein context plays a role in cell-type specificity. For example, a polyglutamine expansion in the Htt protein causes a severe degeneration of the striatum and cortex, while the same size glutamine tract in ataxin-1 causes a degeneration of structures in the cerebellum and brainstem, while sparing the striatum and cortex. This cannot be attributed to gross differences in expression patterns, since both proteins are pan-neural (Fig. 7).

One effect that protein context may have on pathology is to affect subcellular localization. For example, the presence of a functional NLS or NES dictates preferential subcellular localization in the nucleus or cytoplasm. Mice expressing expanded ataxin-1 with a mutated NLS do not develop the SCA1 phenotype, while those without mutated NLSs do. Interaction domains also affect polyglutamine protein toxicity. The ability of expanded polyglutamine proteins to interact with other molecules is a promising avenue in the search for mechanisms of cell-type specificity. Yeast two-hybrid screens and other techniques have yielded interaction partners for a number of polyglutamine proteins. For example, Htt's interactions with transcription factors may prove to be central to its pathological effects. At the same time, it is also evident that polyglutamine tracts are innately toxic. Pure polyglutamine tracts cause toxicity in cell culture. Mice that express a glutamine tract of 150 amino acids in the Hprt protein, which does not naturally contain any such tracts, exhibit progressive neurological deficits and NIs. Thus, it appears that pathology in each polyglutamine disease is due to a gain of function of the glutamine tract that is then modulated by the protein context in which it resides.

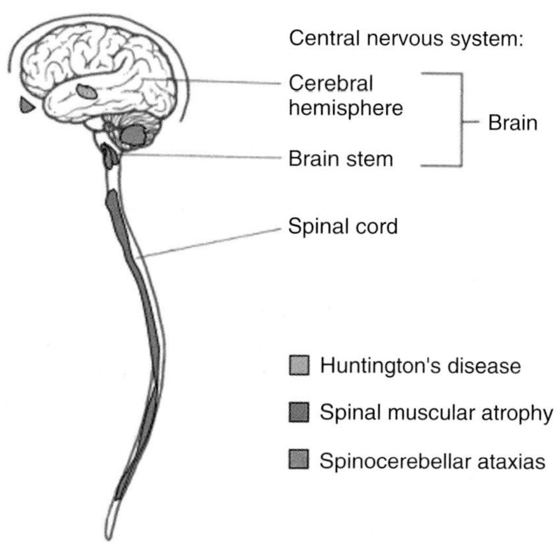

Fig. 7 The conundrum of cell-type specificity in the polyglutamine diseases. Although the different polyglutamine disease proteins are expressed throughout the central nervous system, only select populations of neurons degenerate in the different disorders. The principal regions of selective vulnerability are shown for certain of the polyglutamine diseases.

3.12
Transcriptional Dysregulation

Transcriptional dysregulation is one theory of polyglutamine toxicity that has gained support from many lines of evidence in several disorders, and it appears increasingly likely that it is one of the fundamental causes of pathogenesis. Microarray technology and other genomic techniques are facilitating rapid advances in this area of the polyglutamine field. The majority of polyglutamine disorders are caused by proteins that are either transcription factors/cofactors or interact closely with transcription factors. TBP and AR are transcription factors, atrophin-1 is a transcriptional corepressor, ataxin-7 is part of a transcriptional coactivator complex, and ataxin-1, htt, and ataxin-3 all interact with various transcription factors. It is therefore reasonable to suspect that an abnormally long polyglutamine tract could interfere with the transcriptional activity of these proteins and/or their interactors.

CBP is one of the most studied transcription factors in the polyglutamine field. It is a transcription activator that mediates part of the cellular response to cAMP, and it can interact with numerous polyglutamine proteins in the soluble or insoluble phase, including huntingtin, ataxin-3, ataxin-7, and AR. Expanded htt represses CBP-regulated genes, and overexpression of CBP causes a considerable rescue of the cell death phenotype in HD and SBMA cell culture models. Postnatal mice lacking CBP's upstream activator CREB and its homolog CREM display a profound degeneration of the striatum and hippocampus. Thus, interference with CBP appears to be a common theme in polyglutamine pathology.

Another common theme in the transcription dysregulation equation is the tendency for the polyglutamine disease proteins to alter transcription factor/coactivator-mediated histone acetyltransferase (HAT) activity. The ability of a gene to be transcribed depends upon its chromatin structure, and thus upon the degree of histone acetylation in its vicinity. This is because acetylated histones cause chromatin to be in an "open," transcription-friendly conformation. Acetylation status depends upon the balance between the activity of HATs and their countervailing counterparts, the HDACs. CBP, p300, and p300/CBP-associated factor (PCAF) are all HATs that are inhibited by polyglutamine-expanded htt exon 1 peptide and ataxin-3. In cell culture, expression of a mutant htt fragment reduces global histone acetylation.

Finally, according to a very recent study, ataxin-7 directly interacts with GCN5 as part of the STAGA complex, and upon polyglutamine expansion, mutant ataxin-7 causes a dominant-negative effect upon GCN5 HAT activity. This results in CRX transcription interference that may contribute to the SCA7 retinal degeneration phenotype. Inhibiting HDACs with drugs known as *HDAC inhibitors* (HDAC Is) attenuates the phenotype of HD and SBMA in mouse and fly models, presumably by shifting the cells' acetylation status. HDAC Is such as sodium butyrate and especially suberoylanilide hydroxamic acid (SAHA) have thus emerged as possible candidates for therapeutic trials in human polyglutamine disease patients.

3.13
Proteolytic Cleavage

The occurrence of proteolytic cleavage in polyglutamine diseases first became apparent when it was shown in HD that htt can be cleaved by extracts derived

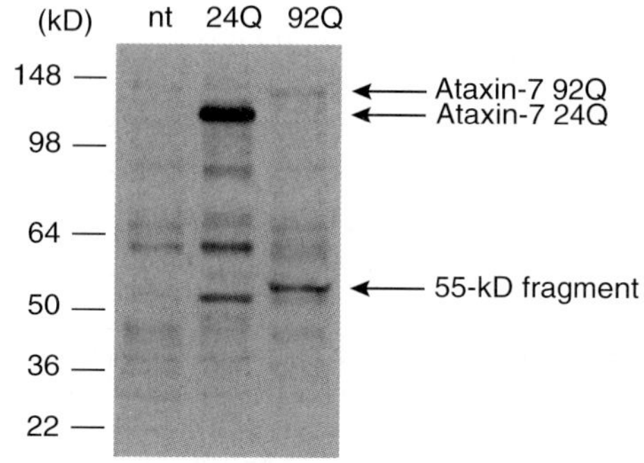

Fig. 8 Proteolytic cleavage in the polyglutamine diseases. Production of amino-terminal truncated proteins is observed in many of the polyglutamine diseases, as is shown here for SCA7. In this experiment, nuclear fractions of retinal lysates from age- and sex-matched nontransgenic (nt), ataxin-7 CAG-24 (24Q), and ataxin-7 CAG-92 (92Q) mice were prepared and immunoblotted with an antibody directed against the amino-terminal region of the protein ataxin-7. As shown here, in addition to soluble full-length ataxin-7 (which is typically reduced in cells where it is aggregating into insoluble inclusions), an ~50–60 kD fragment is detected. (Adapted from Garden, G.A., Libby, R.T., Fu, Y.H., Kinoshita, Y., Huang, J., Possin, D.E., Smith, A.C., Martinez, R.A., Fine, G.C., Grote, S.K., et al. (2002) Polyglutamine-expanded ataxin-7 promotes noncell-autonomous Purkinje cell degeneration and displays proteolytic cleavage in ataxic transgenic mice, *J. Neurosci.* **22**, 4897–4905. Used with permission of the *Journal of Neuroscience*.)

from apoptotic cells. Subsequent publications showed that only amino-terminal epitopes of htt protein are detectable in aggregates. SBMA, DRPLA, and SCA7 were then added to the list of diseases in which aggregates are only immunoreactive for a glutamine-containing fragment of the disease protein. Since then, evidence of proteolytic cleavage has been published for nearly all known polyglutamine disorders. Experimental studies revealed that inhibition of caspase cleavage reduces aggregate formation and toxicity in a cell culture model of HD, underscoring the relevance of proteolytic cleavage in polyglutamine disease pathogenesis. Cleavage promotes aggregation and/or toxicity in SCA3 and SCA7 (Fig. 8). Polyglutamine tract containing htt fragments are more toxic in cell culture than full-length htt, and the most widely used HD mouse model expresses only exon 1 of the *htt* gene.

There are several mechanisms by which cleavage may modulate polyglutamine neurotoxicity. Abnormal proteolysis of polyglutamine proteins may lead to a toxic species that can cause damage in the soluble or insoluble phase. Alternatively, proteolysis may be a normal event in protein turnover and the inability to clear cleaved protein may be the problem. Another possibility is that soluble polyglutamine proteins may cause the activation of proteases such as caspases, which then cleave the disease protein and send the cell on a path to degeneration and ultimately apoptosis. As we will see, each theory has support and not all are mutually exclusive.

Abnormal proteolysis of polyglutamine disease proteins is one possible mechanism of toxicity. Some studies have shown that polyglutamine repeat length modulates susceptibility to proteolytic cleavage, but results have been inconclusive. Few studies have been performed using human brain tissue, however. Considerable evidence exists to show that polyglutamine tracts themselves are resistant to degradation by mammalian proteasomes. The 20S and 26S eukaryotic proteasomes are incapable of cutting within polyglutamine tracts longer than 9 amino acids *in vitro*. The inability of the proteasome to digest glutamine tracts may contribute to the toxicity and aggregation of expanded polyglutamine peptide fragments.

Proteolysis may be a normal part of the turnover of wild-type and mutant htt. One group published evidence that caspase-3 cleaves both mutant and wild-type htt in the cytoplasm of HD brains, suggesting that this particular cleavage is a normal event in its turnover. Since this fragment was located exclusively in the cytoplasm, it also implies that further cleavage occurs before the nuclear translocation of htt. Another study suggested that pepstatin-sensitive aspartyl endopeptidases such as the presenilins and cathepsins D and E may be involved in the normal turnover of htt. This process, which may normally aid in maintaining the balance between htt synthesis and degradation, could generate a toxic, highly aggregation-prone (and hence intermediate oligomer/protofibril-prone) species when it cleaves mutant htt.

Another possible source of toxic cleavage products is the ubiquitin-proteasome system, which is responsible for labeling and digesting misfolded proteins, among other things. Polyglutamine tracts beyond a threshold of about 35 glutamines adopt an abnormal β-pleated sheet conformation, which may cause the host protein to be ubiquitinated and thus targeted for degradation. In support of this theory, a common feature of polyglutamine diseases is the presence of ubiquitin-positive inclusions. Arguing against this idea is an experiment showing that in a cell culture model of SCA1, increased proteasome degradation due to the introduction of a degradation signal reduces aggregates and toxicity. Another potential source of toxic fragments is the autophagy pathway, which is responsible for the bulk degradation of cellular components. According to one study, degradation of htt by autophagy and the autophagosome-associated cathepsin D creates toxic htt fragments.

There are many proteases that are suspected to play a role in the proteolytic cleavage of polyglutamine proteins, the most studied of which are the caspases. Caspases are cysteine- dependent aspartyl proteases that play a crucial role in apoptosis, but are increasingly being investigated for their nonapoptotic functions. In healthy cells, caspases are present predominantly as inactive proenzymes, which must be cleaved for full activity. Caspase-3 is the only protease that has been shown to cleave htt in HD patient's material, although caspases 1 and 6 have also been implicated in cell culture and *in vivo*. Inhibiting the cleavage of htt by any of these three proteases is protective, and an HD mouse model expressing dominant-negative caspase-1 showed significantly delayed disease onset. Caspase-3 has also been implicated in the proteolytic cleavage of atrophin-1, ataxin-3 and AR. Caspases are activated in response to a number of cellular insults, and in postmitotic cells such as neurons where inhibitors of apoptosis proteins (IAPs) are highly expressed, this may cause cellular damage that does not result in classical apoptosis.

Calpains are another family of proteases implicated in the cleavage of polyglutamine proteins. They are calcium-activated cysteine proteases that exist predominantly as proenzymes. As mouse models of HD and material from HD patients exhibit abnormal mitochondrial calcium regulation, some investigators have proposed that altered calcium flux is the mechanism of calpain activation. Indeed, several cell culture studies support the role of calpain proteases I, II, and "m" in htt cleavage. As mentioned previously, other aspartyl proteases such as the Alzheimer's disease-associated presenilins and autophagy-associated cathepsins have also been implicated in htt cleavage.

4 Type 2: the Loss-of-function Repeat Diseases

4.1 Fragile X Syndrome

Fragile X syndrome (FRAXA) is an X-linked disorder with a prevalence of 1 in 4000 in males and 1 in 8000 in females, making it the most common form of inherited mental retardation. Neurological presentation frequently includes mild to severe mental retardation, hyperactivity, poor eye contact, high-pitched speech, and flapping or biting hand movements. Physical signs in males include long, prominent ears, and jaws, macrocephaly, postpubescent macroorchidism, and occasionally, connective tissue abnormalities. The mutation responsible for FRAXA is typically an expansion of a polymorphic (CGG)*n* repeat found in the 5′-untranslated region of the fragile X mental retardation-1 (FMR1) gene. Expanded chromosomes have a folate-sensitive fragile site at Xq27.3 that can be viewed under a light microscope under special cell culture conditions. Normal alleles contain 6–53 triplets punctuated by one or more AGGs, which are considered to have a stabilizing influence on the repeat. Disease alleles contain expansions beyond 200 and up to 2000 repeats, with no AGG interruptions. Pathogenic expansions result exclusively from maternal transmission. FRAXA appears to be a loss-of-function disorder, since deletion of FMR1 and loss-of-function point mutations can also cause FRAXA. This view is further supported by the FMR1 knockout mouse, which reproduces certain aspects of the human disorder.

FRAXA patients have reduced levels of the *FMR1* gene product, FMRP, and there is a linear correlation between reduced protein levels and IQ test scores. Expansion of the disease allele results in hypermethylation of the (CGG)*n* tract, which spreads to a nearby CpG island in the FMR1 promoter region. Some of the proposed secondary structures formed by the FRAXA repeat contain C–C mispairs, which are good targets for human DNA methyltransferase. Another theory invokes the RNAi protein Dicer's ability to cleave CGG repeat RNA, postulating that the resulting siRNAs recruit DNA methyltransferases to the FMR1 locus. Affected alleles also display condensed chromatin, loss of histone acetylation and increased histone methylation. These data support a process whereby DNA methylation recruits transcription silencing machinery, which subsequently suppresses FMR1 transcription in the nearby promoter region. Interestingly, while premutation carriers had been viewed as perfectly normal for decades, recent work indicates that some premutation carriers (60–200 uninterrupted CGG repeats) display a phenotype distinct from FRAXA. Females have a predisposition to premature ovarian failure (POF), and males may develop late-onset ataxia and tremor with the presence of neuronal intranuclear inclusions (NIs) that consist of RNA. (This new FXTAS is discussed in a separate section.) Expression studies have shown that premutation carriers may express up to seven times more FMR1 mRNA than normal individuals. This upregulation is probably not due to compensation for loss of function, because individuals with point mutations resulting in loss of function of FMRP do not have higher levels of the transcript. Abnormal transcript levels due to premutations or full mutations are thought to be as a result of the expansion's effect on transcription initiation rather than on mRNA stability.

The FMR1 promoter lacks a functional TATA box and initiator sequence. Four functional transcription factor binding sites have been identified in normal individuals: two sites binding USF1/USF2 and Nrf-1, and two GC boxes that bind members of the "Sp" family of transcription factors. Sp1 and Nrf-1 binding is disrupted in cells from FRAXA patients. The *FMR1* gene encodes the widely expressed FMRP, which is an RNA-binding protein most highly expressed in the brain and testes. It is 60% homologous to two other proteins, FXR1P and FXR2P, with which it interacts. The protein is thought to bind approximately 4% of all brain mRNA transcripts, through its RGG box domain. It selectively recognizes RNA-containing hairpin or tetraplex secondary structures ("G-quartet"), and shuttles into and out of the nucleus and associates with polyribosomes in messenger ribonucleoprotein complexes (mRNPs). The FMRP-containing mRNP complexes also contain Pur α and mStaufen, proteins involved in the transport of neuronal granules. These granules, which contain RNA and associated proteins, are transported to dendritic spines in a metabotropic glutamate receptor 5 (mGluR5)-dependent manner. FMRP can suppress the translation of certain transcripts *in vitro* and *in vivo*. FMRP may thus regulate the transport, localization, and translation of certain mRNAs in an activity-dependent manner (Fig. 9). In Drosophila, FMRP interacts with components of the RNAi machinery, which is involved in gene silencing and thus translational control. Although FMRP does not seem to affect the siRNA pathway, it may regulate microRNAs (miRNAs), which are noncoding RNAs thought to control the translation of mRNAs by binding to their 3′-untranslated region. FMRP associates with miRNAs, and proteins in miRNA-containing complexes in mammals, interactions that may be relevant to FMRPs regulation of translation.

FMRP is important for the development of dendritic spines and synaptic plasticity. It regulates the expression of MAP1B, an important regulator of microtubule stability. FMRP knockout mice have abnormally high levels of MAP1B, resulting in increased microtubule stability. This may affect the development of dendrites and/or dendritic spines. The absence of FMRP in hippocampal neurons results in immature dendritic spine morphology and delayed synaptic connections, perhaps contributing to the neurological phenotype observed in FRAXA. FMRP may also affect mGluR5-dependent long term depression (LTD). LTD is a process by which neuronal activity can cause a lasting desensitization of neurons to depolarization. LTD-associated protein synthesis at synapses may be enhanced by mGluR5 activation, while it appears to be suppressed by FMRP. This is supported by evidence of enhanced LTD in the FMR-1 knockout mouse. FMRP suppression and mGluR5 activation of LTD-dependent local protein synthesis may be opposing forces that are out of balance in FRAXA, perhaps accounting for part of the cognitive abnormalities.

4.2 Fragile XE Mental Retardation

Fragile XE mental retardation (FRAXE) is an X-linked disorder with a prevalence of about 1–4% that of FRAXA. It accounts for approximately two-thirds of families with nonspecific X-linked mental retardation (MRX), a classification in which mental retardation is the only consistent

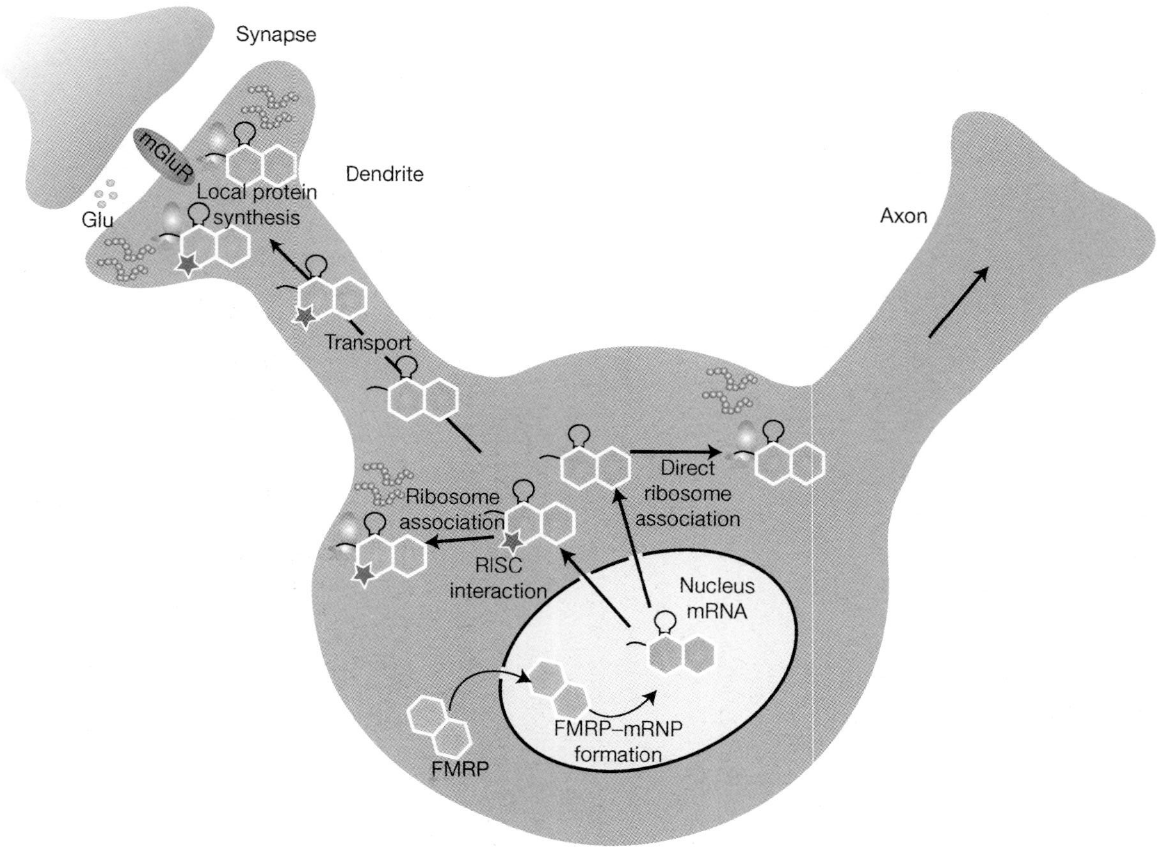

Fig. 9 Advances in the understanding of FMRP's normal function. FMRP binds to recently transcribed messenger RNAs in the nucleus to form a ribonucleoprotein complex (mRNP). The FMRP–mRNP complex moves out of the nucleus and then either directly associates with ribosomes *or* interacts with the RNA-induced silencing complex (RISC). The FMRP–mRNP complex is also transported to dendrites. Whether in the perinuclear cytosol or in dendrites, FMRP is believed to regulate protein translation. (From Jin, P., Alisch, R.S., Warren, S.T. (2004) RNA and microRNAs in fragile X mental retardation, *Nat. Cell Biol.* **6**, 1048–1053. Used with permission of *Nature Cell Biology*.).

clinical feature. Other characteristics are variable and may include hyperactivity, mild facial hypoplasia and nasal abnormalities. FRAXE is caused by the expansion of a (GCC)*n* repeat in the promoter region of *FMR2*, a gene 600 kb downstream of *FMR-1*. It also overlaps with the putative promoter of a gene called *FMR3*, transcribed in the opposite direction. Both maternal and paternal transmission can result in expansion. Normal individuals bear 6–35 GCC repeats, premutation carriers bear 61–200 repeats, and full mutation carriers bear >200 repeats. Pathogenic expansion results in a folate-sensitive fragile site at the disease locus.

In mice, FMR2 mRNA is expressed in adult and fetal brain, kidney, lung, and placenta, with highest brain levels in the hippocampus and amygdala. Transcripts of both FMR2 and FMR3 are reduced beyond detection in FRAXE patients. This may be due to a methylation-dependent silencing process similar to FRAXA. The presence of the repeat within the preinitiation region of the FMR2 promoter may also suggest a more direct disruption of transcription. FMR2 mRNA is highly expressed in the hippocampus, an area critical for learning and memory. Its paralogs AF4 and LAF4 are both transcription transactivator proteins, and the FMR2 protein seems to be a potent transcription activator itself. Furthermore, the FMR2 protein is nuclear, consistent with its proposed role in transcription regulation. The Drosophila ortholog of FMR2/AR4, Lilliputian, is essential for proper organ development, and its loss of function is lethal. FMR2 knockout mice display impairment in the conditioned fear test and enhanced LTP (long-term potentiation) in the hippocampus. The role of FMR3 in FRAXE, if any, is unknown. The mechanism of FRAXE pathogenesis may therefore involve silencing of the human *FMR2* gene resulting in altered transcription in many parts of the developing nervous system and mature brain. The neurological deficits characteristic of FRAXE may involve effects of altered transcription on the hippocampus, amygdala, and possibly other brain structures.

4.3 Friedreich's Ataxia

With a prevalence of about one per 50 000 individuals in the Caucasian population, Friedreich's ataxia (FRDA) is the most common inherited ataxia in this ethnic group. It is a multisystem degenerative disease that is unusual among the triplet repeat disorders due to its autosomal recessive inheritance. Neurological symptoms include gait, limb and truncal ataxia, loss of position and vibration senses, diminished tendon reflexes, and dysarthria. Neuropathology changes include degeneration of the posterior columns of the spinal cord, loss of large primary sensory neurons in the dorsal root ganglia (DRG), and mild, late-onset degeneration of the cerebellar cortex. Other common clinical features are cardiomyopathy, diabetes mellitus, scoliosis, and other skeletal abnormalities. Patients often present with symptoms in childhood, become wheelchair-bound by their late teens or early twenties, and have reduced lifespans due to cardiac failure. Adult presenting patients, however, can have nearly normal lifespans with more protracted progression and less severe nonneuronal involvement. Indeed, until the identification of the causal mutation, many of these adult-onset cases went undiagnosed.

FRDA results from the expansion of a polymorphic (GAA)*n* repeat in the first intron of the gene *X25*, now known as frataxin. FRDA patients have at least one expanded allele. For the disease to occur, the second allele must contain either an expanded repeat, or rarely, a copy of frataxin containing a loss-of-function mutation. No patients have been described with loss-of-function allele mutations in both *FRDA* genes. This is probably due to prenatal lethality, as frataxin-null mice die in utero. Disease severity and onset age are determined by the size of the (GAA)*n* repeat of the smaller expansion allele. Unaffected individuals carry at least one allele between 7–38 repeats, while affected individuals carry two alleles of 66–1700 repeats. Interruptions of the (GAA)*n* repeat result in later onset and attenuated presentations. GAA expansion at the FRDA locus can result in enormous repeats in a single generation. Maternal transmission can result in expansion or contraction, while paternal transmission results primarily in contraction.

FRDA patients exhibit reduced frataxin RNA and protein levels, and evidence suggests a defect in transcription or RNA maturation. Current models propose that frataxin RNA elongation may be disturbed by triplex structure formation between expanded DNA strands during transcription. DNA triplexes are formed when one strand of a double-stranded DNA molecule folds back upon itself, and interacts with two previously annealed strands. This creates a local structure comprising three DNA strands held together by hydrogen bonds. Certain sequences favor this process, among them extended (GAA)*n* repeats. Triplex formation in the first intron would presumably affect transcript elongation but not initiation. In support of this model are *in vitro* transcription experiments demonstrating no effect of repeat length on transcript initiation.

Frataxin is a 210 amino acid protein that is well conserved from prokaryotes to mammals. It contains mitochondrial targeting signal sequences and localizes to the mitochondrial matrix. Frataxin expression occurs at the primary sites of pathology: dorsal root ganglia, spinal cord, sensory nerves, heart, and pancreas. These are tissues that rely upon high levels of oxidative metabolism and consequently are rich in mitochondria. In addition, such cell types are often postmitotic, meaning that most dividing cell types are spared in FRDA patients.

Studies with FRDA patient material have demonstrated an increased heart iron content and a deficiency in iron–sulfur cluster-containing proteins, including aconitase, a protein involved in iron homeostasis. In addition, fibroblasts derived from FRDA patients are abnormally sensitive to iron and hydrogen peroxide induced stress. Ablation of the yeast frataxin homolog results in respiratory dysfunction, abnormal accumulation of mitochondrial iron, impaired biogenesis of iron–sulfur proteins, and increased sensitivity to oxidative stress. Conditional knockout of frataxin in striated muscle results in a heart-specific phenotype resembling the cardiac abnormalities in human FRDA (Fig. 10). This supports a genetic mechanism involving loss of function of the disease protein. Frataxin may be part of a complex that delivers iron to Iron–sulfur clusters (ISCs), which are cofactors essential to the activity of many important cellular proteins. The accumulation of iron outside the mitochondria, although probably not central to FRDA pathogenesis, supports a deficiency in iron delivery to ISCs. Among ISC-dependent cofactors are proteins involved in mitochondrial electron transport and thus respiration. Disturbances in oxidative metabolism are often associated with increases in the production of

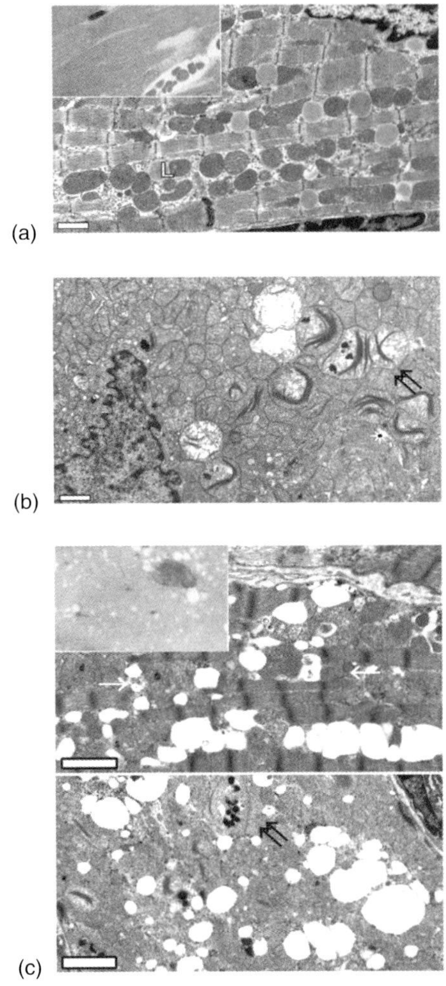

Fig. 10 Friedreich's ataxia mice display cardiac muscle pathology. (a) Transmission electron microscopy of mice lacking expression of frataxin in only their muscle reveals the abnormal accumulation of lipid droplets (L) in cardiac muscle at 4 weeks of age. (b) By 7 weeks of age, mitochondria appear abnormal and iron deposits are visible (arrows). (c) With further progression, large vacuoles emerge (inset-top) as mitochondria undergo continued prominent degeneration. Ultimately, the mitochondria become engorged with electron-dense material consistent with iron deposits. (From Seznec, H., Simon, D., Monassier, L., Criqui-Filipe, P., Gansmuller, A., Rustin, P., Koenig, M., Puccio, H. (2004) Idebenone delays the onset of cardiac functional alteration without correction of Fe-S enzymes deficit in a mouse model for Friedreich ataxia, *Hum. Mol. Genet.* **13**, 1017–1024, used with permission of *Human Molecular Genetics*).

toxic reactive oxygen species (ROS). FRDA therefore is thought to result from reduced frataxin levels, leading to abnormal iron–sulfur metabolism, mitochondrial dysfunction, oxidative stress, and tissue degeneration. FRDA is thus caused by the dynamic mutation of a nuclear-encoded mitochondrial protein. The disease shares features with classic mitochondrial disorders such as MELAS (mitochondrial myopathy, encephalopathy, lactic acidosis, and strokelike episodes syndrome) and MERRF (myoclonus epilepsy with ragged red fibers), which are caused by stable mutations to mitochondrial-encoded proteins. FRDA appears to be a classic mitochondrial disorder with an unusual genetic basis – dynamic mutation.

Given the data suggesting a role for oxidative metabolism and iron transport in FRDA pathogenesis, antioxidants and iron transport molecules are being evaluated as treatments for this disease. Several clinical trials have been conducted in FRDA patients using idebenone, a synthetic analog of coenzyme Q10 and a potent antioxidant. Some studies have shown substantial improvements in the heart function of patients, although the lack of randomized placebo controls and the variability of the disorder render the results controversial. A recent conditional knockout mouse model of FRDA also supports the use of idebenone as a treatment for the disorder, making treatment of FRDA patients with this antioxidant a distinct possibility.

4.4 Progressive Myoclonus Epilepsy Type 1

Progressive myoclonus epilepsy type 1 (EPM1) is a rare, autosomal recessive neurological disorder most prevalent in Finland and parts of North Africa. Its most prominent symptoms are progressive photosensitive myoclonus and tonic-clonic epilepsy. Some patients also experience a progressive cerebellar ataxia and cognitive decline. Neurodegeneration occurs in the thalamus, spinal cord, and the Purkinje and granular neurons of the cerebellum. Typical age of onset is 6 to 16 years of age. EPM1 is sometimes caused by missense mutations in the *cystatin B* gene (CSTB); however, analysis of affected patients lacking such mutations revealed a dodecamer repeat expansion upstream of CSTB. The dodecamer repeat is located between 66 and 77 bp 5′ of two putative transcription start sites, and has the sequence CCCCGC-CCCGCG. Unaffected individuals have a repeat number of 2–3, premutation carriers have 12–17 repeats, and affected individuals carry 30–150 repeat alleles.

CSTB is a highly conserved gene in the cystatin family of cysteine protease inhibitors. CSTB mRNA is ubiquitous, with high transcript levels in the hippocampus. The protein binds to and inhibits lysosomal proteases such as cathepsins B, H, L, and S. Pathogenic expansion of the repeat causes a reduction in CSTB transcription in some cell types. Cstb knockout mice develop symptoms similar to those seen in EPM1 patients, and missense mutations causing the disease disrupt the ability of CSTB to bind cysteine proteases. Together, these data suggest a loss-of-function mechanism for EPM1. Lowered inhibition of cysteine proteases may result in neuronal damage, causing the phenotype observed in EPM1.

There are several ways in which the repeat might reduce CSTB expression. The first is that reduced gene expression may be due to an increase in the distance between promoter elements and the transcription start site. AP-1 binding sites have been located upstream of the repeat. One study demonstrated a two- to fourfold reduction in promoter activity when the repeat is expanded. An equivalent reduction also occurred when similarly sized heterologous DNA was used in place of the expanded repeat. Independent studies have confirmed a large reduction in promoter activity due to repeat expansion. Another way the repeat could affect CSTB expression is through alterations in chromatin structure. The repeat region is G-C rich and might exclude nucleosomes or promote the formation of secondary structures. Abnormal DNA secondary structures have been observed in the EPM1 repeat region. These alterations could affect the expression of *CSTB* or other genes in the vicinity. A third possible mechanism for expression changes lies in the sequence of the repeat region. Each dodecamer repeat contains a GC box, which could act as an Sp1 binding site *in vivo*. In cases of abnormally large repeat numbers, increased Sp1 binding could contribute to the suppression of CSTB transcription.

5 Type 3: the RNA Gain-of-function Repeat Diseases

5.1 Myotonic Dystrophy Type 1

Myotonic dystrophy or DM is an autosomal dominant, multisystem disorder with a prevalence of 1 in 100 000 worldwide, and 1 in 8000 in European and North American Caucasian populations. Patients typically present with proximal or distal muscle dysfunction including weakness, pain, and myotonia (failure of muscle relaxation). DM exhibits a combination of characteristic symptoms: cardiac conduction abnormalities, subcapsular iridescent cataracts, and unusual endocrine changes. Other features include testicular atrophy, type II diabetes, and late-onset cognitive impairment. In its most severe form, congenital DM, mental retardation, craniofacial deformities, and other developmental abnormalities are present. It exhibits both paternal and maternal transmission, although there is an almost exclusive maternal transmission in congenital DM. Myotonic dystrophy type 1 (DM1) is caused by a (CTG)*n* expansion in the 3′-untranslated region of the gene, dystrophica myotonica protein kinase (DMPK), on chromosome 19. Unaffected individuals carry 5–37 repeats, while affected individuals carry 50–4000 repeats. DM exhibits pronounced anticipation and dramatic somatic instability. Interestingly, the clinical phenomenon of anticipation (worsening severity as a disease gene is transmitted from one generation to the next) was first described nearly a century ago in a family segregating DM1. Although the anticipation phenomenon was dismissed as an artifact of ascertainment by geneticists of the mid twentieth century, anticipation is now known to be a genuine feature of dynamic mutation diseases. Anticipation results from the tendency of disease repeats to expand and the inverse correlation between repeat length and age of disease onset.

Several theories have been advanced to explain the molecular basis of DM1, including haploinsufficiency of DMPK, local chromatin effects on neighboring genes, and gain of function exerted by expanded RNAs. DMPK is proposed to have many functions, some of which could relate to the disease, such as modulation of skeletal muscle sodium channels, RNA metabolism, calcium homeostasis, and the cell stress response. Initial expression studies reported a decrease in DMPK RNA and protein levels, but a DMPK knockout mouse designed to test the haploinsufficiency theory exhibited only mild myopathy that was inconsistent with the DM phenotype. The (CTG)*n* tract is a strong nucleosome assembly site, and it was therefore hypothesized to have trans-effects on the expression of neighboring genes. According to this model, the myriad symptoms of DM1 are caused by disturbances in the expression of multiple nearby genes due to the expansion – making DM1 a "contiguous gene syndrome." There are several genes in close proximity to the *DMPK* gene (i.e. <5 kb) whose roles in DM1 were considered; the most studied of these genes has been SIX5. The SIX5 homolog in Drosophila is required for normal eye development and its mouse homolog is involved in regulating distal limb muscle development. Expression data on SIX5 in

DM1 patients has been inconsistent, however. A Six5 knockout mouse developed cataracts, although they were not of the type observed in DM1. Another gene implicated by this model is *DMWD*, which is expressed in the testis and suspected to be involved in male infertility. Studies indicated that the expression of DMWD is not altered.

The third theory for the molecular basis of DM proposes that an RNA gain-of-function mechanism is responsible. RNA foci, which are accumulations of expanded transcripts, accumulate in the nuclei of patient cells (Fig. 11). The RNA gain-of-function theory was buttressed by the discovery that DM2, a disorder with symptoms almost identical to DM1, is caused by a (CCTG)n expansion in intron 1 of the zinc finger 9 protein (ZNF9). The two genes responsible for DM1 and DM2 are unrelated and reside on different chromosomes. Genes near the two loci bear no obvious resemblances. The only striking parallels between the two expansions are: (1) they both contain CTG triplets; and (2) they both occur in transcribed but noncoding regions of the genome. More support for the RNA gain-of-function model came from a mouse model generated by Mankodi et al. in 2000, which contained a (CTG)n expansion in the 3′ untranslated region of the skeletal actin (*HSA*) gene. This mouse exhibited myopathy typical of DM1, although the skeletal muscle-restricted expression pattern of HSA precludes broader conclusions. Thus, there are several lines of evidence suggesting that CUG expansion-containing RNAs are capable of causing DM.

RNA-binding proteins and transcription factors colocalize with the RNA foci found in DM, potentially altering nuclear processes. Elevated levels of CUG-containing RNA has been shown to alter gene splicing in specific transcripts which could be relevant to DM: cardiac troponin T (cTNT), involved in cardiomyopathy; Insulin Receptor (IR), involved in diabetes; and Clc-1, the main chloride channel in muscle. Abnormal splicing of cTNT, IR, and ClC-1 are proposed to account for the cardiac abnormalities, insulin insensitivity and myotonia observed in DM.

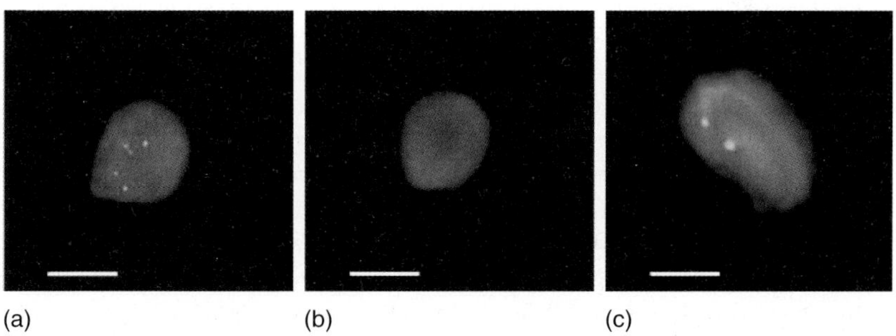

Fig. 11 *In situ* hybridization of muscle sections with fluorescently labeled antisense oligonucleotide probes reveals accumulation of mutant RNA in DM1 and DM2. (a) Probing of DM2 muscle with a CAGG probe indicates that multiple RNA foci are present. (b) Probing of normal muscle with a CAGG probe demonstrates absence of RNA foci. (c) Probing of DM1 muscle with a CAG probe yields prominent RNA foci. (From Liquori, C.L., Ricker, K., Moseley, M.L., Jacobsen, J.F., Kress, W., Naylor, S.L., Day, J.W., Ranum, L.P. (2001) Myotonic dystrophy type 2 caused by a CCTG expansion in intron 1 of ZNF9, *Science* **293**, 864–867, used with permission of *Science*).

Splicing abnormalities cause a reduction in the membrane concentration of ClC-1 and reduce chloride conductance to levels consistent with myotonia. These spicing alterations are thought to be due to the repeat-expanded RNA's effects on two families of RNA-binding proteins: "CUG-BP1 and ETR-3-like factors" (CELF) and "muscleblind-like proteins" (MBNL). CELF proteins regulate pre-mRNA splicing in cTNT, IR, and ClC-1. In patient tissue and cell culture, CUG-BP1 levels and activity are increased in response to elevated levels of CUG-containing RNA. This may be due to a lengthening of the protein's half-life. MBNL proteins were named after their Drosophila ortholog *muscleblind*, which is required for photoreceptor and muscle differentiation in flies. The three known proteins in this family (MBNL1 (MBNL), MBNL2 (MBLL), and MBNL3 (MBXL)) are splicing regulators that are thought to act antagonistically to CELF family proteins. They colocalize with RNA foci *in vivo*, and a mouse model lacking specific isoforms of MBNL1 recapitulates the myotonia, cataracts, and splicing dysregulation observed in DM.

The splicing alterations seen in DM patients are consistent with loss of function of MBNL proteins or an increase in CELF protein activity in muscle and brain. Because of the colocalization of MBNL proteins with RNA foci, it has been proposed that sequestration and subsequent depletion of these proteins from the cellular milieu is responsible for the symptoms of DM. Although this process probably plays a critical role in the disease, recent evidence suggests it is not solely responsible. The most plausible current theory explaining DM pathogenesis proposes an imbalance between the antagonistic MBNL and CELF proteins, resulting in specific splicing abnormalities that cause DM's diverse range of symptoms.

5.2 Myotonic Dystrophy Type 2

Myotonic dystrophy type 2 (DM2) is an autosomal dominant, multisystem disorder very similar to DM1. The majority of its symptoms resemble DM1: progressive weakness, myotonia, cardiac disturbances, iridescent cataracts, and insulin insensitivity. There are some notable differences, however. DM2, unlike DM1, predominantly affects proximal muscles at its onset, which is why many cases were originally classified as proximal myotonic myopathy (PROMM). Other interesting differences are that mental retardation is not observed in DM2, DM2 patients show increased sweating, and DM2 congenital forms have not been observed. DM2 is caused by a (CCTG)n tetranucleotide expansion in the first intron of the zinc finger protein 9 (*ZNF9*) gene on chromosome 3q21.3. The tetranucleotide repeat can expand to stunning lengths, with the longest cases comprising 44 kb of DNA, making them the longest tracts observed in the repeat expansion disorders. The DM2 locus exhibits marked somatic instability. Over the course of a patient's lifetime, the average repeat length increases substantially, as judged by blood drawings.

The similarities between DM1 and DM2 are not restricted to the clinical presentation. The expansions are both large, CTG-containing tracts that are transcribed, but not translated. Both disorders cause nuclear RNA foci that sequester specific RNA-binding proteins (Fig. 11), including CELF and MBNL family members. Similar splicing abnormalities are also observed in DM2. Since the genes associated with the repeat expansions in DM1 and DM2 are unrelated, it is likely that the cause of both disorders is a toxic gain of function of long tracts of CUG-containing RNA. A mouse model generated by Mankodi et al. in 2000, which contained a (CTG)*n* expansion in the 3′ untranslated region of the skeletal actin (*HSA*) gene, exhibited myopathy typical of DM. Thus, even outside the context of the DM1 and DM2 loci, CTG-containing expansions can cause DM-like symptoms. This supports the RNA gain-of-function theory, as transcribed CTG-containing expansions are inherently capable of causing disease. Differences between DM1 and DM2 may be due to regional or temporal expression patterns or differences in the affinity of RNA-binding proteins for CTG or CCTG tracts.

5.3 Spinocerebellar Ataxia Type 8

Spinocerebellar ataxia type 8 (SCA8) is a dominantly inherited cerebellar ataxia. Affected individuals suffer from late-onset, slowly progressing gait ataxia as well as dysarthria, oculomotor incoordination, spasticity, and decreased vibration sense. The cerebellar cortices and vermis undergo a slowly progressing but dramatic atrophy, while the brainstem exhibits little evidence of degeneration. Patients may become wheelchair-bound as early as their fourth decade of life. Using a direct method for cloning expanded triplet repeats (i.e. RAPID cloning), a gene containing an expanded CTG tract was identified in a large family (MN-A) segregating the SCA8 phenotype. Interestingly, this CTG is contained within a gene on the long arm of chromosome 13 that is transcribed, but apparently not translated into a protein product. Thus, it was proposed that the production of an RNA transcript containing this expanded CTG repeat tract is the cause of SCA8. Numerous subsequent studies, however, have indicated that possession of the expanded CTG repeat tract appears necessary, but not sufficient, for the production of the SCA8 phenotype. Thus, reduced penetrance is viewed as a key feature of SCA8 at this time. On the basis of the available genetic data, normal individuals always carry fewer than 70 CTG repeats, while affected SCA8 patients can carry anywhere from 71 to >1000 CTG repeats. As extremely large CTG repeat alleles (>800 triplets) were shown to not cause disease in early reports, this was initially attributed to lack of stability of the mutant RNA product. However, further work has revealed considerable overlap between disease-causing CTG repeat expansions and nonpenetrant CTG repeat alleles, indicating that a secondary factor – either a trans-acting genetic factor or an environmental factor – must be present to yield the SCA8 disease phenotype. In the case of the original MN-A family, evidence for a *cis* modifier appears responsible for the extremely high penetrance in this large pedigree. Thus, while the causality of the CTG repeat expansion in SCA8 has been somewhat controversial, review of the current literature suggests that the CTG repeat expansion is directly involved in SCA8, but may not alone be sufficient to produce the disorder.

Another unique feature of the SCA8 CTG repeat is its extreme and unusual genetic instability. Paternal transmissions generally result in contractions, while maternal transmissions generally result in expansions. Expansions of up to 600 repeats have occurred in one generation through maternal transmission. Large deletions in expanded alleles often occur in sperm cells, offering an explanation for the paternal contraction bias. Pathogenic expansions often have 5′ triplet interruptions, the role of which is unclear.

The molecular basis of how the expanded SCA8 RNA causes disease remains unclear. The transcript has been detected exclusively in the brain, and appears to be transcribed in the CTG orientation. Translation of a polyglutamine tract from the CAG-containing transcript in the opposite direction is precluded by the existence of stop codons flanking the repeat. The longest transcript identified to date contains 6 exons, and many alternatively spliced forms of the gene have been detected, although all these isoforms are expressed at very low levels. None contain significant open reading frames. In 1999, Koob et al. reported that a gene partially overlaps the 5′ end of the *SCA8* gene locus on the antisense strand. This gene, called *Kelch-like 1 (KLHL1)*, for its homology to the Drosophila *KELCH* gene, is highly conserved and predicted to encode an actin-binding protein of 748 amino acids. Its expression overlaps with that of the SCA8 transcript, suggesting that the *SCA8* gene may produce an antisense RNA whose function is to regulate KLHL1 expression. Whether or how this is occurring remains uncertain at this time.

With the discovery of the DM RNA toxic gain-of-function pathway, an emerging theory for SCA8 pathology holds that the production of a CUG-expanded transcript results in RNA gain-of-function toxicity within the restricted neuronal populations where the *SCA8* gene is expressed. In support of this hypothesis, transgenic mice derived with a human bacterial artificial chromosome containing the entire *SCA8* gene with a 118 CTG repeat expansion develop neurological disease. The severity of their phenotype depends on the expression level of the transgene. In the more moderate expressing lines, the SCA8 mice display a slowly progressive gait ataxia reminiscent of the human disease. While no protein product has been detected, 1C2 and ubiquitin antibody staining reveal intranuclear inclusions in cerebellar neurons, consistent with the RNA-containing inclusions observed in FXTAS brains. An alternative interpretation is that translation of CTG into a polyleucine tract is occurring. Further work will be needed to distinguish these two possibilities.

Studies of the SCA8 CTG repeat expansion in *D. melanogaster* have also been informative. Whether directing expression of the *SCA8* gene to fly retina with a normal CTG repeat tract or an expanded CTG repeat tract, a neurodegenerative eye phenotype results. Modifier screens using fly stocks carrying mutations in genes encoding RNA-binding protein yielded a number of genes that could either enhance or suppress this retinal degeneration phenotype. Interestingly, Drosophila *muscleblind*, whose mammalian counterpart has been implicated in the DM RNA toxicity pathway, modified the retinal degeneration caused by expanded SCA8 CTG repeat expression more so than the retinal degeneration caused by normal SCA8 CTG repeat expression. Such data supports the hypothesis that the SCA8 CTG repeat expansion is producing neurotoxicity by altering the function of RNA-binding proteins as in DM. Although it would be premature to definitively conclude that SCA8 is an RNA gain-of-function repeat disease, provisional classification in this category is appropriate.

5.4
The Fragile X Tremor – Ataxia Syndrome (FXTAS)

Fragile X tremor-ataxia syndrome (FXTAS) is a rare and unusual disorder that is associated with premutation of the FRAXA locus in males. Patients develop a progressive ataxia and intention tremor. This is sometimes accompanied by dementia, parkinsonism, and autonomic dysfunction. Neuropathological changes include degeneration of the cerebellum and ubiquitin-positive intranuclear inclusions in neurons and glia. FXTAS is associated with the expansion of a $(CGG)n$ tract in the 5′-untranslated region of the *FMR1* gene, found on chromosome Xq27.3. Expansion of this tract beyond 200 repeats causes FRAXA, a disorder resulting from a reduction in the expression of FMR1. Males with tracts between 55 and 200 CGG repeats are considered to be in the "premutation range," and are now considered to be at risk for FXTAS beyond middle age.

In contrast to FRAXA, FXTAS patients show increased expression of the FMR1 transcript, with longer repeats corresponding to higher transcript levels. Premutation carriers with more than 100 repeats have an average of five times more FMR1 mRNA than individuals in the normal repeat range. Despite the increase in transcript level, the protein product of the *FMR1* gene, FMRP, is slightly reduced in FXTAS patients. Since elevated transcript

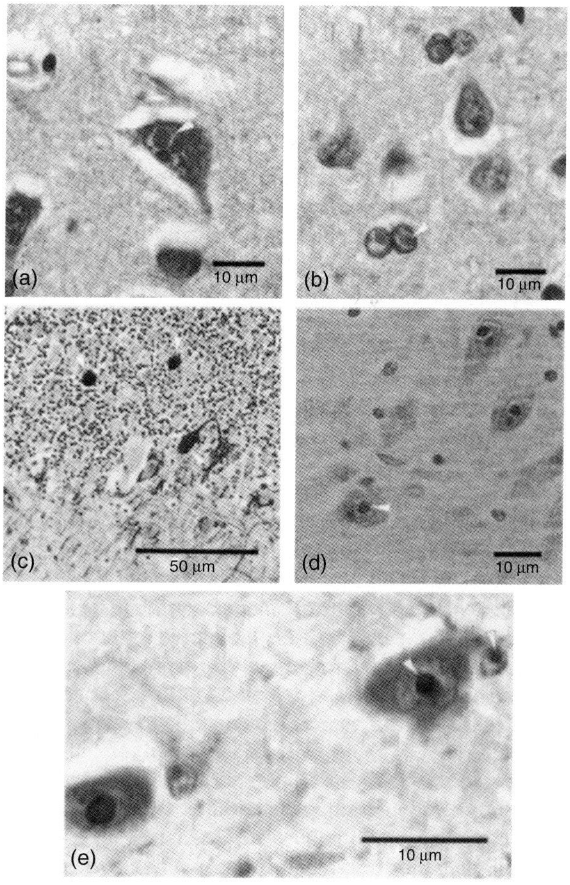

Fig. 12 Intranuclear inclusions are present in the brains of FXTAS patients. (a) Hematoxylin & eosin staining of cerebral neurons reveals a refractile, eosiniphilic nuclear inclusion of ~5 μM in diameter (white arrowhead). (b) Hematoxylin & eosin staining of cerebral astrocytes reveals refractile, eosiniphilic nuclear inclusions of ~2 μM in diameter (white arrowhead). (c) Antineurofilament antibody staining of cerebellum demonstrates presence of dystrophic neurites, consistent with ongoing Purkinje cell degeneration. (d) Antiubiquitin antibody staining of cerebral neurons labels intranuclear inclusions. (e) Antiubiquitin positive intranuclear inclusions form in both neurons (white arrowhead; larger cell) and astrocytes (white arrowhead; smaller cell). (From Greco et al. (2002) *Brain* **125**, 1760, used with permission).

levels are present without great alteration in protein levels, this disorder may be due to an RNA gain of function of the FMR1 transcript. Indeed, immunostaining of FXTAS patients' brain sections reveal intranuclear inclusions in neurons and glia, believed to be comprised of accumulated Fmr1 RNA transcripts and various proteins (Fig. 12). To determine the molecular basis of FXTAS, a knockin mouse model of FXTAS was generated using a human 98 CGG repeat, and was noted to produce elevated levels of the Fmr1 transcript and display intranuclear ribonucleoprotein inclusions, containing ubiquitin, Hsp40 and the 20S proteasome subunit. The inclusions may be a response to RNA toxicity or a result of the aggregation of CGG-binding proteins. While the mechanism of FXTAS may be due to a dominant gain of function of triplet-expanded RNA on RNA-binding proteins as in DM, there is currently no direct evidence supporting this theory.

6 Type 4: The Polyalanine Diseases

6.1 Overview

Polyalanine disorders are characterized by small expansions in trinucleotide repeats encoding alanine tracts. Most are rare, autosomal dominant developmental malformations. Although the normal function of the alanine tract is unknown, they tend to be found in transcription factors. The majority of alanine tract expansions associated with disease loci are also located in transcription factors, most of which play a role in development. In contrast to polyglutamine expansions, polyalanine expansions are small and stable, not exceeding 30 triplets. Most are composed of "imperfect" alanine tracts, including combinations of GCG, GCA, GCC, and GCT codons. This composition argues against an expansion mechanism involving single-strand, hairpin-like secondary structures as in other repeat disorders. A more plausible mechanism for polyalanine expansion is unequal crossing-over, in which alleles mispair during meiotic crossing-over, resulting in one expanded and one contracted tract.

6.2 Oculopharyngeal Muscular dystrophy

Oculopharyngeal muscular dystrophy (OPMD) is a predominantly autosomal dominant, late-onset disorder characterized by progressive drooping eyelids, dysphagia, and proximal limb weakness. Certain skeletal muscles in affected patients degenerate and contain nuclear inclusions and rimmed vacuoles. This effect is particularly striking in the levator palpebra and pharyngeal muscles, responsible for lifting the eyelids and swallowing. OPMD is caused by an alanine expansion in the gene polyadenine-binding protein 2 (PABP2). PABP2 normally contains a 10-alanine tract encoded by $(GCG)_6(GCA)_3GCG$. Affected individuals carry 12–17 alanines, probably resulting from unequal crossing-over of the two alleles. Repeat length appears to correlate with disease severity.

PABP2 is an abundant, ubiquitously expressed pre-mRNA-binding protein that plays a role in controlling the formation and length of mRNA polyA tails. NIs immunoreactive for PABP2 are present in the skeletal muscle cells of OPMD patients and sequester polyA-containing transcripts. The aggregates are filamentous and contain PABP2, ubiquitin and proteasome subunits, leading investigators to conclude that expansion of the alanine tract probably causes PABP2 to misfold and aggregate. Aggregation in OPMD has been linked to toxicity in several experimental systems. Polyalanine-expanded PABP2 aggregates and causes cell death in cultured cells and transgenic mice. Indeed, widespread expression of the human *PABP2* gene in transgenic mice yielded polyalanine length-dependent muscle pathology, including rimmed vacuoles, central nuclei, and numerous dystrophic changes (Fig. 13). Deleting the C-terminal oligomerization domain, overexpressing chaperones, or exposing cells to aggregation inhibitors such as Congo red and doxycycline, reduces aggregate formation and toxicity in cell culture. A polyalanine-expanded peptide, similar to the amino-terminal region of PABP2 was shown to adopt a β-sheet conformation, whereas the same peptide with 7 alanines adopted an α-helical conformation. This is reminiscent of polyglutamine proteins, which also adopt a β-sheet conformation when the polyglutamine tract exceeds a threshold of about 35 glutamines. Expanded polyalanine proteins also activate caspase-3 and -8 in cultured cells, another feature reminiscent of polyglutamine toxicity.

In 2004, Wirtschafter et al. proposed a model for the selective vulnerability of

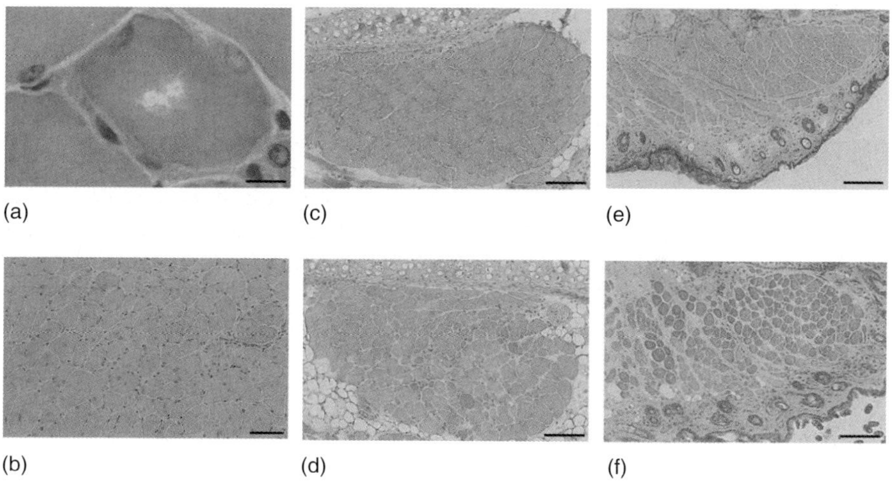

Fig. 13 Mice expressing polyalanine-expanded PABP2 protein display prominent muscle pathology consistent with the OPMD phenotype. Transgenic mice were generated using the pCAGGS expression vector that consists of the chicken beta-actin promoter and a CMV enhancer, and thus ubiquitously express either human PABP2 protein with six alanines (c, e) or nine alanines (a, b, d, f) at roughly comparable levels. Hematoxylin & eosin staining of sections from the soleus muscle (a, b), the pharynx muscle (c, d), and the eyelid muscle (e, f) reveal normal muscle histology in the six-alanine PABP2 expressing mice (c, e). However, mice expressing PABP2 with nine alanines display nonuniform muscle fiber size and prominent connective tissue in pharynx and eyelid musculature (d, f). At high power, cytoplasmic vacuoles, reminiscent of OPMD "rimmed vacuoles" are apparent in soleus muscle from the nine-alanine PABP2 expressing mice (a). Low power examination further indicates frequent central nuclei in the soleus muscle from such mice (b). (From Hino et al. (2004) *Hum. Mol. Genet.* **13**, 181, used with permission).

extraocular muscles in OPMD. Unlike other skeletal muscles, extraocular muscles are not postmitotic and they continually undergo remodeling. This requires the frequent upregulation of genes involved in cell cycling and protein synthesis. Failure of correct mRNA polyadenylation or transport in these cells may have a cumulative toxic effect resulting in the progressive degeneration of these muscles. The mechanism of OPMD could be due to an interference with polyadenylation, disturbances in intracellular trafficking of mRNA, or toxicity due to the aggregation of misfolded and/or aggregated PABP2 species. Further research will be required to distinguish between these possibilities.

6.3 Synpolydactyly (Syndactyly Type II)

Synpolydactyly (SPD) is a rare, autosomal dominant developmental disorder characterized by fused and extra digits (syndactyly and polydactyly, respectively). It is caused by the expansion of a polyalanine tract in the amino-terminal region of the transcription factor HOX-D13. *HOX* genes act in concert to coordinate axial patterning in animals. Tracts of 7–14 alanines in HOX-D13 have been linked to SPD, and disease severity is proportional to repeat length. Multiple studies support a "dominant-negative" role for expanded HOX-D13 protein in SPD. Mice null for Hox-d13 have a phenotype less severe than SPD, while mice with alanine tract expansions have a form that more closely resembles the disorder. Mice lacking Hox11, Hox12, and Hox13 have a phenotype similar to SPD, suggesting that the alanine expansion in SPD antagonizes the function of other *HOX* genes. This is supported by a genetic complementation study. In further support of this theory, humans with suspected loss-of-function mutations in the *HOX-D13* gene do not have a phenotype consistent with SPD.

6.4 Cleidocranial Dysplasia

Cleidocranial dysplasia (CCD) is a rare, autosomal dominant developmental disorder characterized by holes in the skull, dental malformations, absent or hypoplastic clavicles and maxillae, and other skeletal malformations. The primary cause of CCD is thought to be loss-of-function mutations in the gene *RUNX2*. This gene affects osteoblast differentiation and is a member of the Runt family of transcription factors. In one family, phenotypically distinct from classic CCD, an expansion from 17 to 27 alanines in RUNX2 has been detected. This family exhibited brachydactyly and a mild CCD phenotype. The difference in phenotype between the polyalanine expansion mutation family and typical presumed haploinsufficient, loss-of-function CCD patients supports a gain-of-function effect of the expanded alanine tract in this atypical family. Contraction of the tract in RUNX2 is common and does not cause a detectable phenotype.

6.5 Holoprosencephaly

Holoprosencephaly (HPE) is a common developmental malformation resulting in partial or full cyclopia, failure to develop midline structures in the ventral forebrain, and prenatal lethality. In rare cases, HPE is caused by the heterozygous expansion of a 15-amino acid alanine tract in the protein ZIC2. This protein is one member of a family of zinc finger proteins believed to regulate neurulation, left–right axis formation and other developmental processes. Other individuals with heterozygous loss-of-function mutations have a phenotype indistinguishable from those with alanine tract expansions. Partial loss of function of Zic2 in mice similarly causes developmental abnormalities similar to HPE. Therefore, expansion of the alanine tract likely causes loss of function of ZIC2 in the case of HPE.

6.6 Hand-foot-genital Syndrome

Hand-foot-genital syndrome (HFGS) is a rare, dominantly inherited developmental abnormality characterized by malformation of the distal limbs and lower urogenital tract. Short thumbs, short great toes, and abnormal carpals and tarsals are some of its salient features. Some HFGS patients have alanine tract expansions in the protein HOX-A13, which is in the same family as the SPD-associated protein HOX-D13. HOX-A13 contains three alanine tracts, the most C-terminal of which is the most commonly mutated. Expansions enlarge the second or third tract by 6 to 9 alanines. In humans, deletion of HOX-A13 causes a phenotype that is mild in comparison to HFGS caused by alanine tract mutations. Hoxa13-null mice also have a milder phenotype than mice carrying a frameshift deletion suspected to confer gain of function. This evidence suggests a dominant-negative mechanism for HFGS caused by alanine tract expansions.

6.7 Blepharophimosis-ptosis-epicanthus Inversus Syndrome

Blepharophimosis-ptosis-epicanthus inversus syndrome (BPES) is a rare, autosomal dominant developmental disorder resulting in malformation of the upper eyelids and forehead, and occasionally premature ovarian failure in women. It is caused by a number of different mutations in the gene *Forkhead L2* (*Foxl2*), the most common of which is expansion of its carboxy-terminal 14-amino acid alanine tract. FOXL2 is a highly conserved transcription factor whose role in the ovary has been studied most thoroughly. It is expressed in the ovaries and eyelids during development and adulthood. Studies in which Foxl2 have been ablated in mice show that it is important for the differentiation of ovarian granulosa cells, and its absence causes accelerated follicle cell depletion leading to POF. Knockout mice also display craniofacial abnormalities consistent with BPES, suggesting that a dominant-negative mechanism may be responsible in humans.

6.8 Syndromic and Nonsyndromic X-linked Mental Retardation

Several loosely related disorders are associated with alanine tract expansions in the Aristaless related homeobox (ARX) protein on chromosome Xp22.13. One is nonsyndromic X-linked mental retardation (XLMR), a heterogeneous condition in which mental retardation is the main consistent feature. Several syndromic XLMR disorders linked to alanine expansion in ARX include West syndrome (WS) and Partington syndrome (PRTS). WS causes progressive mental retardation with abnormal EEG and infantile seizures, while PRTS causes mental retardation, dysarthria, and dystonic movements of the hands. Brain anatomy appears normal in these disorders.

ARX is a paired-class homeodomain protein that is expressed in the ventricular and marginal zones of the developing mouse brain and continues to be expressed in adult cortex. It is suspected to play a role in neuroepithelial cell differentiation and maintenance of neuronal subtypes in the adult cortex. ARX-null mice have small brains and neuronal migration deficits. Expansions of two different alanine tracts in ARX cause XLMR: a 12-alanine tract at amino acids 144–155, and a second tract at amino acids 100–115. Both sites cause highly variable forms of XLMR, with expansions in the former alanine tract resulting in nonsyndromic XLMR, WS or PRTS. Alanine tract expansions in ARX probably result in partial loss of function of the protein. Heterozygous female carriers of expanded ARX do not exhibit XLMR, suggesting that one normal copy of the protein is sufficient for normal cognitive ability. Also, humans with null mutations of ARX suffer from X-linked lissencephaly with abnormal genitalia (XLAG), a much more severe condition causing major developmental abnormalities in the brain and genitalia. Thus, XLMR due to alanine expansions in ARX may be because of partial loss of function leading to subtle developmental defects and/or failure to maintain specific neuronal populations.

6.9 Congenital Central Hypoventilation Syndrome

Congenital central hypoventilation syndrome (CCHS) is a rare, autosomal dominant disorder causing a failure of autonomic control of breathing. It attenuates or abolishes responses to hypercarbia and hypoxemia. In the majority of cases, it is caused by the expansion of one of two alanine tracts in the protein PHOX-2B. Mutations expand the 20-residue tract to 25–29 alanines. PHOX-2B is a paired-class transcription factor containing a homeodomain. A loss-of-function mutation in murine Phox2b is homozygous lethal, and specifically prevents the development of parasympathetic ganglia. Heterozygous mice show chronic pupil dilation but no parasympathetic or respiratory disturbances. Also, a patient hemizygous for a 5-Mb deletion including Phox2b does not have CCHS. No cases of CCHS have been reported in which PHOX2B is truncated before the homeobox domain. This evidence suggests that the expansion of alanine tracts in PHOX2B may cause a subtle, dominant-negative effect on the development of respiratory control pathways.

7 Unclassified Repeat Diseases Lacking Mechanistic Explanations

7.1 Spinocerebellar Ataxia Type 10

Spinocerebellar ataxia type 10 (SCA10) is an autosomal dominant, progressive ataxia that exhibits nearly pure cerebellar signs. It appears restricted to individuals of Mexican ethnicity. SCA10 patients typically present with gait ataxia, followed by dysarthria, dysphagia, and ocular dysmetria, and most patients also experience recurrent motor seizures. Cerebellar atrophy is the most prominent neuropathological change. SCA10 is caused by the expansion of a highly polymorphic pentanucleotide repeat, (ATTCT)*n*, on chromosome 22. Unaffected individuals carry 10–22 repeats. While expanded alleles have not been successfully PCR amplified, transcript sizes on Northern blots and fragment sizes on Southern blots indicate that disease alleles can range from ~800 to 4500 repeats. Anticipation occurs in SCA10, independently supporting the causal nature of the repeat expansion. The ATTCT repeat is located in intron 9 of the *SCA10* gene, a previously unrecognized 66-kb gene that encodes a novel putative 475 amino acid protein of unknown function with few recognizable motifs or domains. The 2-kb SCA10 transcript is ubiquitously expressed though the highest levels of expression occur in the brain, testis, and adrenal glands. Within the brain, it is most highly expressed in the cerebellum and associated structures. The carboxy-terminal portion of the protein appears to contain "armadillo repeats," which are responsible for the membrane association of β-catenins. Ataxin-10 does not seem to associate with membranes, however. The entire protein is highly conserved between humans and rodents, and potential orthologs exist in Arabidopsis and Drosophila.

There are several proposed mechanisms for SCA10 pathogenesis. The most obvious hypothesis is that the large intronic expansion affects ataxin-10 expression, perhaps by altering local chromatin structure. siRNA knockdown of ataxin-10 in cell culture experiments yields higher rates of cell death in cerebellar neurons than in cortical neurons. However, the expression levels of the SCA10 transcript are not reduced in patient's lymphoblast cells, arguing against simple haploinsufficiency. Consequently, RNA gain-of-function toxicity has been proposed as the potential mechanism; however, there are no further data at this time to support such a hypothesis. Additional work will need to be done to distinguish between these and other possibilities, and will need to account for the cell-type specific pattern of neurodegeneration that occurs in the face of apparently widespread expression of the *SCA10* gene mutation.

7.2
Spinocerebellar Ataxia Type 12

Among the more recent additions to the unstable repeat disease group is spinocerebellar ataxia type 12 (SCA12), a rare, autosomal dominant disorder that may be most prevalent in Indian populations. Its symptoms are distinct from the other SCAs, typically beginning with an action tremor of the upper extremities and progressing to include hyperreflexia, mild cerebellar dysfunction, bradykinesia, increased muscle tone, psychiatric symptoms, and dementia. The brains of SCA12 patients likely undergo a slow, generalized atrophy that is most prominent in the cortex, but also results in loss of Purkinje cells in the cerebellum. Disease onset typically occurs in the third or fourth decade, and a gradually progressive disease course is typical.

SCA12 is caused by a (CAG)n expansion in chromosome 5q31 – q33. The expansion is 5′ to the *PPP2R2B* gene, encoding a regulatory subunit of the protein phosphatase 2A enzyme (PP2A). This gene has many transcription start sites, some of which include the repeat (but many of which do not). GENSCAN predicts an exon including the expansion that would encode a polyserine tract, but this prediction is of low probability and considered unlikely. Other evidence suggests that the expansion is located in the promoter region, and this is currently the most widely accepted view. Unaffected individuals carry 7–32 CAG repeats while affected individuals carry 55–78 CAG triplets. The most common repeat size in unaffected individuals is 10 CAGs. The repeat is fairly stable, with only modest expansions and contractions resulting equally from maternal and paternal transmission. A significant correlation between repeat length and age of onset has not been documented.

The protein PP2A is an essential serine/threonine phosphatase expressed in all known eukaryotic cells. It is involved in diverse cellular functions, including cell growth, differentiation, DNA replication, neurotransmitter release, and apoptosis. PPP2R2B is a brain-specific regulatory subunit of PP2A. The class of regulatory subunits including PPP2R2B may affect PP2A's phosphatase activity for certain substrates, including histone-1, vimentin, and tau. It may also affect PP2A's subcellular localization.

There are several possible explanations for SCA12 pathogenesis. The first is that expanded PPP2R2B may generate a polyamino acid tract-containing protein, resulting in toxicity. Northern blots probing for sequence flanking the CAG repeat did not detect a PPP2R2B transcript, indicating that if a repeat-containing exon exists and is transcribed, it is not present at appreciable levels. Nevertheless, the possibility of polyglutamine, polyserine, or polyalanine toxicity, though unlikely, cannot be completely ruled out at this time. A second possible cause of SCA12 pathogenesis is RNA gain of function, as occurs in DM and FXTAS. The repeat expansions are smaller in SCA12 than in DM, however, and more importantly, CAG tract-containing transcripts appear to be rather scarce. A third possibility is altered splicing of the PPP2R2B transcript. Several different amino-termini are possible, the ratio of which may affect PP2A's subcellular localization. Yet another theory of SCA12 pathogenesis is that the expansion affects PPP2R2B transcript levels. Repeat expansion causes a substantial increase in PPP2R2B expression as measured in reporter assays using a neuroblastoma cell line. Altered levels of the protein could affect PP2A's specificity or subcellular localization. This has the potential to disturb a multitude of processes in the CNS. At this time, all of the above theories of SCA12 CAG repeat expansion neurotoxicity remain plausible.

7.3 Huntington's Disease Like 2 (HDL2)

Perhaps the most exciting and enigmatic recent discovery in the repeat expansion field is that of the mutational basis of a disorder known as Huntington's disease like 2 or HDL2. HDL2 is so named because it is in essence a genocopy of classical HD, as the original HDL2 pedigree was labeled with a diagnosis of HD until HD CAG repeat testing indicated that this family's HD-like disease did not result from a CAG repeat expansion in the *htt* gene. HDL2 patients present with weight loss and diminished coordination, and then develop tremors, dysarthria, hyperreflexia, and rigidity. Patients display psychiatric involvement, chorea, and dystonia, and ultimately become demented. Death occurs 15–25 years after onset, when patients become bedridden as in typical HD cases. MRI findings reveal marked atrophy of the caudate and of the cerebral cortex, making the neuropathology indistinguishable from classic HD. HDL2 patients do not show cerebellar signs or neuropathology.

As soon as it was found that HDL2 patients do not have the HD CAG repeat expansion, direct cloning methods for triplet repeat expansions of the CAG/CTG type were applied to patient samples and an expanded CAG/CTG repeat ($n = 55$) was isolated. Sequence flanking this repeat indicated that HDL2 is caused by CTG repeat expansions in the *junctophilin-3* (*JPH3*) gene, one of a family of structural proteins whose function is to link

Ca++ channels on the ER with voltage sensors on the plasma membrane. Analysis of JPH3 CTG repeat indicates that normal individuals typically carry 7–27 CTG repeats while affected HDL2 patients usually have expansions of 50–60 CTG repeats. Widespread screening of HD-like patients from around the globe suggests that the HDL2 JPH3 CTG repeat expansions are most common in individuals of African ethnicity.

The question of how the JPH3 CTG repeat expansion causes HDL2 remains unknown, but there are at least four possible explanations. These alternative (but not mutually exclusive) theories stem from the documented alternative processing of the *JPH3* gene, which permits the prediction of the CTG repeat tract as: (1) part of intron 1; (2) as part of the 3′ untranslated region; or (3) as encoding either a polyleucine or polyalanine tract. One theory is that haploinsufficiency of the Ca++ regulating brain- and testes-specific junctophilin-3 protein is responsible for HDL2. While JPH3 knockout mice display motor incoordination, no histological abnormalities are found in their brains. Further work with these mice is ongoing to evaluate this hypothesis. Evidence against simple haploinsufficiency has come from study of HDL2 patient's brain material, however. 1C2 and ubiquitin antibody immunostaining reveal intranuclear inclusions in neurons throughout the brains of these patients, with dramatic similarity in distribution to patients with classic HD (Fig. 14). At this time, the molecular basis of HDL2 is unknown; however, the incredible overlap between HD and HDL2 in terms of clinical phenotype and neuropathology strongly suggests that solving HDL2 should have profound implications for our mechanistic understanding of HD.

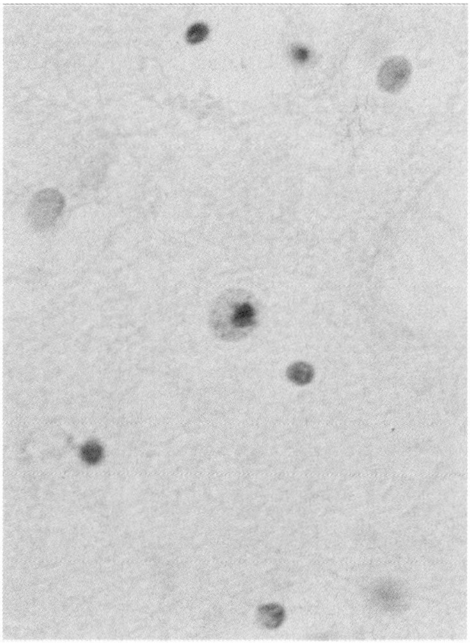

Fig. 14 HDL2 patients have 1C2-positive intranuclear inclusions. 1C2 staining of cerebral cortex (frontal lobe) from an HDL2 patient reveals a prominent intranuclear inclusion that resembles the intranuclear inclusions seen in classic HD patients. As the 1C2 antibody is directed against expanded, misfolded polyglutamine tracts, which are not predicted to be expressed from the causal *HDL2* gene, the explanation for the presence of such nuclear inclusions in HDL2 patients remains unknown. (From Margolis, R.L., O'Hearn, E., Rosenblatt, A., Willour, V., Holmes, S.E., Franz, M.L., Callahan, C., Hwang, H.S., Troncoso, J.C., Ross, C.A. (2001) A disorder similar to Huntington's disease is associated with a novel CAG repeat expansion, *Ann. Neurol.* **50**, 373–380, used with permission of *Annals of Neurology*, and John Wiley & Sons, publisher).

See also Motor Neuron Diseases: Cellular and Animal Models; Motor Neuron Diseases: Molecular Mechanism, Pathophysiology, and Treatments.

Bibliography

Books and Reviews

Brown, L.Y., Brown, S.A. (2004) Alanine tracts: the expanding story of human illness and trinucleotide repeats, *Trends Genet.* **20**, 51–58.

Harper, P.S. (2001) *Myotonic Dystrophy*, Saunders, London, UK.

Harper, P.S., Perutz, M. (2001) *Glutamine Repeats and Neurodegenerative Diseases: Molecular Aspects*, Oxford University Press, New York.

Jin, P., Alisch, R.S., Warren, S.T. (2004) RNA and microRNAs in fragile X mental retardation, *Nat. Cell Biol.* **6**, 1048–1053.

Ranum, L.P., Day, J.W. (2004) Pathogenic RNA repeats: an expanding role in genetic disease, *Trends Genet.* **20**, 506–512.

Wells, R.D., Warren, S.T. (1998) *Genetic Instabilities and Hereditary Neurological Diseases*, Academic Press, San Diego, CA.

Primary Literature

Albrecht, M., Golatta, M., Wullner, U., Lengauer, T. (2004) Structural and functional analysis of ataxin-2 and ataxin-3. *Eur. J. Biochem.* **271**, 3155–3170.

Amiel, J., Laudier, B., Attie-Bitach, T., Trang, H., de Pontual, L., Gener, B., Trochet, D., Etchevers, H., Ray, P., Simonneau, M. et al. (2003) Polyalanine expansion and frameshift mutations of the paired-like homeobox gene *PHOX2B* in congenital central hypoventilation syndrome, *Nat. Genet.* **33**, 459–461.

Arrasate, M., Mitra, S., Schweitzer, E.S., Segal, M.R., Finkbeiner, S. (2004) Inclusion body formation reduces levels of mutant huntingtin and the risk of neuronal death, *Nature.* **431**, 805–810.

Banfi, S., Servadio, A., Chung, M., Capozzoli, F., Duvick, L.A., Elde, R., Zoghbi, H.Y., Orr, H.T. (1996) Cloning and developmental expression analysis of the murine homolog of the spinocerebellar ataxia type 1 gene (*Sca1*), *Hum. Mol. Genet.* **5**, 33–40.

Brais, B., Bouchard, J.P., Xie, Y.G., Rochefort, D.L., Chretien, N., Tome, F.M., Lafreniere, R.G., Rommens, J.M., Uyama, E., Nohira, O. et al. (1998) Short GCG expansions in the *PABP2* gene cause oculopharyngeal muscular dystrophy, *Nat. Genet.* **18**, 164–167.

Brook, J.D., McCurrach, M.E., Harley, H.G., Buckler, A.J., Church, D., Aburatani, H., Hunter, K., Stanton, V.P., Thirion, J.P., Hudson, T. et al. (1992) Molecular basis of myotonic dystrophy: expansion of a trinucleotide (CTG) repeat at the 3′ end of a transcript encoding a protein kinase family member, *Cell.* **69**, 385.

Brown, V., Jin, P., Ceman, S., Darnell, J.C., O'Donnell, W.T., Tenenbaum, S.A., Jin, X., Feng, Y., Wilkinson, K.D., Keene, J.D. et al. (2001) Microarray identification of FMRP-associated brain mRNAs and altered mRNA translational profiles in fragile X syndrome, *Cell.* **107**, 477–487.

Brown, L.Y., Odent, S., David, V., Blayau, M., Dubourg, C., Apacik, C., Delgado, M.A., Hall, B.D., Reynolds, J.F., Sommer, A. et al. (2001) Holoprosencephaly due to mutations in ZIC2: alanine tract expansion mutations may be caused by parental somatic recombination, *Hum. Mol. Genet.* **10**, 791–796.

Burnett, B., Li, F., Pittman, R.N. (2003) The polyglutamine neurodegenerative protein ataxin-3 binds polyubiquitynated proteins and has ubiquitin protease activity, *Hum. Mol. Genet.* **12**, 3195–3205.

Calado, A., Tome, F.M., Brais, B., Rouleau, G.A., Kuhn, U., Wahle, E., Carmo-Fonseca, M. (2000) Nuclear inclusions in oculopharyngeal muscular dystrophy consist of poly(A) binding protein 2 aggregates which sequester poly(A) RNA, *Hum. Mol. Genet.* **9**, 2321–2328.

Chen, S., Peng, G.H., Wang, X., Smith, A.C., Grote, S.K., Sopher, B.L., La Spada, A.R. (2004) Interference of Crx-dependent transcription by ataxin-7 involves interaction between the glutamine regions and requires the ataxin-7 carboxy-terminal region for nuclear localization, *Hum. Mol. Genet.* **13**, 53–67.

Chen, H.K., Fernandez-Funez, P., Acevedo, S.F., Lam, Y.C., Kaytor, M.D., Fernandez, M.H., Aitken, A., Skoulakis, E.M., Orr, H.T.,

Botas, J. et al. (2003) Interaction of Akt-phosphorylated ataxin-1 with 14-3-3 mediates neurodegeneration in spinocerebellar ataxia type 1, *Cell.* **113**, 457–468.

Cummings, C.J., Reinstein, E., Sun, Y., Antalffy, B., Jiang, Y., Ciechanover, A., Orr, H.T., Beaudet, A.L., Zoghbi, H.Y. (1999) Mutation of the E6-AP ubiquitin ligase reduces nuclear inclusion frequency while accelerating polyglutamine-induced pathology in SCA1 mice, *Neuron.* **24**, 879–892.

David, G., Abbas, N., Stevanin, G., Durr, A., Yvert, G., Cancel, G., Weber, C., Imbert, G., Saudou, F., Antoniou, E. et al. (1997) Cloning of the *SCA7* gene reveals a highly unstable CAG repeat expansion, *Nat. Genet.* **17**, 65–70.

Davies, S.W., Turmaine, M., Cozens, B.A., DiFiglia, M., Sharp, A.H., Ross, C.A., Scherzinger, E., Wanker, E.E., Mangiarini, L., Bates, G.P. (1997) Formation of neuronal intranuclear inclusions underlies the neurological dysfunction in mice transgenic for the HD mutation, *Cell.* **90**, 537–548.

Day, J.W., Ricker, K., Jacobsen, J.F., Rasmussen, L.J., Dick, K.A., Kress, W., Schneider, C., Koch, M.C., Beilman, G.J., Harrison, A.R. et al. (2003) Myotonic dystrophy type 2: molecular, diagnostic and clinical spectrum, *Neurology.* **60**, 657–664.

De Baere, E., Beysen, D., Oley, C., Lorenz, B., Cocquet, J., De Sutter, P., Devriendt, K., Dixon, M., Fellous, M., Fryns, J.P. et al. (2003) FOXL2 and BPES: mutational hotspots, phenotypic variability, and revision of the genotype-phenotype correlation, *Am. J. Hum. Genet.* **72**, 478–487.

DiFiglia, M., Sapp, E., Chase, K.O., Davies, S.W., Bates, G.P., Vonsattel, J.P., Aronin, N. (1997) Aggregation of huntingtin in neuronal intranuclear inclusions and dystrophic neurites in brain, *Science.* **277**, 1990–1993.

Dunah, A.W., Jeong, H., Griffin, A., Kim, Y.M., Standaert, D.G., Hersch, S.M., Mouradian, M.M., Young, A.B., Tanese, N., Krainc, D. (2002) Sp1 and TAFII130 transcriptional activity disrupted in early Huntington's disease, *Science.* **296**, 2238–2243.

Ellerby, L.M., Hackam, A.S., Propp, S.S., Ellerby, H.M., Rabizadeh, S., Cashman, N.R., Trifiro, M.A., Pinsky, L., Wellington, C.L., Salvesen, G.S. et al. (1999) Kennedy's disease: caspase cleavage of the androgen receptor is a crucial event in cytotoxicity, *J. Neurochem.* **72**, 185–195.

Fu, Y.H., Kuhl, D.P., Pizzuti, A., Pieretti, M., Sutcliffe, J.S., Richards, S., Verkerk, A.J., Holden, J.J., Fenwick, R.G., Jr., Warren, S.T., et al. (1991) Variation of the CGG repeat at the fragile X site results in genetic instability: resolution of the Sherman paradox, *Cell.* **67**, 1047–1058.

Garden, G.A., Libby, R.T., Fu, Y.H., Kinoshita, Y., Huang, J., Possin, D.E., Smith, A.C., Martinez, R.A., Fine, G.C., Grote, S.K., et al. (2002) Polyglutamine-expanded ataxin-7 promotes noncell-autonomous Purkinje cell degeneration and displays proteolytic cleavage in ataxic transgenic mice, *J. Neurosci.* **22**, 4897–4905.

Gecz, J. (2000) The *FMR2* gene, FRAXE and non-specific X-linked mental retardation: clinical and molecular aspects, *Ann. Hum. Genet.* **64**, 95–106.

Goodman, F.R., Scambler, P.J. (2001) Human *HOX* gene mutations, *Clin. Genet.* **59**, 1–11.

Grabczyk, E., Usdin, K. (2000) The GAA*TTC triplet repeat expanded in Friedreich's ataxia impedes transcription elongation by T7 RNA polymerase in a length and supercoil dependent manner, *Nucleic Acids Res.* **28**, 2815–2822.

Greene, E., Handa, V., Kumari, D., Usdin, K. (2003) Transcription defects induced by repeat expansion: fragile X syndrome, FRAXE mental retardation, progressive myoclonus epilepsy type 1, and Friedreich ataxia, *Cytogenet. Genome Res.* **100**, 65–76.

Hagerman, R.J., Leehey, M., Heinrichs, W., Tassone, F., Wilson, R., Hills, J., Grigsby, J., Gage, B., Hagerman, P.J. (2001) Intention tremor, parkinsonism, and generalized brain atrophy in male carriers of fragile X, *Neurology.* **57**, 127–130.

Helmlinger, D., Hardy, S., Sasorith, S., Klein, F., Robert, F., Weber, C., Miguet, L., Potier, N., Van-Dorsselaer, A., Wurtz, J.M., et al. (2004) Ataxin-7 is a subunit of GCN5 histone acetyltransferase-containing complexes, *Hum. Mol. Genet.* **13**, 1257–1265.

Hillman, M.A., Gecz, J. (2001) Fragile XE-associated familial mental retardation protein 2 (FMR2) acts as a potent transcription activator, *J. Hum. Genet.* **46**, 251–259.

Hodgson, J.G., Agopyan, N., Gutekunst, C.A., Leavitt, B.R., LePiane, F., Singaraja, R., Smith, D.J., Bissada, N., McCutcheon, K., Nasir, J., et al. (1999) A YAC mouse model for Huntington's disease with full-length mutant

huntingtin, cytoplasmic toxicity, and selective striatal neurodegeneration, *Neuron.* **23**, 181–192.
Holmes, S.E., O'Hearn, E., McInnis, M.G., Gorelick-Feldman, D.A., Kleiderlein, J.J., Callahan, C.A., Ingersoll-Ashworth, R.G., Sherr, M., Sumner, A.J. (1999) Expansion of a novel CAG repeat in the 5′ region of a gene encoding a subunit of protein phosphatase 2A is associated with spinocerebellar ataxia type 12, *Am. J. Hum. Genet.* **65**, A4.
Holmes, S.E., O'Hearn, E., Rosenblatt, A., Callahan, C., Hwang, H.S., Ingersoll-Ashworth, R.G., Fleisher, A., Stevanin, G., Brice, A., Potter, N.T., et al. (2001) A repeat expansion in the gene encoding junctophilin-3 is associated with Huntington disease-like 2, *Nat. Genet.* **29**, 377–378.
Huntington's Disease, Collaborative, Research and Group (1993) A novel gene containing a trinucleotide repeat that is expanded and unstable on Huntington's disease chromosomes, *Cell* **72**, 971–983.
Huynh, D.P., Figueroa, K., Hoang, N., Pulst, S.M. (2000) Nuclear localization or inclusion body formation of ataxin-2 are not necessary for SCA2 pathogenesis in mouse or human, *Nat. Genet.* **26**, 44–50.
Ishikawa, K., Fujigasaki, H., Saegusa, H., Ohwada, K., Fujita, T., Iwamoto, H., Komatsuzaki, Y., Toru, S., Toriyama, H., Watanabe, M., et al. (1999) Abundant expression and cytoplasmic aggregations of [alpha]1A voltage-dependent calcium channel protein associated with neurodegeneration in spinocerebellar ataxia type 6, *Hum. Mol. Genet.* **8**, 1185–1193.
Jiang, H., Mankodi, A., Swanson, M.S., Moxley, R.T., Thornton, C.A. (2004) Myotonic dystrophy type 1 is associated with nuclear foci of mutant RNA, sequestration of muscleblind proteins and deregulated alternative splicing in neurons, *Hum. Mol. Genet.* **13**, 3079–3088.
Katsuno, M., Adachi, H., Kume, A., Li, M., Nakagomi, Y., Niwa, H., Sang, C., Kobayashi, Y., Doyu, M., Sobue, G. (2002) Testosterone reduction prevents phenotypic expression in a transgenic mouse model of spinal and bulbar muscular atrophy, *Neuron.* **35**, 843–854.
Kitamura, K., Yanazawa, M., Sugiyama, N., Miura, H., Iizuka-Kogo, A., Kusaka, M., Omichi, K., Suzuki, R., Kato-Fukui, Y., Kamiirisa, K., et al. (2002) Mutation of ARX causes abnormal development of forebrain and testes in mice and X-linked lissencephaly with abnormal genitalia in humans, *Nat. Genet.* **32**, 359–369.
Klement, I.A., Skinner, P.J., Kaytor, M.D., Yi, H., Hersch, S.M., Clark, H.B., Zoghbi, H.Y., Orr, H.T. (1998) Ataxin-1 nuclear localization and aggregation: role in polyglutamine-induced disease in SCA1 transgenic mice, *Cell.* **95**, 41–53.
Koob, M.D., Moseley, M.L., Schut, L.J., Benzow, K.A., Bird, T.D., Day, J.W., Ranum, L.P. (1999) An untranslated CTG expansion causes a novel form of spinocerebellar ataxia (SCA8), *Nat. Genet.* **21**, 379–384.
Kremer, E.J., Pritchard, M., Lynch, M., Yu, S., Holman, K., Baker, E., Warren, S.T., Schlessinger, D., Sutherland, G.R., Richards, R.I. (1991) Mapping of DNA instability at the fragile X to a trinucleotide repeat sequence p(CCG)n, *Science.* **252**, 1711–1714.
La Spada, A.R., Wilson, E.M., Lubahn, D.B., Harding, A.E., Fischbeck, K.H. (1991) Androgen receptor gene mutations in X-linked spinal and bulbar muscular atrophy, *Nature.* **352**, 77–79.
La Spada, A.R., Roling, D.B., Harding, A.E., Warner, C.L., Spiegel, R., Hausmanowa Petrusewicz, I., Yee, W.C., Fischbeck, K.H. (1992) Meiotic stability and genotype-phenotype correlation of the trinucleotide repeat in X-linked spinal and bulbar muscular atrophy, *Nat. Genet.* **2**, 301–304.
La Spada, A.R., Fu, Y., Sopher, B.L., Libby, R.T., Wang, X., Li, L.Y., Einum, D.D., Huang, J., Possin, D.E., Smith, A.C., et al. (2001) Polyglutamine-expanded ataxin-7 antagonizes CRX function and induces cone-rod dystrophy in a mouse model of SCA7, *Neuron.* **31**, 913–927.
Lafreniere, R.G., Rochefort, D.L., Chretien, N., Rommens, J.M., Cochius, J.I., Kalviainen, R., Nousiainen, U., Patry, G., Farrell, K., Soderfeldt, B. et al. (1997) Unstable insertion in the 5′ flanking region of the *cystatin B* gene is the most common mutation in progressive myoclonus epilepsy type 1, EPM1, *Nat. Genet.* **15**, 298–302.
Lalioti, M.D., Antonarakis, S.E., Scott, H.S. (2003) The epilepsy, the protease inhibitor and the dodecamer: progressive myoclonus epilepsy, cystatin b and a 12-mer repeat expansion, *Cytogenet. Genome Res.* **100**, 213–223.

Lalioti, M.D., Scott, H.S., Antonarakis, S.E. (1999) Altered spacing of promoter elements due to the dodecamer repeat expansion contributes to reduced expression of the *cystatin B* gene in EPM1, *Hum. Mol. Genet.* **8**, 1791–1798.

Lin, X., Antalffy, B., Kang, D., Orr, H.T., Zoghbi, H.Y. (2000) Polyglutamine expansion down-regulates specific neuronal genes before pathologic changes in SCA1, *Nat. Neurosci.* **3**, 157–163.

Liquori, C.L., Ricker, K., Moseley, M.L., Jacobsen, J.F., Kress, W., Naylor, S.L., Day, J.W., Ranum, L.P. (2001) Myotonic dystrophy type 2 caused by a CCTG expansion in intron 1 of ZNF9, *Science.* **293**, 864–867.

Loesch, D.Z., Huggins, R.M., Bui, Q.M., Epstein, J.L., Taylor, A.K., Hagerman, R.J. (2002) Effect of the deficits of fragile X mental retardation protein on cognitive status of fragile x males and females assessed by robust pedigree analysis, *J. Dev. Behav. Pediatr.* **23**, 416–423.

Lunkes, A., Lindenberg, K.S., Ben-Haiem, L., Weber, C., Devys, D., Landwehrmeyer, G.B., Mandel, J.L., Trottier, Y. (2002) Proteases acting on mutant huntingtin generate cleaved products that differentially build up cytoplasmic and nuclear inclusions, *Mol. Cell.* **10**, 259–269.

Mangiarini, L., Sathasivam, K., Seller, M., Cozens, B., Harper, A., Hetherington, C., Lawton, M., Trottier, Y., Lehrach, H., Davies, S.W., et al. (1996) Exon 1 of the *HD* gene with an expanded CAG repeat is sufficient to cause a progressive neurological phenotype in transgenic mice, *Cell.* **87**, 493–506.

Mankodi, A., Logigian, E., Callahan, L., McClain, C., White, R., Henderson, D., Krym, M., Thornton, C.A. (2000) Myotonic dystrophy in transgenic mice expressing an expanded CUG repeat, *Science.* **289**, 1769–1773.

Margolis, R.L., O'Hearn, E., Rosenblatt, A., Willour, V., Holmes, S.E., Franz, M.L., Callahan, C., Hwang, H.S., Troncoso, J.C., Ross, C.A. (2001) A disorder similar to Huntington's disease is associated with a novel CAG repeat expansion, *Ann. Neurol.* **50**, 373–380.

Matsuura, T., Fang, P., Lin, X., Khajavi, M., Tsuji, K., Rasmussen, A., Grewal, R.P., Achari, M., Alonso, M.E., Pulst, S.M., et al. (2004) Somatic and germline instability of the ATTCT repeat in spinocerebellar ataxia type 10, *Am. J. Hum. Genet.* **74**, 1216–1224.

Matsuura, T., Yamagata, T., Burgess, D.L., Rasmussen, A., Grewal, R.P., Watase, K., Khajavi, M., McCall, A.E., Davis, C.F., Zu, L., et al. (2000) Large expansion of the ATTCT pentanucleotide repeat in spinocerebellar ataxia type 10, *Nat. Genet.* **26**, 191–194.

McCampbell, A., Taylor, J.P., Taye, A.A., Robitschek, J., Li, M., Walcott, J., Merry, D., Chai, Y., Paulson, H., Sobue, G., et al. (2000) CREB-binding protein sequestration by expanded polyglutamine, *Hum. Mol. Genet.* **9**, 2197–2202.

Monckton, D.G., Cayuela, M.L., Gould, F.K., Brock, G.J., Silva, R., Ashizawa, T. (1999) Very large (CAG)(n) DNA repeat expansions in the sperm of two spinocerebellar ataxia type 7 males, *Hum. Mol. Genet.* **8**, 2473–2478.

Mortlock, D.P., Innis, J.W. (1997) Mutation of HOXA13 in hand-foot-genital syndrome, *Nat. Genet.* **15**, 179–180.

Mosemiller, A.K., Dalton, J.C., Day, J.W., Ranum, L.P. (2003) Molecular genetics of spinocerebellar ataxia type 8 (SCA8), *Cytogenet. Genome Res.* **100**, 175–183.

Mundlos, S., Otto, F., Mundlos, C., Mulliken, J.B., Aylsworth, A.S., Albright, S., Lindhout, D., Cole, W.G., Henn, W., Knoll, J.H., et al. (1997) Mutations involving the transcription factor CBFA1 cause cleidocranial dysplasia, *Cell.* **89**, 773–779.

Muragaki, Y., Mundlos, S., Upton, J., Olsen, B.R. (1996) Altered growth and branching patterns in synpolydactyly caused by mutations in HOXD13, *Science.* **272**, 548–551.

Mutsuddi, M., Marshall, C.M., Benzow, K.A., Koob, M.D., Rebay, I. (2004) The spinocerebellar ataxia 8 noncoding RNA causes neurodegeneration and associates with staufen in Drosophila, *Curr. Biol.* **14**, 302–308.

Nagafuchi, S., Yanagisawa, H., Sato, K., Shirayama, T., Ohsaki, E., Bundo, M., Takeda, T., Tadokoro, K., Kondo, I., Murayama, N., et al. (1994) Dentatorubral and pallidoluysian atrophy expansion of an unstable CAG trinucleotide on chromosome 12p, *Nat. Genet.* **6**, 14–18.

Nakamura, K., Jeong, S.Y., Uchihara, T., Anno, M., Nagashima, K., Nagashima, T., Ikeda, S., Tsuji, S., Kanazawa, I. (2001) SCA17, a novel autosomal dominant cerebellar ataxia caused by an expanded polyglutamine in TATA-binding protein, *Hum. Mol. Genet.* **10**, 1441–1448.

Okazawa, H., Rich, T., Chang, A., Lin, X., Waragai, M., Kajikawa, M., Enokido, Y., Komuro, A., Kato, S., Shibata, M., et al. (2002) Interaction between mutant ataxin-1 and PQBP-1 affects transcription and cell death, *Neuron*. **34**, 701–713.

Oostra, B.A., Willemsen, R. (2003) A fragile balance: FMR1 expression levels, *Hum. Mol. Genet.* **12**, **Spec No 2**, R249–R257.

Orr, H.T., Chung, M.Y., Banfi, S., Kwiatkowski, Jr., T.J., Servadio, A., Beaudet, A.L., McCall, A.E., Duvick, L.A., Ranum, L.P., Zoghbi, H.Y. (1993) Expansion of an unstable trinucleotide CAG repeat in spinocerebellar ataxia type 1, *Nat. Genet.* **4**, 221–226.

Paulson, H.L., Perez, M.K., Trottier, Y., Trojanowski, J.Q., Subramony, S.H., Das, S.S., Vig, P., Mandel, J.L., Fischbeck, K.H., Pittman, R.N. (1997) Intranuclear inclusions of expanded polyglutamine protein in spinocerebellar ataxia type 3, *Neuron*. **19**, 333–344.

Penrose, L.S. (1948) The problem of anticipation in pedigrees of dystrophica myotonica, *Ann. Eugen.* **14**, 125–132.

Perutz, M.F., Johnson, T., Suzuki, M., Finch, J.T. (1994) Glutamine repeats as polar zippers: their possible role in inherited neurodegenerative diseases, *Proc. Natl. Acad. Sci. U. S. A.* **91**, 5355–5358.

Poirier, M.A., Li, H., Macosko, J., Cai, S., Amzel, M., Ross, C.A. (2002) Huntingtin spheroids and protofibrils as precursors in polyglutamine fibrilization, *J. Biol. Chem.* **277**, 41032–41037.

Pulst, S.M., Nechiporuk, A., Nechiporuk, T., Gispert, S., Chen, X.N., Lopes-Cendes, I., Pearlman, S., Starkman, S., Orozco-Diaz, G., Lunkes, A., et al. (1996) Moderate expansion of a normally biallelic trinucleotide repeat in spinocerebellar ataxia type 2, *Nat. Genet.* **14**, 269–276.

Rasmussen, A., Matsuura, T., Ruano, L., Yescas, P., Ochoa, A., Ashizawa, T., Alonso, E. (2001) Clinical and genetic analysis of four Mexican families with spinocerebellar ataxia type 10, *Ann. Neurol.* **50**, 234–239.

Sanchez, I., Mahlke, C., Yuan, J. (2003) Pivotal role of oligomerization in expanded polyglutamine neurodegenerative disorders, *Nature*. **421**, 373–379.

Sanpei, K., Takano, H., Igarashi, S., Sato, T., Oyake, M., Sasaki, H., Wakisaka, A., Tashiro, K., Ishida, Y., Ikeuchi, T., et al. (1996) Identification of the *spinocerebellar ataxia type 2* gene using a direct identification of repeat expansion and cloning technique, DIRECT, *Nat. Genet.* **14**, 277–284.

Satterfield, T.F., Jackson, S.M., Pallanck, L.J. (2002) A Drosophila homolog of the polyglutamine disease gene *SCA2* is a dosage-sensitive regulator of actin filament formation, *Genetics*. **162**, 1687–1702.

Scherzinger, E., Lurz, R., Turmaine, M., Mangiarini, L., Hollenbach, B., Hasenbank, R., Bates, G.P., Davies, S.W., Lehrach, H., Wanker, E.E. (1997) Huntingtin-encoded polyglutamine expansions form amyloid-like protein aggregates in vitro and in vivo, *Cell*. **90**, 549–558.

Schilling, G., Wood, J.D., Duan, K., Slunt, H.H., Gonzales, V., Yamada, M., Cooper, J.K., Margolis, R.L., Jenkins, N.A., Copeland, N.G., et al. (1999) Nuclear accumulation of truncated atrophin-1 fragments in a transgenic mouse model of DRPLA, *Neuron*. **24**, 275–286.

Seznec, H., Simon, D., Monassier, L., Criqui-Filipe, P., Gansmuller, A., Rustin, P., Koenig, M., Puccio, H. (2004) Idebenone delays the onset of cardiac functional alteration without correction of Fe-S enzymes deficit in a mouse model for Friedreich ataxia, *Hum. Mol. Genet.* **13**, 1017–1024.

Sopher, B.L., Thomas, P.S., Jr., LaFevre-Bernt, M.A., Holm, I.E., Wilke, S.A., Ware, C.B., Jin, L.W., Libby, R.T., Ellerby, L.M., La Spada, A.R. (2004) Androgen receptor YAC transgenic mice recapitulate SBMA motor neuronopathy and implicate VEGF164 in the motor neuron degeneration, *Neuron*. **41**, 687–699.

Stevanin, G., Camuzat, A., Holmes, S.E., Julien, C., Sahloul, R., Dode, C., Hahn-Barma, V., Ross, C.A., Margolis, R.L., Durr, A., et al. (2002) CAG/CTG repeat expansions at the Huntington's disease-like 2 locus are rare in Huntington's disease patients, *Neurology*. **58**, 965–967.

Stromme, P., Mangelsdorf, M.E., Shaw, M.A., Lower, K.M., Lewis, S.M., Bruyere, H., Lutcherath, V., Gedeon, A.K., Wallace, R.H., Scheffer, I.E., et al. (2002) Mutations in the human ortholog of Aristaless cause X-linked mental retardation and epilepsy, *Nat. Genet.* **30**, 441–445.

Sutherland, G.R., Baker, E. (1992) Characterisation of a new rare fragile site easily confused with the fragile X, *Hum. Mol. Genet.* **1**, 111–113.

Uda, M., Ottolenghi, C., Crisponi, L., Garcia, J.E., Deiana, M., Kimber, W., Forabosco, A., Cao, A., Schlessinger, D., Pilia, G. (2004) Foxl2 disruption causes mouse ovarian failure by pervasive blockage of follicle development, *Hum. Mol. Genet.* **13**, 1171–1181.

Ueno, S., Kondoh, K., Kotani, Y., Komure, O., Kuno, S., Kawai, J., Hazama, F., Sano, A. (1995) Somatic mosaicism of CAG repeat in dentatorubral-pallidoluysian atrophy (DRPLA), *Hum. Mol. Genet.* **4**, 663–666.

Wallis, J., Williamson, R., Chamberlain, S. (1990) Identification of a hypervariable microsatellite polymorphism within D9S15 tightly linked to Friedreich's ataxia, *Hum. Genet.* **85**, 98–100.

Warren, S.T. (1997) Polyalanine expansion in synpolydactyly might result from unequal crossing-over of HOXD13, *Science.* **275**, 408–409.

Watase, K., Weeber, E.J., Xu, B., Antalffy, B., Yuva-Paylor, L., Hashimoto, K., Kano, M., Atkinson, R., Sun, Y., Armstrong, D.L., et al. (2002) A long CAG repeat in the mouse Sca1 locus replicates SCA1 features and reveals the impact of protein solubility on selective neurodegeneration, *Neuron.* **34**, 905–919.

Wellington, C.L., Ellerby, L.M., Gutekunst, C.A., Rogers, D., Warby, S., Graham, R.K., Loubser, O., van Raamsdonk, J., Singaraja, R., Yang, Y.Z., et al. (2002) Caspase cleavage of mutant huntingtin precedes neurodegeneration in Huntington's disease, *J. Neurosci.* **22**, 7862–7872.

Wellington, C.L., Ellerby, L.M., Hackam, A.S., Margolis, R.L., Trifiro, M.A., Singaraja, R., McCutcheon, K., Salvesen, G.S., Propp, S.S., Bromm, M., et al. (1998) Caspase cleavage of gene products associated with triplet expansion disorders generates truncated fragments containing the polyglutamine tract, *J. Biol. Chem.* **273**, 9158–9167.

Willemsen, R., Hoogeveen-Westerveld, M., Reis, S., Holstege, J., Severijnen, L.A., Nieuwenhuizen, I.M., Schrier, M., van Unen, L., Tassone, F., Hoogeveen, A.T., et al. (2003) The FMR1 CGG repeat mouse displays ubiquitin-positive intranuclear neuronal inclusions; implications for the cerebellar tremor/ataxia syndrome, *Hum. Mol. Genet.* **12**, 949–959.

Wirtschafter, J.D., Ferrington, D.A., McLoon, L.K. (2004) Continuous remodeling of adult extraocular muscles as an explanation for selective craniofacial vulnerability in oculopharyngeal muscular dystrophy, *J. Neuroophthalmol.* **24**, 62–67.

Yoo, S.Y., Pennesi, M.E., Weeber, E.J., Xu, B., Atkinson, R., Chen, S., Armstrong, D.L., Wu, S.M., Sweatt, J.D., Zoghbi, H.Y. (2003) SCA7 knockin mice model human SCA7 and reveal gradual accumulation of mutant ataxin-7 in neurons and abnormalities in short-term plasticity, *Neuron.* **37**, 383–401.

Yvert, G., Lindenberg, K.S., Picaud, S., Landwehrmeyer, G.B., Sahel, J.A., Mandel, J.L. (2000) Expanded polyglutamines induce neurodegeneration and trans-neuronal alterations in cerebellum and retina of SCA7 transgenic mice, *Hum. Mol. Genet.* **9**, 2491–2506.

Zakany, J., Duboule, D. (1996) Synpolydactyly in mice with a targeted deficiency in the HoxD complex, *Nature.* **384**, 69–71.

Zhang, S., Xu, L., Lee, J., Xu, T. (2002) Drosophila atrophin homolog functions as a transcriptional corepressor in multiple developmental processes, *Cell* **108**, 45–46.

Zhuchenko, O., Bailey, J., Bonnen, P., Ashizawa, T., Stockton, D.W., Amos, C., Dobyns, W.B., Subramony, S.H., Zoghbi, H.Y., Lee, C.C. (1997) Autosomal dominant cerebellar ataxia (SCA6) associated with small polyglutamine expansions in the alpha 1A-voltage-dependent calcium channel, *Nat. Genet.* **15**, 62–69.

Zuccato, C., Tartari, M., Crotti, A., Goffredo, D., Valenza, M., Conti, L., Cataudella, T., Leavitt, B.R., Hayden, M.R., Timmusk, T., et al. (2003) Huntingtin interacts with REST/NRSF to modulate the transcription of NRSE-controlled neuronal genes, *Nat. Genet.* **35**, 76–83.

19
Molecular Biology of Neural Tube Defects

Andrew J. Copp
Institute of Child Health, University College London, London, UK

Neurobiology. From Molecular Basis to Disease. Edited by Robert A. Meyers.

ISBN: 978-3-527-32293-0

Keywords

Anencephaly
Exposure of the interior of the brain to the outside, with absence of the skull vault, resulting from failure of neural tube closure in the future brain.

Closed spina bifida (or dysraphism)
Skin-covered spinal lesions in which the lowest portion of the spinal cord fails to become separated from adjacent tissues during embryonic development, leading to spinal cord tethering inside the vertebral canal.

Craniorachischisis
The most severe NTD in which most of the brain and the entire spinal cord remain open, as a result of failure of the initiating event of neurulation (Closure 1).

Exencephaly
An early developmental form of anencephaly, in which the persistently open cranial neural folds appear everted and "overgrown." Later in gestation, degeneration of the exposed neural tissue results in the "flat skull" appearance of anencephaly.

Hydrocephalus
Increase in cerebrospinal fluid pressure within the cavities of the brain, often associated with spina bifida.

Myelomeningocele
Alternative term for open spina bifida. The spinal cord is open (myelo-), and the meninges emerge from the defect in a cystic lesion (meningocele). An open spinal cord without a covering meningeal sac is called *myelocele*.

Neural plate
A layer of ectodermal cells on the back of the early embryo that undergoes folding and fusion to produce the neural tube, the precursor of the entire brain and spinal cord.

Neural tube defects
The group of birth defects comprising spina bifida, anencephaly, and craniorachischisis that result when the neural tube fails to complete closure during embryogenesis.

Neural tube
The tubular structure formed in the early embryo by folding and fusion of the neural plate in the dorsal midline to produce the precursor of the brain and spinal cord.

Neuropore
A transient "holelike" opening in the neural tube at which neural tube closure is undergoing completion.

Neurulation
The embryonic process by which the neural tube is formed. In higher vertebrates, neurulation comprises primary events, characterized by neural folding, and secondary neurulation in which the caudal neural tube forms by hollowing of a cord of cells to create a tubular structure.

Open spina bifida (spina bifida aperta)
Exposure of the interior of the spinal cord to the outside environment. This definition encompasses the pathological terms myelocele and myelomeningocele.

Spina bifida is usually grouped together with the closely related conditions anencephaly and craniorachischisis, to form the category of congenital malformations known as *neural tube defects* (NTDs). NTDs affect approximately 1 in every 1000 pregnancies, equating to a worldwide prevalence of around 120 000 pregnancies with NTDs per year. Hence, NTDs represent one of the commonest, most severe birth defects in humans. The term NTDs reflects the origin of the malformations. In normal development, the *neural plate*, the precursor of the brain and spinal cord in the early embryo, undergoes folding and fusion to form the *neural tube*. If neural tube closure fails in the developing spine, the result is open spina bifida (also called *myelomeningocele or myelocele*; Fig. 1(a)–(d)), a defect in which the interior of the spinal cord remains open to the outside, rather than being covered by the bony vertebrae and skin. The exposed nervous tissue degenerates before birth so that babies born with open spina bifida are often paralysed and lack sensation below the level of the lesion. They may exhibit urinary and faecal incontinence, and have associated defects such as *hydrocephalus* (Fig. 1(e)). If the neural tube fails to close in the embryonic brain, the resulting birth defect is initially described as *exencephaly* (Fig. 1(b)), in which the neural folds appear everted. Later in gestation, degeneration of the neural tissue converts the defect to *anencephaly*, in which the interior of the skull is exposed owing to loss of the open brain and failure of the skull roof to develop. Fetuses with anencephaly do not survive beyond birth, and are often stillborn or die during pregnancy. If the entire neural tube fails to close, from brain to low spine, this is referred to as *craniorachischisis* (Fig. 1(f)–(h)), a uniformly lethal condition. Closed spina bifida (also called *spina bifida occulta*) represents the mild end of the NTD spectrum. The lowest part of the spinal cord fails to separate from adjacent tissues during development, a defect that can produce weakness of the legs or urinary incontinence. Surgical intervention may be required in order to free the spinal cord within the vertebral canal, relieving pressure on the spinal nerve roots.

1 Neurulation: A Multiphase Embryonic Event

NTDs (neural tube defects) result from failure of neurulation, the embryonic event responsible for formation of the neural tube. Neurulation comprises a series of coordinated cellular behavior, beginning with differentiation of neurectoderm from lateral ectoderm. Concomitantly, the neural plate narrows medio-laterally and lengthens rostro-caudally by the process of convergent extension, with neuroepithelial and underlying mesodermal cells moving toward the midline and intercalating along the rostro-caudal axis. Subsequently, neural folds form at the edges of the neural

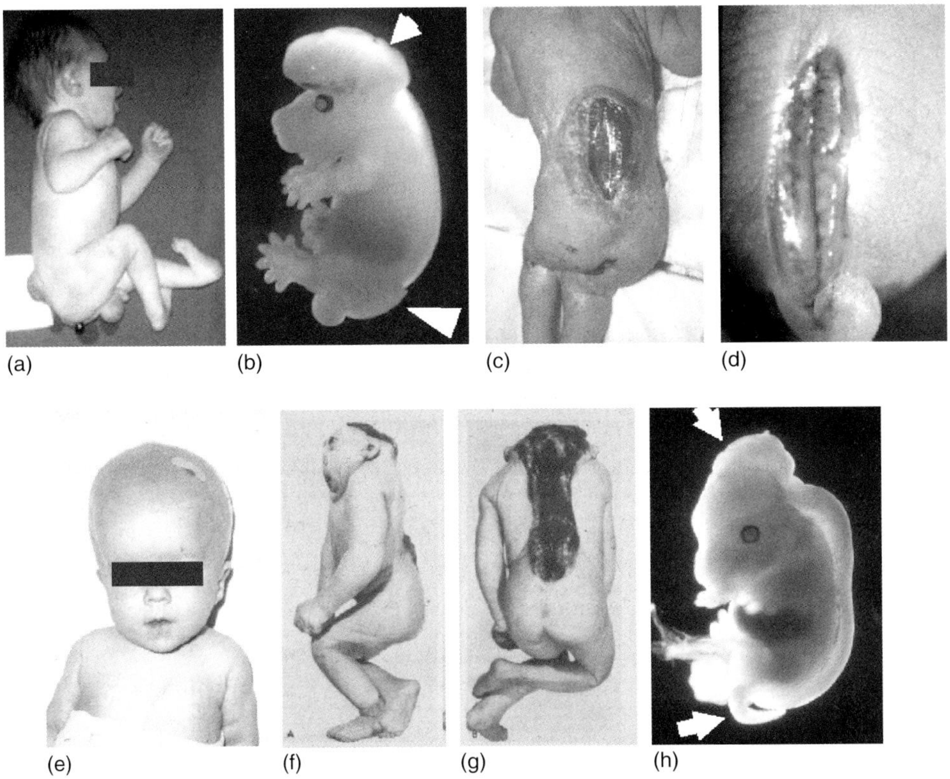

Fig. 1 The spectrum of NTDs in humans and mice. (a) Myelomeningocele in a human neonate; (b) exencephaly (arrow) and open spina bifida (arrowhead) in an E15.5 mouse embryo homozygous for the *curly tail* mutation; (c) myelocele in a human neonate; (d) open spina bifida in an E18.5 mouse fetus. Note the close similarity between these mouse and human NTDs; (e) hydrocephalus, with marked expansion of the cranial vault; (f, g) craniorachischisis in a human fetus from sideview (f) and back view (g). Most of the brain and spine are open; (h) craniorachis-chisis in an E15.5 mouse embryo homozygous for a mutation of *Celsr1*. The neural tube is open between the arrows, affecting most of the brain and the entire spine, as in the human defect. Parts (b, h) modified from Copp, A.J., Greene, N.D.E., Murdoch, J.N. (2003) *Nat. Rev. Genet.* **4**, 784–793; Part (c) from Copp, A.J. (1999) *Encyclopaedia of Life Sciences/www.els.net* Nature Publishing Group, London, UK; Part (d) from Stiefel, D., Shibata, T., Meuli, M., Duffy, P., Copp, A.J. (2003) *J. Neurosurg. (Spine)* **99**, 206–213.

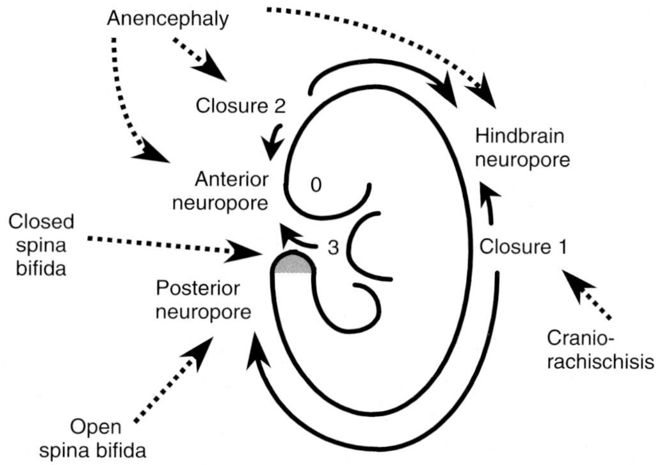

Fig. 2 Schematic summary of the successive initiation events of mouse neurulation (Closures 1, 2, and 3), the direction of spread of closure from the initiation sites (solid arrows), and the sites of completion of closure (anterior, hindbrain, and posterior neuropores). The tail bud region (red shading) is the site of secondary neurulation, which follows immediately after closure of the posterior neuropore. The main types of NTD that result from failure of the different components of neurulation are indicated by the dotted arrows. Modified from Copp, A.J., Bernfield, M. (1994) *Curr. Opin. Pediatr.* **6**, 624–631. (See color plate p. xl).

plate, which then elevate, with bending at precise dorso-ventral locations, until the tips of the neural folds come into contact in the dorsal midline. The apposed neural folds adhere and then fuse, with subsequent remodeling of the epithelia, to create the neural tube and covering surface ectoderm.

1.1 Rostro-caudal Sequence of Neurulation Events

In the mouse, neural tube closure is initiated at the hindbrain/cervical boundary (Closure 1), and then proceeds concurrently in the future brain and spinal regions (Fig. 2). Brain closure is initiated independently at the forebrain/midbrain boundary (Closure 2) and at the extreme rostral end of the forebrain (Closure 3). Closure between these initiation sites occurs simultaneously, culminating in the completion of cranial neurulation at the anterior and hindbrain neuropores. The spread of closure caudally along the spinal axis leads to progressive closure of the posterior neuropore, marking the end of primary neurulation.

While the neurulation process has been studied most closely in the mouse, human neurulation appears quite similar. Closures 1 and 3 certainly occur in human embryos, whereas there is disagreement between authors about the occurrence of an event comparable to mouse Closure 2. In fact, even in the mouse, Closure 2 is a variable event, with some inbred strains exhibiting the closure rather caudally, within the midbrain, whereas other strains undergo Closure 2 more rostrally, within the forebrain. In these latter strains, Closures 2 and 3 may be located extremely close together, making the two events difficult to distinguish from each other. This variant of mouse Closure 2 resembles the human situation, where a distinct Closure 2 event has been inconsistently observed.

2 Causation of Spina Bifida and Other NTDs

NTDs have a multifactorial aetiology, with both genetic and environmental influences playing an important role.

2.1 Genetic Factors

The genetic component of NTD etiology is revealed by the dramatically increased risk among women who have had a previous NTD pregnancy. Compared with the average background prevalence of 0.5 to 2 per 1000 pregnancies, the recurrence risk rises after one affected pregnancy to around 1 in 25 and after two affected pregnancies to about 1 in 10. These average figures relate to live born or therapeutically aborted NTD cases, and may underestimate the true recurrence rates. Significant numbers of NTDs are lost as spontaneous abortions early in pregnancy, suggesting that the true incidence, and perhaps also the recurrence rate, of NTDs may be higher than commonly quoted. The finding that the risk of an NTD pregnancy is also increased in first-degree relatives of an index case, with risk declining in second- and third-degree relatives is further evidence of a significant genetic component to the etiology of NTDs.

In the mouse, NTDs occur as a result of mutations in more than 80 different genes. This suggests that normal functioning of a large number of genes is necessary to ensure normal neural tube development. It predicts, moreover, that human NTDs are unlikely to result from the action of just a single gene, but ultimately may prove to be etiologically related to many genes. Hence, NTDs are probably heterogeneous both morphologically, in terms of the varying level of the body axis that is primarily affected, and also etiologically in terms of their precise genetic causation.

Despite the evidence for a strong genetic etiology, no major causative genes have yet been identified for human NTDs. The strongest link is to the gene encoding 5,10-methylene tetrahydrofolate reductase (MTHFR), an enzyme involved in intracellular folate metabolism. A polymorphic, thermolabile variant of MTHFR (caused by the C677T mutation) has a higher frequency among NTD cases and their families than among normal controls in Dutch and Irish populations. Although a link with NTDs is more difficult to demonstrate in other human populations, a meta-analysis of the worldwide data showed an overall significant association between MTHFR and NTDs. The C677T mutation of MTHFR appears to contribute to the overall risk of NTDs, particularly when combined with low vitamin B_{12} status or with homozygosity for the A66G variant of the enzyme methionine synthase reductase. Hence, genetic variations in folate metabolism may predispose to NTDs.

Several of the genes implicated in mouse NTDs have been examined in human NTD cases, but to date, only rare mutations have been identified. There are several possible reasons for this negative finding. First, most of the mouse NTD–causing mutations act only in homozygous form, while heterozygotes are usually normal. This contrasts with human genetic disease where haplo-insufficiency is the commonest cause of clinical phenotype. With a few exceptions, such as cystic fibrosis, homozygotes for specific genetic defects are usually rare and contribute little to the overall burden of genetically determined human disease. Second, many of the mouse NTD mutants are embryonic lethals. Given the extremely high level of pregnancy wastage in humans, it is probable that most disadvantaged human embryos fail to survive to late gestation. Third, it seems likely that gene–gene interactions, rather than single-gene effects, are most likely to play a major role in the etiology of human NTDs.

Gene–gene interactions have been well documented in the mouse. For example, while the *curly tail* (*ct*) mutation and the unrelated *splotch* (Sp^{2H}) mutation each cause spina bifida in homozygotes, mice doubly heterozygous for both mutations ($ct/+$; $Sp^{2H}/+$) also have NTDs. Embryos of genotype ct/ct; $Sp^{2H}/+$ exhibit particularly severe spina bifida. Translated to the human situation, this suggests that NTDs may result most commonly from the coming together in a single individual of two or more different genetic variants, which provides the interaction necessary to disturb neurulation. If many genes are required for successful neurulation, as in the mouse, a large number of possible combinations of genetic variants might be capable of producing NTDs. This could yield a "sporadic" occurrence of NTDs at a relatively high frequency, as is observed in all human populations.

2.2
Environmental Factors

Epidemiological studies point to a likely role for nongenetic factors, as well as genetic influences, in the etiology of NTDs. For example, the relatively high prevalence of defects among pregnancies in low socioeconomic groupings has led to the suggestion that nutritional factors may be of significance. This idea receives support from the finding of a preventive effect of folic acid supplementation on the occurrence and recurrence of NTDs.

As with the genetic aspects of NTD etiology, rather few environmental factors have been shown definitively to cause human NTDs. It is now widely accepted that exposure to the anticonvulsant sodium valproate during the early weeks of pregnancy, when the neural tube is closing, raises the risk of spina bifida to around 1.5 to 3%, a 15- to 30-fold increase in prevalence compared to background incidence. An elevated risk of NTDs has also been reported with carbamazepine, another anticonvulsant. Apart from these drug associations, the only other well-defined environmental association relates to the uterine milieu associated with maternal diabetes mellitus. The frequency of NTDs (like many other birth defects) is increased severalfold in pregnancies of diabetic mothers.

A variety of exogenous agents, when applied to rodent embryos, exert teratogenic effects with the production of abnormalities including NTDs. The majority of NTD-causing agents are drugs, chemicals, or other substances extraneous to pregnancy. For example, administration of sodium valproate either to the pregnant female or to the embryo *in vitro* produces NTDs although the principal defect produced in rodent embryos is exencephaly, not spina bifida as in humans. Other well-known NTD-producing teratogens include metals such as arsenic and cadmium, drugs including cytochalasins and alkaloids, and physical agents including radiation and hyperthermia. Apart from exogenous teratogenic agents, a number of molecules derived from metabolism cause NTDs, particularly in the context of maternal disease condition. Elevated concentrations of glucose and ketone bodies, as found in diabetes mellitus, increase the risk of NTDs, a finding that parallels the predisposition to NTDs in human diabetic pregnancy. Moreover, an excess or deficiency of retinol (vitamin A), and particularly its active metabolite retinoic acid, are also potent causes of NTDs. Retinoic acid is produced endogenously in the embryo and is essential for a variety of developmental processes. An imbalance in signaling via retinoic acid receptors leads to NTDs, among other defects.

As with gene–gene interactions, environmental influences on nervous system development seem most likely to operate in the context of varying genetic susceptibility. This idea has been particularly well illustrated by studies in which different inbred mouse strains, each representing a unique genetic background, were found to exhibit varying susceptibility to cranial NTDs caused by teratogenic agents, including valproic acid. The genes that mediate this strain-specific variation in NTD predisposition are not yet known, whereas the developmental basis of the strain variation has been traced to a polymorphism in the precise pattern of cranial neurulation. Inbred strains differ in the position along the rostro-caudal axis of the developing brain at which the Closure 2 closure event occurs. While the majority of strains exhibit Closure 2 at the midbrain/forebrain boundary, others have this event occurring more caudally within the midbrain (e.g. DBA/2 strain), or more rostrally within the forebrain (e.g. NZW strain). Experimental studies have shown that strains with a caudal Closure 2 point are relatively resistant to exencephaly (Fig. 3), whether caused by teratogenic or genetic influences, whereas strains with a rostral Closure 2 point are relatively prone to exencephaly. Hence, the precise position of Closure 2, as dictated by the genetic background, sets the risk level of a strain in its interaction with either teratogenic or genetic factors that can cause cranial NTDs.

Teratogenic agents can also interact directly with the single-gene effects to modulate the frequency or severity of NTDs. For example, the low natural frequency of exencephaly in the *curly tail* strain is elevated by the administration of all-trans-RA although, interestingly, the frequency of spina bifida in this strain is actually reduced by low-dose RA. This example demonstrates that environmental agents can exert sharply contrasting effects, either exacerbating or ameliorating, upon interaction with a genetic predisposition.

3
Developmental Mechanisms of Spina Bifida and Other NTDs

Of the more than 80 genes demonstrated to be necessary for successful neurulation, only a few have been studied in sufficient detail to reveal how the genetic requirement is mediated developmentally. Recent studies have led to an improved understanding of two aspects of neurulation: the initial closure event (Closure 1), whose failure leads to the severe NTD, craniorachischisis, and the progressive closure of the spinal neural tube, which yields spina bifida when defective. In both cases, specific signaling pathways have been shown to be critical for successful neurulation. More diverse cellular events appear crucial for cranial neural tube closure,

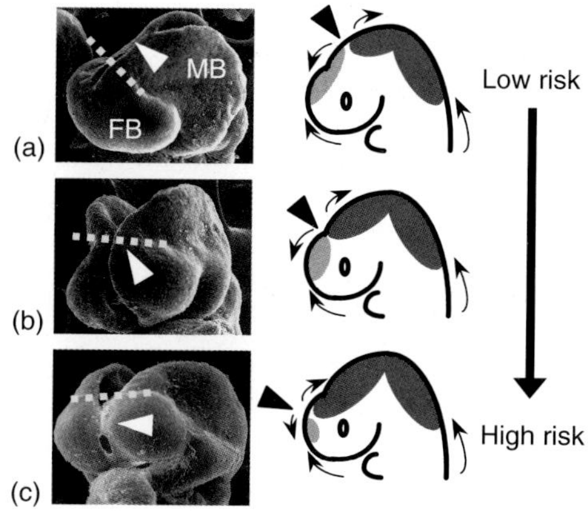

Fig. 3 Variations in position of Closure 2, in the future brain of different inbred mouse strains, in relation to the risk of exencephaly and anencephaly. Left panel shows scanning electron micrographs of embryonic brains from the left frontal view. Dotted lines indicate the forebrain/midbrain (FB/MB) boundary. Arrowheads point to the site of initiation of Closure, which is located caudal to the FB/MB boundary in the DBA/2 strain (a), at the FB/MB boundary in the CD1 strain (b), and rostral to the FB/MB boundary in the NZW strain (c). Middle panel summarizes diagrammatically this variation in Closure 2 morphology. Red shading indicates open neural folds caudal to the site of Closure 2, whereas green shading indicates the anterior neuropore, rostral to Closure 2. Arrows indicate direction of neural tube closure. Right panel shows an increasing risk of exencephaly correlated with a progressively more rostral position of Closure 2. The reason for this correlation is that the midbrain region, which lies at the apex of the cranial flexure, is mechanically the least favorable brain region for neural tube closure, and is aided in its closure by a caudal Closure 2 position (a) but hampered in its closure by a rostral Closure 2 position (c). Modified from Fleming, A., Copp, A.J. (2000) *Hum. Mol. Genet.* **9**, 575–581. (See color plate p. xl).

whose failure results in the development of anencephaly.

3.1
Initiation of Neurulation: Role of Planar Cell Polarity Signaling

The initial closure event (Closure 1) occurs when the neural folds fuse at the level of the future hindbrain/cervical boundary (Fig. 2). If Closure 1 fails, the embryo develops with an extensively open neural tube, from midbrain to low spine (craniorachischisis, Fig. 1(f)–(h)), although the forebrain is closed, as Closure 2 occurs normally in such embryos.

Recently, the cell movements of convergent extension, which occur from the stage of gastrulation onward, have been found to be essential for initiation of neural tube closure. Convergent extension requires signaling via the so-called planar cell polarity pathway, a noncanonical Wnt signaling pathway in which extracellular signals are transduced via cell-surface frizzled receptors and intracellular proteins of the dishevelled family (Fig. 4). Downstream events in this pathway include the activation of cytoskeleton-related proteins such as RhoA and the Jun-N-terminal kinases (JNKs).

In *Xenopus*, misexpression of dishevelled, or its interacting partner protein strabismus, prevents convergent extension, yielding embryos with an abnormally short, broad neural plate that fails to complete neural tube closure. This defect in *Xenopus* closely resembles the NTDs

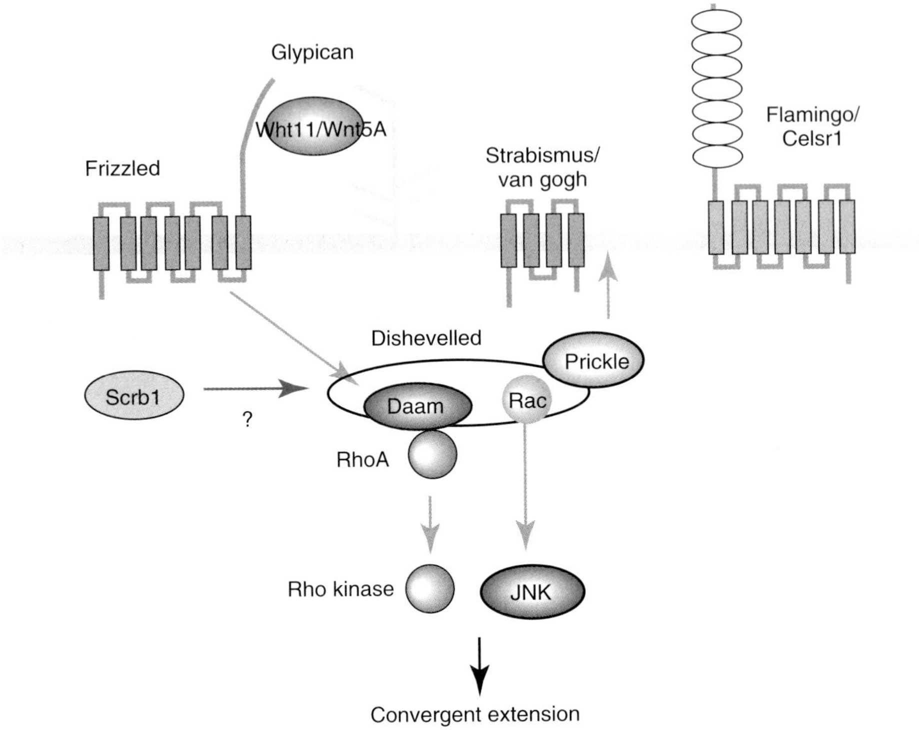

Fig. 4 The noncanonical Wnt (planar cell polarity) signaling pathway, which has been implicated in the regulation of convergent extension cell movements and initiation of neural tube closure (Closure 1). Mutations in the *strabismus/van gogh* (*Vangl2*), *Flamingo/Celsr1*, *Scrb1*, and *Dishevelled* genes all disturb Closure 1 in the mouse embryo, leading to the severe NTD craniorachischisis. Downstream signaling is thought to be via Rho kinase and/or jun-N-terminal (JNK) kinase. Modified from Ueno, N., Greene, N.D.E. (2003) *Birth Defects Res. (Part C)* **69**, 318–324.

in mouse embryos homozygous for the *loop-tail, circletail,* and *crash* mouse mutants, and in embryos lacking function of both *dishevelled1* and *dishevelled2* genes. In each of the mouse mutants, the neural plate is abnormally broad and has a non-bending region intervening between the neural folds, in contrast to the well-defined median hinge point of normal embryos. Although the neural folds elevate relatively normally, they are located too far apart to achieve closure (Fig. 5). Hence, convergent extension is required to establish a neural plate of width that is compatible with the medial bending essential for the initiation of closure.

The critical role of planar cell polarity in mammalian convergent extension and initiation of neurulation has become clear from recent genetic findings. All of the mouse mutants with failure of Closure 1 have proven to harbor mutations in components of the planar cell polarity pathway: *loop-tail* mice are mutant for the *strabismus/Vangl2* gene, *circletail* mice have mutations in *Scrb1*, and *crash* mice are mutant for *Celsr1*, a homologue of the *Drosophila* gene *flamingo*. It is striking,

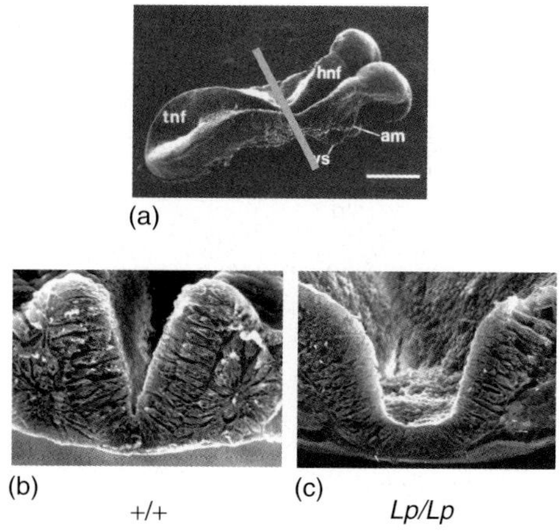

Fig. 5 Initiation of neurulation (Closure 1) in the mouse embryo, as studied by scanning electron microscopy. (a) Whole E8.5 embryo demonstrating the approaching neural folds approximately half way along the body axis, indicating the incipient Closure 1 event. This event occurs at the boundary of the hindbrain and cervical regions. (b, c) Sections transverse to the body axis as indicated by the red line in (a). The wild type embryo (b) shows "Mode 1" morphology with bending at a compact midline hinge point, and straight lateral neural folds. The homozygous *loop-tail* (*Lp/Lp*) embryo (c) shows a disturbance of the midline, which is broader than normal, without a compact bend in the neural plate. Although the neural folds elevate, laterally, they are not able to appose and fuse in the midline, leading to craniorachischisis in mutant embryos. Scale bar represents 0.25 mm (a) and 0.03 mm (b, c). Modified from Copp, A.J., Brook, F.A., Estibeiro, J.P., Shum, A.S.W., Cockroft, D.L. (1990) *Prog. Neurobiol.* **35**, 363–403; Greene, N.D.E., Gerrelli, D., Van Straaten, H.W.M., Copp, A.J. (1998) *Mech. Dev.* **73**, 59–72. (See color plate p. xli).

therefore, that the only group of mouse mutants that develop craniorachischisis all affect the same signaling pathway, suggesting a crucial role for the establishment of planar cell polarity in ensuring the successful initiation of neural tube closure.

3.2
Regulation of Neural Plate Bending: Mechanisms of Spina Bifida

In birds and mammals, bending of the neuroepithelium occurs at two specific sites in the neural plate: at the ventral midline (the median hinge point, MHP) and at paired dorsolateral bending sites near the junction of the neural plate and surface ectoderm (the dorsolateral hinge points, DLHPs). MHP bending creates the "neural groove," with a V-shaped cross-section, while DLHP bending creates longitudinal furrows that bring the lateral aspects of the neural plate toward each other in the dorsal midline.

Closure in the spinal region of the mouse embryo involves the progressive "zippering" of the neural tube down the spinal axis, until the posterior neuropore completes its closure at the future upper sacral level at E10. As the upper spine is forming, at embryonic day (E) 8.5, the closing neural folds exhibit an MHP but no DLHPs. By early E9.5, the neural plate begins to bend at DLHPs in addition to the MHP and, at the lowest spinal levels (at E10), the MHP disappears and the neural plate bends only at DLHPs. Hence, there is a complete switch from closure based on midline bending to closure dominated by dorsolateral bending during the course of mouse spinal neurulation (Fig. 6(a)–(d)).

Dorso-ventral molecular interactions have been implicated in the regulation of neural plate bending along the spinal region of the mouse embryo. A concentration-dependent influence of Shh, originating from the notochord and

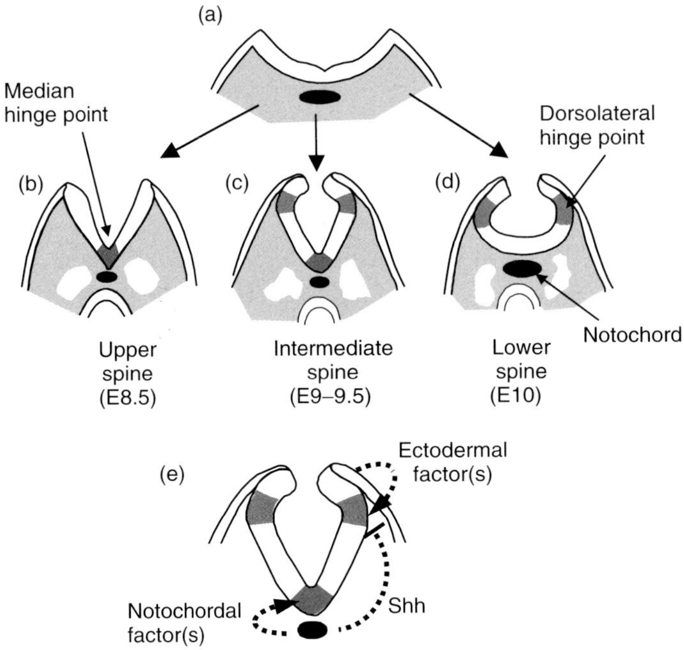

Fig. 6 Changing morphology of neurulation as neural tube closure progresses along the spinal region of the mouse embryo. The neural folds elevate from an initially flat morphology (a). In the upper spine (b), the neural plate bends only in the midline (Mode 1), at the median hinge point (MHP; see also Fig. 5(b)). In the intermediate spinal axis (c), both midline and dorsolateral bending are observed (Mode 2), whereas in the low spine (d), midline bending is lost and only dorsolateral hinge points (DLHPs) are observed (Mode 3). This changing pattern of neurulation is regulated by dorso-ventral patterning molecules (e). Notochordal factor(s) induce the MHP, while Shh, a secreted notochordal factor, inhibits DLHP formation. Factor(s) from the surface ectoderm induce DHLPs. As neurulation progresses along the spinal axis, the strength of notochordal signaling diminishes, with consequent loss of MHP-mediated bending and the emergence of DLHP-mediated neural tube closure. Modified from Ybot-Gonzalez, P., Cogram, P., Gerrelli, D., Copp, A.J. (2002) *Development* **129**, 2507–2517.

floor plate, patterns neuronal differentiation events along the dorso-ventral axis of the neural tube. BMPs (bone morphogenetic proteins) and Wnts, from the dorsal ectoderm, oppose Shh influence and promote dorsal patterning events. The MHP is induced by signals from the notochord, as demonstrated by experiments in which notochordal development was suppressed, with prevention of MHP formation. Shh, emanating from the notochord, may participate in the induction of the MHP, although midline bending is not completely abolished in Shh null embryos, suggesting a role for other, as yet unknown, notochordal inductive factors. Interestingly, although suppression of the notochord prevents MHP formation, the neural tube still closes, suggesting the existence of a "default" mechanism of neural tube closure that operates in the absence of Shh influence (Fig. 6(e)).

An explanation for this observation has come from experimental studies with cultured mouse embryos, which show that Shh emanating from the notochord serves to inhibit DLHP formation in the upper spine of normal embryos. This inhibitory Shh influence counteracts a positive effect of the surface ectoderm that induces bending of the dorsal neural plate; in the absence of surface ectoderm, DLHPs are not formed. BMPs are possible mediators of this ectodermal influence, although this is speculative at present. As the wave of spinal neurulation passes down the body axis, the strength of Shh signaling from the underlying notochord lessens, enabling DLHP formation to "break through" at lower spinal levels. DLHP formation in the lower spine can be inhibited by local release of Shh peptide, demonstrating the negative regulation of dorsolateral bending by Shh. In the absence of Shh, as in notochordless embryos or in embryos genetically null for Shh, DLHP formation is de-inhibited and ensures closure of the neural tube at all levels of the body axis.

Hence, the precise pattern of bending of the neural plate during spinal neurulation is dictated by mutually inhibitory signals from the notochord and surface ectoderm. This conclusion is supported by the finding that mutations that increase the strength of Shh signaling (e.g. knockout of the *Patched1* gene) inhibit neural tube closure, presumably owing to an increased strength of Shh-mediated inhibition of DLHP formation, which is incompatible with the completion of closure of the posterior neuropore.

3.3 Molecular Regulation of Cranial Neurulation

Cranial closure is more susceptible to disturbance than spinal closure in the mouse, as evidenced by the increased occurrence of exencephaly compared to spina bifida among knockouts and mutants with NTD phenotypes. Similarly, more teratogenic agents induce mouse cranial NTDs than spinal NTD. Studies of gene knockout mice with cranial NTDs suggest several cell and molecular processes that appear crucial for the successful closure of the cranial neural folds. These include expansion of the cranial mesenchyme, function of the actin cytoskeleton, emigration of the cranial neural crest, and regulated balance between cell proliferation and cell death (Fig. 7). Knockout mice in which these events are disrupted have cranial but not spinal NTDs, suggesting that spinal neurulation is relatively independent of these mechanisms.

3.3.1 Cranial Mesenchyme

As the cranial neural folds elevate, there is a marked expansion of the cranial mesenchyme, with cell proliferation and a striking increase in extracellular space. This causes the neural folds to take on a bulging (biconvex) appearance, particularly in the midbrain. Mice with loss of function of the *Twist* or *Cart1* genes have cranial NTDs in which the principal defect appears to be a reduction in the proliferation and expansion of the cranial mesenchyme. Both genes are expressed in the mesenchyme, and chimeric analysis in the *Twist* mutant demonstrates that exencephaly develops only if there is a high proportion of *Twist*$^{-/-}$ cells in the cranial mesenchyme.

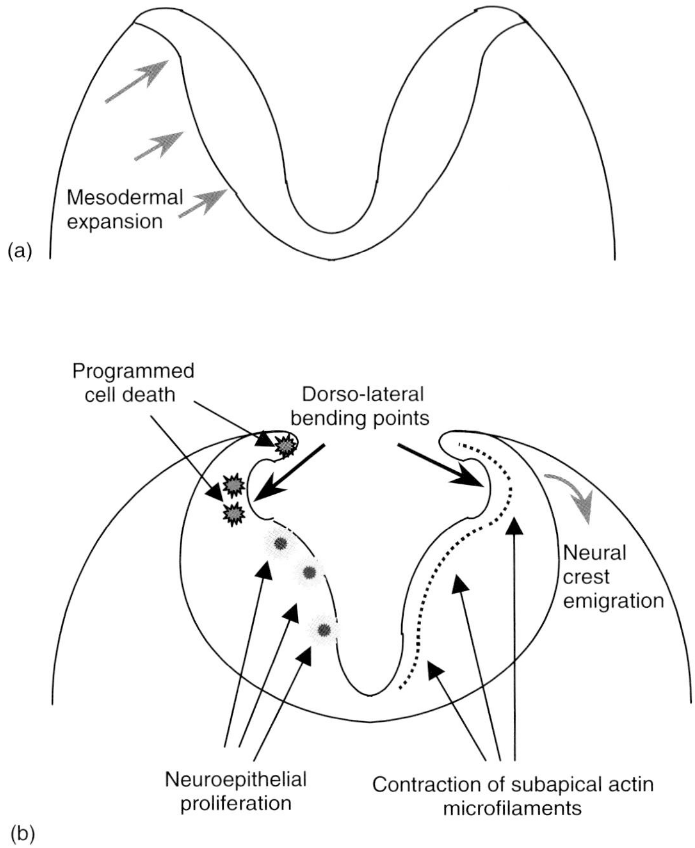

Fig. 7 Summary of the cellular and morphogenetic processes that are required for successful neural tube closure in the future brain. Cranial neural folds in the midbrain of the mouse embryo elevate initially with a biconvex morphology (a). Subsequently, dorsolateral bending occurs, causing the tips of the neural folds to converge on the midline (b), ensuring fusion and completion of cranial neurulation. Mesodermal expansion appears to be the most important morphogenetic factor in the initial elevation of the neural folds (a). However, the later stages of cranial neurulation require a number of additional processes including a functional actin cytoskeleton, successful initiation of neural crest emigration, precisely regulated programmed cell death and continuation of regulated neuroepithelial cell proliferation, with postponement of neuronal differentiation until after closure is complete (b). Defects in each of these processes are seen in one or more of the 80 gene knockout mice with NTDs. In each case, cranial neural tube closure fails and embryos develop exencephaly/anencephaly. Modified from Copp, A.J., Greene, N.D.E., Murdoch, J.N. (2003) *Nat. Rev. Genet.* **4**, 784–793.

3.3.2 Actin Cytoskeleton

In the second phase of cranial neural tube closure, the edges of the biconvex neural folds "flip around" and adopt a biconcave morphology, approaching the dorsal midline where fusion occurs to complete brain closure. This second neurulation phase is highly dependent on the actin cytoskeleton. Mice lacking the filamentous actin-associated protein shroom develop exencephaly, revealing a dependence of cranial neurulation on the actin microfilaments. This finding is consistent with *in vitro* studies in which rodent embryos are treated with cytochalasins, drugs that disassemble actin microfilaments. Cranial neural tube closure is inhibited by cytochalasins, whereas spinal neurulation proceeds to completion in the presence of these drugs. Hence, both genetic and nongenetic interventions demonstrate that the requirement for actin microfilaments is specific to cranial neural tube closure.

3.3.3 Neural Crest Emigration

The transition from biconvex to biconcave neural plate morphology also seems dependent on the initiation of cranial neural crest migration. In the midbrain and hindbrain, neural crest cells detach from the apices of the neural folds and start migration well in advance of neural tube closure, in contrast to the spinal region where neural crest emigration follows completion of neural tube closure. Mice overexpressing the gap junction protein connexin 43 exhibit marked cranial neural crest migration defects, and also develop exencephaly. Similarly, neural crest emigration and dorsolateral bending of the cranial neural folds are delayed in cultured rat embryos treated with chondroitinase ABC, to digest preexisting chondroitin sulphate. These findings suggest that neural crest emigration is a prerequisite for cranial neural tube closure, perhaps by increasing the flexibility of the lateral neural plate.

3.3.4 Programmed Cell Death

There is growing evidence to implicate a precise regulation of programmed cell death (apoptosis) in cranial neurulation. Dying cells occur within the dorsolateral regions of the cranial neural plate and at the tips of the fusing neural folds. Moreover, studies of knockout mice demonstrate that either an increase or decrease in the amount of this apoptosis is detrimental for cranial neurulation. Exencephaly develops in knockouts such as the *Apaf-1*, *caspase 9*, and *p53* genes, which have reduced apoptotic cell death, and in knockouts of *AP-2*, *bcl10*, and *Tulp1*, which show increased apoptosis in the cranial neural folds. The precise role played by cell death in cranial neurulation is not clear. Dorsolateral cell death may synergize with neural crest cell emigration, to "loosen" the cranial neuroepithelium, thereby allowing the conversion from a biconvex to biconcave neural fold morphology. Additionally, apoptosis at the tips of the neural folds may play a quite different role. Once the apposing neural folds have contacted and adhered, midline epithelial remodeling is needed to break the continuity between neuroepithelium and surface ectoderm on each side, enabling the neural tube roof and its covering ectoderm to be formed at the midline. Inhibition of apoptosis using the peptide Zvad-fmk produces spinal NTDs in the chick embryo, probably by preventing this dorsal midline remodeling. Hence, apoptosis may play a dual role in cranial neurulation.

3.3.5 Neuronal Differentiation

The onset of neuronal differentiation occurs only after neural tube closure is complete, during normal development, and this temporal sequence appears essential for successful cranial neurulation. Gene knockouts such as *RBP-Jκ, Hes1,* and *Numb,* which affect the Notch and related signaling pathways, exhibit failure of cranial neurulation in association with the premature onset of neuronal differentiation in the neural plate. The early appearance of differentiating neuroepithelial cells in these mutants might interfere with neurulation by rendering the neural plate mechanically inflexible, by interfering with the release of neural crest cells, or by inhibiting adhesion of the neural folds.

3.3.6 **Cell Proliferation**

Excessive cell proliferation has been implicated in some NTD mutants, such as mice with overexpression of Notch3, which produces increased numbers of neural progenitor cells in association with exencephaly. Neuroepithelial "overgrowth" has been suggested in relation to the exencephalic brains of other mutants but, when analyzed in detail, a normal or even lengthened cell cycle has generally been observed in exencephalic brains. It remains to be determined whether a disturbance of cell proliferation *per se* is a common cause of cranial NTDs.

4 Prevention of NTDs by Folic Acid and Other Vitamins

Much interest was stimulated by the finding that supplementation with folic acid can prevent a proportion of NTDs, estimated at around 70%, in the UK Medical Research Council Vitamin Research Group randomized clinical trial of NTD recurrence. Folic acid prevented 40 to 85% in a nonrandomized supplementation study in China, and there were no cases of NTD among 2394 pregnancies that received a folic acid–containing multivitamin supplement, compared with six out of 2310 unsupplemented pregnancies, in a randomized controlled trial of NTD occurrence in Hungary. Fortification of bread flour with folic acid in the United States has been associated with a 19% reduction in the prevalence of NTD. The variation in the proportion of cases prevented by folic acid in the various studies may relate to the differing doses of folic acid used. Indeed, NTD risk appears related to the log (serum folate level) in humans. Nevertheless, reports of successive NTD pregnancies in a single family despite high-dose folate intake suggest that a proportion of NTD cases may be resistant to folate, perhaps reflecting heterogeneity of etiology, with some subtypes of NTD being sensitive to folic acid whereas others are folate resistant.

4.1 Animal Models of Folate-preventable NTDs

Supporting evidence for the existence of folic acid–sensitive and folic acid–resistant NTDs comes from studies of mouse genetic mutant strains with NTDs. In some mutants, NTDs can be prevented by folic acid (e.g. *splotch, crooked tail, Cart1, Cited2*). Moreover, exencephaly has been reported in mice lacking the folate transport protein encoded by the gene *Folbp1*. This gene knockout causes lethality around the time of neural tube closure, hampering an assessment of whether *Folbp1* is essential for neural tube closure. However, treatment of the pregnant dams with folic acid rescues survival in these mice, and a proportion of the surviving homozygous knockouts exhibit exencephaly later in gestation. It is interesting that spina bifida is not part of the *Folbp1* knockout phenotype, even though folic acid appears to prevent both spina bifida and anencephaly among human NTDs. Moreover, analysis of the folate receptors α and β have revealed no association with the NTD phenotype in a series of human pregnancies.

Other genes that are involved in folate transport and metabolism have not yet been associated with neural tube closure defects in mice. For example, MTHFR, which has been implicated as a risk factor in human NTDs (see Section 4.1), does not cause NTDs when inactivated in mice.

4.2 Folate-resistant NTDs and Inositol as an Adjunct Therapy

While there is ample evidence from experimental studies to support a preventive effect of folic acid on NTDs, some mouse NTD models have proven insensitive to folic acid therapy, for example, the *curly tail, axial defects,* and *ephrin-A5* mutants. *Curly tail* is the best-characterized mouse model of folate-resistant NTD, with information available on the embryonic mechanism of neural tube closure failure. Recently, *Ghl3* was identified as a strong candidate for the gene mutated in *curly tail.* The spinal NTDs in this mutant are not altered in frequency by administration of folic acid, folinic acid (5-formyl tetrahydrofolate), Pregnavite Forte F (a multivitamin preparation active in preventing human NTDs), or methionine. Interestingly, however, a significant reduction in frequency of spinal NTD, with a particular reduction in the incidence of the most severe defect lumbosacral spina bifida, is seen after maternal administration of *myo*-inositol and *D-chiro*-inositol. *Curly tail* embryos are also at increased risk of developing exencephaly under conditions of inositol deficiency, compared to nonmutant embryos.

The preventive effect of inositol on spinal NTD in *curly tail* mice is mediated via intracellular signaling through the inositol phospholipid cycle, by stimulation of protein kinase C (PKC) activity. A role for PKC is demonstrated by the ability of the PKC agonist, 12-O-tetradecanoylphorbol-13-acetate (TPA), to substitute for *myo*-inositol in preventing the NTD. On the other hand, the broad spectrum inhibitor of PKC action, GF109203X, blocks prevention both by *myo*-inositol and TPA. Recent studies have identified PKCβI, γ or ζ, but not other isoforms of PKC, as essential for mediating the preventive effect of inositol on NTDs. The PKCβI isoform is required for stimulation of cell proliferation in the hindgut endoderm of the caudal region of *curly tail* embryos, which counteracts the genetically determined cell proliferation defect, and rescues the phenotype.

Hence, the pathway that mediates the preventive action of *myo*-inositol on NTD in the *curly tail* mouse is becoming understood. The finding of inositol as an alternative to folic acid for prevention of mouse NTDs raises the possibility of a novel future adjunct therapy to folic acid for use in human pregnancy.

5 Conclusions

Spina bifida is a member of the group of congenital malformations known as NTDs. A combination of studies of humans and animal models is contributing to our increasing knowledge of the etiology of NTDs, with identification of key genetic and nongenetic influences, and a growing appreciation of the types of interaction that promote or ameliorate the risk of NTDs. Study of mouse models is shedding light on the developmental mechanisms that underlie the different categories of NTDs, with marked variations in mechanism along the body axis. Primary prevention of NTDs is now a reality with introduction of folic acid, which has a preventive effect on a proportion of NTD cases. Folate-resistant NTDs require other strategies for prevention, and inositol appears promising in studies of one-mouse model.

See also Brain Development.

Bibliography

Books and Reviews

Bock, G., Marsh, J. (Eds.) (1994) *Neural Tube Defects*, Ciba Foundation Symposium 181. John Wiley & Sons, Chichester, UK.

Copp, A.J., Brook, F.A., Estibeiro, J.P., Shum, A.S.W., Cockroft, D.L. (1990) The embryonic development of mammalian neural tube defects, *Prog. Neurobiol.* **35**, 363–403.

Copp, A.J., Greene, N.D.E., Murdoch, J.N. (2003) The genetic basis of mammalian neurulation, *Nat. Rev. Genet.* **4**, 784–793.

Harding, B.N., Copp, A.J. (2002) Congenital Malformations, in: Graham, D.l., Lantos, P.L. (Eds.) *Greenfield's Neuropathology*, 7th edition, Arnold, London, pp. 357–483.

Schoenwolf, G.C., Smith, J.L. (1990) Mechanisms of neurulation: Traditional viewpoint and recent advances, *Development* **109**, 243–270.

Primary Literature

Barber, R.C., Shaw, G.M., Lammer, E.J., Greer, K.A., Biela, T.A., Lacey, S.W., Wasserman, C.R., Finnell, R.H. (1998) Lack of association between mutations in the folate receptor-α gene and spina bifida, *Am. J. Med. Genet.* **76**, 310–317.

Berry, R.J., Li, Z., Erickson, J.D., Li, S., Moore, C.A., Wang, H., Mulinare, J., Zhao, P., Wong, L.Y.C., Gindler, J., Hong, S.X., Correa, A., China-US Collaborative Project Neu. (1999) Prevention of neural-tube defects with folic acid in China, *N. Engl. J. Med.* **341**, 1485–1490.

Botto, L.D., Moore, C.A., Khoury, M.J., Erickson, J.D. (1999) Neural-tube defects, *N. Engl. J. Med.* **341**, 1509–1519.

Carter, C.O. (1974) Clues to the aetiology of neural tube malformations, *Dev. Med. Child Neurol.* **16**(()Suppl. 32), 3–15.

Cavalli, P., Copp, A.J. (2002) Inositol and folate-resistant neural tube defects, *J. Med. Genet.* **39**, e5.

Chen, W.-H., Morriss-Kay, G.M., Copp, A.J. (1994) Prevention of spinal neural tube defects in the curly tail mouse mutant by a specific effect of retinoic acid, *Dev. Dyn.* **199**, 93–102.

Chen, Z.T., Karaplis, A.C., Ackerman, S.L., Pogribny, I.P., Melnyk, S., Lussier-Cacan, S., Chen, M.F., Pai, A., John, S.W.M., Smith, R.S., Bottiglieri, T., Bagley, P., Selhub, J., Rudnicki, M.A., James, S.J., Rozen, R. (2001) Mice deficient in methylenetetrahydrofolate reductase exhibit hyperhomocysteinemia and decreased methylation capacity, with neuropathology and aortic lipid deposition, *Hum. Mol. Genet.* **10**, 433–443.

Chen, Z.-F., Behringer, R.R. (1995) *Twist* is required in head mesenchyme for cranial neural tube morphogenesis, *Genes Dev.* **9**, 686–699.

Cockroft, D.L., Brook, F.A., Copp, A.J. (1992) Inositol deficiency increases the susceptibility to neural tube defects of genetically predisposed (curly tail) mouse embryos in vitro, *Teratology* **45**, 223–232.

Cogram, P., Hynes, A., Dunlevy, L.P.E., Greene, N.D.E., Copp, A.J. (2004) Specific isoforms of protein kinase C are essential for prevention of folate-resistant neural tube defects by inositol, *Hum. Mol. Genet.* **13**, 7–14.

Cogram, P., Tesh, S., Tesh, J., Wade, A., Allan, G., Greene, N.D.E., Copp, A.J. (2002) D-*chiro*-inositol is more effective than *myo*-inositol in preventing folate-resistant mouse neural tube defects, *Hum. Reprod.* **17**, 2451–2458.

Copp, A.J., Greene, N.D.E., Murdoch, J.N. (2003) Dishevelled: linking convergent extension with neural tube closure, *Trends Neurosci.* **26**, 453–455.

Creasy, M.R., Alberman, E.D. (1976) Congenital malformations of the central nervous system in spontaneous abortions, *J. Med. Genet.* **13**, 9–16.

Curtin, J.A., Quint, E., Tsipouri, V., Arkell, R.M., Cattanach, B., Copp, A.J., Fisher, E.M., Nolan, P.M., Steel, K.P., Brown, S.D.M., Gray, I.C., Murdoch, J.N. (2003) Mutation of *Celsr1* disrupts planar polarity of inner ear hair cells and causes severe neural tube defects in the mouse, *Curr. Biol.* **13**, 1–20.

Czeizel, A.E., Dudás, I. (1992) Prevention of the first occurrence of neural-tube defects by periconceptional vitamin supplementation, *N. Engl. J. Med.* **327**, 1832–1835.

Erickson, C.A., Weston, J.A. (1983) AN SEM analysis of neural crest migration in the mouse, *J. Embryol. Exp. Morphol.* **74**, 97–118.

Estibeiro, J.P., Brook, F.A., Copp, A.J. (1993) Interaction between splotch (*Sp*) and curly tail (*ct*) mouse mutants in the embryonic development of neural tube defects, *Development* **119**, 113–121.

Finnell, R.H., Spiegelstein, O., Wlodarczyk, B., Triplett, A., Pogribny, I.P., Melnyk, S., James, J.S. (2002) DNA methylation in Folbp1 knockout mice supplemented with folic acid during gestation, *J. Nutr.* **132**, 2457S–2461S.

Fleming, A., Copp, A.J. (2000) A genetic risk factor for mouse neural tube defects: defining the embryonic basis, *Hum. Mol. Genet.* **9**, 575–581.

Greene, N.D.E., Copp, A.J. (1997) Inositol prevents folate-resistant neural tube defects in the mouse, *Nat. Med.* **3**, 60–66.

Greene, N.D.E., Copp, A.J. (2002) Animal Models of Folate-Related Neural Tube Defects, in: Massaro, E.J., Rogers, J.M. (Eds.) *Folate and Human Development*, Humana Press, Totowa, NJ, pp. 1–26.

Greene, N.D.E., Gerrelli, D., Van Straaten, H.W.M., Copp, A.J. (1998) Abnormalities of floor plate, notochord and somite differentiation in the *loop-tail* (*Lp*) mouse: a model of severe neural tube defects, *Mech. Dev.* **73**, 59–72.

Hall, J.L., Harris, M.J., Juriloff, D.M. (1997) Effect of multifactorial genetic liability to exencephaly on the teratogenic effect of valproic acid in mice, *Teratology* **55**, 306–313.

Hildebrand, J.D., Soriano, P. (1999) Shroom, a PDZ domain-containing actin-binding protein, is required for neural tube morphogenesis in mice, *Cell* **99**, 485–497.

Holmes, L.B., Harvey, E.A., Coull, B.A., Huntington, K.B., Khoshbin, S., Hayes, A.M., Ryan, L.M. (2001) The teratogenicity of anticonvulsant drugs, *N. Engl. J. Med.* **344**, 1132–1138.

Honein, M.A., Paulozzi, L.J., Mathews, T.J., Erickson, J.D., Wong, L.Y.C. (2001) Impact of folic acid fortification of the US food supply on the occurrence of neural tube defects, *JAMA* **285**, 2981–2986.

Jessell, T.M. (2000) Neuronal specification in the spinal cord: inductive signals and transcriptional codes, *Nat. Rev. Genet.* **1**, 20–29.

Juriloff, D.M., Harris, M.J., Tom, C., MacDonald, K.B. (1991) Normal mouse strains differ in the site of initiation of closure of the cranial neural tube, *Teratology* **44**, 225–233.

Kucera, J. (1971) Rate and type of congenital anomalies among offspring of diabetic women, *J. Reprod. Med.* **7**, 61–70.

Lardelli, M., Williams, R., Mitsiadis, T., Lendahl, U. (1996) Expression of the Notch 3 intracellular domain in mouse central nervous system progenitor cells is lethal and leads to disturbed neural tube development, *Mech. Dev.* **59**, 177–190.

Lee, K.J., Jessell, T.M. (1999) The specification of dorsal cell fates in the vertebrate central nervous system, *Annu. Rev. Neurosci.* **22**, 261–294.

Morriss, G.M., Solursh, M. (1978) Regional differences in mesenchymal cell morphology and glycosaminoglycans in early neural-fold stage rat embryos, *J. Embryol. Exp. Morphol.* **46**, 37–52.

Morriss-Kay, G., Tan, S.-S. (1987) Mapping cranial neural crest cell migration pathways in mammalian embryos, *Trends Genet.* **3**, 257–261.

Morriss-Kay, G.M. (1981) Growth and development of pattern in the cranial neural epithelium of rat embryos during neurulation, *J. Embryol. Exp. Morphol.* **65**(Suppl.), 225–241.

Morriss-Kay, G.M., Tuckett, F. (1985) The role of microfilaments in cranial neurulation in rat embryos: effects of short-term exposure to cytochalasin D, *J. Embryol. Exp. Morphol.* **88**, 333–348.

Murdoch, J.N., Doudney, K., Paternotte, C., Copp, A.J., Stanier, P. (2001) Severe neural tube defects in the *loop-tail* mouse result from mutation of *Lpp1*, a novel gene involved in floor plate specification, *Hum. Mol. Genet.* **10**, 2593–2601.

Murdoch, J.N., Henderson, D.J., Doudney, K., Gaston-Massuet, C., Phillips, H.M., Paternotte, C., Arkell, R., Stanier, P., Copp, A.J. (2003) Disruption of *scribble* (*Scrb1*) causes severe neural tube defects in the *circletail* mouse, *Hum. Mol. Genet.* **12**, 87–98.

Nakatsu, T., Uwabe, C., Shiota, K. (2000) Neural tube closure in humans initiates at multiple sites: evidence from human embryos and implications for the pathogenesis of neural tube defects, *Anat. Embryol.* **201**, 455–466.

O'Rahilly, R., Müller, F. (2002) The two sites of fusion of the neural folds and the two neuropores in the human embryo, *Teratology* **65**, 162–170.

Piedrahita, J.A., Oetama, B., Bennett, G.D., Van Waes, J., Kamen, B.A., Richardson, J., Lacey, S.W., Anderson, R.G.W., Finnell, R.H. (1999) Mice lacking the folic acid-binding protein Folbp1 are defective in early embryonic development, *Nat. Genet.* **23**, 228–232.

Sadler, T.W. (1980) Effects of maternal diabetes on early embryogenesis: II. Hyperglycemia-induced exencephaly, *Teratology* **21**, 349–356.

Schluter, G. (1973) Ultrastructural observations on cell necrosis during formation of the neural tube in mouse embryos, *Z. Anat. Entwickl.-Gesch.* **141**, 251–264.

Seller, M.J. (1981) Recurrence risks for neural tube defects in a genetic counselling clinic population, *J. Med. Genet.* **18**, 245–248.

Seller, M.J., Embury, S., Polani, P.E., Adinolfi, M. (1979) Neural tube defects in curly-tail mice. II. Effect of maternal administration of vitamin A., *Proc. R. Soc. Lond. B* **206**, 95–107.

Shenefelt, R.E. (1972) Morphogenesis of malformations in hamster caused by retinoic acid: relation to dose and stage at treatment, *Teratology* **5**, 103–108.

Shum, A.S.W., Copp, A.J. (1996) Regional differences in morphogenesis of the neuroepithelium suggest multiple mechanisms of spinal neurulation in the mouse, *Anat. Embryol.* **194**, 65–73.

Ting, S.B., Wilanowski, T., Auden, A., Hall, M., Voss, A.K., Thomas, T., Parekh, V., Cunningham, J.M., Jane, S.M. (2003) Inositol and folate-resistant neural tube defects in mice lacking the epithelial-specific factor Grhl-3, *Nat. Med.* **9**, 1513–1519.

Van der Put, N.M.J., Eskes, T.K.A.B., Blom, H.J. (1997) Is the common 677C→T mutation in the methylenetetrahydrofolate reductase gene a risk factor for neural tube defects? A meta-analysis, *Q. J. Med.* **90**, 111–115.

Van Straaten, H.W.M., Copp, A.J. (2001) Curly tail: a 50-year history of the mouse spina bifida model, *Anat. Embryol.* **203**, 225–237.

Wald, N., Sneddon, J., Densem, J., Frost, C., Stone, R., MRC Vitamin Study Res Group. (1991) Prevention of neural tube defects: results of the medical research council vitamin study, *Lancet* **338**, 131–137.

Wald, N.J., Law, M.R., Morris, J.K., Wald, D.S. (2001) Quantifying the effect of folic acid, *Lancet* **358**, 2069–2073.

Weil, M., Jacobson, M.D., Raff, M.C. (1997) Is programmed cell death required for neural tube closure, *Curr. Biol.* **7**, 281–284.

Wilson, A., Platt, R., Wu, Q., Leclerc, D., Christensen, B., Yang, H., Gravel, R.A., Rozen, R. (1999) A common variant in methionine synthase reductase combined with low cobalamin (vitamin B12) increases risk for spina bifida, *Mol. Genet. Metab.* **67**, 317–323.

Wilson, D.B. (1980) Cellular proliferation in the exencephalic brain of the mouse embryo, *Brain Res.* **195**, 139–148.

Ybot-Gonzalez, P., Cogram, P., Gerrelli, D., Copp, A.J. (2002) Sonic hedgehog and the molecular regulation of neural tube closure, *Development* **129**, 2507–2517.

Ybot-Gonzalez, P., Copp, A.J. (1999) Bending of the neural plate during mouse spinal neurulation is independent of actin microfilaments, *Dev. Dyn.* **215**, 273–283.

Zhao, Q., Behringer, R.R., De Crombrugghe, B. (1996) Prenatal folic acid treatment suppresses acrania and meroanencephaly in mice mutant for the Cart1 homeobox gene, *Nat. Genet.* **13**, 275–283.

20
Receptor, Transporter and Ion Channel Diseases

J. Jay Gargus
University of California, Irvine, CA, USA

Neurobiology. From Molecular Basis to Disease. Edited by Robert A. Meyers.

ISBN: 978-3-527-32293-0

Keywords

Action potential
An all-or-none transient electrical depolarization propagated in excitable tissues.

ADH
Antidiuretic hormone or vasopressin; peptide hormone from the pituitary that controls plasma osmolarity by regulating aquaporins in the apical membrane of renal Principle cells.

Aldosterone
Adrenal steroid hormone that controls plasma volume and salt balance by regulating EnaC channels in the apical membrane of renal Principle cells.

Allele
The specific sequence of a given gene. Each possible variation of that sequence, functional or not, reflects a different allele.

Apical membrane
The specialized face of an epithelial cell that generally faces into the lumen of the gland and often is folded into a brush border with microvilli.

Aquaporin
A transmembrane protein that allows the transport of water free of solvated ions.

Ataxia
Inability to coordinate voluntary muscle activity, generally due to defect in cerebellum or spinal cord.

Channel
Transmembrane protein that allows the conductive transmembrane transport of small molecules, most commonly ions and water.

Channelopathy
A disease originating from dysfunction of a channel, or channel-like membrane protein.

Depolarization
A change in membrane potential tending to make the cell interior more positive.

Diuretic
A drug that serves to enhance urine flow, generally by acting to inhibit a renal sodium transport mechanism.

Dominant Allele
One copy of a mutation is sufficient to produce the mutant phenotype, such that the heterozygote expresses the phenotype.

Dominant Negative Mutation
A dominant mutation that results in less activity of the gene product than would be produced in a heterozygote with one wild-type and one null allele. Most commonly occurs because gene product functions as an inhibitory subunit in protein multimer.

EKG
Electrocardiogram; standardized clinical recording of extracellular electrical activity generated by the heart.

EMG
Electromyogram; standardized clinical recording of electrical activity generated by muscle contraction.

Environmental Disease
A disease caused by an agent external to the host.

Epilepsy
A chronic condition of seizure susceptibility, characterized by paroxysmal brain dysfunction and synchronized neuronal depolarizations.

Epithelium
A coupled sheet of polarized cells.

Excitation-contraction Coupling
The process by which electrical depolarization of a muscle membrane leads to activation of contractile fibers and cell contraction.

Gain-of-function Mutation
A mutation that increases the activity of the gene product to any degree.
Syn = hypermorphic.

Gate
Portions of channel protein that close access to permeability pore and is operated in response to ligand binding, voltage, or other stimulus.

Gene Family
An assembly of genes related by sequence homology, likely reflecting decent from a common ancestor followed by divergence.

G-protein Coupled Receptor
A large family of membrane proteins having seven transmembrane spans, the prototype being rhodopsin, which serve in sensory transduction by coupling to a signaling cascade through trimeric GTP binding proteins.

Haploinsufficiency
A mechanism of dominant inheritance whereby a heterozygote with one wild-type and one null mutant allele manifests the mutant phenotype. This is an atypical situation since rarely is 50% of wild-type activity insufficient for physiological function.

Heteromultimer
A combination of protein product subunits encoded by distinct genes uniting to produce a functional protein assembly.

Heterozygote
An individual having two different alleles of a gene. The default use implies that one copy is wild-type and the other mutant; however, this can also be an individual having two different mutant alleles, in which case they are specified.

Homozygote
An individual having identical copies for both alleles of a gene. These can both be wild-type or mutant.

Hypermorphic
A mutation that increases the activity of the gene product to any degree. Syn = gain of function.

Hyperpolarization
A change in membrane potential tending to make the cell interior more negative.

Hypomorphic Mutation
A mutation that inactivates the gene product to any degree. Syn = loss of function.

Inactivation
A nonconducting conformation of an ion channel from which it is incapable of carrying out an activating transition into a conducting conformation.

Ion channel
A transmembrane protein that allows the selective transport of ions freed from solvating water.

Ligand-gated Ion Channel
Binding a chemical compound to a site on the channel protein operates the gating mechanism to open or close substrate access to the permeability pore.

Loss-of-function Mutation
A mutation that inactivates the gene product to any degree. Syn = hypomorphic.

Loss of Heterozygosity
Regional loss of the chromosome carrying a normal copy of a gene, exposing clinical manifestations of a recessive mutation that is otherwise carried silently on the other copy of the chromosome.

Mendelian Disease
The same as a monogenic disease. Its inheritance follows Mendel's laws and can be either dominant or recessive.

Migraine
A condition characterized by paroxysmal onset of severe headache, generally unilateral, often associated with vertigo, nausea and photophobia and proceeded by a sensory aura.

Miniature Endplate Potentials (MEPPs)
Spontaneous depolarizations recorded at the motor endplate, which represent postsynaptic neuroreceptor response to spontaneous quantal release of neurotransmitter vesicles into the synapse.

Monogenic Disease
A disease caused by a major-effect gene mutation, such that the presence of this one mutation alone is highly predictive of the disease phenotype.

Motor Endplate
Postsynaptic membrane of specialized synapse of the motor neuron onto the skeletal muscle sarcolemma.

Myotonia
Delayed relaxation of skeletal muscle.

Neomorphic Mutation
A mutation that conveys to the gene product a new activity not present in wild-type.

Nernst Equilibrium Potential
The electrical potential difference at which no net ionic flux is observed across a permeable membrane despite the presence of a concentration difference between the sides. $E_K = [RT/nF]\ln[K_o/K_i]$ gives the Nernst potential for potassium, E_K, where R is the gas constant, T the absolute temperature Kelvin, n the valance, F the Faraday constant, and K_o and K_i the extracellular and intracellular potassium concentration, respectively.

Neuromuscular Junction
Specialized synapse of the motor neuron onto the skeletal muscle.

Null Mutation
A total loss of gene function, the extreme limit of a hypomorphic mutation, can include deletion or disruption (as in a knockout allele) of the gene itself.

Pacemaker
Endogenous source of rhythmic activity. In the heart, this is the sinoatrial node and the rhythm arises via spontaneous depolarization of the membrane.

Paracellular
Pathways across an epithelial sheet that pass between cells.

Paralysis
Inability to contract muscles.

Patch Electrode
Fire-polished glass electrode with 1-μm tip diameter that forms a molecularly-tight high resistance seal when applied to cell membrane, allowing recording from single channel molecules.

Pathophysiological Mechanism
Alteration in biological function leading to a disease process. This can be resolved from the molecular to the organismal and ecological levels.

Permeability Pore
Pathway though the channel protein taken by the conducted substrate.

Pharmacogenetic Disease
A disease caused by the interaction of an inciting "triggering" drug and an individual with a susceptible "at risk" genotype.

Plateau Phase
Segment of the cardiac myocyte action potential that coincides with the QT interval as observed on the EKG.

Polygenic Disease
A disease caused by an assembly of minor-effect gene mutations, such that the presence of any one of the mutations contributes only a small additional susceptibility to the disease.

Primary Active Transport
A transmembrane transport process that is capable of moving a substrate against its electrochemical potential gradient by directly utilizing the chemical energy of ATP hydrolysis.

QT Interval
The phase of cardiac ventricular repolarization as observed on the EKG. This coincides with the plateau phase of the cardiac myocytes action potential.

Recessive Allele
Carrying one copy of the mutation, the heterozygote appears like wild-type, not mutant. Only when both gene copies are mutant is the mutant phenotype observed.

Resorptive Epithelium
An epithelial sheet designed to create a flow of salt and water out of the lumen and into the blood.

Sarcoplasmic Reticulum
The specialized endoplasmic reticulum of skeletal muscle cells.

Secondary Active Transport
A transmembrane transport process that is capable of moving a substrate against its electrochemical potential gradient by coupling to the electrochemical potential of another chemical species, often sodium.

Secretory Epithelium
An epithelial sheet designed to create a flow of salt and water into the lumen and out of the blood.

Segregation
The disjunction of homologous chromosomes at meiosis such that offspring receive only one copy of each parental chromosome. This process underlies the characteristic Mendelian patterns of inheritance.

Seizure
Paroxysmal brain dysfunction associated with synchronized neuronal depolarizations.

Selective Advantage
An attribute of an allele that evolution favors, reflected by the survival of an excess of its descendants in subsequent generations.

Selectivity Filter
Portion of channel protein that restricts conductance to only specific substrates.

Splice Variant
Gene products that result from the use of alternative exons, often by exon skipping.

Subcellular Compartment
Membrane enclosed regions within eukaryotic cells. Examples include the endoplasmic reticulum, Golgi, mitochondria, and lysosomes.

Toxin
A natural biomolecule with potent biological activity. Ion channel toxins typically act to block or open ion channels at picomolar concentrations.

Transcellular
Pathways across an epithelial sheet that go through the cytosol.

T-tubules
Specialized invaginations of sarcolemma that play specialized role in excitation-contraction coupling of muscle fiber contraction.

Voltage Clamp
A feedback amplifier that adjusts externally applied current across a membrane such that the membrane potential is kept at a constant value even though its permeability character is varying.

Voltage-gated Ion Channel
Ion channel that is gated in response to changes in the membrane potential.

Wild-type
The allele of a gene most abundantly found in nature, for simplicity this can be considered the normal, nonmutant version of the gene.

Ion channels reflect an important but still very new mechanism of disease, first formalized in 1989 with the discovery of mutations in the *CFTR* ion channel gene causing cystic fibrosis, thereby carving out a new category of genetic inborn error, the channelopathy. Now, however, ion channel disorders are recognized to cover the gamut of medical disciplines and to cause significant pathology in virtually every organ system. The opening of this era was driven largely by the success of positional cloning strategies in isolating the genes underlying the rare monogenic disorders of ion channels and the explosive growth in basic ion channel physiology and biophysics that rendered these pathogenic mutant alleles interpretable, but in addition crystal structures have recently become available. Ion channels are a large family of over 400 related proteins representing over 1% of our genetic endowment, and many of the features shared among the rare monogenic ion channel diseases provide a solid foundation to begin to target these mechanisms for the development of novel therapeutics. They also render members of this gene family important "functional candidates" in the much more common complex polygenic diseases.

1 Introduction

Ion channel diseases reflect a relatively new category of genetic disease, or inborn error. Monogenic Mendelian ion channel diseases, or *channelopathies,* were first recognized in 1989 with the isolation of pathogenic alleles causing cystic fibrosis. The field explosively advanced as positional cloning revealed new pathogenic loci and alleles, and patch electrode electrophysiology allowed detailed characterization of the functional effects of mutations. This work most powerfully converged in the dissection of components of the action potential of excitable tissues – nerve, muscle and heart and while it remains true that diseases of excitable tissue still most clearly illustrate this family of diseases, ion channel disorders now cover the gamut of medical disciplines, causing significant pathology in virtually every organ system, producing a surprising range of often unanticipated symptoms, and providing valuable targets for pharmacological intervention.

2 Physiological Function of Ion Channels

Ion channels are a large family of related transmembrane proteins that provide ions, predominantly K^+, Na^+, Ca^{++}, and Cl^-, a passive pathway through which they can rapidly diffuse down their electrochemical gradient across the hydrophobic barrier of the plasma membrane. The standing electrochemical gradients that drive ion movements though channels are established by ion pumps – such as the Na/K-ATPase, and ion carriers – such as the Na-K-Cl cotransporter and the Na-H antiporter. While classical kinetic studies portray channels, pumps, and carriers as radically different transport mechanisms, with channels conducting ions 4 orders of magnitude faster and seemingly quite differently than the one-at-a-time transmembrane turnover of the pumps and carriers, we now recognize much similarity in their molecular mechanisms. Most cells maintain a high intracellular K^+, and a low intracellular Na^+ and Ca^{++}, compared with the surroundings, while Cl^- is often passively distributed at its *Nernst equilibrium*, which in a cell with a typical −60 mV membrane potential (inside negative) would be 10-fold lower on the inside (i.e. $E_{Cl^-} = (RT/nF) \log [Cl^-]_i/[Cl^-]_o \sim -59$ mV). The K^+ gradient, created at the cost of ATP hydrolysis by the ion pumps, is typically 10-fold in mammalian cells, and since the K^+ permeability, P_{K^+}, of most membranes overwhelms that of other ionic species, the K^+ diffusion potential predominates, setting the cell membrane potential, and giving rise to the typical −60 mV resting membrane potential near E_{K^+}. *Note that the physiological ions both create and respond to electrical phenomena across the cell membrane.*

In many ways, channels act like highly selective water filled pores that can be opened and closed in a controlled fashion to allow a specific ion species to flow, a flow that in turn can be perceived as both a miniscule chemical flux, with the use of isotopic tracer, or, more commonly, as an electrical current that changes the membrane potential toward the Nernst potential of the conducted ion, an interior positive *depolarization* created by opening Na^+ and Ca^{++} channels, an increasingly negative *hyperpolarization* created by opening K^+ channels, and a stabilization of the membrane potential by opening Cl^- channels. Note that unlike enzymes, channels carry out no biochemical transformations; the product and substrate ions differ only in regard to the side of the membrane on which they are found. Ion channels are further distinguished from enzymes in that rarely is the transported ion itself of any physiological consequences, since the flux through channels minimally changes the ion's concentration across the cell membrane; it is predominantly the electrical consequences of the ion current flow that underlies the physiology of the ion channels. Calcium is often an important exception to this rule, since it plays a critical role in coupling electrical activity to biochemical pathways. A second exception is that the chemical consequences of ion channel flux often predominates their function in epithelial cells.

We now recognize that ion channel proteins have structures that impose the three cardinal functions of a channel mentioned above: a *permeability pore* that provides ions but not water a passageway across the bilayer, a *selectivity filter* that allows some ions access while excluding others, and one or more *gates* that opens and closes the passageway. Three eras of Nobel Prize-winning technological and conceptual breakthroughs have led to our current level of understanding channel function (Fig. 1). All three key features of channel function were first defined, as abstract entities modeled with mathematical exactitude, in the era of the pioneers in membrane biophysics, an era that was opened up by *voltage clamp* technology fifty years ago and crowned by the Nobel Prize-winning classic squid axon studies of Hodgkin and Huxley (Fig. 1a). However, recent Nobel Prize-winning crystallographic work from the lab of Rod MacKinnon has now defined all three aspects in exquisite molecular detail for at least a few prototype channels (Fig. 1c). An essential era linking the mathematical and structural eras was that brought about by the breakthrough of *patch electrode* electrophysiology and crowned by the Nobel Prize-winning work of Neher and Sakmann. This era was critical in defining channels as biochemical mechanisms, and worked hand-in-glove with the molecular genetics of gene isolation and *in vitro* expression studies of cloned wild type and mutant channels, a breakthrough enabling paradigm pioneered in the lab of Shosaku Numa. Since the patch electrode would seal onto the plasma membrane so tightly that ions, and therefore ionic currents, could not leak between the two structures, magically tight gigaohm seals could be achieved that allowed electronic amplifiers the sensitivity to monitor in real time a single ion channel molecule in its millisecond stochastic dance between conformations: "closed," where it is nonconducting, and "open," where it passes on the order of 10^7 ions s^{-1}, producing a current of 10^{-12} amps, a picoamp, with stochastic transitions between the two states giving rise to the beautiful square wave traces of single-channel current records that are emblematic of the era (Fig. 1b).

One can separate the greater family of ion channels into various large classes based upon the cardinal channel features of selectivity and gating. One major distinction is based upon the nature of the selectivity filter and the predominant permeant ion species: hence, since the era of Hodgkin and Huxley, there have been K^+ channels, Na^+ channels and so on (Fig. 1a). Another major classification is based upon the gating mechanism. One large class of channels gate in response to their direct binding of ligand – the *ligand-gated ion channels.* Many of these ligands are classical neurotransmitters: hence, there are acetylcholine-gated ion channels, dopamine-gated ion channels and so on. But a wide range of other types of extracellular and intracellular ligands are also able to directly gate ion channels through binding. In addition, a large family of ion channels is indirectly gated by ligands, many by the same neurotransmitters mentioned above, but in this case neurotransmitter binding occurs to a heptahelical *G-protein coupled receptor* (GPCR) and the channel is activated by a second messenger ligand or a covalent modification, such as a protein

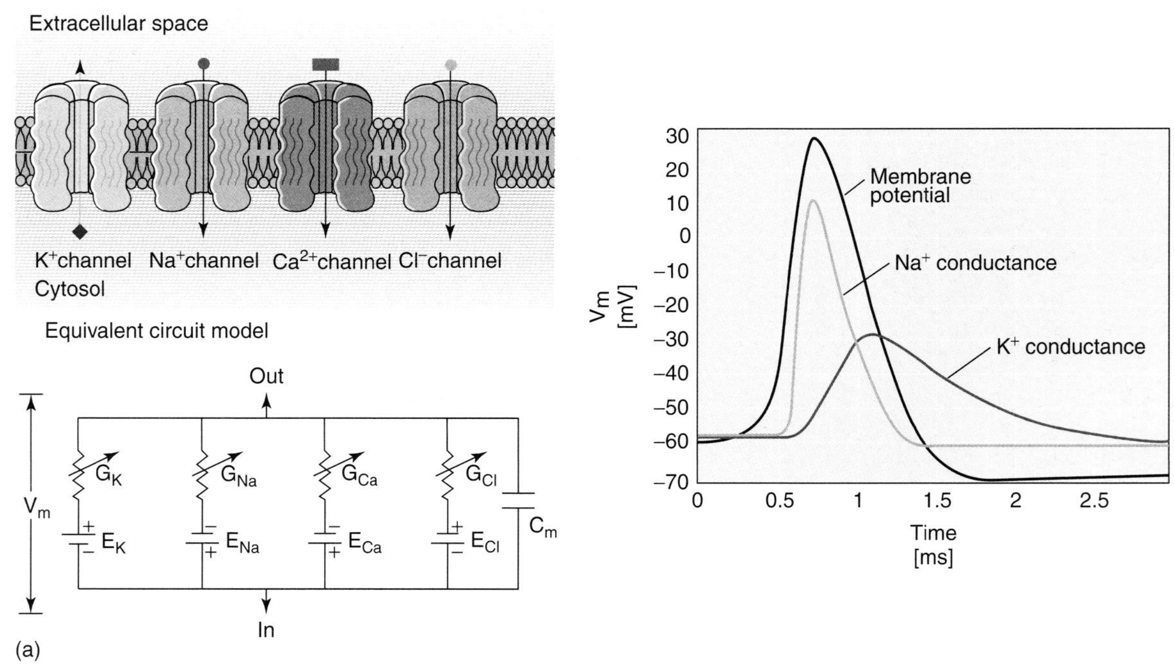

Fig. 1 3 ERAS of the ion channel. (a) Hodgkin–Huxley era, (b) Patch-clamp era, (c) Crystallography era.

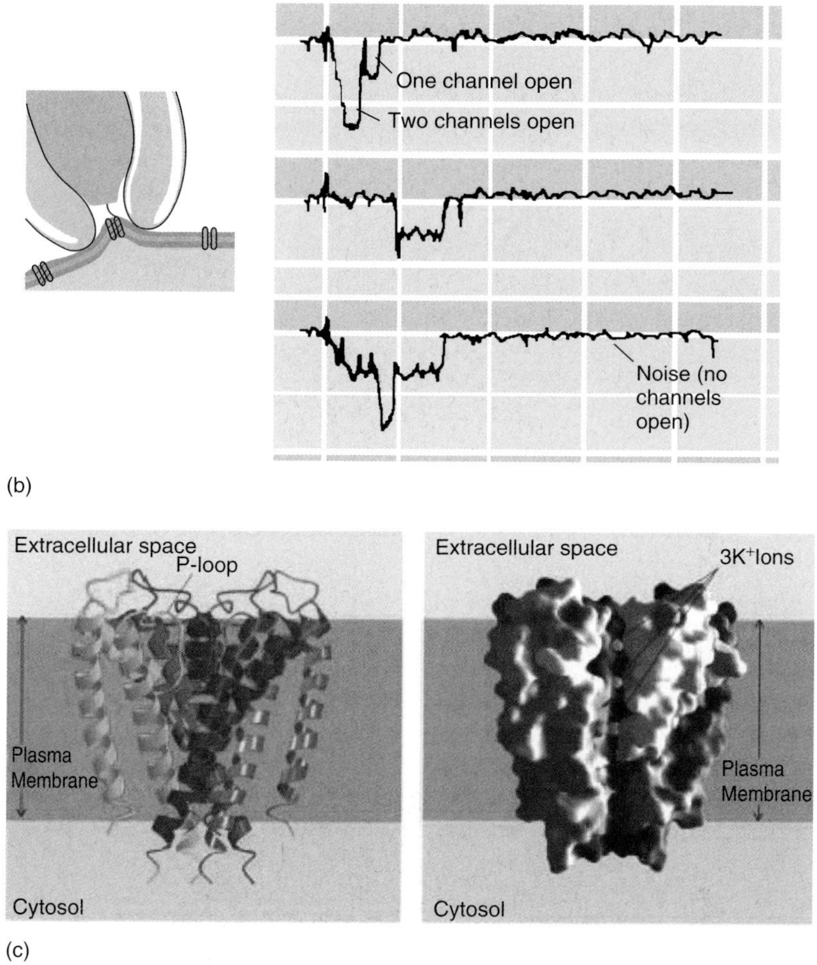

Fig. 1 *(Continued)*

phosphorylation or dephosphorylation reaction. Finally, a very large class of ion channels gate in response to changes in the electrical potential across the membrane, the *voltage-gated ion channels*. You can imagine that these are particularly tricky to study since the channel's own ionic current that is being controlled by the membrane potential *itself* alters the membrane potential, a vicious cycle that could only be broken by the voltage clamp amplifier. This experimentally vexing behavior of voltage-gated ion channels is, however, critical to their function in the perpetuation of a *propagating action potential* (AP). As a patch of membrane begins to depolarize, often because of the activation of a ligand-gated channel, voltage-gated Na^+ and Ca^{++} channels begin to respond to the voltage change by undergoing a conformational change from the "closed" into the "open" conformation (Fig. 2), increasing the membrane's permeability to sodium and calcium (P_{Na^+} and $P_{Ca^{++}}$) and hence

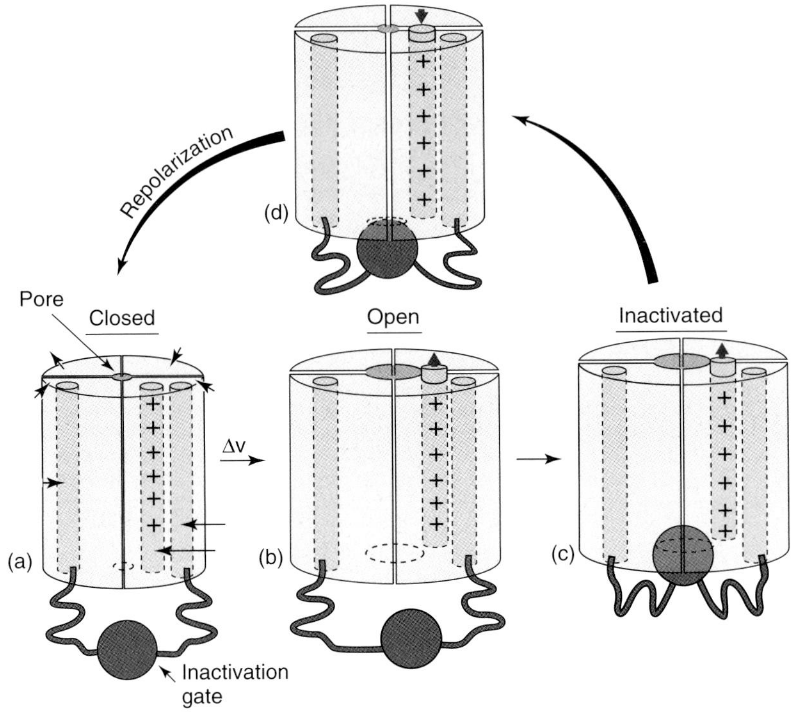

Fig. 2 Three states of the channel.

driving the membrane potential toward *their* inside-positive Nernst potential, a depolarization that serves to further activate more channels in surrounding patches of membrane in an explosive self-reinforcing cycle, recognized as a propagating AP, a depolarizing wave that rapidly spreads through an excitable tissue until all of its channels have been activated. Those channels next spontaneously enter an *inactive*, nonconducting conformation that is distinctly different from the "closed" state, since while "inactive," a channel cannot be opened and the tissue is rendered nonexcitable. Only by restoring the resting membrane potential does the channel's conformation become reset to the "closed" state that has the potential to open, and excitability is restored to the tissue. This resetting and restoring job is carried out by the more slowly activating voltage-activated K^+ channels. Since cells have but one membrane potential, it is always integrating inputs from all channel mechanisms in the membrane, and for this reason it is the great integrator in signaling pathways. Also, unlike biochemical pathways whose overall flux is predominated by a single rate-limiting step, membrane potential is intrinsically a continuous variable fully reflecting subtle changes in the entire host of contributing channels.

3 Biochemical Structure of Ion Channels

Ion channel proteins share many structural features in common that today allow

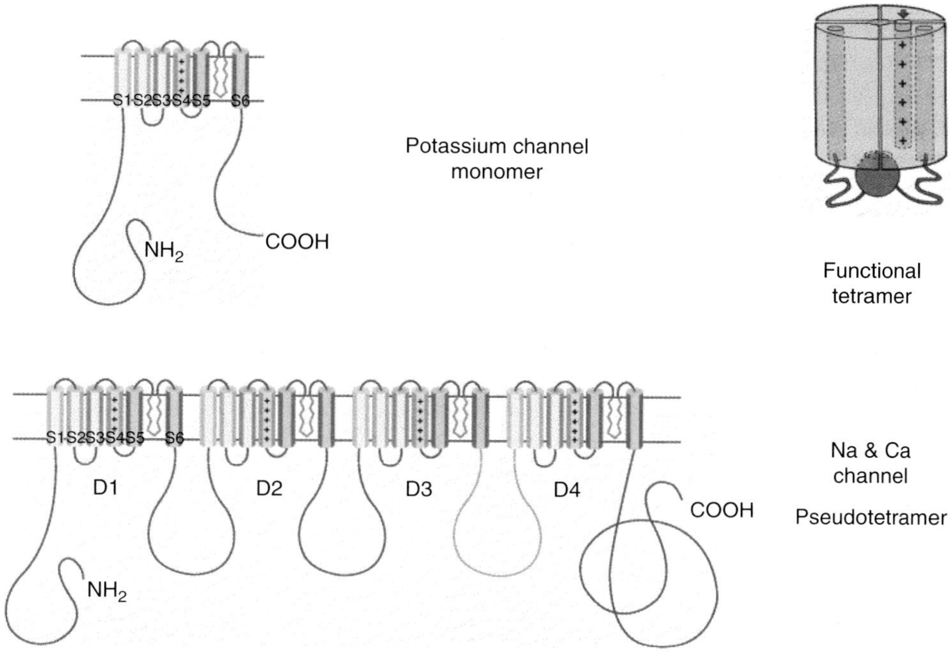

Fig. 3 Voltage-gated channels.

us to model their consensus structures and functions based upon the few prototype channel structures that have emerged from crystallography (Fig. 1c). For example, the large superfamily of voltage-gated potassium channels share a monomer structure of 6 transmembrane helices, each helix characteristically 20 to 25 amino acids in length, and both the N- and C-termini facing the cytosol (Fig. 3). The N-terminus forms a tethered "ball and chain" motif that gates the channel under the influence of electrostatics and the altered membrane potential its activation has brought about. It does this by swinging into the inner mouth of the pore and thereby effects the "opened" to "inactive" transition of the channel. The forth helix characteristically contains a cluster of 4 to 8 positively charged residues, converting that helix into a dipole that allows it to move under the influence of the membrane potential. This allows it to serve as the voltage sensor that gates the channel from the "closed" to the "opened" conformation. Between the fifth and sixth helix is found the heart of channel function, a signature pore domain characterized by a hairpin motif that dips partially through the membrane to form the conductive pathway used by the ion (Fig. 4). Some families of channels, such as the inward rectifiers, contain only this pore and the two flanking helices (Fig. 5). This pore domain contains the K^+ selectivity filter, a nonhelical motif conveyed by the amino acid sequence, TVGYG. It is, however, the strongly dipolar carbonyl groups of the α-carbons in the peptide backbone of this amino acid sequence, not the specialized side chains, that line the pore and create, through electrostatics and geometry, the selectivity filter that allows potassium ions

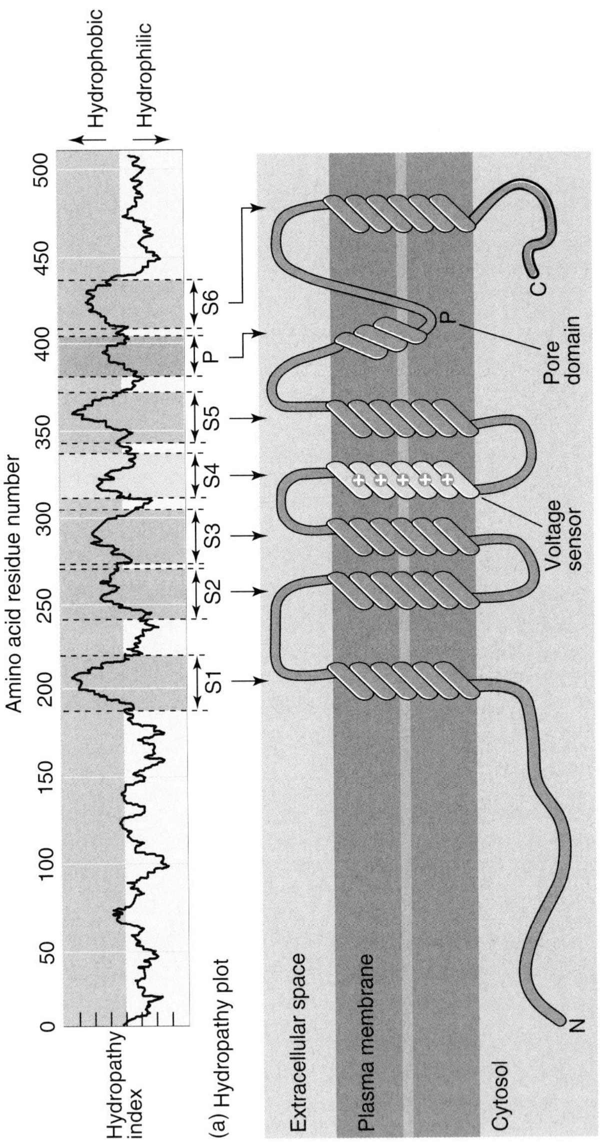

Fig. 4 Hydropathy plot and secondary structure prediction of amino acid sequence of potassium channel monomer.

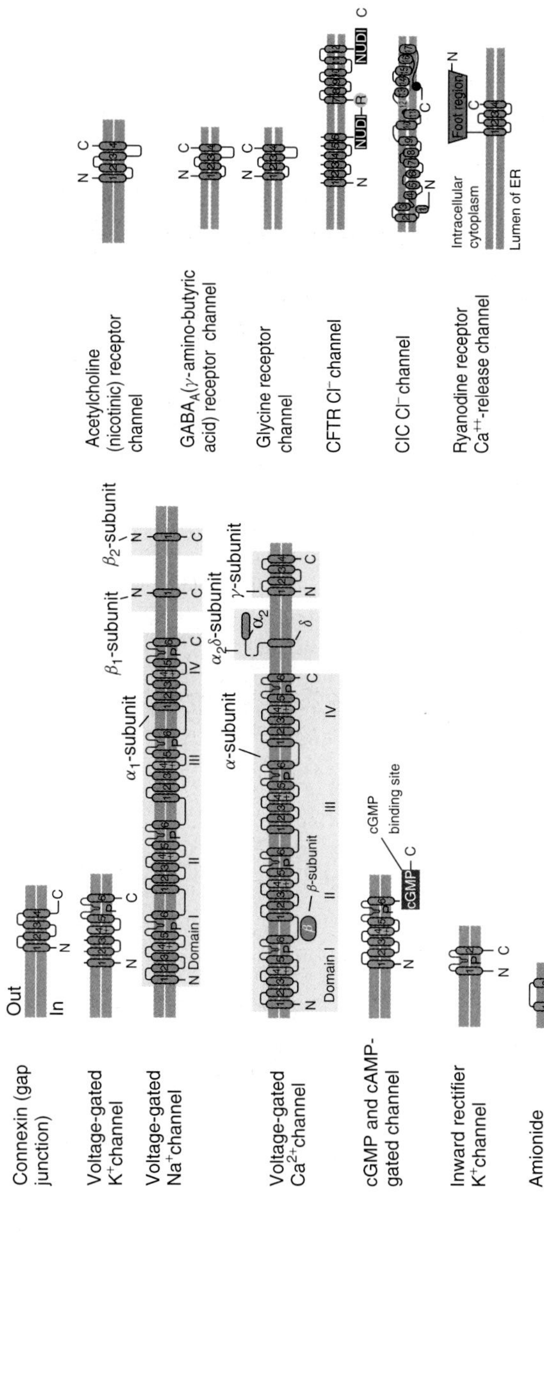

Fig. 5 Secondary structure of channel families.

entrance, but excludes sodium and other ions. The functional potassium channel that creates the transmembrane pore is a tetramer and the overall structure of the tetrameric assembly is that of an inverted teepee created by the tertiary structure of the fifth through sixth helices of the four interacting subunits, with the large vestibule of the teepee facing outward and lined by the selectivity filter residues from each monomer (Fig. 1c). Helices 1 to 4 of the monomers surround this core structure and present a surface that interacts with the lipid bilayer. Upon channel opening, the potassium ions move in a single file fashion along the pore, rapidly hopping between interactions with each of the critical residues of the filter motif, thousands of ions flowing during each channel's opening.

Conserved features that families of ion channels share in common have also proven useful in the identification of their genes in the genome. The human genome contains over 400 channel genes representing between 1 to 2% of our genetic endowment, although the substrate transported by most remains to be identified. While there are already known several large families of ion channels that apparently carry out extremely similar biochemical functions (Figs. 5 and 6), for instance, over 100 voltage-gated channels, the large number of channel genes almost certainly does *not* reflect redundancy. Since their job is not primarily contingent on their ability to transport ions from one side of the membrane to the other, but rather on the shape of the electrical signal they produce during conductive ion flow, it is apparently this that is under evolutionary selective pressure and it suggests that physiology requires the diverse vocabulary provided by multiple members of a gene family, which carry out nearly indistinguishable transport reactions with subtly differing kinetics. Furthermore, it is becoming increasingly clear that gene-family diversity is additionally greatly expanded through alternative exon splicing, allowing one channel gene locus to produce multiple splice isoforms and hence multiple functionally distinct protein products. Finally, for at least a handful of channels, and best characterized in the glutamate receptor channels, the exotic complex process of RNA editing is carried out to change single amino acid residues in the permeability pathway that critically tune ion selectivity and other aspects of channel function. Clearly, very subtle differences in channel function are important since evolution seems to have gone to very great lengths to preserve and enhance channel gene diversity.

For the potassium channels, diversity is further tremendously amplified through combinatorial monomer assembly into the functional tetramer (Fig. 3). This predominantly arises because channel protein subunits encoded by all members of a channel gene *family*, not just the allelic splice-variant protein products, coassemble to create a functional channel that differs in biophysical behavior as a function of the stoichiometry of the constituent subunits in the assembly. The panoply of combinatorial heteromultimers formed as monomers from family members coassemble, creates tremendous functional diversity in the potassium currents. This finely tunes the behavior of the population of expressed functional channels. This diversity conveys the essence of potassium channel function, allowing them to be tuned to modulate signal processing as they interact and jointly shape the membrane potential, thereby "deciding" the cell's aggregate signaling output.

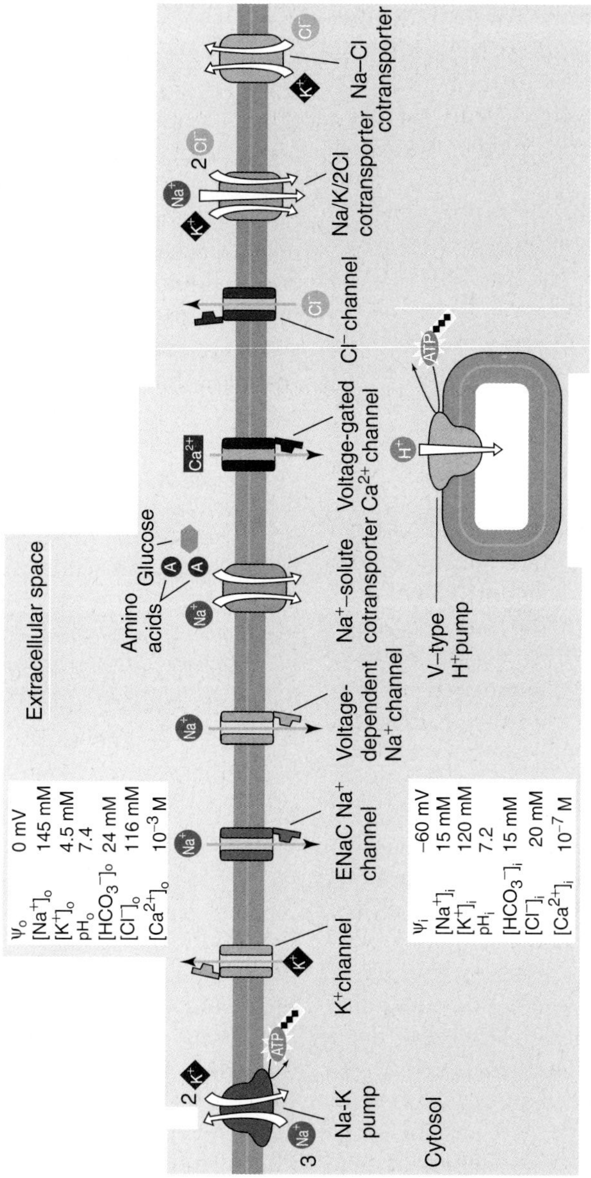

Fig. 6 Ion pumps, carriers, and channels.

In contrast to the broad heterogeneity of composition and function of the "contemplative" potassium channels, the voltage-gated sodium and calcium channels that are the "do-ers," able to execute the result of signal processing, are extremely homogeneous in their structure and function in a given tissue. While they can be perceived to share the structure of the potassium channel monomer, in their case, where a crisp uniformity of channel function is required within a tissue, a fixed pseudo-tetramer structure is hard-wired into the large endoduplicated genes that encode a pseudotetrameric functional monomer (Figs. 3 and 5). Tissue-specific diversity in the function of these channels is conveyed by tissue-specific isoforms of the gene family of voltage-gated sodium and calcium channels.

In addition to the major, α, channel subunits that directly contribute to the conducting pore of ion channels, discussed above, there are also a host of auxiliary subunits, each represented by gene families often labeled $\beta, \gamma, \delta, \varepsilon$, that modify channel function (Fig. 5). This obviously adds another enormous combinatorial explosion of channel diversity. Much of this diversity allows channel function to be tuned for its specific use in a specific tissue, since many channel genes are expressed in a remarkably tissue-selective fashion. On the other hand other channels may be found to be broadly expressed, but to predominate in the physiology of only a few tissues. As will become clear below, both of these features allow channels to produce tissue-selective disease. This tissue selectivity also points out that ion channels should offer highly focused targets for therapeutic intervention, perhaps devoid of the cross-tissue "side effects" that complicate many current receptor-based drug targets.

4 Ion Channels as a Mechanism of Disease

4.1 Contribution of Genes and Environment to Disease

While classically diseases are recognized to arise from a host of extrinsic environmental causes, such as infectious agents, or from intrinsic causes, most simply a reflection of our genes, we now recognize that genes and the environment are inextricably connected in all disease, and in most we see various contributions from a small number of "major-effect" genes, smaller contributions from a large number of "minor-effect" genes, and variable contributions from a host of factors in the environment. While recognizing this interconnectedness of genes and environment, we still tend to refer to diseases as being "caused" by the most salient of these three factors, the *monogenic* diseases, the *polygenic* diseases, and the environmental diseases. *Most parsimoniously, disease reflects a mismatch between genotype and environment.*

4.2 Monogenic Ion Channel Disease

Rare mutations arise *de novo* in the population each generation, breaking the molecular mechanism encoded by every gene in the genome in every possible way, most mutations being either neutral or deleterious to the gene's function; should the effect disrupt a critical physiological pathway, they are "disease-causing." *Dominant* mutations are expressed in a phenotype even in the presence of one normal allele, so the phenotype occurs in the individual in whom they first arise; those producing a phenotype incompatible with reproduction being immediately eliminated from the population, but being constantly replenished though new mutations. Thus, dominant mutations that are maintained and transmitted in the population tend to produce a later onset or milder course of disease. Recessive mutations that arise *de novo* are carried silently in a *heterozygote*, only rarely producing a disease on those rare occasions when two copies of the mutant gene are contributed to a *homozygous* offspring of two rare heterozygotes, this likelihood tremendously enhanced in a consanguineous mating. As most recessive alleles in the population are carried in heterozygotes, even lethal homozygote phenotypes have little effect on the disease-allele's frequency. On the other hand some "disease-causing" alleles are now very abundant in a given population, exceeding 1% of the population's alleles, frequencies typical of *polymorphic variants* that are commonly considered to be of "neutral" functional significance. An example of such a common polymorphism, carried by approximately 1 in 20 Caucasians, is the mutation that causes the most common genetic disease in Caucasians and the first recognized genetic ion channel disease, cystic fibrosis (CF). To understand the apparent anomaly of a lethal mutation rising to such abundance, a historical perspective is required. Often such high "deleterious" allele frequencies arise because of some *selective advantage* conveyed to the heterozygote carrier. In other instances, it simply appears that by chance heterozygotes for a rare mutation were among the founders who established an isolated population; with inbreeding, this allele became abundant and the disease frequency in this isolate became appreciable. Many common disease-causing mutations seem to have served in our evolutionary past as primitive "inoculations" protecting heterozygous carriers, who possess most of the "disease-alleles" in the population, against specific disease agents. The best-known example is the assent of the globin sickle cell and thalassemia hemolytic anemia alleles in the malaria belt, but CF alleles appear to be particularly protective against the infectious secretory diarrheas that imposed a high infant mortality in our recent past. While carrying two copies of the common CF allele is sufficient to produce a disease phenotype, independent of the contribution of other genes or environmental agents, and hence is recognized as a monogenic disease, the rare, nonpolymorphic mutant alleles of genes more commonly cause the rare monogenic diseases, such as the Long QT syndrome ion channel disorder, or channelopathy, to be discussed below.

Diseases such as those discussed above, in which there is a major single gene contribution to the disease phenotype are classically recognized as monogenic or *Mendelian* genetic diseases. The classical Mendelian pattern of recessive disease arises most commonly from *loss of function*, *null*, or *hypomorphic*, mutations, predominantly in a situation where sufficient gene product is available from the sole remaining copy of an essential gene for homeostasis. The dominant pattern of inheritance commonly requires a *gain of function*, such that a phenotype is created even in the presence of a normal copy of the gene; this can be an enhanced normal *hypermorphic* activity, a *neomorphic* activity not present in the native protein, or an inhibitory activity, *negative-dominance*, generally arising from an inactivating multimerization with the native protein. An alternative mechanism of dominance is *haploinsufficiency*, the less common situation where 50% of the physiological level of the gene product is *not* enough for homeostasis and in such cases a loss-of-function allele can be dominant. Even for these "monogenic" diseases we recognize that other genes in the genome and the environment still contribute to the disease phenotype, and these effects are acknowledged under the classical terms of *penetrance* and *expressivity*; two individuals carrying exactly the same two alleles of a major-effect gene may have vastly different phenotypes with one showing the classical disease and the other ranging from having such minor manifestations that no aspect of the phenotype reveals the presence of the "disease genotype," in which case he is called *nonpenetrant*, to perhaps manifesting only a minor, yet still diagnostic, feature of the disease, in which case he would demonstrate the variable expressivity of the disease.

4.3 Polygenic Ion Channel Disease

The situation described above is magnified in the polygenic diseases, common diseases of ill-defined pathogenesis, such as hypertension, diabetes, and epilepsy, where typically one mutant gene is insufficient to create the disease phenotype, most of the genes still remain to be defined, and the pattern of inheritance observed is simply familial clustering, not the clear segregation patterns of recessive or dominant inheritance seen in the Mendelian disorders. While many gene loci contribute together to produce the common polygenic diseases, both common polymorphisms and rare mutant alleles seem to participate, and while major-effect genes play a very minor role in the population, not uncommonly rare Mendelian forms of the disease phenotype do occur, their etiological gene providing a peak into the pathogenic mechanisms likely underlying the common form of the disease, such as the monogenic channelopathies recognized to cause epilepsy in benign familial neonatal convulsions (BFNC), hypertension in Liddle syndrome, and diabetes in permanent neonatal diabetes (all to be discussed in detail below). Most polygenic diseases additionally have a significant environmental component, hence their more comprehensive title of *complex polygenic diseases*. The environmental component of these diseases can be recognized since even identical monozygotic twins, sharing all of their genes in common, are discordant for the disease phenotype a high percentage of the time.

4.4 Pharmacogenetic Ion Channel Disease

Finally, some diseases appear to be caused by environmental agents. However, we now recognize that only in the face of an appropriate "at risk" genotype is an appropriate physiological target available such that the agent can be a toxin. Should the toxin be fairly rare in the environment, such as a general anesthetic, it may appear initially to cause an idiosyncratic environmental disease, such as the lethal rare adverse reaction to anesthetics referred to as *malignant hyperthermia* (MHS). Much later it was recognized that specific heritable ion channel mutations cause one to be vulnerable to this syndrome, an individual inheriting an MHS allele being perfectly healthy, just at risk for malignant hyperthermia, and not developing the lethal disease phenotype until exposed to the inducing anesthetic. The recognition of major genetic and environmental components in the disease classifies it as a *pharmacogenetic* disease, and likely many of the idiosyncratic drug reactions we observe, either toxic or therapeutic, reflect this same phenomenon. On the other hand, should the environmental agent be ubiquitous, such as phenylalanine, an amino acid found in all foods, a disease caused by an environmental agent may appear to be a genetic disease, in this case phenylketonuria (PKU), the archetypical "genetic" disease for which every newborn is screened, but which produces no disease phenotype if phenylalanine exposure is controlled.

4.5 Ion Channel Toxins

That ion channels are exquisitely sensitive, finely tuned, evolutionarily highly conserved mechanisms essential to an organism's physiology has not been lost to the combatants on Evolution's fierce battleground. Nearly all of the most potent toxins in Nature target ion channels, most with nanomolar to picomolar affinity. Tetrodotoxin from the puffer fish, saxitoxin from red tide dinoflagellates, and a huge number of scorpion and coelenterate peptide toxins target the voltage-gated sodium channels; ω-conotoxin from the cone snails and a host of spider venom toxins target the voltage-gated calcium channels; a number of scorpion, insect, and coelenterate toxins target families of potassium channels; α-bungarotoxin, from snake venom, and the Amazon blow-dart alkaloid toxin curare target the nicotinic acetylcholine receptor and the plant alkaloid picrotoxin targets the GABA receptor channel. Most toxins act by relying upon intimate, high-affinity contacts within the channel molecule, and for this reason many have proven to be essential tools in defining classes of ion channels and their mechanisms of action. Since most of these toxins are quite rare in the environment and none seem to have participated in major human evolutionary bottlenecks, nearly all human genotypes are vulnerable to these toxins, despite the fact that in the laboratory mutations can be selected to convey toxin resistance. For this reason, a huge array of toxins can be recognized to produce quite pure forms of environmental or toxic channelopathies.

4.6 Mechanism of Ion Channel Disease Pathogenesis

Ion channel diseases span the gamut of pathogenic mechanisms discussed above. Furthermore, the pathophysiological mechanisms that lead from a defective ion channel protein to a disease phenotype are complex and multifaceted. That Nature requires a genetic investment in the huge array of subtly differing channels leads one to expect that subtle changes in ion channels matter, foreshadowing the wide range of phenotypic abnormalities different mutants of the same gene might produce. It is therefore not surprising that ion channel diseases as apparently distinct as migraine and ataxia, paralysis, and myotonia or even as having dominant versus recessive inheritance, are found to be allelic, caused by different mutations in the same gene.

A *periodic* disturbance of rhythmic function is the cardinal feature of ion channel disease of excitable tissues. In the heart, this produces a fatal *arrhythmia*, observed as a loss of the normal synchronous contraction of the muscle composing the pumping chambers secondary to a disruption in the rhythmic action potentials arising in this electrically coupled tissue. In skeletal muscle, abnormal ion channel function produces periodic alterations in contractility, ranging from the inability to contract – *paralysis*, to the inability to relax – *myotonia*. And in the central nervous system (CNS), the rhythmic disturbance is observed as an abnormally synchronous electrical discharge, a seizure. The remarkable finding in all of these syndromes is the absence of an overt functional abnormality the vast majority of the time. In the extreme, this results in the tragic situation where no phenotype is obvious until the moment of premature death. While many insights have been garnered as to how a variety of stresses serve to create the decompensations that allow these periodic phenotypes to become manifest, the mechanisms that serve to compensate for a constitutionally defective channel such that healthful homeostatic function is maintained the majority of the time, remain to be defined.

Ion channel phenotypes in nonexcitable tissues are more diverse because of the very different roles channels play in such tissues, participating in signaling involved in endocrine secretion, the function of cytosolic compartments and complex epithelial secretory and resorptive function, which in the kidney, produce secondary changes in systemic electrolyte balance and blood pressure.

It is simplest to develop the pathophysiology of ion channel disease by starting with the cardiac phenotype of arrhythmia and the Long QT (LQT) syndromes, demonstrating how the ion channels in the membrane work together in tissue function and how component parts of the membrane potential are dissected by the genetic lesions. It further provides specific paradigms for how subunit interactions can be manifest, how loss-of-function and gain-of-function phenotypes occur and how dominance can arise through either haploinsufficiency, gain-of-function, or in a "dominant negative" fashion. The muscle and CNS phenotypes, while not lending themselves to as simple a dissection of function are, however, easily built by extrapolation from this paradigm. They further illustrate the power of gene families in identifying candidate disease genes and reinforce the intrinsic polygenic nature of ion channel phenotypes. Finally, the nonexcitable tissue phenotypes can be used to show the range of functions ion channels carry out.

5
Ion Channel Diseases of the Heart

5.1
Function of Ion Channels in the Heart

Cardiac cells possess an intrinsic instability of their membrane potential, such that spontaneous depolarizing action potentials arise (as described at the end of Sect. 2 above), that can be observed as contractions in dissociated cardiac cells in culture. Muscle contraction is directly coupled to the propagation of an action potential, a phenomenon referred to as *excitation-contraction* (EC) *coupling,* since the calcium ions that enter the cell, amplified by other calcium sources inside the cell, produce the biochemical activation of the contractile proteins. Briefly, this entails calcium binding to troponin and releasing tropomyosin's block on myosin's cycling cross-bridge formation with actin. The electrical *pacemaker* function in a healthy heart only has the opportunity to manifest itself at the sinoatrial (SA) node, the physiological pacemaker of the heart. Since all of the other cardiac cells have a slower endogenous rate of spontaneous depolarization (pacing), and since all cells in the heart are electrically coupled, the fastest spontaneously depolarizing pacemaker, the SA node, is the origin of a propagated action potential that drives though the remaining cells of the organ,

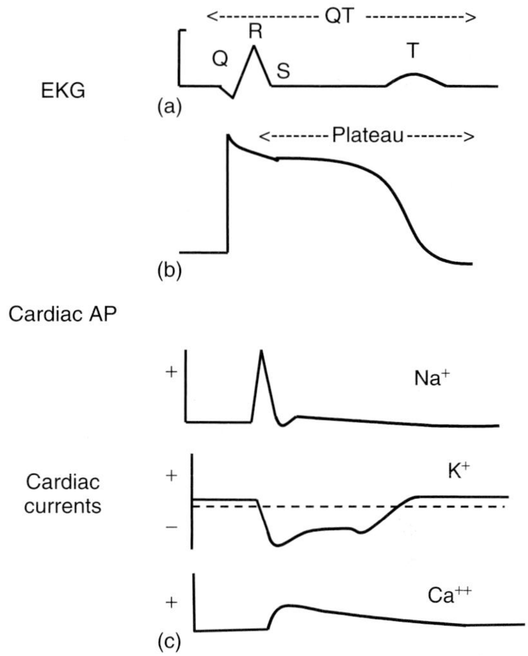

Fig. 7 Synchronized EKG, cardiac action potential and component currents. (a) EKG trace, (b) cardiac ventricle myocyte action potential, (c) resolved ionic currents underlying ventricular myocyte action potential. From Gargus, J.J. (2003) Unraveling monogenic channelopathies and their implications for complex polygenic disease.

forcing them to depolarize before they would do so spontaneously. Should the fastest pacemaker cells be lost, the next fastest pacemakers progressively take over, a very reassuring redundancy in a complicated process that must be repeated several billion times, failure rapidly resulting in death. The directional flow of the propagated action potential through the heart muscle creates the vector of the surface current revealed by the EKG traces, and produces a synchronous rhythmic coordinated contraction of the muscle cells composing the pumping chambers of the heart. If the cardiac cells of a pumping chamber all contract synchronously, the chamber volume rapidly shrinks, forcefully pumping blood that is directed by the cardiac valves. Without synchrony there is no pumping, only ineffectual fibrillation of individual cardiocytes and death to the individual. The cardiac pump cycle thus can be followed with the EKG. The QT interval measured on the EKG is the long interval of the cardiac cycle during which the ventricles, the major pumping chambers, repolarize, their conducted depolarizing action potential being reflected by the QRS waves (Fig. 7).

5.2 Ion Channel Dysfunction in Arrhythmia

A design feature engineered into the ventricular action potential by the mix of ion channels in its membrane is a prolonged plateau phase of depolarization, a phase *not* present in the classic rapid action potential of nerve (Fig. 1a). LQT syndrome is caused by mutations in the cardiac ion channel genes contributing to this component of the cardiac action potential. The plateau phase is a time during which calcium enters ventricular myocytes to produce contraction, but as importantly, the depolarized plateau phase serves to hold sodium channels in the "inactive" nonexcitable state until the inciting depolarizing wave has spread to all of the electrically coupled myocardium. In this manner, it can be assured that all ventricular myocytes will repolarize and regain the ability to again depolarize in synchrony. If this mechanism fails, and one group of cells fires an action potential that cannot be conducted through a still-recovering inactive area, those inactive cells may subsequently have sufficient time to develop their endogenous pacemaker-like depolarization and asynchrony in the tissue is initiated as an endlessly looping futile depolarization wave spreads through any newly excitable domains of the tissue. Such asynchrony is lethal since unless the entire ventricular muscle depolarizes and contracts together, it does not pump blood, but only fibrillates. During the plateau phase, the membrane potential struggles between the depolarizing influence of sodium and calcium currents and the repolarizing influence of potassium currents, with the slowly developing potassium currents finally overcoming the depolarizing currents and restoring the resting membrane potential (Fig. 7). Mutations that serve to either enhance the sodium current or to reduce the potassium currents predictably prolong the plateau phase and are associated with LQT syndrome. A long plateau phase is a dangerous situation since prolonging it opens the opportunity for asynchrony between the cells and susceptibility to arrhythmia, as described above.

5.3 Long QT Syndrome

Classically the LQT syndromes are divided into those with a dominant pattern of inheritance and a phenotype limited to the heart, the Romano–Ward syndrome (RW), and those having recessive inheritance and a phenotype that includes sensorineural hearing loss, the Jervell and Lange-Nielsen syndrome (JLN). In addition, there are now recognized acquired forms of the disease, primarily developing in a failing heart, and pharmacogenetic forms, a serious complicating side effect for a wide spectrum of drugs. As the name suggests, LQT prolongs the QT interval measured on the EKG and predisposes to a fatal arrhythmia; however, typically the pedigree may only reveal early sudden death of no known etiology. There are no diagnostic pathological findings and commonly no symptoms before death such that the disease is commonly not diagnosed until repeated deaths in a family have occurred. Diagnosis of the syndrome, on the other hand, does not depend upon a clinically recognized phenotype; rather, it depends only upon the EKG findings. There is a characteristic context of death observed in the syndrome that further illuminates the critical balancing act performed by multiple cardiac channels in normal physiology. The context is that of excess adrenergic outflow, as occurs in high emotion or exertion. This is a risk since the heart rate dramatically increases during the "fight or flight" responses these transmitters mediate. While we may initially focus on this response in terms of the β-adrenergic receptor increasing the rate at which the cardiac pacemaker depolarizes and the heart contracts, it is clear that to depolarize more rapidly, it must also *repolarize* more rapidly, and in fact specific cAMP-dependent channel regulatory mechanisms downstream of the β-adrenergic receptor assure this coordination, effectively shortening the QT as the heart rate rises. This explains why the rational way to evaluate a QT is in the context of the underlying heart rate, the QTc. In a patient with an intrinsic LQT defect that lengthens the QT, this adjustment of the rate of repolarization can not keep pace, the QT can not sufficiently shorten at high heart rates, and an arrhythmia ensues.

5.4
Nature of LQT Genetic Loci and Alleles

LQT mutations have been found in six cardiac ion channel genes. One LQT gene encodes the cardiac sodium channel and five encode the primary and auxiliary subunits of three different potassium channels (Fig. 8). While this provides a nearly complete dissection of the major channels involved during the critical plateau phase, one major actor, the cardiac calcium channel, encoded by *CACNA1C*, has yet to yield a pathogenic allele, perhaps being an early embryonic lethal. However, the most recently discovered LQT locus, LQT4, encoding ankyrin-B, while being unique in *not* encoding an ion channel protein, regulates a complex of transporters in the cardiac membrane mediating calcium homeostasis. *LQT1*, the first LQT gene identified, was found using a strategy involving linkage and positional cloning. All of the other LQT genes were identified as functional candidates within a mapped chromosomal interval.

5.5
The Dominant LQT Alleles

LQT1 was located on chromosome 11p 15.5, and it was ultimately demonstrated that mutations in a potassium channel α-subunit gene, *KCNQ1*, were causal. When expressed *in vitro*, *KCNQ1* produces a current that is not found in the heart; however, when expressed together with its β-subunit, encoded by *KCNE1*, they can be recognized to produce the I_{Ks} channel underlying the slow delayed rectifier K^+ current that participates in repolarization (Fig. 8). Over 35 different pathogenic alleles have been reported in this gene and they are the most common cause of RW syndrome. They primarily confer a dominant-negative phenotype when studied *in vitro*. Defective subunits coassemble with wild type copies producing, through combinatorials, a supermajority of defective channel tetramers, and hence reduced repolarizing current.

LQT2 was the first LQT gene cloned, taking advantage of a candidate gene approach within the region of chromosome 7q35–36 where a second locus was identified in families not mapping to chromosome 11. The gene, *KCNH2*, was a strong functional candidate based upon its homology to a fly gene with a proven ability to create a rhythm disorder phenotype, ether-a-go-go. Like *KCNQ1*, coexpression of *KCNH2* with its β-subunit, *KCNE2*, was required to produce a current that can be recognized in the heart. The channel they form is I_{Kr}, the rapidly activating delayed rectifier K^+ current (Fig. 8). Over 16 alleles of *KCNH2* have been reported, and like *KCNQ1*, many confer a "dominant-negative" phenotype when expressed *in vitro*. However, other dominant alleles appear to be simple loss-of-function alleles, suggesting that the membrane current conducted by this channel is so finely tuned that a haploinsufficiency mechanism is adequate to produce dominance.

LQT3 is the only sodium channel locus involved in the disease, and it is caused by mutations in the *SCN5A* gene found at chromosome 3p21–24. It encodes the cardiac-specific voltage-gated sodium channel that underlies the rapid depolarization phase of the ventricular action potential that produces the QRS complex and ventricular contraction (Fig. 8). Over 30 alleles of this gene have been

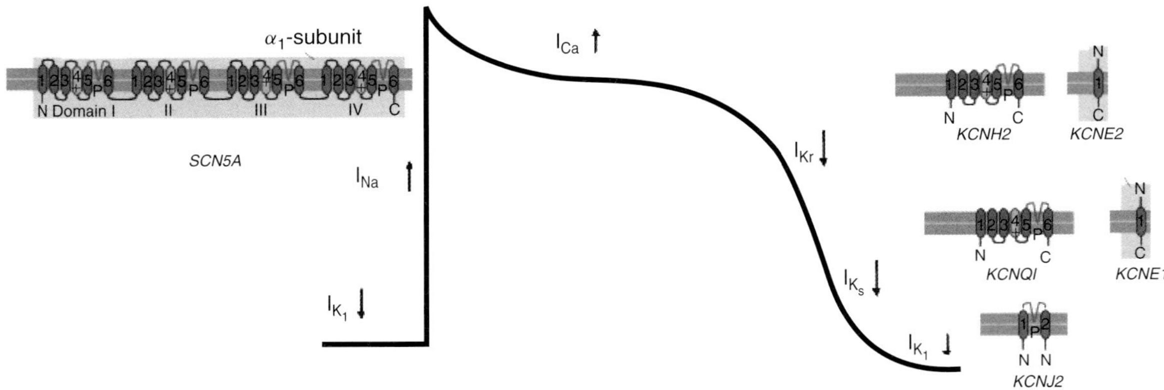

Fig. 8 Cardiac ion channel LQT genes. Each channel subunit gene product shown carries known LQT mutations. ↑ signifies depolarizing current, excess leading to LQT; ↓ signifies hyperpolarizing current, deficiency leading to LQT.

reported. Channels encoded by pathogenic alleles have delayed or decreased inactivation after opening, leaving excess inward depolarizing current during the plateau phase, delaying the time at which potassium currents can bring about repolarization. They are thus dominant "gain-of-function" mutations.

A distinct set of *SCN5A* alleles produce the *opposite* effect on inactivation from the *LQT3* alleles; they produce a rapid recovery from inactivation. They produce a different dominant arrhythmia syndrome, Brugada syndrome. The myocardium of these individuals contains a mixed population of sodium channels that are no longer locked in synchrony by a common period of inactivation, forming an ideal substrate for arrhythmia. Some alleles, surprisingly, are capable of producing either sodium channel phenotype within a family, demonstrating the delicate balancing act between excitation and inactivation carried out by the sodium channel that will be more clearly illustrated in the skeletal muscle disorders, and hinting at modifier genes that alter this balance, and therefore polygenic disorders.

LQT4 is the exception that proves the rule that LQT is an ion channel disease. The *LQT4* locus, found at 4q25 encodes ankyrin-b, a scaffold protein that colocalizes and regulates the Na/K-ATPase, the Na/Ca exchanger and the IP3 receptor. It is not a channel, but pathogenic loss-of-function alleles disturb calcium homeostasis through the functional loss of these critical mechanisms. It thus creates a dominant arrhythmia syndrome that includes LQT, but is not restricted to LQT.

LQT5 and *LQT6* are two adjacent loci on chromosome 21q22.1, and their genes, *KCNE1* and *KCNE2*, respectively, encode the two highly homologous potassium channel β-subunits discussed above as the accessory subunits that interact with their α-subunit partner to form I_{Ks} and I_{Kr}, respectively (Fig. 8). Both are proteins with only a single transmembrane α-helix and both function only to modify the behavior of the α-subunit with which they multimerize. Dominant pathogenic alleles are missense and effectively achieve dominance by altering channel gating to produce less current.

LQT7 is an alternative name for Andersen syndrome, a dominant multisystem disorder that includes long QT, but also extracardiac findings such as periodic paralysis and, more surprisingly, dysmorphology. Over nine different dominant-negative alleles of the potassium channel gene *KCNJ2* are responsible for this syndrome. The *KCNJ2* channel subunits multimerize to form the inwardly rectifying potassium channel that governs the resting membrane potential of the cardiac myocyte (Fig. 8). Current through this channel participates at the very end of repolarization, and hence dominant-negative loss of its function prolongs the process. The extracardiac findings imply that it also contributes in a significant way to membrane signaling in muscle and perhaps to the developmental processes underlying the dysmorphology.

5.6 The Recessive LQT Alleles

The recessive JLN syndrome differs from the dominant RW syndrome not only in its pattern of inheritance but also in its extracardiac manifestation of a constitutive sensorineural hearing loss. While in both syndromes the cardiac arrhythmia phenotype is lethal and intermittent, the striking differences between the two syndromes would never lead you to guess they might be allelic. In fact, the JLN alleles are strong loss-of-function or null alleles in the same genes that cause dominant *LQT1* and *LQT5*, *KCNQ1* and *KCNE1*, the two channel genes encoding the I_{Ks} channel (Fig. 8). Surprisingly the same channel is expressed selectively in the heart and the inner ear, where its abundant expression is involved in the production of the potassium-rich endolymph fluid that fills this chamber and participates in the transduction of auditory sensation. Only complete loss of the channel significantly disturbs this secretory process. It is of note that another inner ear *LQT1* homolog, encoded by *KCNQ4*, has dominant alleles that produce only deafness without LQT (*DFNA2*).

5.7 Acquired LQT Channelopathies

While heart failure is the very common end-stage phenotype of a dilated and poorly contractile myocardial pump, and is produced by a variety of common insults to the heart, most deaths in this disorder arise from a fatal arrhythmia, and it likely reflects one of the most common forms of the LQT syndrome. The failing myocytes downregulate expression of potassium currents, perhaps in an attempt to enhance contractility by prolonging the period of calcium entry, the QT interval, but doing so predisposes them to the arrhythmia discussed above.

5.8 Pharmacogenetic and Polygenic Arrhythmia Syndromes

Initial studies suggested that LQT was a highly penetrant Mendelian phenotype with "strong" alleles that, while clinically silent, would dependably be revealed by EKG. Not surprisingly, it has become clear that there are both "weak" and "subclinical" alleles of these genes as well and likely polygenic interactions between them. Additionally, there is clearly an environmental component to the disease, seen as a pharmacogenetic syndrome, which greatly broadens the scope of the diseases involving these cardiac ion channel genes. These more subtle features of the disease became most apparent in studies on families initially considered to have a case of "sporadic" LQT. While half of the "sporadic" probands in the study had the predicted new dominant mutation in one of the LQT genes, the other half were found *not* to have new mutations, but rather to have families segregating a "weak" pathogenic LQT allele. This allele in these families had many silent carriers with no EKG abnormality, or they had carriers who only expressed the abnormality once taking a medication with potassium channel blocking activity. The silent carriers presumably lacked a phenotype because they lacked other unidentified susceptibility alleles at other loci carried by the proband. It is easy to imagine, based upon the descriptions above of the interactions between the multiple normally and abnormally functioning cardiac ion channels in maintaining and terminating the plateau phase of the cardiac action potential, how minor functional variants in these same genes could sum with one another to produce a major alteration in the plateau phase and hence produce a polygenic LQT, arrhythmia or sudden death phenotype. It is likewise easy to imagine how they might sum with a drug effect that acts like one of these mutations by blocking a potassium channel.

The pharmacological induction of the phenotype in an individual such as those in the Priori study suggests that environmental or genetic liability can sum with that contributed by the weak LQT alleles. Since the drugs that unmask the phenotype produce potassium channel block, similar to known LQT mutation, so presumably do the hidden unmasking liability alleles. The pharmacogenetic LQT syndrome is clinically quite important; causing the recall of several otherwise valuable drugs, and is now recognized as a potentially lethal side effect of many common medications. These include antihistamines, such as Seldane, antibiotics and cisapride, as well as the more predictable antiarrhythmics. Most of the implicated drugs share a common mechanism of action that involves the block of I_{Kr} potassium channels, and a number of cases now explicitly demonstrate that weak or subclinical alleles in the LQT genes, encoding both Na^+ and K^+ channels, are contributory to the pharmacogenetic disease.

A special consideration of pathogenic weak LQT alleles is raised by the contribution of these genes to sudden infant death syndrome (SIDS). In a landmark prospective study spanning an 18-year period, Schwartz and coworkers from nine large maternity hospitals performed EKGs on all healthy newborns between the third to fourth day of life, studying a total of over 34 000 neonates. At the 1-year follow-up evaluation, it could be observed that infants who had died of SIDS had a longer QTc interval *at birth* than did survivors or infants dying of other causes, even though none had a positive family history for LQT. Half of the infants dying of SIDS (12 of 24) had a QTc at birth greater than two standard deviations above the mean for the cohort, a finding that alone predicted greater than a 40-fold increase in the odds of dying of SIDS. However, because SIDS is so rare, despite the finding's dramatic odds ratio, only 2% of the neonates found to have a long QT die of SIDS. More recent studies have begun to add a molecular dimension to these cases, and for a few cases, a new mutation in one of the LQT genes, producing an allele previously seen in a family with the classical syndrome, can be identified. It remains to be determined whether weak inherited mutations, or even polymorphisms, in two or more of the LQT genes contribute to the rest.

6
Ion Channel Diseases of Muscle

6.1
Function of Ion Channels in Muscle

In many ways, the skeletal muscle AP is similar to the initial and final portions of that in the heart, but lacks the long plateau phase (Fig. 7). It is a fast depolarization, driven by sodium and calcium currents, followed immediately by a rapid repolarization driven by potassium currents, and in this regard is much like that of nerve (Fig. 1a). Instead of arising endogenously within the tissue, as at the cardiac sinoatrial node, the inciting muscle depolarization arrives *synaptically* via the quantal release of a chemical neurotransmitter, acetylcholine (ACh), from a motor nerve terminal bouton that forms a specialized neuromuscular junction with the *motor endplate* of the muscle plasma membrane, or *sarcolemma*. As the motor nerve's propagating AP reaches its terminus, cytosolic calcium rises, triggering a few hundred *synaptic vesicles*, each holding several thousand ACh molecules, to fuse into the presynaptic neuronal membrane (Fig. 9). As they fuse, each vesicle releases, in a quantal fashion, all its contents. This released ACh diffuses, in a matter of milliseconds, across the *synaptic cleft* to the muscle, where it is degraded by acetyl

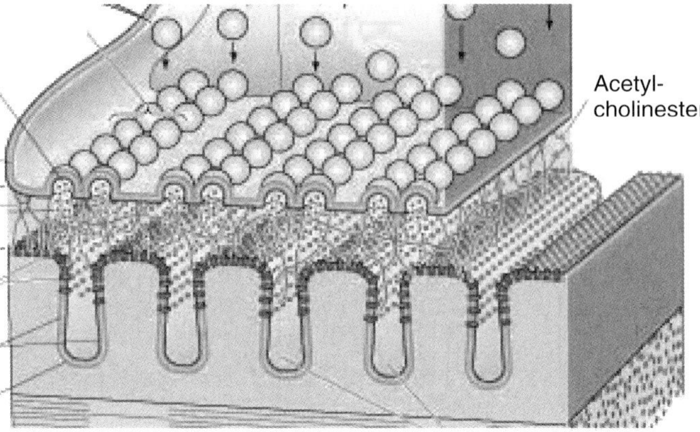

Fig. 9 Neuromuscular junction.

cholinesterase. A ligand-gated cation channel found on the postjunctional folds of the sarcolemmal endplate membrane, the heteropentameric nicotinic ACh receptor (nAChR), is activated by ACh binding (Figs. 5 and 9). Each receptor activated progressively brings about an initial incremental muscle membrane depolarization, and these events, referred to as *miniature endplate potentials* (MEPPs), were critical in uncovering the quantal nature of synaptic transmission. This receptor is the target of autoimmune and genetic myasthenia syndromes, and these quantal synaptic events are disturbed by these diseases, which are discussed below. In the physiological condition, a large number of tiny receptor-mediated MEPPs summate, producing an endplate potential that brings the membrane potential to the *threshold* at which a few sodium channels begin to open (Fig. 10). The sodium channels in turn further depolarize the membrane, causing more adjacent sodium channels to open in a reinforcing cycle until all sodium channels in the muscle have explosively opened while rapidly spreading a depolarizing wave across the surface of the muscle. At this point, all of the sodium channels are rendered "inactive" and nonexcitable, beginning the process of repolarization, a process completed when more slowly activating potassium channels restore the resting potential and return the sodium channels to the "closed" state. Unlike the situation in cardiac muscle, the rapid recovery of skeletal muscle excitability allows rapid repetitive contractions, an advantage in performing fine precision movements. The major subunit of muscle sodium channels is the muscle-specific isoform $\alpha 4$, encoded by *SCN4A*, a relative of the gene involved in *LQT3*. Alleles of this one gene produce the diverse array of muscle phenotypes that will be discussed below.

As the depolarization wave of the muscle AP spreads across the sarcolemmal plasma membrane, it extends into specialized membrane invaginations, called *T-tubules*, which dive deep into the cytoplasm and ramify near the contractile proteins (Fig. 9). Their function is to facilitate calcium-dependent activation of contraction throughout the large muscle fiber. In the muscle T-tubule, as in the motor nerve bouton, the depolarization wave activates the voltage-gated calcium channel, which is similar in structure to

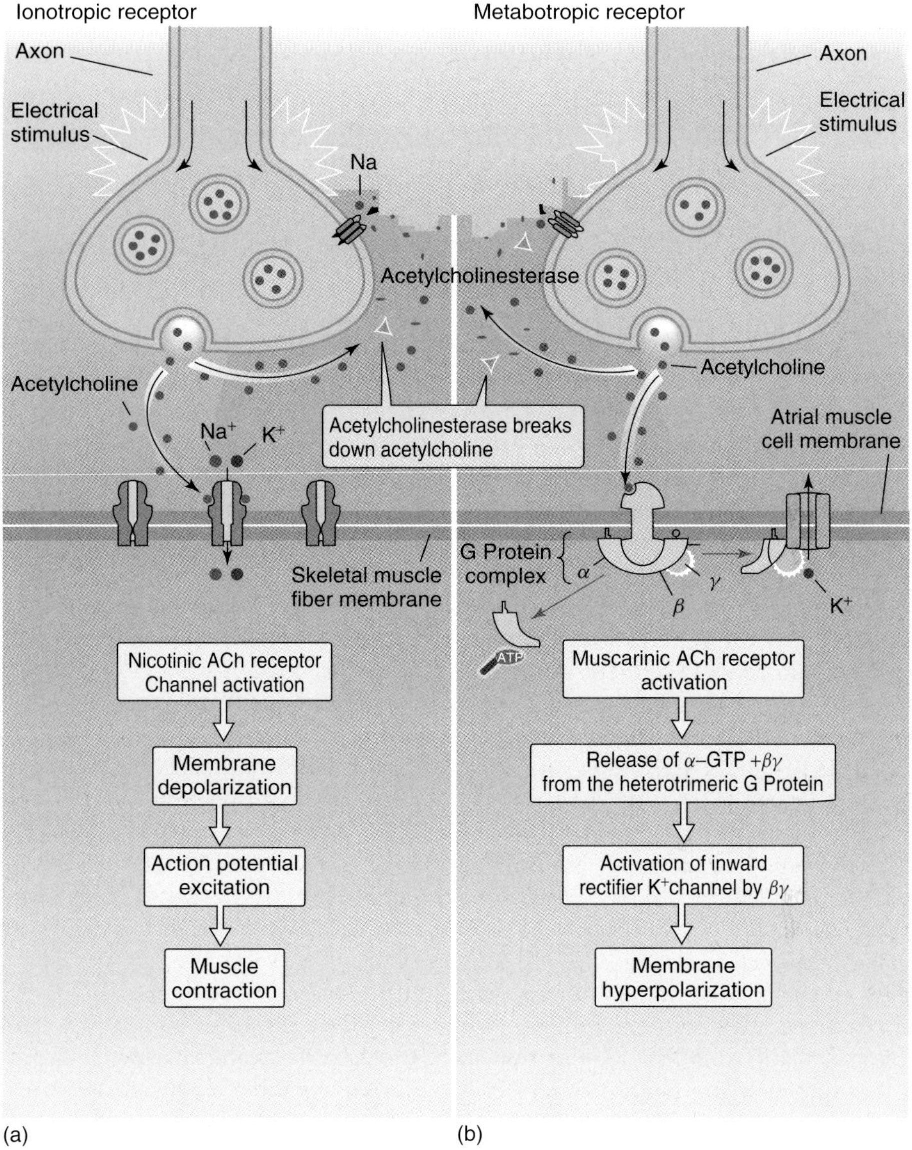

Fig. 10 Mechanisms of acetylcholine receptor activation.

the sodium channel, discussed above. The major, α_1, subunit is pseudotetrameric and contains the ion-conducting pore. Its function is modified by auxiliaryβ, γ and α_2/δ subunits, the skeletal muscle calcium channel's isoforms being encoded by *CACNA1S, CACNB1, CACNG1,* and *CACNA2D1,* respectively (Fig. 5). This channel is responsible for rapidly immersing the contractile proteins throughout the large muscle cell in elevated concentrations of ionized calcium, thereby

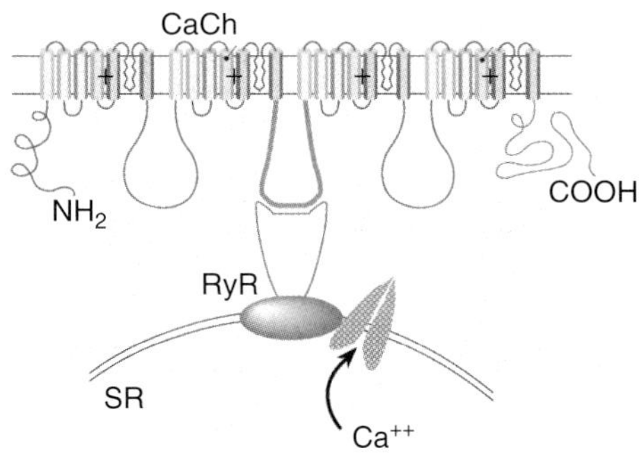

Fig. 11 Ca channel and Ryanodine receptor.

bringing about concerted contraction, much as was described for the cardiac contractile machinery. It does this in part by allowing the passage of extracellular calcium through its pore, down its electrochemical gradient into the cytoplasm. But additionally a large cytosolic loop in the skeletal muscle α_1-subunit contacts a *different* calcium channel located in a closely opposed *different* membrane compartment, the sarcoplasmic reticulum (SR) (Figs. 5 and 11). The SR membrane contains calcium pumps that rapidly sequester intracellular calcium after each AP, creating a releasable calcium store and restoring muscle relaxation. Through the physical interaction with its cytosolic loop, the voltage-gated calcium channel directly gates the SR calcium release channel. The release channel is called the *ryanodine receptor*, encoded by *RYR1*, and opening it spills the SR intracellular calcium stores into the cytoplasm. After activation, both the sodium and calcium voltage-gated channels spontaneously enter the nonconducting "inactive" conformation, beginning the process of membrane repolarization, a process completed by potassium channels that hyperpolarize the membrane, resetting the sodium and calcium channels into the "closed" conformation. The sodium and calcium channels then are kept closed by the large stabilizing current of the chloride channels, encoded by *CLCN1* (Fig. 5).

6.2
Ion Channel Dysfunction in Paralysis and Myotonia

The two major skeletal muscle phenotypes produced by ion channel disorders reflect either an inability to contract the muscle, ranging from weakness to *paralysis*, or an inability to relax the muscle, *myotonia*. While a third muscle phenotype, malignant hyperthermia (MHS), appears quite different, it is surprisingly closely related to the paralysis syndromes, in fact, being allelic. Two major themes in the ion channel diseases begin to develop in discussing the skeletal muscle diseases. First, it becomes clear that the LQT syndrome provides a powerful paradigm for understanding the types of alleles that produce disease and the mechanisms by which they act. Second, the proclivity of certain ion channel gene families to produce disease begins to emerge as does their ability to produce distinctly different clinical phenotypes with subtly differing alleles (Table 1).

While the pathophysiology of paralysis and hyperthermia is complex, that of myotonia is fairly straightforward. Since the high chloride conductance of *CLCN1* in the muscle membrane serves to stabilize

Tab. 1 Channelopathy gene families.

Gene family	***Tissue***	***Disease phenotype***	***Pathogenic mutation***
KCNE1	Heart/ear	LQT5/JLN	Dominant negative/recessive loss-of-function
KCNE2 (100%)[a]	Heart	LQT6	Dominant negative
KCNE3	Muscle	HOKPP	Dominant negative
KCNQ1	Heart/ear	LQT1/JLN	Dominant negative/recessive loss-of-function
KCNQ2	CNS	BFNC1	Dominant loss-of-function
KCNQ3 (80%)[a]	CNS	BFNC2	Dominant loss-of-function
KCNQ4	Ear	DFNA2	Dominant missense
KCNQ5	CNS	6q14	
RYR1	Muscle	MHS1	Missense
RYR2 (66%)[a]	Heart	Ventricular tachycardia	Missense
RYR3	CNS	15q14	
SCN1A	CNS	GEFS + /SMEI	Gain-of-function/loss-of-function
SCN2A	CNS	GEFS+	Gain-of-function
SCN3A	CNS	2q24	
SCN4A (40%)[a]	Muscle	HYPP/HOKPP/MHS/etc.	Gain-of-function
SCN5A	Heart	LQT3/Brugada S	Gain-of-function
SCN6A and *SCN7A*	Neuronal	2q23	
SCN8A	Motor endplate	12q13	
SCN9A	Neuroendo	2q24	
SCN10A	Nerve/muscle	3p22	
SCN11A and *SCN12A*	Sensory neurons	3p24	
SCN1B	CNS	GEFS+	Gain-of-function
SCN2B	Neuronal	11q23	
CACNA1S	Muscle	HOKPP/MHSS	Missense
CACNA1A	CNS	FHM1/EA2/SCA6/seizures	Many types
CACNA1F	Retina	Night blindness	Hemizygous loss-of-function
CHRNA1	Muscle	SCCMS	Gain-of-function

Sources: Gargus, J.J. (2003) Unraveling monogenic channelopathies and their implications for complex polygenic disease, *Am. J. Hum. Genet.* **72**(4), 785–803.
[a]Percent of gene family members already associated with disease.

the membrane potential, thereby preventing "echo" contractions after an induced contraction, its functional loss is associated with the inability to normally relax the muscle, and hence the slow muscle relaxation and muscle stiffness of myotonia. Both dominant and recessive mutations of this gene, which decrease channel current, produce myotonia and a characteristic "dive bomber" EMG record of repetitive contractions. The repetitive volleys of contraction can arise in skeletal muscle because unlike in the heart, where sodium channels are held in the inactive state for a prolonged period to prevent arrhythmia, in skeletal muscle, sodium channels must be ready to accept a rapid train of controlled activations from the motor nerve. Problems arise, however, in the absence of a strong chloride stabilizing current, since instability of the membrane potential during recovery from a depolarization leaves a high probability that a secondary "echo" depolarization will reach threshold and trigger a repeated activation of the sodium channels.

The recessive loss-of-function alleles of *CLCN1* produce Becker myotonia, with a phenotype characterized by stiffness that warm-up exercise improves, and paradoxically weak hypertrophied muscles. The dominant-negative alleles leave more residual current and produce Thomsen myotonia, having a similar but milder phenotype without weakness. The hypertrophy arises because the muscles are constantly "exercising" even at rest secondary to their electrical instability, and the weakness arises because of incomplete recovery from sodium channel inactivation, as will be described for the periodic paralysis syndromes below.

6.3 Periodic Paralysis Syndromes

To understand how the array of periodic paralysis syndromes arise, it is essential to understand how very subtle "tweaks" in sodium channel function give rise to macroscopic phenotypes that are quite distinct – superficially the antithesis of one another – yet are so closely related that a given patient can transition from one state to the other. All of these pathogenic muscle sodium channel alleles appear to share with the cardiac sodium channel *LQT3* alleles a dominant gain-of-function mechanism of pathogenesis – *delayed and incomplete inactivation*. The essential biophysical feature of sodium channel function that makes this interpretable is the fact that while a depolarization will make it more probable that the voltage-sensitive gate of the closed channel will open, when sustained, depolarization also leaves behind more of the channels in an "inactive" nonexcitable conformation. As the sodium channel is the major mechanism for propagating APs in muscle (as it is in most tissues), this delicate balancing act has the potential to move these tissues from a state of hyperexcitability to a state of inexcitability, manifested as weakness and paralysis. Plasma potassium levels are commonly altered during these attacks of paralysis, and, while incompletely understood, probably has to do with the fact that the vast majority of the potassium in the body, >95%, is found in the muscle. The anxiety that surrounds clinical management of such changes is that very small amounts of potassium dramatically change the plasma potassium concentration and this is a critical determinant of the potassium Nernst potential (hence the membrane potential of all cells) and such changes in the heart pose the risk of fatal arrhythmia.

The ranges of phenotypes gain-of-function *LQT3*-like incompletely inactivating *SCN4A* alleles produce include: (1) hyperkalemic periodic paralysis (HYPP), (2) hypokalemic periodic paralysis (HOKPP), (3) paramyotonia congenita, a cold exacerbated myotonia and (4) potassium-activated myotonia.

HYPP alleles are the most common, and the disorder is characterized by short, mild frequent attacks of profound weakness beginning in infancy and provoked by rest just after exercise or stress. Commonly, the individual performs significant physical exertion, then sits to rest and finds he is unable to move for an hour or more (although respiration is not compromised). There is often myotonia between attacks. During an attack, plasma potassium levels rise to pathological levels, probably via release from the depolarized muscle. The alleles are characteristically missense mutations altering amino acids in the transmembrane spans of the channel protein. Andersen syndrome, discussed above as caused by *LQT7* alleles, additionally has the muscle phenotypes of HYPP and that of MHS, discussed below, suggesting an intrinsic relation between the two. Furthermore, some HYPP alleles of *SCN4A* additionally produce MHS, and the gene is probably also the *MSH2* locus, a rare cause of the phenotype.

HOKPP is characterized by an onset in the second decade of infrequent, profound painless episodes of weakness that can last for days. During an attack, the plasma potassium can fall to dangerous levels, presumably being driven into the muscles. The factors known to incite the attacks are known to physiologically have this activity. They include insulin or glucose intake (which stimulates insulin release) and the hormonal changes that occur upon awakening. Two other loci, *CACNA1S* and *KCNE3*, also have alleles that produce HOKPP. Missense dominant alleles of *CACNA1S*, encoding the major α_1, subunit of the muscle voltage-gated calcium channel, were the first recognized and remain the most common cause of hypokalemic periodic paralysis. Most recently mutations in one of the skeletal muscle potassium channels involved in the rapid repolarization of the membrane, encoded by *KCNE3*, a relative of the *LQT5* and *LQT6* genes also have been shown to cause HOKPP. With this discovery, all of the members of the fully ascertained *KCNE* gene family are proven pathogenic loci, suggesting such familial relationships may be useful in promoting candidate genes for evaluation in disease association (Table 1).

6.4 Malignant Hyperthermia

The pathophysiology of malignant hyperthermia is typical of a pharmacogenetic syndrome; it involves an inciting agent, typically an inhalation anesthetic such as halothane, and an individual with a predisposing genotype. The pathogenic mechanism arises from a dramatic rise in muscle cytoplasmic calcium as the hypersensitized SR stores are discharged. This then secondarily produces contracture of the muscles and a hyper-metabolic state of glycogenolysis, ATP depletion, and lactic acidosis driven by the attempt to resequester calcium. This process evolves to produce hyperthermia (not fever) from activity of the contractile machinery and mitochondrial respiration, and disruption of the muscle membranes, releasing myoglobin, creatine kinase, and potassium, and leading to renal failure, disseminated intravascular coagulation and death The predisposing genotypes enhance the sensitivity of the calcium release process, which primarily involves the sarcolemmal calcium channel and the SR calcium release channel. Missense alleles altering either the cytosolic loop of the *CACNA1S*-encoded α_1 calcium channel subunit or its contact domains on the ryanodine receptor render this complex hypersensitive and capable of triggering massive calcium release, muscle activation, and malignant hyperthermia (Fig. 11). Over 50% of the cases are caused by dominant *MHS1* alleles of *RYR1* and the next most common cause is dominant *MHS5* alleles of *CACNA1S*. Some of the HYPP alleles of *SCN4A* additionally seem to produce this syndrome, and it probably reflects *MHS2*. Three other *MHS* loci have been mapped but susceptibility alleles have yet to be identified. Additionally, this phenotype is also a component of Andersen syndrome, discussed above, affecting the potassium channel that controls the resting membrane potential in muscle and heart, *KCNJ2* (Fig. 8). Finally, alleles of *RYR1* also cause central core disease (CCD), a congenital myopathy characterized by nonprogressive proximal muscle weakness and distinctive muscle fiber histology. While individuals with CCD do have a susceptibility to anesthetic-induced MHS, the remainder of the disease phenotype is constitutive and does not require pharmacological induction, and therefore the disorder is illustrative of the continuum between genetic and environmental disease.

6.5 Other Monogenic Muscle Ion Channel Syndromes

Dominant gain-of-function mutations in four of the five subunits that compose the nAChR found at the neuromuscular junction and encoded by *CHRNA1*, *CHRNB1*, *CHRND* and *CHRNE* produce slow-channel congenital myasthenic syndrome (SCCMS), a phenotype characterized by muscle weakness. Acetylcholine dissociates slowly from these mutant receptors, leaving them persistently activated, depolarizing the membrane. One might guess that the prolonged receptor activation the mutant alleles produce will give a hyperexcitable muscle; however, as was discussed above in the periodic paralysis syndromes, the phenotype seen is produced by the loss of sodium channels into the inactive state in this chronically depolarized tissue.

6.6 Autoimmune Channelopathies in Muscle

Myasthenia gravis (MG) is one of the classic autoimmune diseases, first placed on a molecular footing in the early 1970s, well before the era of the monogenic channelopathies. It develops in mid-life, produces facial, respiratory, and limb muscle weakness and is associated with tumors of the thymus and antibodies to the nAChR. It is clear that the antibodies produce the disease, since they can passively transfer the symptoms and their removal is therapeutic. The autoantibodies accelerate the destruction of the nAChR, reducing their abundance in the synapse and hence the efficacy of synaptic transmission. This was first observed as a decrease in the size, but not number, of MEPPs, suggesting that the ACh-filled vesicles were normally releasing quanta of neurotransmitter, but that they had a smaller postsynaptic effect, causing weakness. Enhancing the synaptic concentration of ACh by inhibiting acetyl cholinesterase rapidly improves strength, and is used diagnostically and therapeutically (Fig. 9). Note that the muscle phenotype is very similar to that created by SCCMS alleles of the nAChR subunit genes, discussed above, but that those genetic syndromes do not have thymus tumors or autoantibodies, confirming that receptor dysfunction does not cause those findings in MG, the tumor appearing to be primary and causal in autoimmunity.

Lambert-Eaton myasthenia syndrome (LEMS) is another autoimmune disorder that produces a clinical picture very similar to that of MG, but preferentially targeting proximal limb muscles and being associated with small-cell carcinoma of the lung. It additionally produces symptoms referable to the autonomic nervous system, such as dry mouth, so it is clear that the epitope targeted by the antibodies is not strictly limited to muscle. In fact, the antibodies are directed at the P/Q-type voltage-gated calcium channels of the presynaptic neuron. This channel is encoded by *CACNA1A* and is similar to its family members expressed in heart and skeletal muscle, but it participates in excitation-*secretion* coupling (not EC coupling), serving as a calcium entry pathway that couples the propagated AP to the biochemical pathways underlying neurotransmitter release at the presynaptic bouton, to be discussed below. It is distinguished from its family members by its high affinity for ω-conotoxin. Antibodies do not cross the blood-brain barrier, so symptoms in the autoimmune disorders are confined to effects in the periphery. As will be seen below, very different symptoms arise from genetic lesions in *CACNA1A* where CNS symptoms predominate.

7 Ion Channel Diseases of Nerve

7.1 Function of Ion Channels in Neurons

The neuronal AP is very similar to that of muscle: typically, activation is initiated across a *synapse,* neuroreceptors and their endogenous or coupled channels are activated, the supra-threshold summated postsynaptic currents in turn trigger a rapidly spreading depolarization wave by voltage-gated sodium and calcium channel currents, this is rapidly followed by a repolarization driven by potassium currents, and the AP couples to biochemical effector pathways through a rise in the cytosolic calcium concentration. Despite this gross oversimplification of all of the diversity of neuronal structure and function, and their vastly more complex geometry, the major distinctions from muscle lie at the afferent and efferent ends of the neuronal transduction pathway, not in the propagated AP in between.

First, neurons have a vastly more complex variety of activating, as well as inhibitory, neuroreceptors and synapses than muscle, and they are primarily found at the branching *dendritic,* or afferent, ends of the prototypical neuron (Fig. 12). The receptors produce both depolarizing, excitatory, as well as hyperpolarizing, inhibitory, *postsynaptic potentials* (EPSPs and IPSPs) that, like the MEPPs of muscle, summate over time, but also through the complex spacial geometry of the arborized dendritic tree. This summation serves the "computational" function of the neuron; some combinations of dendritic inputs will drive the soma, or neuronal cell body, to initiate an AP that is propagated down the axon, while others that differ slightly in timing, spacial geometry of the activated synapses, or the mix of incoming neurotransmitters will not.

Second, when the neuron does "decide" to produce a propagated AP, the elevated cytosolic calcium wave does not participate in EC coupling, but rather in excitation-secretion (ES) coupling, the controlled release of neurotransmitter into another synapse (Fig. 10), as was briefly discussed for LEMS, above. In the neuron, the calcium-coupled neurosecretion events occur after the AP has been conducted down the *axon,* the long cellular projection that serves as the efferent and effector side of the neuron. There a wide variety of small-molecule neurotransmitters can be released into a synapse to initiate in the downstream neuron the process just described above (Fig. 12).

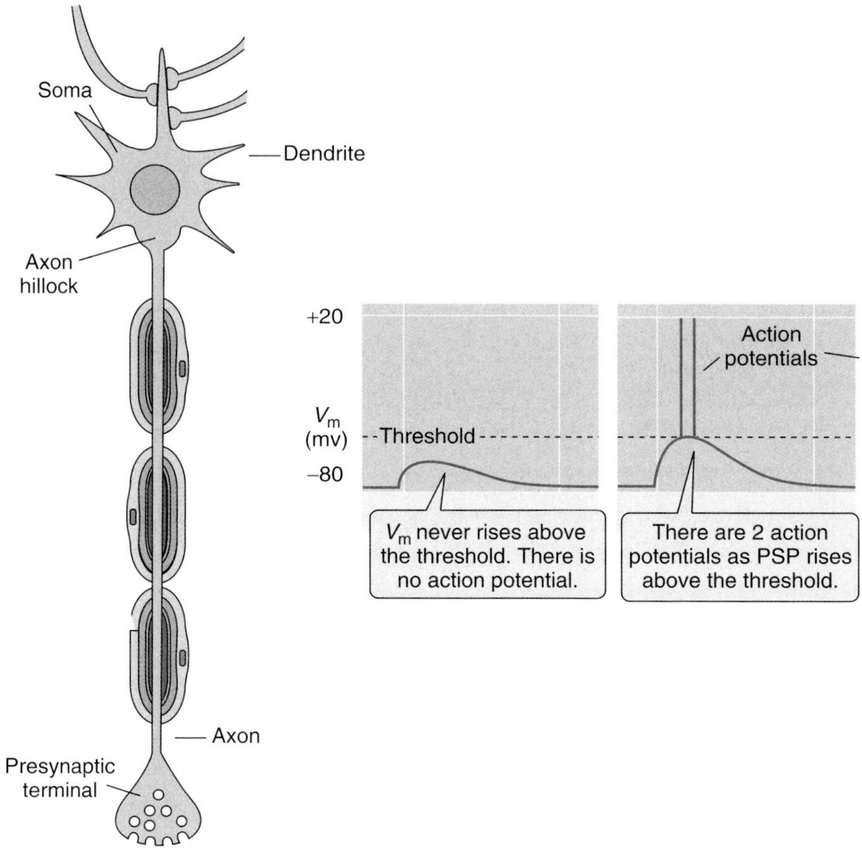

Fig. 12 Neuron structure and initiation of neuronal AP.

7.2
Ion Channel Dysfunction in Seizures

As opposed to the situation in heart and muscle, synchrony of electrical activity in the CNS is actively prevented. An abnormally synchronous discharge of neurons in the CNS produces a *seizure*, a pathological state characterized by a stereotyped alteration in behavior. Seizures are the most common monogenic ion channel phenotype in the CNS and in many ways are similar to the arrhythmia phenotype that arises in the hyperexcitable LQT heart. Both are periodic disorders in constitutionally hyperexcitable tissues in which the normal rhythmic electrical activity of the tissue is temporarily lost. However, whereas the heart is typified by its extremely homogeneous set of responding cells, the CNS displays maximal tissue complexity both in terms of the number of different cell types present and in the number of different ways they connect with and influence one another in stimulatory or inhibitory fashions. Therefore, while one can visualize the very specific and individual contribution each channel type makes to components of an action potential perturbed in LQT in the heart or in the skeletal muscle syndromes, one can much less specifically interpret how

Tab. 2 Unlike heart, must still guess WHERE neuronal channels act.

Seizures:		
BFNC1	*KCNQ2*	Haploinsufficiency, decrease I_K
BFNC2	*KCNQ3*	Haploinsufficiency, decrease I_K
ADNFLE	*CHRNA4*	GOF increase activation, increase I_{Na}
ADNFLE	*CHRNB2*	GOF increase activation, increase I_{Na}
GEFS+	*SCN1B*	GOF decrease inactivation, increase I_{Na}
GEFS+	*SCN1A*	GOF decrease inactivation, increase I_{Na}
GEFS+	*SCN2A*	GOF decrease inactivation, increase I_{Na}
GEFS+	*GABRG2*	Dominant missense, decrease I_{Cl}
JME	*GABRA1*	Dominant missense, decrease I_{Cl}
JME	*CACNB4*	Dominant missense, I_{Ca}
SMEI	*SCN1A*	Haploinsufficiency, decrease I_{Na}
...[f]	*KCNA1*	Haploinsufficiency, decrease I_K
...[f]	* *CACNA1A*	Dominant negative, decrease I_{Ca}
Ataxia		
...[f]	* *CACNA1A*	Dominant negative, decrease I_{Ca}
SCA6	* *CACNA1A*	Dominant CAG expansion, decrease I_{Ca}
EA2	* *CACNA1A*	Haploinsufficiency, decrease I_{Ca}
EA1	*KCNA1*	Haploinsufficiency, decrease I_K
Migraine:		
FHM1	* *CACNA1A*	Dominant missense, decrease I_{Ca}
FHM2	*ATP1A2*	Dominant loss-of-function

Sources: Gargus, J.J. (2003) Unraveling monogenic channelopathies and their implications for complex polygenic disease, *Am. J. Hum. Genet.* **72**(4), 785–803.
* Different manifestations of P/Q channel dysfunction;
f – case reports, discussed in text.

or even where the channel dysfunction underlying seizures occurs. Also, rhythm disorders in the heart are often a once in a lifetime event; in the CNS they are usually a lifelong chronic condition, *epilepsy*. Epilepsy is also a useful phenotype to serve as a springboard to understanding the other major monogenic CNS channelopathy phenotypes, *ataxia* and *migraine*, as they can be allelic.

While the complexity of neuroanatomy obscures functional analysis of the mutations causing monogenic epilepsy syndromes, extrapolation from the cardiac and muscle syndromes provides a simplified predictive framework of molecular pathology: K^+ and Cl^- channels, which physiologically stabilize excitable tissue, will likely have pathological lesions that diminish their current, and Na^+ and Ca^{++} channels, which physiologically excite the tissue, will likely have gain-of-function lesions (Table 2).

7.3 Hereditary Epilepsy Syndromes

Nearly 40% of all epilepsy is associated with no macroscopic pathological lesion, and is therefore classically referred to as *idiopathic epilepsy*, a common disorder that affects ~1% of the population. It is overwhelmingly a genetic disease, with monozygotic twins greater than 95% concordant for the phenotype; however, nearly all cases are polygenic and most of those genes still remain to be identified. Obviously, since ion channels control electrical activity in the CNS, as in the heart and muscle, they are strong functional candidates for this disorder. In man, they were the first identified, and remain nearly the exclusive, proven genetic causes of epilepsy. Only a handful of rare monogenic epilepsy syndromes are recognized, but predictably success at pathogenic allele definition has thus far come only from these rare monogenic syndromes.

7.4 LQT-like Loci and Alleles in Neuronal Syndromes

The first rare monogenic seizure syndrome for which the major etiological genes were identified was benign familial neonatal convulsions (BFNC), a rare autosomal dominant disorder characterized by a brief period of seizures in the neonatal period, generally resolving in weeks, but 10% having persistent adult epilepsy. The BFNC loci were mapped to chromosomes 20 and 8. The *BFNC1* gene was positionally localized within chromosome 20q13.3 taking advantage of a family with the syndrome and a microdeletion chromosome. The interrupted locus contained a promising candidate gene, *KCNQ2*, a member of the *LQT1* gene family that was predominantly expressed in neurons. Unlike the *LQT1* alleles that demonstrate dominant-negative interactions *in vitro*, this first *BFNC1* allele was a null, and most subsequent alleles are simply loss-of-function. The relevant neuronal currents mediated by this channel must be so critically tuned that pathogenic alleles can achieve dominance simply via haploinsufficiency. Heterozygous null mice display a milder inducible pharmacogenetic syndrome, showing no basal seizures, but only an increased sensitivity to seizure-inducing drugs. Presumably, the strain's polygenic background raises their seizure threshold making them appear much like silent-carrier family members of the "sporadic" LQT syndrome probands who have disease produced by weak alleles.

Since the *KCNQ* family had the demonstrated ability to produce disease, other family member genes were sought by homology. *KCNQ3* was thus identified, and mapping to chromosome 8q, was a tempting candidate for the second BFNC locus, a hypothesis proven by finding loss-of-function alleles segregating in a family. While both KCNQ subunits produced channels when expressed *in vitro*, neither subunit alone produced a recognizable current. However, KCNQ2 and KCNQ3 were subsequently recognized to heteromultimerize to form the "M current," a long-sought signature potassium current activated by muscarinic acetylcholine receptors (mAChR) (Fig. 10). Therefore, both their gene-family relationship and their functional subunit interaction help explain the common phenotype that mutations in either gene produce. The final member of this *KCNQ* family, *KCNQ5*, mapping to chromosome 6q14, and the only member still to be associated with a disease, is expressed in neurons and can also interact with KCNQ3 to produce M current, making it an extremely strong functional candidate disease gene.

7.5 Other Monogenic Ion Channel Epilepsy Syndromes

The gene *CHRNA4* was the first ion channel gene demonstrated to contribute pathogenic alleles to an epilepsy syndrome. It is found on chromosome 20q13 and encodes the most abundant neuronal isoform of the major subunit, α_4, of the nAChR. Unlike the mAChR, discussed immediately above, which couples to channel gating via G-protein activation (a G-protein coupled receptor, GPCR), the nAChR, like that found in muscle, is a heteropentameric receptor/channel that is *itself* a ligand-gated nonselective cation channel, its activation depolarizing the membrane (Fig. 10). It is closely related to the muscle nAChR, which participates in the phenotype of SCCMS congenital myasthenia. In the nerve, it is composed of two α- and three β-homologous subunits, each with four transmembrane α-helices (Fig. 5). The rare seizure syndrome, autosomal dominant nocturnal frontal lobe epilepsy (ADNFLE), producing a phenotype with brief clusters of frontal seizures that occur at night, was mapped to this region in a family and missense alleles identified. To date, three alleles have been recognized, all altering the channel pore region and receptor function but having no obvious common effect on *in vitro* channel behavior. More recently, mutations in the most abundant neuronal β-subunit, β_2, were also recognized to cause this syndrome, suggesting that the channel isoform relevant to this phenotype is an $\alpha_4\beta_2$ pentamer. These missense alleles in *CHRNB2*, located on chromosome 1q21, produced the common *in vitro* effect of receptor activation, suggesting that some depolarizing gain of function is probably the mechanism of dominant pathogenesis, like that found in the SCCMS muscle pathogenic homologs. The vulnerable cell in the CNS whose function is perturbed by these mutations, however, remains to be defined.

Fever, a common accompaniment of many illnesses, lowers the seizure threshold so it is not surprising that febrile seizures are by far the most common polygenic seizure disorder, affecting 3% of children worldwide. While these susceptibility genes remain to be defined, a rare Mendelian dominant seizure syndrome that includes febrile seizures provides the first lead to their identification. This febrile seizure syndrome additionally evolves to include a variety of afebrile seizures, and is called *generalized epilepsy with febrile seizures plus* (GEFS+). The first locus responsible for GEFS+ was mapped in a large family to chromosome 19q13.1 and a missense mutation was identified in an auxiliary subunit of the voltage-gated sodium channel. This subunit, the β_1 subunit encoded by the gene *SCN1B*, has a single transmembrane helix and functions only to modify the activity of the large α-subunit (Fig. 5). Subsequently, mutations producing GEFS+ or closely related syndromes were identified in two different adjacent neuronal sodium channel α-subunit genes, *SCN1A* and *SCN2A*, on chromosome 2q24. Not until the α_1 subunit alleles were coexpressed with their auxiliary β-subunits did a reproducible picture of defective channel inactivation and sustained sodium current appear. This is a picture similar to that observed with the *LQT3* and HYPP and HOKPP alleles of the gene-family members *SCN5A* and *SCN4A*, respectively. As in the heart, most simply, a persistent sodium channel current can be viewed as a gain-of-function dominant lesion that favors a hyperexcitable state; however, the periodic paralysis lesions caution that hyperexcitability can easily transition into inexcitability, through the inactivation mechanism, and this dichotomous nature of sodium channel lesions is reinforced by the recent observations that a rare, very severe dominant seizure syndrome, initially associated with febrile seizures, called *severe myoclonic epilepsy of infancy* (SMEI), proved to be allelic with GEFS+. All of dozens of known alleles were shown to be new mutations in *SCN1A*, and their nature (e.g. frameshift, nonsense) suggests they are predominantly functional nulls. While the physiological interpretation of these findings is clear, together they are puzzling; haploinsufficiency for the major subunit of the sodium channel produces a very severe dominant *loss*-of-function phenotype via some critically tuned cell type; a milder syndrome being caused by the dominant *gain*-of-function GEFS+ alleles.

The remaining GEFS+ locus, *GABRG2*, found within a cluster of GABA receptor genes on chromosome 5q31, encodes the γ_2 subunit of the inhibitory $GABA_A$ receptor. The $GABA_A$ receptor, like the nAChR discussed above, is a pentameric ligand-gated channel, but it differs in that instead of conducting a depolarizing cationic current, it conducts a stabilizing chloride current (Fig. 5). This receptor is the major mechanism through which GABA, the principle *inhibitory* neurotransmitter in the brain, functions. It is at this site that the benzodiazepine drugs act to treat seizures, by potentiating GABA effects, and that toxins, such as picrotoxin, act to induce seizures, by blocking GABA. So physiologically the inhibitory, neuronal silencing, activity of GABA participates in preventing seizures. The GEFS+ allele produces a decrease in GABA-induced chloride current *in vitro*. Therefore, like the seizure-inducing toxins, seizure-inducing mutations alter receptor/channel function to produce *less* GABA-induced stabilizing current.

While some human "seizure" genes have not recapitulated that phenotype in mice (for instance, mice homozygous for null alleles of *CHRN4A* and *CHRNB2* – the two loci thus far recognized with alleles causing ADNFLE in man – have a phenotype that alters the pain response, not seizures), spontaneous mutations producing a seizure phenotype in mice have led to the recognition of an important new category of ion channel seizure disorders in man, those altering voltage-gated calcium channel function. More importantly, those loci have proven to be particularly informative for their ability to tie the various neurological phenotypes together in an interpretable fashion. As discussed above, the voltage-gated calcium channels are similar to the voltage-gated sodium channel, and the major pore-containing α_1 subunit that dictates the channel's subtype is pseudotetrameric with its function modified by the auxiliary β, γ, and α_2/δ subunits (Fig. 5). Alleles in all four different types of neuronal voltage-gated calcium channel subunit genes produce a monogenic absence seizure phenotype in mice. The "lethargic," "stargazer," and "ducky" mice are recessive phenotypes produced by mutations altering neuronal calcium channel auxiliary subunits, *CACNB4* encoding a β-subunit, *CACNG2* encoding a γ-subunit, and *CACNA2D2* encoding a protein cleaved to produce the α_2 and δ-subunit, respectively. Of these three genes, a human phenotype has been reported only for mutations in the first. The human *CACNB4* gene maps to human chromosome 2q22-q23 and two dominant alleles, a truncation and missense allele, were found in families with a dominant seizure syndrome. The human *CACNG2* maps to chromosome 22q13.1 and the human *CACNA2D2* gene maps to 3p21.3, both remaining only promising seizure candidate genes in man. The "tottering" and "leaner" mice are recessive seizure phenotypes produced by alleles of *CACNA1A*, the gene encoding the major α_1 subunit of the neuronal P/Q-type calcium channel. The P/Q-type calcium channel plays a major role in calcium entry underlying synaptic release of the major *excitatory* neurotransmitter, glutamate, and while this process is decreased in the mutants, suggesting they are hypomorphic alleles, it is unclear how this produces a hyperexcitable seizure phenotype.

7.6 Monogenic Ataxia and Migraine Syndromes

While the "tottering" and "leaner" mice are recessive seizure phenotypes produced by alleles of *CACNA1A*, a seizure phenotype is *not* the major phenotype caused by mutations in this gene in man. Human mutations in *CACNA1A*, which maps to chromosome 19p13, have been shown to cause three apparently distinct and different late-onset neurological disease phenotypes: episodic ataxia type 2 (*EA2*), familial hemiplegic migraine type 1 (*FHM1*), and spinocerebellar ataxia type 6 (*SCA6*). The *EA2* alleles are predominantly truncations, with over 22 known alleles, including frameshift and splice site mutations, but also five missense alleles, some altering conserved pore residues producing complete loss of function. It is therefore presumably a haploinsufficiency dominant loss-of-function syndrome. Familial hemiplegic migraine is caused by at least nine known missense alleles producing no obvious uniform functional change in calcium current *in vitro*; therefore, the relevant functional change they share in common is presumably subtle. Finally, SCA6 is a progressive degenerative phenotype primarily caused by trinucleotide expansion alleles. The polymorphic CAG repeat encodes a C-terminal polyglutamine repeat of 5 to 20 residues in unaffected individuals. Pathogenic alleles encode 21 to 30 glutamines. The only point mutant allele recognized to produce the syndrome is G293R, which changes a conserved pore residue. An intermediate-length CAG repeat allele with only 20 repeats produced the milder phenotype of EA2.

Thus, the four allelic *CACNA1A* diseases in man and mouse are clearly closely related at the molecular level. The phenotypes also can blend one into the other, for instance, many individuals with FHM have ataxia as do most of the murine seizure mutants, and recently a dominant-negative truncation allele has been reported to produce a syndrome of progressive ataxia and seizures in a child. In man, the mildest alleles appear to be the missense mutations producing FHM1. On a phenotypic continuum, these appear to be hypomorphic since the truncations, pore mutations, and intermediate repeat expansions produce EA2, apparently a haploinsufficiency syndrome. Expanding the polyglutamine repeat by only a few additional residues yields the potent long-repeat alleles producing the progressive SCA6 syndrome, which can also be produced by one specific missense pore mutation. While the mechanism of action of all pathogenic polyglutamine expansion alleles are controversial, if the three human *CACNA1A* phenotypes are truly the continuum suggested by the EA2 intermediate repeat phenotype, these repeat expansion alleles should be dominant negative to eliminate more of the P/Q current than haploinsufficiency and to keep the progression of allele potency parallel to the severity of the symptoms and pathology. While all of the human alleles produce the phenotype in heterozygotes, the mouse seizure alleles, which are missense, like FHM1, and demonstrated to be hypomorphic, produce no phenotype in heterozygotes, including knockout null alleles; decreased synaptic glutamate neurotransmitter release and seizures being produced only in the homozygotes.

Another strong clue of the relationship between ataxia and seizures comes from the other known episodic ataxia locus, *EA1*. EA1 is produced by dominant loss-of-function alleles of perhaps the most "famous" potassium channel, "shaker", named after its mutant phenotype in fly, and the paradigmatic first K^+ channel cloned. It is a voltage-gated potassium channel formally called *KCNA1*. Knockout null alleles in the mouse produce a recessive seizure disorder, with the heterozygotes lacking a phenotype, just like the calcium channel mutants, but like them showing a reduced threshold for induced seizures, hinting at a potential mechanism for pharmacogenetic neurological syndromes in man. While the vast majority of EA1 patients have only ataxia, a few cases have been reported to additionally have seizures.

Just as examining both loci that produce episodic ataxia, *EA1* and *EA2*, informs the underlying pathophysiology of the indistinguishable diseases they produce, so too does examining both loci that cause familial hemiplegic migraine, *FHM1* and *FHM2*. Whereas *FHM1* alleles are subtle missense mutations of the calcium channel-encoding *CACNA1A* gene, recently *FHM2* alleles were discovered to be mutations in *ATP1A2*, the gene encoding the $\alpha 2$ subunit of the Na/K-ATPase, one of the major ion pump mechanisms responsible for establishing the ion gradients tapped by ion channels (Fig. 6). This is the first disease that is established to be caused by mutations in the Na/K pump, a central physiological mechanism. All FHM2 alleles identified to date are missense and alter narrowly conscribed domains of the pump protein. While loss-of-function and haploinsufficiency mechanisms are more suggestive of deletions and other forms of more readily produced null alleles than the observed constrained missense, this has been suggested to be the mechanism of some pathogenic alleles; however, detailed enzyme kinetics have more recently shown others to achieve dominance by gain of function, producing altered kinetics of the pump. It is still not clear what biophysical property of the neuron is perturbed in common by all of the FHM lesions, although a common denominator seems to be membrane depolarization and elevated cytosolic calcium.

7.7 Other Monogenic Neuronal Ion Channel Syndromes

One of the first monogenic CNS ion channel disorders for which the underlying pathogenic alleles were identified is the rare disorder hyperekplexia (HE), more commonly called *Startle syndrome*. The phenotype includes hypertonia and a severe startle reaction that simulates a drop seizure. It is produced by dominant and recessive mutations in *GLRA1*, the gene encoding the $\alpha 1$ subunit of the glycine receptor/channel. Since like the inhibitory GABA – and excitatory nAChR – receptor/channels, described above, the chloride-conducting inhibitory glycine receptor is a ligand-gated ion channel, sharing with other members of this channel family a subunit structure with four membrane-spanning helices and a characteristic heteropentameric composition (Fig. 5), it is not surprising that recessive mutations in the *GLRB* gene, encoding the glycine receptor β-subunit, also cause the same HE phenotype. Further, it is interesting that pathogenic alleles in all three types of ligand-gated channels, causing the discussed syndromes of HE, SCCMS, ADNFLE, and GEFS+, are missense mutations primarily altering residues in the extracellular loop between the second and third transmembrane helix that controls gating, suggesting a related molecular pathogenesis of these macroscopically distinct diseases.

Many of the mutations that underlie hereditary defects in the special senses, such as vision and hearing, are found in ion channel genes. X-linked congenital stationary night blindness type 2 is caused by defects in the retinal-specific close relative of both the neuronal calcium channel "seizure-ataxia-migraine" gene *CACNA1A* and the skeletal muscle "HOKPP periodic paralysis" gene *CACNA1S*, discussed above. The syndrome is caused by several loss-of-function mutations in *CACNA1F*. Color vision requires, in addition to the three cone opsin pigment genes, mutations in which cause loss of color detected by one cone type, only the genes encoding the two subunits of a specific cyclic nucleotide-gated heteromultimeric ion channel found in the photoreceptor membrane (Fig. 5). Defects in these channel subunits cause color blindness due to dysfunction of all cones. Rod monochromacy is produced by at least 10 recessive missense alleles of *CNGA3* and Achromatopsia type 3 is produced by missense loss-of-function alleles in *CNGB3*.

Nonsyndromic deafness (NSD) is produced by mutations in dozens of genes, including those potassium channel genes that contribute alleles to produce Jervell and Lange-Nielsen recessive LQT syndrome, discussed above. Additionally, the genes encoding subunits of a very different type of channel, the connexin gap junctions (Figs. 5 and 13), are the most common source of alleles for NSD, being polymorphic in many populations. Gap junctions serve to connect all types of cells via a large nonselective pore that mediates the transit between cells of not only ions but also small macromolecules and metabolites. They are functionally dimers contributed by each member of a pair of cells, each cell membrane having a hexameric hemichannel that seals together with its partner. Each hemichannel is composed of the family of connexin subunits. Alleles of *GJB2*, encoding connexin 26, are the most common cause of NSD. There are a few dominant alleles, but several dozen recessive alleles, including one specific polymorphic pathogenic allele, 35delG, which accounts for as many as 4% of the European Caucasian alleles. Alleles of *GJB6* and *GJB3* also contribute to this syndrome. The mechanism by which they act appears to be similar to that of the Jervell and Lange-Nielsen LQT alleles; they perturb the formation of the potassium-rich endolymph fluid in the inner ear that is required for signal transduction. Curiously, some alleles of all three connexin genes also cause distinct hyperkeratotic skin syndromes with or without deafness.

7.8
Acquired CNS Ion Channel Syndromes

Just as cardiac failure produces an altered pattern of ion channel gene expression, and thereby creates an acquired channelopathy arrhythmia syndrome (see Sect. 5.7), experimental forms of induced temporal lobe epilepsy, the most common form of adult epilepsy, greatly reduce expression of the *KCND2* potassium channel gene, producing a downregulation of the neuronal A-type potassium current and a reduced threshold for seizure initiation in the hippocampus, a common origin for such seizures. Thus, seizure activity produces a reduced threshold for subsequent seizures.

7.9 Polygenic Syndromes of Altered CNS Excitability

While most common seizure, migraine, and ataxia syndromes behave as complex polygenic traits, <u>not</u> as the simple monogenic Mendelian traits discussed above, and none of the genes involved in the polygenic forms of the disease have yet been convincingly demonstrated, a definition of the shared pathophysiology of these monogenic syndromes is likely to provide the best leads to the nature of those genes and to novel therapeutics for these common diseases. Likewise, while the definition of specific pharmacogenetic syndromes of altered neural excitability is only in its infancy, largely limited at this time to the definition of those individuals with altered rates of metabolism of seizure or antipsychotic medications, one would anticipate a dramatic impact on neuropsychopharmacology just by defining the genotypic basis of the wide variance now observed in patient response to current neuropsychiatric drugs. Additionally, it is likely that polygenic ion channel disorders produce a wider range of phenotypes in the CNS than those already revealed by the monogenic disorders. The salient feature of the CNS ion channel disorders, as well as those in the heart and muscle, was that they arose from a *periodic* disturbance in normal rhythmic activity. They were intrinsically episodic disorders, although many progressed to become constant. Many of the common complex polygenic neuropsychiatric disorders share this character. Further, several of the drugs used to treat the recognized channel phenotypes are additionally used to treat these disorders (e.g. valproate is used to treat both seizures and bipolar disease). Ion channel genes are thus important "functional candidates" to pursue within the broad chromosomal regions shown through linkage or linkage disequilibrium to be associated with these diseases, and are particularly intriguing should alleles be recognized that are similar in activity to those seen in the monogenic channel disorders.

8 Ion Channel Diseases of Epithelia

8.1 Structure and Function of Epithelial Tissues

Epithelia are sheets of *polarized* cells (sometimes consisting of many layers but only one cell layer of which is functionally significant) that participate in the transport of ions, water, and metabolites, not merely in and out of the cell, but fully *across* the entire cell sheet (Fig. 13). The epithelial cells are not excitable, like heart, muscle, and nerve, but they are electrically and metabolically coupled horizontally by gap junctions, described above, and the epithelial sheet therefore functions nearly as though it was two sheets of parallel membrane with cytosol between. One side of the epithelial sheet, called the *basolateral membrane*, faces the blood and is relatively unspecialized; the other, called the *apical membrane*, lines a nonblood compartment, often the lumen of a tubule, and the membrane is often elaborated into a brush border and microvilli (Fig. 13). It is primarily in the apical membranes that are found the many special channels and transporters that distinguish one epithelium from another. The point of demarcation between the two faces of

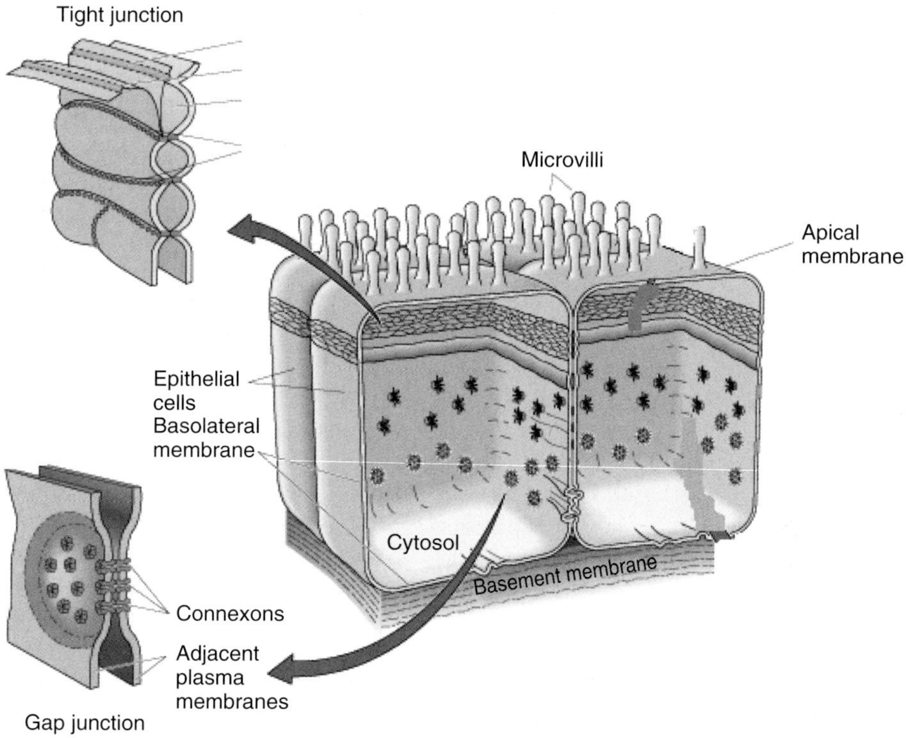

Fig. 13 Structure of epithelia.

an epithelial cell is the *tight junction*, the special barrier-forming connection made between adjacent cells in the epithelium (Fig. 13).

Epithelia serve two essential functions in physiology that can be perturbed in disease: to transport salt, water and other substances from one compartment to another, and to maintain gradients of salt, water, or other substances between compartments. While all epithelia share a fundamental organization and have in common a basic set of transport functions, various epithelia are optimized to perform one or the other of these two principle functions, and this is primarily conferred by the nature of the tight junctions between cells in the sheet. Those optimized for rapid transepithelial transport have "leaky" tight junctions and large *paracellular* pathways between cells providing a shunt parallel to the *transcellular* pathway across the sheet, and those optimized to form large electrical or chemical gradients across the membrane sheet have "tight" tight junctions that do not allow the passage of ions and water except through the cells.

8.2 Function of Ion Channels in Epithelia

Even though various epithelia transport a wide spectrum of compounds, there are a number of basic membrane properties that they all share, and these shared properties even extend to most nonpolarized mammalian cells, as introduced in Sect. 2. This generalization is most particularly true for epithelial basolateral membranes, which are much less specialized and more like those of other cells. First there is a Na/K-ATPase ion pump located in the basolateral membrane (Figs. 6 and 14). This ATPase is oriented so as to extrude Na^+ from the cytoplasm in exchange for the uptake of K^+. Second, the electrical potential difference across the basolateral membrane is electrically negative (cell interior with respect to the blood) and on the order of 50–60 mV, like most mammalian cells. Third, the intracellular ion concentrations reflect the activity of the Na/K pump, with a high internal K^+, a low internal Na^+ and a Cl^- concentration near its Nernst electrochemical equilibrium potential (Fig. 6). Fourth, the passive permeability of the basolateral membrane is predominated by that for K^+, with low permeabilities to Na^+ and Cl^-. Since K^+ is not in electrochemical equilibrium across the K^+-permeable basolateral membrane, the −60 mV resting membrane potential is then a K^+ diffusion potential attenuated by low finite Na^+ and Cl^- permeabilities. Finally, the properties of the apical membrane of epithelia are very diverse, but commonly are predominated by a high Na^+ permeability. The specifics of these apical Na^+ permeabilities and their properties, however, vary such that broad generalizations can not be made. Major breakthroughs in understanding epithelial function occurred as Ussing defined how simply opening a passive Na^+ channel in the apical membrane of an epithelial sheet with the five cardinal characteristics described above allowed transepithelial transport of sodium, and later Curran defined how such vectorial transport of sodium could be coupled to the iso-osmotic vectorial transepithelial flow of water and other solutes. To allow these mechanisms to create a net flux, anion had to be allowed to passively accompany the transported Na^+ since macroscopic electroneutrality had to be maintained on both sides of the epithelium, as it required an enormous amount of energy to separate charge across a membrane. As will be discussed below, if there is no pathway for chloride movement, as occurs in cystic fibrosis, transepithelial salt and water flow are greatly impeded.

8.3 Ion Channel Dysfunction and Diseases in Epithelia

On one hand, the macroscopic phenotypes produced by defects in epithelial channels can be seen to reflect the wide range of specific uses to which Nature has put epithelia: failure of lung, kidney, digestive and other internal organs, deafness, systemic fluid and electrolyte abnormalities, and even abnormalities of morphogenesis. On the other hand, these phenotypes all ultimately can be recognized to arise from defects in either of the two cardinal functions of epithelia, transepithelial transport, or maintenance of transepithelial gradients.

8.4 Cystic Fibrosis

Cystic fibrosis is the most common simple genetic disease in Caucasians, nearly 5% of the members of those populations carrying one pathogenic allele for this lethal autosomal recessive disease, with one common allele, Δ F508, reflecting 70%, but at least four other pathogenic alleles also reaching polymorphic gene frequencies (~1%) and over 1000 alleles now being known. It is also the first genetic disease for which the etiological gene and pathogenic alleles were discovered strictly on the basis of only its chromosomal position, without benefit of a known protein or a pathogenic allele created by a chromosomal rearrangement. It was additionally a landmark since it represented the first genetic channelopathy. The clinical phenotype of this disease – predominated by failure of lung, pancreas, and intestinal function – suggests nothing to bring to mind ion channel processes in the context discussed for the excitable tissues, and only slowly is the full molecular pathophysiology of this disease coming to light.

The gene, *CFTR*, standing for "cystic fibrosis transmembrane regulator," encodes a large transmembrane protein with 12 transmembrane helical spans, two cytosolic ATP-binding domains, and a large cytoplasmic regulatory domain that has multiple phosphorylation sites (Fig. 5). It is itself an ATP-regulated chloride channel, activated by the combination of both direct ATP binding and ATP-dependent phosphorylation via a cAMP-dependent kinase.

How does loss of function of this epithelial chloride channel produce the multifaceted phenotype of CF? The key features of the disease reflect epithelial dysfunction and include very thick dehydrated mucus secretions in the respiratory tree, which become purulent with chronic infections and progress to destruction of the lung parenchyma and therefore chronic respiratory failure. Additionally, a majority of patients have malabsorption caused by pancreatic insufficiency, caused by plugging of the secretory ducts resulting in activation of the digestive enzymes within the ducts and eventually pancreatic fibrosis (hence CF's old names "mucoviscidosis" and "cystic fibrosis of the pancreas"). As medical treatment allowed a longer lifespan, it was recognized that nearly all males at adulthood have infertility caused by plugged atretic vasa deferens and that a few percent of the patients first presented as newborns with meconium ileus, an obstruction of the newborn's intestines caused by cement-like intestinal contents. The physiology of these complex secretory processes was insufficiently understood to lead to an understanding of this disease or its diagnosis. However, the modern scientific underpinnings of CF began during a New York heat wave in 1953, when it was astutely observed that many of the children who were hospitalized for dehydration and salt depletion came to be diagnosed with CF. The dehydration arose because they secreted a salt-rich sweat. This observation led di Sant'Agnese and coworkers to the development of the diagnostic "sweat chloride test" based upon the recognition that the high concentration of chloride in sweat of CF homozygotes, well before they became symptomatic, could be distinguished from unaffected controls or carriers. Not until 30 years later did patch-clamp electrophysiology studies on the sweat gland by Quinton and coworkers establish that a defect in the function of an epithelial chloride channel prevented the physiological resorption of salt from sweat during its transit through the duct. Therefore, the cardinal aspects of transepithelial transport, salt and water movements, were impeded by the CF chloride channel defect. As the sweat gland was a *resorptive epithelium*, salt resorption was prevented, so salt was wasted in sweat. The same molecular lesion in the *secretory epithelium* of the respiratory mucus gland, the pancreatic acinus, and the vas deferens resulted in concretions of dehydrated product that plugged and scarred the ducts, ultimately destroying the tissues.

Despite the complicated name of the CFTR channel it is clear that this protein *itself* functions as an ion channel, not just an indirect regulator of other channels. This point is most simply made by classes of mutant alleles that alter ion selectivity and single-channel conductance, functions intrinsic to a channel and *not* subject to its regulation. There are also informative classes that alter channel gating; however, this *is* a process subject to regulation, so such alleles do not prove intrinsic channel activity. Likewise, the vast majority of the mutant alleles are even less informative in that they merely alter the membrane protein's biosynthesis, maturation, and stability, telling little about the protein's function, but perhaps pointing to vulnerabilities that can be targeted in therapy.

8.5 Renal Channelopathies

The range of the diversity of apical sodium entry pathways available in epithelial designs is reflected in channelopathies of the kidney tubules, but mutations in the subunits of the heterotrimeric EnaC epithelial sodium channels, encoded by *SCNN1A, SCNN1B*, and *SCNN1G*, provide the simplest entry to the pathophysiology of these diseases. The α-, β-, and γ-subunits of this channel are homologous to each other and to the degenerins that participate as an olfaction and neurodegeneration channel in *Caenorhabditis elegans*. Each gene encodes a simple ion channel subunit that contains cytoplasmic termini and only two transmembrane helices (Figs. 5 and 6). Expression of the α-subunit alone produced ion channel activity *in vitro*, but it did not reflect a recognized native current. Neither the β- nor γ-subunits alone produced any channel activity. All three subunits expressed together, however, could be recognized to produce the amiloride-sensitive, aldosterone-regulated sodium channel best characterized in the principal cell of the distal nephron (Fig. 14d), but also found in the colon, lung, and exocrine glands, many places CFTR was expressed. Amiloride is a diuretic drug, and drugs we call diuretics are in fact naturetics in that they cause sodium wasting by inhibiting a renal sodium resorption mechanism. Each class of clinically useful diuretic targets a different renal epithelial apical sodium entry pathway, and, remarkably, mutations have been found targeting each of these pathways, simulating the effects of the diuretic drugs furosemide, thiazide, and amiloride.

Just as without a rudimentary understanding of transepithelial transport one would be unlikely to guess the phenotype of CFTR mutations, it takes at least a superficial appreciation of renal physiology to derive the phenotype produced by these mutations: alterations in blood pressure. These days we tend to forget that salt is a precious essential ingredient for life and that loss of body salt (much more than loss of body water, as will be discussed below) results in blood volume depletion and a syndrome of shock similar to that produced by a severe hemorrhage. While the sweat gland is a relatively simple structure that does a fair job of resorbing some salt from the small volume of blood ultrafiltrate that gives rise to the excreted fluid, the kidneys are Olympic performers in this field in that they cope with fully one-fifth of the cardiac output, sending one-fifth of that across the glomerular filter (G) into the epithelial nephron tubules, where in excess of 99.5% of the salt can be reabsorbed by crossing the luminal apical membrane and then the basolateral membrane back into the blood, leaving by default unspecified waste to be lost in the urine (Fig. 14). The epithelial sheet that forms the nephron tubule changes in character along its length, and classical renal tubule

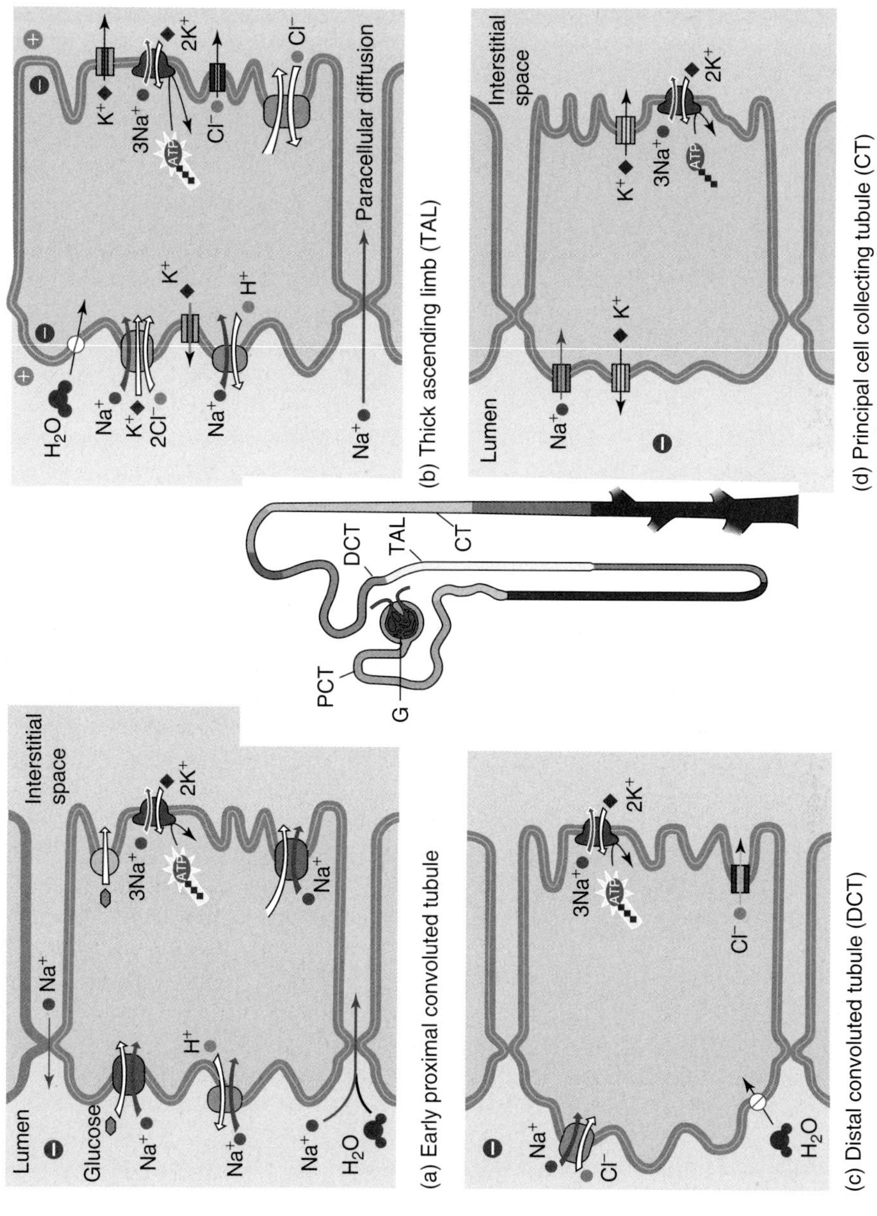

Fig. 14 Renal epithelia.

segment names have now come to be associated with specific molecular transport mechanisms. For instance, 66% of the filtered salt is resorbed in first segment, the proximal tubule (PCT), 25% in the thick ascending limb (TAL) of the loop of Henle-via the furosemide-sensitive apical Na-K-Cl cotransporter, 7% in the distal convoluted tubule (DCT) via the thiazide-sensitive apical Na-Cl cotransporter, and the final 2% via the amiloride-sensitive, aldosterone-regulated heterotrimeric EnaC epithelial sodium channels in the collecting tubule (CT) (Fig. 14).

The kidney is capable of regulating sodium resorption over a wide range to compensate for changes in intake, using only the final 2% of the filtered load. This is all achieved by control of the abundance of the EnaC epithelial sodium channels in the principal cells of the CT by the adrenal steroid hormone aldosterone (Fig. 14d). Aldosterone is secreted in response to angiotensin II, which in turn is produced from angiotensin I by angiotensin converting enzyme (ACE), an important pharmacological target for the control of blood pressure. The entire cascade leading to aldosterone is initiated in response to sensors that detect aspects of blood volume, critical ones being the intrarenal and central baroreceptors that trigger specialized cells in the nephron to release an enzyme, called *renin*, which cleaves the inactive angiotensinogen precursor to angiotensin I. Therefore, hemorrhage causes a low blood volume and pressure to be sensed, triggering renin release, leading to the production of angiotensin and aldosterone, the insertion of excess EnaC epithelial sodium channels in the principal cells of the CT, enhanced distal nephron sodium resorption into the blood and a physiological trend to restoring blood volume. On the other hand, artificially producing high blood pressure will homeostatically decrease renin release, leaving angiotensin and aldosterone levels low, reducing the abundance of EnaC epithelial sodium channels and distal nephron sodium resorption, wasting sodium in urine and diminishing blood volume. Because the transepithelial electrical potential across the "tight" CT epithelium is used to drive the secretion of potassium and hydrogen into the nephron lumen, and that electrical potential results from EnaC sodium channel activation, a low abundance of EnaC sodium channels results in reduced K^+ and H^+ wasting in the urine, and hence elevated plasma levels of both. Conversely, high levels of EnaC sodium channels enhance K^+ and H^+ wasting and low plasma levels of both.

8.6 Pseudohypoaldosteronism and Liddle Syndrome

Loss-of-function mutations, primarily produced by deletions, in *SCNN1A*, *SCNN1B*, and *SCNN1G* genes, encoding subunits of the EnaC channel, produce the phenotype of autosomal recessive pseudohypoaldosteronism type 1. This is a severe disease that presents in infancy with life-threatening failure to thrive caused by salt wasting and hypovolemia. They additionally have hyperkalemia and metabolic acidosis. As the name implies, the phenotype appears like that produced by an absence of aldosterone; however, in this case aldosterone (as well as angiotensin and renin) levels are very high. One allele, *SCNN1B G37S*, replaces a conserved glycine residue of the channel with serine, thereby altering gating and reducing the channel open probability. This establishes that channel function *itself* is critical to the phenotype, not some other aspect of the protein. This phenotype powerfully defines the physiology: all of the complex and apparently multifaceted actions of aldosterone can be understood to emanate from increasing the EnaC channel current, since loss of the hormone and loss of the channel appear indistinguishable. The sodium current conducted by the channel is the effector of aldosterone action; *a classical endocrine and hormonal pathway is truly a complex means of ion channel regulation, its diseases truly channelopathies.*

Gain-of-function dominant mutations in *SCNN1B* and *SCNN1G*, altering the C-terminus of the β- or γ-subunits of EnaC channels, cause a pseudohyperaldosterone state called *Liddle syndrome*. The phenotype is hypertension, with low renin, angiotensin and aldosterone levels, suggesting primary excess sodium retention and volume expansion. They also have hypokalemia and metabolic alkalosis, secondary to excessive renal K^+ and H^+ wasting. The mutations mainly result in increased numbers of EnaC channels in the CT; however, some alleles do increase the channel open probability, again showing the sodium current, and not some other aspect of the protein, to be critical. Since the diuretic amiloride blocks the EnaC channel, the drug is useful in treating the disease. Spironolactone, a drug that inhibits aldosterone action, however, has no effect, since the mutant channels are constitutively active and no longer under hormonal control. Not only does amiloride block the excessive sodium resorption, allowing sodium wasting, volume loss, and a normalization of blood pressure, but it also indirectly blocks excessive K^+ and H^+ wasting and corrects those plasma disorders. EnaC does *not* itself transport K^+ or H^+. Blocking the sodium channel in the CT prevents K^+ and H^+ secretion there via effects on the CT transepithelial membrane potential. For this reason, amiloride is a "potassium-sparing" diuretic, in contrast to all of the others to be discussed below.

8.7 Bartter and Gitelman Syndrome

Amiloride, with its ability to block EnaC channels and therefore resorption of only the final 2% of the filtered sodium load, is a relatively weak diuretic. The most potent diuretics, such as furosemide, target the apical Na-K-Cl cotransporter found in the TAL where 25% of the filtered sodium load is resorbed (Fig. 14b). The cotransporter is a solute carrier protein with 12 transmembrane helical spans encoded by *SLC12A1*. It has the ability to couple the electroneutral movement of four ions across the membrane, $1Na^+$, $1K^+$, and $2Cl^-$ (Fig. 6). The carrier's movement is passive in that no ATP energy is directly consumed in transport, but because the flows of all four ions are coupled and must take place in concert for any movement to occur, flux coupling, as was mentioned above, allows the cotransporter to capture the potential energy available in the sodium electrochemical gradient to move K^+ and Cl^- ions along with it as it enters the TAL cell across the apical membrane. This happens despite the fact that K^+ and Cl^- move uphill against their individual electrochemical gradients since it is the net electrochemical potential of the entire complex that dictates the direction of the coupled movements. This is a process called *secondary active transport*, a bioenergetic process coupled to an ion's electrochemical potential, and it is contrasted to *primary active transport*, a process coupled to the chemical potential of ATP hydrolysis and exemplified by the Na/K-ATPase, discussed above.

Loss-of-function alleles of *SLC12A1* were first recognized to cause the severe autosomal recessive disease phenotype of Bartter syndrome (BS), characterized by prenatal polyhydramnios, infantile polyuria, and polydipsia, renal Na^+, K^+, and H^+ wasting with subsequent hypokalemia and hypochloremic metabolic alkalosis, and very high renin, angiotensin, and aldosterone levels. It looked in every way as if these infants were on high doses of furosemide since conception, with the massive renal salt wasting producing the polyhydramnios, polyuria, and volume depletion that the homeostatic aldosterone signaling pathway was mightily struggling to restore. The high aldosterone was forcing a maximal EnaC channel response, but this low-capacity system could not handle the huge load of sodium the dysfunctional TAL had left behind, so sodium continued to be wasted. However, the electrical effect of maximal EnaC channel activation in the CT is present, hence the massive K^+ and H^+ wasting. Just as the EnaC channel has a major impact on the transepithelial electrical potential in the CT, and hence the flux of other ions in that renal segment, the SLC12A1 cotransporter has a major role in creating the TAL transepithelial potential, a very atypical potential in that it is lumen positive, not negative like nearly every other epithelial sheet (Fig. 14b). Physiologically, the lumen positive potential, created by the secondary active transport of Cl^- by the cotransporter, repels cations, particularly the divalent cations Ca^{++} and Mg^{++}, favoring their paracellular resorption and being the major site in the nephron at which they are resorbed. With the loss of SLC12A1 function, the divalent cations are wasted in the urine and the blood levels fall.

Since the SLC12A1 cotransporter carries an electroneutral load of two cations and two anions, it appears puzzling it should be able to create the unusual lumen positive condition. This arises because of two types of critical ion channels in the TAL apical and basolateral membranes (Fig. 14b). Mutations in these channels also produce the BS phenotype, BS type 1 used to specifically designate the disease caused by *SLC12A1* mutations. Since the TAL cells have a high internal K^+, the energetics of cotransport are favored by allowing K^+ to futilely recycle back into the lumen across the apical membrane. This apical potassium channel, the ATP-sensitive ROMK channel encoded by *KCNJ1*, a member of the channel family causing Andersen syndrome (Figs. 5 and 8), was shown to be essential to TAL transepithelial transport since loss-of-function mutations in this gene produced the Bartter syndrome phenotype, BS type 2. The high abundance of ROMK channels in the apical membrane of the TAL contributed half of the explanation for the unusual positive transepithelial potential; abundant K^+ channels gave the apical membrane a hyperpolarized membrane potential nearer the K^+ Nernst potential (Fig. 14b). The last part of the puzzle is the presence of abundant Cl^- channels in the basolateral membrane (Fig. 14b). They depolarize the basolateral membrane potential, and therefore accentuate the difference in potential between the two faces of the TAL cell. They are there to provide a pathway for Cl^- to leave the cell and passively reenter the blood, accompanying Na^+ and effecting net salt resorption in the TAL – the classical "diluting segment" of the tubule where hypotonic luminal fluid is produced. The basolateral Cl^- channel is composed of subunits from the large CLC channel family (Fig. 5) that includes the muscle chloride channel underlying Becker and Thomsen myotonia, discussed above. A crystal structure has recently been obtained for this family of chloride channels, showing them to function as "double-barreled" dimers with each subunit, composed of 18 helical segments and both termini on the cytoplasmic face, having its own ion conduction pathway. The specific subunits found in the kidney are encoded by *CLCNKB* and *CLCNKA*, which are inserted in the membrane accompanied by barttin, their small β-subunit having two transmembrane helices and encoded by *BSND*. Recessive loss-of-function mutations in *CLCNKB* cause classical Bartter syndrome, BS type 3. Bartter syndrome, type 4 is caused by homozygosity for loss-of-function alleles of *BSND*, but this phenotype additionally includes sensorineural deafness. This same set of phenotypes is produced as a *digenic* syndrome with loss-of-function alleles of *both CLCNKB* and *CLCNKA*.

Channelopathies as a cause of deafness were briefly mentioned in the context of recessive cardiac LQT syndromes in Sect. 5.6. Surprisingly, the same channel is expressed selectively in the heart and the inner ear epithelium, where its abundant expression is involved in the production of the potassium-rich endolymph fluid that fills this chamber and participates in the transduction of auditory sensation. Only complete loss of the channel significantly disturbs this secretory process. It is of note that another inner ear *LQT1* homolog, encoded by *KCNQ4*, has dominant alleles that produce only deafness without LQT (*DFNA2*). Just as even a small amount of LQT channel function, such as provided by the dominant LQT alleles, was sufficient for production of the KCl-rich fluid of the inner ear compartment – deafness only occurring with a profound homozygous loss of function – only complete loss of inner ear epithelium chloride channel function causes deafness. The chloride that accompanies potassium in endolymph production passes through the *CLCNKB*- and *CLCNKA*-encoded channel. Without barttin, neither the *CLCNKB*- nor *CLCNKA*-encoded channel can be inserted into either the TAL or inner ear epithelial basolateral membrane. Either channel alone is sufficient to maintain the endolymph production for hearing, but only the *CLCNKB*-encoded channel can support the higher demands of TAL function.

Gitelman syndrome is the result of homozygosity for loss-of-function mutations in *SLC12A3*. It encodes a close relative of the Na-K-Cl cotransporter, called the *thiazide-sensitive Na-Cl cotransporter* to reflect its sensitivity to the thiazide class of diuretics and the K^+ insensitivity of the transport mechanism (Figs. 6 and 14c). This apical cotransporter is found in the DCT of the nephron and is responsible for the resorption of 7% of the filtered sodium load (Fig. 14c). Like Bartter syndrome, the disease presents with renal Na^+ wasting and a secondary wasting of K^+ and H^+ as well as resultant hypokalemia and hypochloremic metabolic alkalosis, as aldosterone and the EnaC channels struggle to correct the volume loss, which ultimately is only minor. Renin, angiotensin, and aldosterone are all elevated; however, the disease is much milder than Bartter syndrome, since the DCT resorbs a much smaller fraction of filtered sodium. Therefore, it classically presents in adolescence, but the presenting picture is *not* one of volume depletion, it is rather a picture predominated by consequences of the chronic plasma electrolyte abnormalities and their impact on excitable tissues. This is commonly manifested as muscle weakness, abdominal pain, and neuromuscular irritability. Patients commonly have hypomagnesemia and accompanying tetany, although the pathogenesis of this abnormality remains to be elucidated.

It should be noted that one would expect hypomorphic alleles in the five genes contributing to Bartter and Gitelman syndrome and the three genes contributing to the EnaC channel to be protective against hypertension. Conversely, hypermorphic alleles in these same eight genes have the potential to contribute to hypertension. Therefore, polymorphic variants of these genes are promising candidates for evaluation in the heritable polygenic trait of blood pressure.

8.8 Hypomagnesemia

Autosomal recessive primary hypomagnesemia has played an important role in beginning to reveal the mechanisms underlying the paracellular pathways of epithelial ion transport, revealing them to be far more complex than simply an unmediated leak between cells (Fig. 14b). The disease presents in early childhood with symptoms similar to Gitelman syndrome, but additionally polyuria and renal stones. It is caused by homozygosity for loss-of-function alleles of *CLDN16*. This gene encodes Paracellin-1, a transmembrane protein with four helical spans found in the TAL and DCT tight junctions where it appears to contribute to a calcium and magnesium selective paracellular path. This gene family appears to encode proteins designated for such paracellular epithelial transport pathways since *CLDN4* and *CLDN15* also influence ion selectivity in other tight junctions.

Autosomal dominant hypomagnesemia appears to arise by a very different mechanism. It was shown in a large extended family to cosegregate with a heterozygous missense mutation in *FYXD2*, the gene encoding the small single membrane-spanning proteolipid γ-subunit of the Na/K-ATPase, a modulatory subunit found associated with the functional $\alpha\beta$ subunits only when the pump is expressed in basolateral membranes of the distal nephron. This subunit alters the transport kinetics of the ion pump, the prime mover setting the gradients and energetics for all coupled transport in these tight epithelia. The pathological *FYXD2 G41R* allele must function as a dominant negative gain-of-function allele, since haploinsufficiency produced by heterozygous deletions spanning the locus does not produce the phenotype. The mutation alters charge in the middle of the transmembrane helix and causes the γ-subunit to fail to traffic beyond the golgi, allowing functional $\alpha\beta$ subunit insertion into the membrane alone. As this is the only pathogenic allele reported to date, the mechanism by which mutations in this locus produce magnesium wasting remains unclear.

Hypomagnesemia with secondary hypocalcemia is caused by homozygosity for loss-of-function mutations in the *TRPM6* gene, over a dozen pathogenic alleles already known. This gene, and other members of its large extended family, are cousins of the genes encoding the pore-containing hexahelical channel subunits of voltage-gated K^+ channels, described in Sect. 3; however, the TRP branch of the family encode an array of selective and nonselective tetrameric cation channels that are coming to be recognized as playing important roles in sensory transduction and epithelial function (Fig. 5). The TRPM6 and TRPM7 channel subunits, each of which contains a C-terminal protein kinase domain, appear to heteromultimerize to form a Mg^{++} channel and pathogenic alleles of *TRPM6* prevent subunit association and channel function. No pathogenic alleles, however, have yet been reported in *TRPM7*.

8.9
Polycystic Kidney Disease

Autosomal dominant polycystic kidney disease (ADPKD) is one of the most common genetic conditions, affecting over 1/800 births and responsible for 10% of chronic dialysis. Mutations – primarily null alleles caused by deletions, nonsense, frameshifts or splicesite alterations – in a relative of the TRP family of genes, PKD2 – encoding polycystin-2 – and in PKD1 – encoding polycystin-1, its homologous heteromultimerizing partner – cause this disease, and produce cystic dilation of kidney tubules, hepatic cysts, and aneurysms in blood vessels. Most of the nephrons appear normal, suggesting a "two-hit" pathogenic model reflecting a germline loss-of-function mutation and probabilistic somatic loss-of-heterozygosity (LOH), similar to that which has proved so powerful in understanding dominant cancers. Coexpression of PKD1 and PKD2 *in vitro* produce functional calcium-permeable nonselective cation channels, but expression of either PKD1 or PKD2 alone does not, and pathogenic alleles prevent heteromultimerization and channel function. The channels are found localized in association with the cilium, a structure of the apical membrane, perhaps transducing mechanical sensory signals it detects. While it is clear that calcium signaling is induced by the cilium and perturbed by the mutant alleles, it is unclear how either alters morphogenesis producing cyst formation. However, recently a clear example of morphogenetic pattern formation, left-right axis asymmetry, was shown to require epithelial PKD2 function.

8.10
Glucosuria

The notion that energy conserved in the sodium gradient across the apical membrane of an epithelial cell can be used in secondary active transport was introduced above in discussing Bartter and Gitelman syndrome. Sodium-coupled transport, however, is used in the gut and nephron in the resorption of many uncharged metabolites, not only ions (Fig. 6). A clear picture of how such energy coupling occurs is provided by two autosomal recessive glucose-wasting disorders. The sodium gradient across the apical membrane arises as a consequence of Na/K pump activity at the basolateral membrane. Secondary active transport of a substance, in this case glucose, occurs "uphill," against its electrochemical gradient by being obligatorily coupled on a carrier protein to the movement of sodium down its electrochemical gradient into the cell, such that the net movement of the coupled system obeys thermodynamics and remains "downhill" (Fig. 14a). A tremendous amount of energy is stored in the sodium gradient. A simple estimation reveals that cellular sodium is about 1/10th that found in the renal tubular fluid and that the membrane potential is $\sim$−60 mV – the Nernst equation revealing that translates to the equilibrium diffusion potential set up by a 10-fold gradient of a monovalent ion (*see* Sect. 2). Both the 10-fold chemical gradient and the equivalent 10-fold electrical gradient favor movement of Na^+ into the cell, and those potentials are multiplicative. This should allow 1 : 1 coupling of sodium:glucose to drive cellular glucose to a concentration 100 times that found in tubular fluid. Should one obligatorily couple the movement of two sodium ions to cotransport with one glucose, the multiplicative properties suggest one could achieve the 100-fold enhancement for *each* Na^+ site, and hence a 10 000-fold gradient favoring glucose absorption. These are the properties of SLC5A2, the low affinity 1 : 1 apical sodium-glucose cotransporter and SLC5A1, the high affinity 2 : 1 apical sodium-glucose cotransporter, respectively.

Isolated renal glucosuria is caused by homozygous loss-of-function mutations in *SLC5A2*. The high-capacity, low-affinity cotransporter it encodes is glucose-selective and is abundantly expressed in the apical membrane of the initial segment of the proximal tubule of the kidney, the site where the majority of the glucose filtered across the glomerulus is reabsorbed (Fig. 14a). Physiologically, no glucose remains in the urine unless blood levels become so high that they overwhelm the transport capacity of this transporter, as occurs in diabetes mellitus. Therefore, the symptoms of this transport defect are massive glucosuria, and the resultant diuresis of water obligated by this large osmotic load.

Homozygous loss-of-function alleles of *SLC5A1* cause a severe neonatal dehydration syndrome; however, this is because of diarrhea, not the renal loss it produces. The disease caused by these mutations is glucose-galactose malabsorption, and it is characterized by loss of high affinity glucose transport in the kidney and gut. This high-affinity cotransporter is expressed in the apical membrane of the final segment of the proximal tubule where it scavenges the last of the glucose missed by the SLC5A2 cotransporter. Its low capacity, however, is easily overwhelmed, so it cannot compensate for loss of *SLC5A2*. Unlike SLC5A2, it transports galactose in addition to glucose, and is the mechanism that predominates intestinal absorption of these sugars – essential for a newborn dependent on the lactose of milk. Just as sugar left behind in the lumen of the nephron osmotically obligates water loss, so too does this unresorbed intestinal sugar; however, providing the infant a formula of fructose, a sugar that uses a different transporter, is life-saving.

8.11 Nephrogenic Diabetes Insipidus

Ancient physicians recognized that not only could diabetes, characterized by excessive urination and thirst – polyuria and polydipsia – be accompanied by sugary-sweet tasting urine – diabetes mellitus – but could alternatively be accompanied by tasteless urine – diabetes insipidus. Autosomal nephrogenic diabetes insipidus (NDI) results from defects in water channels, called *aquaporins*, and a recent Nobel Prize was awarded for the solution of their nature. Aquaporins reflect a large family of water channels and their function – allowing the flow of water while excluding that of ions, is the mirror image of that of ion channels, which allow the flow of ions without water. Loss-of-function recessive and dominant-negative alleles of *AQP2* result in NDI and in a loss of regulated water channels from the apical membrane of the principal cell of the CT (Fig. 14d), the site of regulated water resorption in the kidney. The peptide hormone alternatively called *vasopressin* or antidiuretic hormone (ADH), and long known to be responsible for the production of concentrated urine is tonically secreted by the pituitary in response to small changes in plasma osmolarity. It was shown to control insertion of the

AQP2 channel into the apical membrane of the principal cell; in the absence of hormone, the aquaporin was sequestered in vesicles beneath the membrane and hormone receptor signaling through cAMP triggered insertion of the channels. The receptor receiving the ADH signal, called the *V2 vasopressin receptor*, is encoded by the X-linked gene *AVPR2* and mutations in this gene were shown to cause the X-linked, more common form of nephrogenic diabetes insipidus. The phenotype in NDI, lethargy progressing to coma, results from effects on excitable tissues of euvolumic hypernatremic hyperosmolarity, as water is wasted by the kidney in excess of salt. Note that both major hormones involved in salt and water balance act independently by directing the insertion of different apical channels into the principal cell of the CT: aldosterone regulates sodium balance and hence plasma volume via the insertion of epithelial sodium channels and ADH regulates water balance and hence plasma osmolarity via the insertion of aquaporin-2.

8.12 Hereditary Kidney Stone

Kidney stone formation, nephrolithiasis, is a common polygenic trait with a lifetime risk of up to 10%. Stones are crystal-like structures that precipitate from urine within the kidney. Their origin is now recognized to be an abnormal urine environment, and recently defects in renal ion channels have been shown to contribute to this phenotype. Above it was briefly mentioned that stone formation was a component of the phenotype in primary hypomagnesemia. Nephrolithiasis is most commonly associated with hypercalciuria and calcium salts are a key component of most stones. Loss-of-function alleles of *CLCN5*, encoding a voltage-gated chloride channel within the family of those causing Bartter syndrome and myotonia (Fig. 5), described above, were recognized to cause a Mendelian X-linked form of hypercalciuria and stone formation, called *Dent disease*, a phenotype that additionally included proteinuria and progressed to renal failure. Additional alleles in the same gene were found to cause two other diseases that had not previously been recognized to be related, X-linked nephrolithiasis and X-linked hypophosphatemic rickets – a phenotype characterized by demineralization of bone. This chloride channel is localized to endosomes within epithelial cells of the proximal tubule, *demonstrating that ion channels play a critical part not only in the function of cellular membranes but also of subcellular compartments.* Defects in the function of the endosomal Cl^- channel impair the process of endocytosis from the apical membrane, secondarily altering the regulation of various apical transporters, but the exact mechanism of pathogenesis remains to be defined. It is of note that homozygous loss-of-function and heterozygous dominant-negative mutations in its close relative *CLCN7*, while causing a distinctly different phenotype, osteopetrosis, ultimately produce the phenotype as a result of a failure in endosomal function. Further, homozygous loss-of-function alleles in *TCIRG1*, encoding a subunit of the endosomal vacuolar ATPase proton pump (Fig. 6), also produce osteopetrosis. Osteopetrosis is a phenotype that includes very dense bone since bone remodeling by the degradative bone osteoclasts fails because endosomes fail to function since they can not acidify. Together, the two osteopetrosis diseases loci suggest that the endosomal Cl^- channels provide a shunt pathway that allows the endosomal H^+-ATPase to pump a proton gradient without stalling because of electrostatic charge accumulation, and suggest a similar role for *CLCN5* in the kidney.

9 Ion Channel Disease of Other Nonexcitable Tissue

In addition to major phenotypes produced by defects directly assignable to abnormalities in the function of heart, muscle, nerve, and epithelia, ion channel defects are already known to produce a wide array of ancillary phenotypes, some of which were mentioned above as additional components to a syndrome, for instance, the skin findings produced by alleles producing deafness in the three gap junction loci, mentioned at the end of Sect. 7.7. Further, new channel phenotypes are coming to be recognized as arising from a failure of internal membranous organelle function, and these organelles play similar functions in most cell types. In these other tissues, channels seem to play pathophysiological roles similar to those played in diseases of excitable tissues or epithelia. For instance, the mirror-image pancreatic β-cell endocrine channelopathies of hyperinsulinism in infancy (HI) and permanent neonatal diabetes (PND) are very similar to neuronal receptor/channel disease, once the special mechanism of glucose sensing by the ATP-sensitive sulfonylurea receptor/channel is taken into account. This receptor/channel contains K^+ channel subunits, encoded by *KCNJ11* (from the family of those involved in Andersen and Bartter syndromes, discussed above) (Figs. 5 and 8), and a regulatory ATP-binding cassette subunit, encoded by *ABCC8*, named *SUR* for the sulfonylurea (SU) drugs that interact with it. Blocking this K^+ channel depolarizes the membrane, activating voltage-activated Ca^{++} channels and a calcium influx that triggers excitation-secretion coupling (as described in Sect. 7.1) with the fusion of insulin-laden vesicles into the β-cell membrane, and hence insulin secretion into the blood. Glucose induces insulin release through its metabolism in the β-cell, the ATP produced serving to block the channel; the SU drugs induce insulin release by directly blocking the channel's conductance. Not surprisingly, HI, which simulates the effect of SU, is caused by homozygosity for loss-of-function mutations in either of the two channel subunits. PND, where insulin can not be released, results from heterozygous dominant gain-of-function missense alleles of *KCNJ11*, which produce constitutively active channels that cannot respond to physiological ATP block induced by glucose, but can respond with insulin secretion when blocked by an SU. Notice how these alleles and the mirror-image phenotypes they produce are reminiscent of the mirror-image "endocrine" EnaC renal channelopathies, discussed in Sect. 8.6. As a second example, osteopetrosis, a defect in bone morphogenesis caused by defects in Cl^- channels and ultimately endosomal function, was briefly mentioned above as an endosomal channelopathy, an initial case of a disease caused by defects in membrane compartments *within* cells. Other membrane compartment channelopathies are beginning to emerge: endoplasmic reticulum channelopathies and MHS, caused by alleles of *RYR1*, discussed above in Sect. 6.4; lysosomal channelopathies and the lysosomal storage disease mucolipidosis type IV, caused by loss-of-function alleles in *MLN1*, encoding a lysosomal TRP-like nonselective cation channel; and mitochondrial channelopathies and severe obesity with diabetes, caused by loss-of-function alleles of *UCP3*, encoding a mitochondrial proton channel that shuts protons across the membrane and thereby uncouples respiration from ATP synthesis, physiologically making mitochondria less efficient, and the disease appearing to result from hyper-efficient mitochondria.

10 Ion Channels as Targets for Drug Therapy in Common Diseases

For the discovery of a pathogenic allele to have full impact in modern medicine, it is necessary to understand its function and to find small molecules able to alter that function – such "rationally designed" novel pharmaceuticals being one of the major hopes for the postgenomic era. While often positionally identified disease genes have been slow to yield either their function or a "rational drug," ion channels are highly amenable and proven targets. For example, discovering ion channel participants in a disease, such as the Mendelian seizure disorder caused by alleles of *KCNQ2* and *KCNQ3*, suggests not only a novel *type* of target for therapy (e.g. suggesting use of acetazolamide – a drug proven effective in a wide range of ion channel diseases), but these channels themselves can further serve as the guides to *finding* entirely novel selective pharmaceuticals. Because the function of channel mechanisms is so well understood, and it is now clear that the *full* range of pathogenic abnormalities in the huge ensemble of channel diseases discussed above is either too much or too little conductance through the channel (not *anything* else), one does not need to guess what has to be changed to remedy the gene defect. It is clear that the therapeutic goal must be to discover a channel closer/blocker or a channel opener/activator, and a number of drugs of both types are already in current use. *In vitro* expression assays of channel function can be used in high-throughput drug screening and in drug optimization. An example of this kind of breakthrough is retigabine, a novel class of seizure medication proven to activate the KCNQ2/KCNQ3 "M current" channel suggested as a seizure drug target by the role of these genes in BFNC. Since the channel mechanisms perturbed in such disorders are likely to be critical participants in producing the relevant membrane potentials leading to disease vulnerability, the drugs targeting them should be useful for treating the disease even in those *not* having an intrinsic defect in the targeted mechanism (i.e. even if a patient has the disease because of a pathologically depolarized critical membrane because of some *other* mechanism, hyperpolarizing the implicated membrane via *any* mechanism should be helpful). Another advantage of pharmacology targeted to channels is the expectation, discussed above, that *a narrow range of tissues express the channel in a physiologically important context, therefore drug specificity is achieved and unwanted side effects are minimized.* Therefore, in this postgenomic world rich in potential therapeutic targets, the potential channels hold for rapid novel drug discovery warrants their prioritization.

Most of the diseases discussed in this chapter are rare channelopathies and despite being promising targets for pharmacological intervention would likely remain "orphan" diseases were it not for the fact that they produce a *phenocopy* of a common disease, and perhaps provide the only molecular lead to a new drug urgently needed by a major market. However, it is likely that many of the loci critical to the development of the common complex polygenic diseases *cannot* be revealed by major-effect Mendelian alleles, such alleles perhaps being early embryonic lethals. An increasingly important approach to the identification of genes underlying complex polygenic disease is to identify "functional candidates" within broad chromosomal regions implicated in the disease and to test disease association (i.e. linkage disequilibrium) with polymorphic markers within these candidates. Ion channel candidates in such diseases could be suggested by the known physiology, pharmacology and pathophysiology, their appropriate or extremely narrow tissue distribution, their membership in a demonstrably pathogenic gene family, contribution to a demonstrably vulnerable component of the action potential, or a salient disease-associated motif. The existence of alleles sharing features common to known pathogenic alleles underlying the monogenic channelopathies, such as dominant-negative inhibition of expressed channel function or incomplete inactivation, would be particularly suggestive.

See also Receptor Biochemistry.

Bibliography

Books and Reviews

Agre, P., Kozono, D. (2003) Aquaporin water channels: molecular mechanisms for human diseases, *FEBS Lett.* **555**(1), 72–78.

Gargus, J.J. (2003) Unraveling monogenic channelopathies and their implications for complex polygenic disease, *Am. J. Hum. Genet.* **72**(4), 785–803.

Hille, B. (2001) *Ionic Channels of Excitable Membranes.* 2nd edition, Sinauer Associates, Sunderland, Massachusetts.

Jurkat-Rott, K., Lerche, H., Lehmann-Horn, F. (2002) Skeletal muscle channelopathies, *J. Neurol.* **249**, 1493–1502.

Lang, B., Vincent, A. (2003) Autoantibodies to ion channels at the neuromuscular junction, *Autoimmun. Rev.* **2**(2), 94–100.

Lerche, H., Jurkat-Rott, K., Lehmann-Horn, F. (2001) Ion channels and epilepsy, *Am. J. Med. Genet.* **106**, 146–159.

Lifton, R.P., Gharavi, A.G., Geller, D.S. (2001) Molecular mechanisms of human hypertension, *Cell* **104**, 545–556.

MacKinnon, R. (2003) Potassium channels, *FEBS Lett.* **555**(1), 62–65. '

Marban, E. (2002) Cardiac channelopathies, *Nature* **415**, 213–218.

Schultz, S.G. (1998) A century of (epithelial) transport physiology: from vitalism to molecular cloning, *Am. J. Physiol.* **274**(1 Pt 1), C13–C23.

Vander, A.J. (1995) *Renal Physiology*, 5th edition, McGraw-Hill, New York.

Primary Literature

Abbott, G.W., Butler, M.H., Bendahhou, S., Dalakas, M.C., Ptacek, L.J., Goldstein, S.A.N. (2001) MiRP2 forms potassium channels in skeletal muscle with Kv3.4 and is associated with periodic paralysis, *Cell* **104**, 217–231.

Abbott, G.W., Sesti, F., Splawski, I., Buck, M.E., Lehmann, M.H., Timothy, K.W., Keating, M.T., Goldstein, S.A. (1999) MiRP1 forms IKr potassium channels with HERG and is associated with cardiac arrhythmia, *Cell* **97**, 175–187.

Argyropoulos, G., Brown, A.M., Willi, S.M., Zhu, J., He, Y., Reitman, M., Gevao, S.M., Spruill, I., Garvey, W.T. (1998) Effects of mutations in the human uncoupling protein 3 gene on the respiratory quotient and fat oxidation in severe obesity and type 2 diabetes, *J. Clin. Invest.* **102**, 1345–1351.

Barclay, J., Balaguero, N., Mione, M., Ackerman, S.L., Letts, V.A., Brodbeck, J., Canti, C., Meir, A., Page, K.M., Kusumi, K., Perez-Reyes, E., Lander, E.S., Frankel, W.N., Gardiner, R.M., Dolphin, A.C., Rees, M. (2001) Ducky mouse phenotype of epilepsy and ataxia is associated with mutations in the Cacna2d2 gene and decreased calcium channel current in cerebellar Purkinje cells, *J. Neurosci.* **21**, 6095–6104.

Barhanin, J., Lesage, F., Guillemare, E., Fink, M., Lazdunski, M., Romey, G. (1996) K(v)LQT1 and IsK (minK) proteins associate to form the I(Ks) cardiac potassium current, *Nature* **384**, 78–80.

Baulac, S., Huberfeld, G., Gourfinkel-An, I., Mitropoulou, G., Beranger, A., Prud'homme, J.-F., Baulac, M., Brice, A., Bruzzone, R., LeGuern, E. (2001) First genetic evidence of GABA(A) receptor dysfunction in epilepsy: a mutation in the gamma-2-subunit gene, *Nat. Genet.* **28**, 46–48.

Bennett, P.B., Yazawa, K., Makita, N., George, A.L. Jr. (1995) Molecular mechanism for an inherited cardiac arrhythmia, *Nature* **376**, 683–685.

Bernard, C., Anderson, A., Becker, A., Poolos, N.P., Beck, H., Johnston, D. (2004) Acquired dendritic channelopathy in temporal lobe epilepsy, *Science* **305**(5683), 532–535.

Birkenhager, R., Otto, E., Schurmann, M.J., Vollmer, M., Ruf, E.-M., Maier-Lutz, I., Beekmann, F., Fekete, A., Omran, H., Feldmann, D., Milford, D.V., Jeck, N., Konrad, M., Landau, D., Knoers, N.V.A.M., Antignac, C., Sudbrak, R., Kispert, A., Hildebrandt, F. (2001) Mutation of BSND causes Bartter syndrome with sensorineural deafness and kidney failure, *Nat. Genet.* **29**, 310–314.

Browne, D.L., Gancher, S.T., Nutt, J.G., Brunt, E.R., Smith, E.A., Kramer, P., Litt, M. (1994) Episodic ataxia/myokymia syndrome is associated with point mutations in the human potassium channel gene, KCNA1, *Nat. Genet.* **8**, 136–140.

Bulman, D.E., Scoggan, K.A., van Oene, M.D., Nicolle, M.W., Hahn, A.F., Tollar, L.L., Ebers, G.C. (1999) A novel sodium channel mutation in a family with hypokalemic periodic paralysis, *Neurology* **53**, 1932–1936.

Burgess, D.L., Jones, J.M., Meisler, M.H., Noebels, J.L. (1997) Mutation of the Ca(2+) channel beta subunit gene Cchb4 is associated with ataxia and seizures in the lethargic (lh) mouse, *Cell* **88**, 385–392.

Caddick, S.J., Wang, C., Fletcher, C.F., Jenkins, N.A., Copeland, N.G., Hosford, D.A. (1999) Excitatory but not inhibitory synaptic transmission is reduced in lethargic (Cacnb4(lh)) and tottering (Cacna1atg) mouse thalami, *J. Neurophysiol.* **81**, 2066–2074.

Canessa, C.M., Horisberger, J.D., Rossier, B.C. (1993) Epithelial sodium channel related to proteins involved in neurodegeneration, *Nature* **361**(6411), 467–470.

Canessa, C.M., Schild, L., Buell, G., Thorens, B., Gautschi, I., Horisberger, J.D., Rossier, B.C. (1994) Amiloride-sensitive epithelial Na^+ channel is made of three homologous subunits, *Nature* **367**(6462), 463–467.

Cascio, M. (2004) Structure and function of the glycine receptor and related nicotinicoid receptors, *J. Biol. Chem.* **279**(19), 19383–19386.

Chandy, K.G., Fantino, E., Wittekindt, O., Kalman, K., Tong, L.L., Ho, T.H., Gutman, G.A., Crocq, M.A., Ganguli, R., Nimgaonkar, V., Morris-Rosendahl, D.J., Gargus, J.J. (1998) Isolation of a novel potassium channel gene hSKCa3 containing a polymorphic CAG repeat: a candidate for schizophrenia and bipolar disorder? *Mol. Psychiatry.* **3**, 32–37.

Chang, S.S., Grunder, S., Hanukoglu, A., Rosler, A., Mathew, P.M., Hanukoglu, I., Schild, L., Lu, Y., Shimkets, R.A., Nelson-Williams, C., Rossier, B.C., Lifton, R.P. (1996) Mutations in subunits of the epithelial sodium channel cause salt wasting with hyperkalaemic acidosis, pseudohypoaldosteronism type 1, *Nat. Genet.* **12**, 248–253.

Charlier, C., Singh, N.A., Ryan, S.G., Lewis, T.B., Reus, B.E., Leach, R.J., Leppert, M. (1998) A pore mutation in a novel KQT-like potassium channel gene in an idiopathic epilepsy family, *Nat. Genet.* **18**, 53–55.

Chen, Q., Kirsch, G.E., Zhang, D., Brugada, R., Brugada, J., Brugada, P., Potenza, D., Moya, A., Borggrefe, M., Breithardt, G., Ortiz-Lopez, R., Wang, Z., Antzelevitch, C., O'Brien, R.E., Schulze-Bahr, E., Keating, M.T., Towbin, J.A., Wang, Q. (1998) Genetic basis and molecular mechanism for idiopathic ventricular fibrillation, *Nature* **392**, 293–295.

Chubanov, V., Waldegger, S., Mederos y Schnitzler, M., Vitzthum, H., Sassen, M.C., Seyberth, H.W., Konrad, M., Gudermann, T. (2004) Free in PMC Disruption of TRPM6/TRPM7 complex formation by a mutation in the TRPM6 gene causes hypomagnesemia with secondary hypocalcemia, *Proc. Natl. Acad. Sci. U.S.A.* **101**(9), 2894–2899.

Claes, L., Del-Favero, J., Ceulemans, B., Lagae, L., Van Broeckhoven, C., De Jonghe, P. (2001) De novo mutations in the sodium-channel gene SCN1A cause severe myoclonic epilepsy of infancy, *Am. J. Hum. Genet.* **68**, 1327–1332.

Cleiren, E., Benichou, O., Van Hul, E., Gram, J., Bollerslav, J., Singer, F.R., Beaverson, K.,

Aledo, A., Whyte, M.P., Yoneyama, T., deVernejou, M.-C., Van Hul, W. (2001) Albers-Schonberg disease (autosomal dominant osteopetrosis, type II) results from mutations in the ClCN7 chloride channel gene, *Hum. Mol. Genet.* **10**, 2861–2867.

Cummings, C.J., Zoghbi, H.Y. (2000) Trinucleotide repeats: mechanisms and pathophysiology, *Annu. Rev. Genomics Hum. Genet.* **1**, 281–328.

Curran, M.E., Splawski, I., Timothy, K.W., Vincent, G.M., Green, E.D., Keating, M.T. (1995) A molecular basis for cardiac arrhythmia: HERG mutations cause long QT syndrome, *Cell* **80**, 795–803.

Curran, P.F., McIntosh, J.R. (1962) A model system for biological water transport, *Nature* **193**, 347–348.

De Fusco, M., Marconi, R., Silvestri, L., Atorino, L., Rampoldi, L., Morgante, L., Ballabio, A., Aridon, P., Casari, G. (2003) Haploinsufficiency of ATP1A2 encoding the Na^+/K^+ pump alpha-2 subunit associated with familial hemiplegic migraine type 2, *Nat. Genet.* **33**, 192–196.

Deen, P.M.T., Verdijk, M.A.J., Knoers, N.V.A.M., Wieringa, B., Monnens, L.A.H., van Os, C.H., van Oost, B.A. (1994) Requirement of human renal water channel aquaporin-2 for vasopressin-dependent concentration of urine, *Science* **264**, 92–94.

Delmas, P. (2004) Polycystins: from mechanosensation to gene regulation, *Cell* **118**(2), 145–148.

Di sant'agnese, P.A., Darling, R.C., Perera, G.A., Shea, E. (1953) Abnormal electrolyte composition of sweat in cystic fibrosis of the pancreas; clinical significance and relationship to the disease, *Pediatrics* **12**(5), 549–563.

Doyle, D.A., Morais Cabral, J., Pfuetzner, R.A., Kuo, A., Gulbis, J.M., Cohen, S.L., Chait, B.T., MacKinnon, R. (1998) The structure of the potassium channel: molecular basis of K^+ conduction and selectivity, *Science* **280**, 69–77.

Drachman, D.B. (1978) Myasthenia gravis (first of two parts), *N. Engl. J. Med.* **298**(3), 136–142.

Dror, V., Shamir, E., Ghanshani, S., Kimhi, R., Swartz, M., Barak, Y., Weizman, R., Avivi, L., Litmanovitch, T., Fantino, E., Kalman, K., Jones, E.G., Chandy, K.G., Gargus, J.J., Gutman, G.A., Navon, R. (1999) hKCa3/KCNN3 potassium channel gene: association of longer CAG repeats with schizophrenia in Israeli Ashkenazi Jews, expression in human tissues and localization to chromosome 1q21, *Mol. Psychiatry.* **4**, 254–260.

Ducros, A., Denier, C., Joutel, A., Cecillon, M., Lescoat, C., Vahedi, K., Darcel, F., Vicaut, E., Bousser, M.G., Tournier-Lasserve, E. (2001) The clinical spectrum of familial hemiplegic migraine associated with mutations in a neuronal calcium channel, *N. Engl. J. Med.* **345**, 17–24.

Duggal, P., Vesely, M.R., Wattanasirichaigoon, D., Villafane, J., Kaushik, V., Beggs, A.H. (1998) Mutation of the gene for IsK associated with both Jervell and Lange-Nielsen and Romano-Ward forms of long-QT syndrome, *Circulation* **97**, 142–146.

Dunne, M.J., Cosgrove, K.E., Shepherd, R.M., Aynsley-Green, A., Lindley, K.J. (2004) Hyperinsulinism in infancy: from basic science to clinical disease, *Physiol. Rev.* **84**(1), 239–275.

Dutzler, R., Campbell, E.B., Cadene, M., Chait, B.T., MacKinnon, R. (2002) X-ray structure of a ClC chloride channel at 3.0 a reveals the molecular basis of anion selectivity, *Nature* **415**(6869), 287–294.

Dutzler, R., Campbell, E.B., MacKinnon, R. (2003) Gating the selectivity filter in ClC chloride channels, *Science* **300**(5616), 108–112.

Engel, A.G., Ohno, K., Wang, H.-L., Milone, M., Sine, S.M. (1998) Molecular basis of congenital myasthenic syndromes: mutations in the acetylcholine receptor, *The Neuroscientist* **4**, 185–194.

Escayg, A., De Waard, M., Lee, D.D., Bichet, D., Wolf, P., Mayer, T., Johnston, J., Baloh, R., Sander, T., Meisler, M.H. (2000a) Coding and noncoding variation of the human calcium-channel beta(4)-subunit gene CACNB4 in patients with idiopathic generalized epilepsy and episodic ataxia, *Am. J. Hum. Genet.* **66**, 1531–1539.

Escayg, A., MacDonald, B.T., Meisler, M.H., Baulac, S., Huberfeld, G., An-Gourfinkel, I., Brice, A., LeGuern, E., Moulard, B., Chaigne, D., Buresi, C., Malafosse, A. (2000b) Mutations of SCN1A, encoding a neuronal sodium channel, in two families with GEFS + 2, *Nat. Genet.* **24**, 343–345.

Estevez, R., Boettger, T., Stein, V., Birkenhager, R., Otto, E., Hildebrandt, F., Jentsch, T.J. (2001) Barttin is a Cl- channel beta-subunit crucial for renal Cl- reabsorption and inner ear K^+ secretion, *Nature* **414**, 558–561.

Eunson, L.H., Rea, R., Zuberi, S.M., Youroukos, S., Panayiotopoulos, C.P., Liguori, R., Avoni,

P., McWilliam, R.C., Stephenson, J.B., Hanna, M.G., Kullmann, D.M., Spauschus, A. (2000) Clinical, genetic, and expression studies of mutations in the potassium channel gene KCNA1 reveal new phenotypic variability, *Ann. Neurol.* **48**, 647–656.

European Polycystic Kidney Disease Consortium. (1994) The polycystic kidney disease 1 gene encodes a 14 kb transcript and lies within a duplicated region on chromosome 16, *Cell* **77**, 881–894.

Fambrough, D.M., Drachman, D.B., Satyamurti, S. (1973) Neuromuscular junction in myasthenia gravis: decreased acetylcholine receptors, *Science* **182**(109), 293–295.

Fletcher, C.F., Lutz, C.M., O'Sullivan, T.N., Shaughnessy, J.D. Jr., Hawkes, R., Frankel, W.N., Copeland, N.G., Jenkins, N.A. (1996) Absence epilepsy in tottering mutant mice is associated with calcium channel defects, *Cell* **87**, 607–617.

Forbush, B. III, Kaplan, J.H., Hoffman, J.F. (1978) Characterization of a new photoaffinity derivative of ouabain: labeling of the large polypeptide and of a proteolipid component of the Na, K-ATPase, *Biochemistry* **17**(17), 3667–3676.

Frattini, A., Orchard, P.J., Sobacchi, C., Giliani, S., Abinun, M., Mattsson, J.P., Keeling, D.J., Andersson, A.-K., Wallbrandt, P., Zecca, L., Notarangelo, L.D., Vezzoni, P., Villa, A. (2000) Defects in TCIRG1 subunit of the vacuolar proton pump are responsible for a subset of human autosomal recessive osteopetrosis, *Nat. Genet.* **25**, 343–346.

Gloyn, A.L., Pearson, E.R., Antcliff, J.F., Proks, P., Bruining, G.J., Slingerland, A.S., Howard, N., Srinivasan, S., Silva, J.M., Molnes, J., Edghill, E.L., Frayling, T.M., Temple, I.K., Mackay, D., Shield, J.P., Sumnik, Z., van Rhijn, A., Wales, J.K., Clark, P., Gorman, S., Aisenberg, J., Ellard, S., Njolstad, P.R., Ashcroft, F.M., Hattersley, A.T. (2004) Activating mutations in the gene encoding the ATP-sensitive potassium-channel subunit Kir6.2 and permanent neonatal diabetes, *N. Engl. J. Med.* **350**(18), 1838–1849.

Gomez, C.M., Maselli, R.A., Vohra, B.P., Navedo, M., Stiles, J.R., Charnet, P., Schott, K., Rojas, L., Keesey, J., Verity, A., Wollmann, R.W., Lasalde-Dominicci, J. (2002) Novel delta subunit mutation in slow-channel syndrome causes severe weakness by novel mechanisms, *Ann. Neurol.* **51**(1), 102–112.

Goodenough, D.A., Paul, D.L. (2003) Beyond the gap: functions of unpaired connexon channels, *Nat. Rev. Mol. Cell Biol.* **4**(4), 285–294.

Grant, A.O., Carboni, M.P., Neplioueva, V., Starmer, C.F., Memmi, M., Napolitano, C., Priori, S. (2002) Long QT syndrome, Brugada syndrome, and conduction system disease are linked to a single sodium channel mutation, *J. Clin. Invest.* **110**, 1201–1209.

Grunder, s., Firsov, D., Chang, S.S., Jaeger, N.F., Gautschi, I., Schild, L., Lifton, R.P., Rossier, B.C. (1997) A mutation causing pseudohypoaldosteronism type 1 identifies a conserved glycine that is involved in the gating of the epithelial sodium channel, *EMBO J.* **16**, 899–907.

Hamill, O.P., Marty, A., Neher, E., Sakmann, B., Sigworth, F.J. (1981) Improved patch-clamp techniques for high-resolution current recording from cells and cell-free membrane patches, *Pflugers Arch.* **391**(2), 85–100.

Hanaoka, K., Qian, F., Boletta, A., Bhunia, A.K., Piontek, K., Tsiokas, L., Sukhatme, V.P., Guggino, W.B., Germino, G.G. (2000) Coassembly of polycystin-1 and -2 produces unique cation-permeable currents, *Nature* **408**, 990–994.

Hansson, J.H., Schild, L., Lu, Y., Wilson, T.A., Gautschi, I., Shimkets, R., Nelson-Williams, C., Rossier, B.C., Lifton, R.P. (1995) A de novo missense mutation of the beta subunit of the epithelial sodium channel causes hypertension and Liddle syndrome, identifying a proline-rich segment critical for regulation of channel activity, *Proc. Natl. Acad. Sci. U.S.A.* **92**, 11495–11499.

Hansson, J.H., Nelson-Williams, C., Suzuki, H., Schild, L., Shimkets, R., Lu, Y., Canessa, C., Iwasaki, T., Rossier, B., Lifton, R.P. (1995) Hypertension caused by a truncated epithelial sodium channel gamma subunit: genetic heterogeneity of Liddle syndrome, *Nat. Genet.* **11**, 76–82.

Hevers, W., Luddens, H. (1998) The diversity of GABAA receptors. Pharmacological and electrophysiological properties of GABAA channel subtypes, *Mol. Neurobiol.* **18**, 35–86.

Hodgkin, A.L., Huxley, A.F. (1952) The components of membrane conductance in the giant axon of Loligo, *J. Physiol.* **116**(4), 473–496.

Ishikawa, K., Tanaka, H., Saito, M., Ohkoshi, N., Fujita, T., Yoshizawa, K., Ikeuchi, T., Watanabe, M., Hayashi, A., Takiyama, Y., Nishizawa, M., Nakano, I., Matsubayashi, K., Miwa, M.,

Shoji, S., Kanazawa, I., Tsuji, S., Mizusawa, H. (1997) Japanese families with autosomal dominant pure cerebellar ataxia map to chromosome 19p13.1-p13.2 and are strongly associated with mild CAG expansions in the spinocerebellar ataxia type 6 gene in chromosome 19p13.1, *Am. J. Hum. Genet.* **61**, 336–346.

Jentsch, T.J. (2000) Neuronal KCNQ potassium channels: physiology and role in disease, *Nat. Rev. Neurosci.* **1**, 21–30.

Jentsch, T.J. (2002) Chloride channels are different, *Nature* **415**(6869), 276–277.

Jentsch, T.J., Stein, V., Weinreich, F., Zdebik, A.A. (2002) Molecular structure and physiological function of chloride channels, *Physiol. Rev.* **82**, 503–568.

Jiang, Y., Ruta, V., Chen, J., Lee, A., MacKinnon, R. (2003) The principle of gating charge movement in a voltage-dependent K^+ channel, *Nature* **423**(6935), 42–48.

Jiang, Y., Lee, A., Chen, J., Cadene, M., Chait, B.T., MacKinnon, R. (2002) The open pore conformation of potassium channels, *Nature* **417**(6888), 523–526.

Jodice, C., Mantuano, E., Veneziano, L., Trettel, F., Sabbadini, G., Calandriello, L., Francia, A., Spadaro, M., Pierelli, F., Salvi, F., Ophoff, R.A., Frants, R.R., Frontali, M. (1997) Episodic ataxia type 2 (EA2) and spinocerebellar ataxia type 6 (SCA6) due to CAG repeat expansion in the CACNA1A gene on chromosome 19p, *Hum. Mol. Genet.* **6**, 1973–1978.

Jouvenceau, A., Eunson, L.H., Spauschus, A., Ramesh, V., Zuberi, S.M., Kullmann, D.M., Hanna, M.G. (2001) Human epilepsy associated with dysfunction of the brain P/Q-type calcium channel, *Lancet* **358**, 801–807.

Jun, K., Piedras-Renteria, E.S., Smith, S.M., Wheeler, D.B., Lee, S.B., Lee, T.G., Chin, H., Adams, M.E., Scheller, R.H., Tsien, R.W., Shin, H.S. (1999) Ablation of P/Q-type Ca(2+) channel currents, altered synaptic transmission, and progressive ataxia in mice lacking the alpha(1A)-subunit, *Proc. Natl. Acad. Sci. U.S.A.* **96**, 15245–15250.

Kaunisto, M.A., Harno, H., Vanmolkot, K.R., Gargus, J.J., Sun, G., Hamalainen, E., Liukkonen, E., Kallela, M., van den Maagdenberg, A.M., Frants, R.R., Farkkila, M., Palotie, A., Wessman, M. (2004) A novel missense ATP1A2 mutation in a Finnish family with familial hemiplegic migraine type 2, *Neurogenetics* **5**(2), 141–146.

Keating, M., Atkinson, D., Dunn, C., Timothy, K., Vincent, G.M., Leppert, M. (1991) Linkage of a cardiac arrhythmia, the long QT syndrome, and the Harvey RAS-1 gene, *Science* **252**, 704–706.

Koefoed-Johnsen, V., Ussing, H.H. (1958) The nature of the frog skin potential, *Acta Physiol. Scand.* **42**(3–4), 298–308.

Kohl, S., Marx, T., Giddings, I., Jagle, H., Jacobson, S.G., Apfelstedt-Sylla, E., Zrenner, E., Sharpe, L.T., Wissinger, B. (1998) Total colourblindness is caused by mutations in the gene encoding the alpha-subunit of the cone photoreceptor cGMP-gated cation channel, *Nat. Genet.* **19**, 257–259.

Konrad, M., Weber, S. (2003) Recent advances in molecular genetics of hereditary magnesium-losing disorders, *J. Am. Soc. Nephrol.* **14**(1), 249–260.

Konrad, M., Schlingmann, K.P., Gudermann, T. (2004) Insights into the molecular nature of magnesium homeostasis, *Am. J. Physiol. Renal Physiol.* **286**(4), F599–F605.

Kornak, U., Kasper, D., Bosl, M.R., Kaiser, E., Schweizer, M., Schulz, A., Friedrich, W., Delling, G., Jentsch, T.J. (2001) Loss of the ClC-7 chloride channel leads to osteopetrosis in mice and man, *Cell* **104**, 205–215.

Kors, E.E., van den Maagdenberg, A.M., Plomp, J.J., Frants, R.R., Ferrari, M.D. (2002) Calcium channel mutations and migraine, *Curr. Opin. Neurol.* **15**, 311–316.

Kubisch, C., Schroeder, B.C., Friedrich, T., Lutjohann, B., El-Amraoui, A., Marlin, S., Petit, C., Jentsch, T.J. (1999) KCNQ4, a novel potassium channel expressed in sensory outer hair cells, is mutated in dominant deafness, *Cell* **96**, 437–446.

Langman, C.B. (2004) The molecular basis of kidney stones, *Curr. Opin. Pediatr.*. **16**(2), 188–193.

Leppert, M., Anderson, V.E., Quattlebaum, T., Stauffer, D., O'Connell, P., Nakamura, Y., Lalouel, J.M., White, R. (1989) Benign familial neonatal convulsions linked to genetic markers on chromosome 20, *Nature* **337**, 647–648.

Lerche, C., Scherer, C.R., Seebohm, G., Derst, C., Wei, A.D., Busch, A.E., Steinmeyer, K. (2000) Molecular cloning and functional expression of KCNQ5, a potassium channel subunit that may contribute to neuronal M-current diversity, *J. Biol. Chem.* **275**, 22395–22400.

Letts, V.A., Felix, R., Biddlecome, G.H., Arikkath, J., Mahaffey, C.L., Valenzuela, A., Bartlett, F.S.I.I., Mori, Y., Campbell, K.P., Frankel, W.N. (1998) The mouse stargazer gene encodes a neuronal Ca(2+)-channel gamma subunit, *Nat. Genet.* **19**, 340–347.

Lewis, T.B., Leach, R.J., Ward, K., O'Connell, P., Ryan, S.G. (1993) Genetic heterogeneity in benign familial neonatal convulsions: identification of a new locus on chromosome 8q, *Am. J. Hum. Genet.* **53**, 670–675.

Lin, S.H., Cheng, N.L., Hsu, Y.J., Halperin, M.L. (2004) Intrafamilial phenotype variability in patients with Gitelman syndrome having the same mutations in their thiazide-sensitive sodium/chloride cotransporter, *Am. J. Kidney Dis.* **43**(2), 304–312.

Lipscombe, D., Pan, J.Q., Gray, A.C. (2002) Functional diversity in neuronal voltage-gated calcium channels by alternative splicing of Ca(v)alpha1, *Mol. Neurobiol.* **26**(1), 21–44.

Liu, B.A., Juurlink, D.N. (2004) Drugs and the QT interval – caveat doctor, *N. Engl. J. Med.* **351**(11), 1053–1056.

Lloyd, S.E., Pearce, S.H.S., Fisher, S.E., Steinmeyer, K., Schwappach, B., Schelnman, S.J., Harding, B., Bolino, A., Devoto, M., Goodyer, P., Rigden, S.P.A., Wrong, O., Jentsch, T.J., Craig, I.W., Thakker, R.V. (1996) A common molecular basis for three inherited kidney stone diseases, *Nature* **370**, 445–449.

Lossin, C., Wang, D.W., Rhodes, T.H., Vanoye, C.G., George, A.L. Jr. (2002) Molecular basis of an inherited epilepsy, *Neuron* **34**, 877–884.

Makita, N., Horie, M., Nakamura, T., Ai, T., Sasaki, K., Yokoi, H., Sakurai, M., Sakuma, I., Otani, H., Sawa, H., Kitabatake, A. (2002) Drug-induced long-QT syndrome associated with a subclinical SCN5A mutation, *Circulation* **106**, 1269–1274.

Marubio, L.M., del Mar Arroyo-Jimenez, M., Cordero-Erausquin, M., Lena, C., Le Novere, N., de Kerchove d'Exaerde, A., Huchet, M., Damaj, M.I., Changeux, J.-P. (1999) Reduced antinociception in mice lacking neuronal nicotinic receptor subunits, *Nature* **398**, 805–810.

McClatchey, A.I., McKenna-Yasek, D., Cros, D., Worthen, H.G., Kuncl, R.W., DeSilva, S.M., Cornblath, D.R., Gusella, J.F., Brown, R.H. Jr. (1992) Novel mutations in families with unusual and variable disorders of the skeletal muscle sodium channel, *Nat. Genet.* **2**, 148–152.

McGrath, J., Somlo, S., Makova, S., Tian, X., Brueckner, M. (2003) Two populations of node monocilia initiate left-right asymmetry in the mouse, *Cell* **114**(1), 61–73.

McRory, J.E., Hamid, J., Doering, C.J., Garcia, E., Parker, R., Hamming, K., Chen, L., Hildebrand, M., Beedle, A.M., Feldcamp, L., Zamponi, G.W., Snutch, T.P. (2004) The CACNA1F gene encodes an L-type calcium channel with unique biophysical properties and tissue distribution, *J. Neurosci.* **24**(7), 1707–1718.

Meij, I.C., Koenderink, J.B., van Bokhoven, H., Assink, K.F.H., Groenestege, W.T., de Pont, J.J.H.H.M., Bindels, R.J.M., Monnens, L.A.H., van den Heuvel, L.P.W.J., Knoers, N.V.A.M. (2000) Dominant isolated renal magnesium loss is caused by misrouting of the Na^{+},K(+)-ATPase gamma-subunit, *Nat. Genet.* **26**, 265–266.

Meisler, M.H., Kearney, J., Ottman, R., Escayg, A. (2001) Identification of epilepsy genes in human and mouse, *Annu. Rev. Genet.* **35**, 567–588.

Meyer-Kleine, C., Steinmeyer, K., Ricker, K., Jentsch, T.J., Koch, M.C. (1995) Spectrum of mutations in the major human skeletal muscle chloride channel gene (CLCN1) leading to myotonia, *Am. J. Hum. Genet.* **57**, 1325–1334.

Miller, M.J., Rauer, H., Tomita, H., Rauer, H., Gargus, J.J., Gutman, G.A., Cahalan, M.D., Chandy, K.G. (2001) Nuclear localization and dominant-negative suppression by a mutant SKCa3 N-terminal channel fragment identified in a patient with schizophrenia, *J. Biol. Chem.* **276**, 27753–27756.

Mochizuki, T., Wu, G., Hayashi, T., Xenophontos, S.L., Veldhuisen, B., Saris, J.J., Reynolds, D.M., Cai, Y., Gabow, P.A., Pierides, A., Kimberling, W.J., Breuning, M.H., Constantinou Deltas, C., Peters, D.J.M., Somlo, S. (1996) PKD2, a gene for polycystic kidney disease that encodes an integral membrane protein, *Science* **272**, 1339–1342.

Mohler, P.J., Splawski, I., Napolitano, C., Bottelli, G., Sharpe, L., Timothy, K., Priori, S.G., Keating, M.T., Bennett, V. (2004) A cardiac arrhythmia syndrome caused by loss of ankyrin-B function, *Proc. Natl. Acad. Sci. U.S.A.* **101**(24), 9137–9142.

Mohler, P.J., Schott, J.J., Gramolini, A.O., Dilly, K.W., Guatimosim, S., duBell, W.H., Song, L.S., Haurogne, K., Kyndt, F., Ali, M.E., Rogers, T.B., Lederer, W.J., Escande, D., Le

Marec, H., Bennett, V. (2003) Ankyrin-B mutation causes type 4 long-QT cardiac arrhythmia and sudden cardiac death, *Nature* **421**(6923), 634–639.

Monnier, N., Procaccio, V., Stieglitz, P., Lunardi, J. (1997) Malignant-hyperthermia susceptibility is associated with a mutation of the alpha-1-subunit of the human dihydropyridine-sensitive L-type voltage-dependent calcium channel receptor in skeletal muscle, *Am. J. Hum. Genet.* **60**, 1316–1325.

Morais-Cabral, J.H., Zhou, Y., MacKinnon, R. (2001) Energetic optimization of ion conduction rate by the K^+ selectivity filter, *Nature* **414**, 37–42.

Morimoto, T., Nagao, H., Yoshimatsu, M., Yoshida, K., Matsuda, H. (1993) Pathogenic role of glutamate in hyperthermia-induced seizures, *Epilepsia* **34**, 447–452.

Moslehi, R., Langlois, S., Yam, I., Friedman, J.M. (1998) Linkage of malignant hyperthermia and hyperkalemic periodic paralysis to the adult skeletal muscle sodium channel (SCN4A) gene in a large pedigree, *Am. J. Med. Genet.* **76**, 21–27.

Mulders, S.M., Bichet, D.G., Rijss, J.P.L., Kamsteeg, E.-J., Arthus, M.-F., Lonergan, M., Fujiwara, M., Morgan, K., Leijendekker, R., van der Sluijs, P., van Os, C.H., Deen, P.M.T. (1998) An aquaporin-2 water channel mutant which causes autosomal dominant nephrogenic diabetes insipidus is retained in the Golgi complex, *J. Clin. Invest.* **102**, 57–66.

Nielsen, S., Chou, C.-L., Marples, D., Christensen, E.I., Kishore, B.K., Knepper, M.A. (1995) Vasopressin increases water permeability of kidney collecting duct by inducing translocation of aquaporin-CD water channels to plasma membrane, *Proc. Natl. Acad. Sci. U.S.A.* **92**, 1013–1017.

Noda, M., Takahashi, H., Tanabe, T., Toyosato, M., Kikyotani, S., Furutani, Y., Hirose, T., Takashima, H., Inayama, S., Miyata, T., Numa, S. (1983) Structural homology of Torpedo californica acetylcholine receptor subunits, *Nature* **302**, 528–532.

Ohmori, I., Ouchida, M., Ohtsuka, Y., Oka, E., Shimizu, K. (2002) Significant correlation of the SCN1A mutations and severe myoclonic epilepsy in infancy, *Biochem. Biophys. Res. Commun.* **295**, 17–23.

Ophoff, R.A., Terwindt, G.M., Vergouwe, M.N., van Eijk, R., Oefner, P.J., Hoffman, S.M.G., Lamerdin, J.E., Mohrenweiser, H.W., Bulman, D.E., Ferrari, M., Haan, J., Lindhout, D., van Ommen, G.-J.B., Hofker, M.H., Ferrari, M.D., Frants, R.R. (1996) Familial hemiplegic migraine and episodic ataxia type-2 are caused by mutations in the Ca(2+) channel gene CACNL1A4, *Cell* **87**, 543–552.

Parkinson, N., Brown, S.D. (2002) Focusing on the genetics of hearing: you ain't heard nothin' yet, *Genome Biol.*. **3**(6), COMMENT2006.

Patsalos, P.N. (2000) Antiepileptic drug pharmacogenetics, *Ther. Drug Monit.* **22**(1), 127–130.

Phillips, H.A., Favre, I., Kirkpatrick, M., Zuberi, S.M., Goudie, D., Heron, S.E., Scheffer, I.E., Sutherland, G.R., Berkovic, S.F., Bertrand, D., Mulley, J.C. (2001) CHRNB2 is the second acetylcholine receptor subunit associated with autosomal dominant nocturnal frontal lobe epilepsy, *Am. J. Hum. Genet.* **68**, 225–231.

Piwon, N., Gunther, W., Schwake, M., Bosl, M.R., Jentsch, T.J. (2000) ClC-5 Cl(-)-channel disruption impairs endocytosis in a mouse model for Dent's disease, *Nature* **408**, 369–373.

Preisig-Muller, R., Schlichthorl, G., Goerge, T., Heinen, S., Bruggemann, A., Rajan, S., Derst, C., Veh, R.W., Daut, J. (2002) Heteromerization of Kir2.x potassium channels contributes to the phenotype of Andersen's syndrome, *Proc. Natl. Acad. Sci. U.S.A.* **99**, 7774–7779.

Priori, S.G., Napolitano, C., Schwartz, P.J. (1999) Low penetrance in the long-QT syndrome: clinical impact, *Circulation* **99**, 529–533.

Ptacek, L.J., George, A.L. Jr., Barchi, R.L., Griggs, R.C., Riggs, J.E., Robertson, M., Leppert, M.F. (1992) Mutations in an S4 segment of the adult skeletal muscle sodium channel cause paramyotonia congenita, *Neuron* **8**, 891–897.

Ptacek, L.J., George, A.L. Jr., Griggs, R.C., Tawil, R., Kallen, R.G., Barchi, R.L., Robertson, M., Leppert, M.F. (1991) Identification of a mutation in the gene causing hyperkalemic periodic paralysis, *Cell* **67**, 1021–1027.

Ptacek, L.J., Tawil, R., Griggs, R.C., Engel, A.G., Layzer, R.B., Kwiecinski, H., McManis, P.G., Santiago, L., Moore, M., Fouad, G., Bradley, P., Leppert, M.F. (1994) Dihydropyridine receptor mutations cause hypokalemic periodic paralysis, *Cell* **77**, 863–868.

Pu, H.X., Scanzano, R., Blostein, R. (2000) Distinct regulatory effects of the Na,K-ATPase gamma subunit, *J. Biol. Chem.* **277**(23), 20270–20276.

Quane, K.A., Healy, J.M., Keating, K.E., Manning, B.M., Couch, F.J., Palmucci, L.M., Doriguzzi, C., Fagerlund, T.H., Berg, K., Ording, H., Bendixen, D., Mortier, W., Linz, U., Muller, C.R., McCarthy, T.V. (1993) Mutations in the ryanodine receptor gene in central core disease and malignant hyperthermia, *Nat. Genet.* **5**, 51–55.

Quinton, P.M., Bijman, J. (1983) Higher bioelectric potentials due to decreased chloride absorption in the sweat glands of patients with cystic fibrosis, *N. Engl. J. Med.* **308**(20), 1185–1189.

Ratjen, F., Doring, G. (2003) Cystic fibrosis, *Lancet* **361**(9358), 681–689.

Raychowdhury, M.K., Gonzalez-Perrett, S., Montalbetti, N., Timpanaro, G.A., Chasan, B., Goldmann, W.H., Stahl, S., Cooney, A., Goldin, E., Cantiello, H.F. (2004) Molecular pathophysiology of mucolipidosis type IV: pH dysregulation of the mucolipin-1 cation channel, *Hum. Mol. Genet.* **13**(6), 617–627.

Rees, M.I., Lewis, T.M., Kwok, J.B.J., Mortier, G.R., Govaert, P., Snell, R.G., Schofield, P.R., Owen, M.J. (2002) Hyperekplexia associated with compound heterozygote mutations in the beta-subunit of the human inhibitory glycine receptor (GLRB), *Hum. Mol. Genet.* **11**, 853–860.

Rho, J.M., Szot, P., Tempel, B.L., Schwartzkroin, P.A. (1999) Developmental seizure susceptibility of kv1.1 potassium channel knockout mice, *Dev. Neurosci.* **21**, 320–327.

Riordan, J.R., Rommens, J.M., Kerem, B., Alon, N., Rozmahel, R., Grzelczak, Z., Zielenski, J., Lok, S., Plavsic, N., Chou, J.L., et al. (1989) Identification of the cystic fibrosis gene: cloning and characterization of complementary DNA, *Science* **245**, 1066–1073.

Rosenthal, W., Seibold, A., Antaramian, A., Lonergan, M., Arthus, M.-F., Hendy, G.N., Birnbaumer, M., Bichet, D.G. (1992) Molecular identification of the gene responsible for congenital nephrogenic diabetes insipidus, *Nature* **359**, 233–235.

Roux, A.F., Pallares-Ruiz, N., Vielle, A., Faugere, V., Templin, C., Leprevost, D., Artieres, F., Lina, G., Molinari, N., Blanchet, P., Mondain, M., Claustres, M. (2004) Molecular epidemiology of DFNB1 deafness in France, *BMC Med. Genet.* **5**(1), 5.

Rowntree, R.K., Harris, A. (2003) The phenotypic consequences of CFTR mutations, *Ann. Hum. Genet.* **67**(Pt 5), 471–485.

Sanders, D.B. (2003) Lambert-eaton myasthenic syndrome: diagnosis and treatment, *Ann. N.Y. Acad. Sci.* **998**, 500–508.

Sanguinetti, M.C., Curran, M.E., Spector, P.S., Keating, M.T. (1996) Spectrum of HERG K^+-channel dysfunction in an inherited cardiac arrhythmia, *Proc. Natl. Acad. Sci. U.S.A.* **93**, 2208–2212.

Sansone, V., Griggs, R.C., Meola, G., Ptacek, L.J., Barohn, R., Iannaccone, S., Bryan, W., Baker, N., Janas, S.J., Scott, W., Ririe, D., Tawil, R. (1997) Andersen's syndrome: a distinct periodic paralysis, *Ann. Neurol.* **42**, 305–312.

Schlingmann, K.P., Konrad, M., Jeck, N., Waldegger, P., Reinalter, S.C., Holder, M., Seyberth, H.W., Waldegger, S. (2004) Salt wasting and deafness resulting from mutations in two chloride channels, *N. Engl. J. Med.* **350**(13), 1314–1319.

Schlingmann, K.P., Weber, S., Peters, M., Niemann Nejsum, L., Vitzthum, H., Klingel, K., Kratz, M., Haddad, E., Ristoff, E., Dinour, D., Syrrou, M., Nielsen, S., Sassen, M., Waldegger, S., Seyberth, H.W., Konrad, M. (2002) Hypomagnesemia with secondary hypocalcemia is caused by mutations in TRPM6, a new member of the TRPM gene family, *Nat. Genet.* **31**(2), 166–170.

Schultz, S.G., Curran, P.F. (1970) Coupled transport of sodium and organic solutes, *Physiol. Rev.* **50**(4), 637–718.

Schwartz, P.J., Priori, S.G., Dumaine, R., Napolitano, C., Antzelevitch, C., Stramba-Badiale, M., Richard, T.A., Berti, M.R., Bloise, R. (2000) A molecular link between the sudden infant death syndrome and the long-QT syndrome, *N. Engl. J. Med.* **343**, 262–267.

Schwartz, P.J., Priori, S.G., Bloise, R., Napolitano, C., Ronchetti, E., Piccinini, A., Goj, C., Breithardt, G., Schulze-Bahr, E., Wedekind, H., Nastoli, J. (2001) Molecular diagnosis in a child with sudden infant death syndrome, *Lancet* **358**, 1342–1343.

Schwartz, P.J., Stramba-Badiale, M., Segantini, A., Austoni, P., Bosi, G., Giorgetti, R., Grancini, F., Marni, E.D., Perticone, F., Rosti, D., Salice, P. (1998) Prolongation of the QT interval and the sudden infant death syndrome, *N. Engl. J. Med.* **338**, 1709–1714.

Seeburg, P.H., Single, F., Kuner, T., Higuchi, M., Sprengel, R. (2001) Genetic manipulation of key determinants of ion flow in glutamate receptor channels in the mouse, *Brain Res.* **907**(1–2), 233–243.

Segall, L., Scanzano, R., Kaunisto, M.A., Wessman, M., Palotie, A., Gargus, J.J., Blostein, R. (2004) Kinetic alterations due to a missense mutation in the Na,K-ATPase alpha 2 subunit cause familial hemiplegic migraine type 2, *J. Biol. Chem.* **279**(42), 43692–43696.

Sesti, F., Abbott, G.W., Wei, J., Murray, K.T., Saksena, S., Schwartz, P.J., Priori, S.G., Roden, D.M., George, A.L. Jr., Goldstein, S.A. (2000) A common polymorphism associated with antibiotic-induced cardiac arrhythmia, *Proc. Natl. Acad. Sci. U.S.A.* **97**, 10613–10618.

Shakkottai, V.G., Regaya, I., Wulff, H., Fajloun, Z., Tomita, H., Fathallah, M., Cahalan, M.D., Gargus, J.J., Sabatier, J.M., Chandy, K.G. (2001) Design and characterization of a highly selective peptide inhibitor of the small conductance calcium-activated K^+ channel, SkCa2, *J. Biol. Chem.* **276**(46), 43145–43151.

Shiang, R., Ryan, S.G., Zhu, Y.Z., Hahn, A.F., O'Connell, P., Wasmuth, J.J. (1993) Mutations in the alpha 1 subunit of the inhibitory glycine receptor cause the dominant neurologic disorder, hyperekplexia, *Nat. Genet.* **5**(4), 351–358.

Shieh, C.C., Coghlan, M., Sullivan, J.P., Gopalakrishnan, M. (2000) Potassium channels: molecular defects, diseases, and therapeutic opportunities, *Pharmacol. Rev.* **52**, 557–594.

Shimkets, R.A., Warnock, D.G., Bositis, C.M., Nelson-Williams, C., Hansson, J.H., Schambelan, M., Gill, J.R. Jr., Ulick, S., Milora, R.V., Findling, J.W., Canessa, C.M., Rossier, B.C., Lifton, R.P. (1994) Liddle's syndrome: heritable human hypertension caused by mutations in the beta subunit of the epithelial sodium channel, *Cell* **79**, 407–414.

Simon, D.B., Karet, F.E., Hamdan, J.M., Di Pietro, A., Sanjad, S.A., Lifton, R.P. (1996a) Bartter's syndrome, hypokalemic alkalosis with hypercalciuria, is caused by mutations in the Na-K-2Cl cotransporter NKCC2, *Nat. Genet.* **13**, 183–188.

Simon, D.B., Karet, F.E., Rodriguez-Soriano, J., Hamdan, J.H., DiPietro, A., Trachtman, H., Sanjad, S.A., Lifton, R.P. (1996b) Genetic heterogeneity of Bartter's syndrome revealed by mutations in the K^+ channel, ROMK, *Nat. Genet.* **14**, 152–156.

Simon, D.B., Nelson-Williams, C., Bia, M.J., Ellison, D., Karet, F.E., Molina, A.M., Vaara, I., Iwata, F., Cushner, H.M., Koolen, M., Gainza, F.J., Gitelman, H.J., Lifton, R.P. (1996c) Gitelman's variant of Bartter's syndrome, inherited hypokalaemic alkalosis, is caused by mutations in the thiazide-sensitive Na-Cl cotransporter, *Nat. Genet.* **12**, 24–30.

Simon, D.B., Lu, Y., Choate, K.A., Velazquez, H., Al-Sabban, E., Praga, M., Casari, G., Bettinelli, A., Colussi, G., Rodriguez-Soriano, J., McCredle, D., Milford, D., Sanjad, S., Lifton, R.P. (1999) Paracellin-1, a renal tight junction protein required for paracellular Mg(2+) resorption, *Science* **285**, 103–106.

Simon, D.B., Bindra, R.S., Mansfield, T.A., Nelson-Williams, C., Mendonca, E., Stone, R., Schurman, S., Nayir, A., Alpay, H., Bakkaloglu, A., Rodriguez-Soriano, J., Morales, J.M., Sanjad, S.A., Taylor, C.M., Pilz, D., Brem, A., Trachtman, H., Griswold, W., Richard, G.A., John, E., Lifton, R.P. (1997) Mutations in the chloride channel gene, CLCNKB, cause Bartter's syndrome type III, *Nat. Genet.* **17**, 171–178.

Singh, N.A., Charlier, C., Stauffer, D., DuPont, B.R., Leach, R.J., Melis, R., Ronen, G.M., Bjerre, I., Quattlebaum, T., Murphy, J.V., McHarg, M.L., Gagnon, D., Rosales, T.O., Peiffer, A., Anderson, V.E., Leppert, M. (1998) A novel potassium channel gene, KCNQ2, is mutated in an inherited epilepsy of newborns, *Nat. Genet.* **18**, 25–29.

Smart, S.L., Lopantsev, V., Zhang, C.L., Robbins, C.A., Wang, H., Chiu, S.Y., Schwartzkroin, P.A., Messing, A., Tempel, B.L. (1998) Deletion of the K(V)1.1 potassium channel causes epilepsy in mice, *Neuron* **20**, 809–819.

Splawski, I., Timothy, K.W., Vincent, G.M., Atkinson, D.L., Keating, M.T. (1997a) Molecular basis of the long-QT syndrome associated with deafness, *New. Eng. J. Med.* **336**, 1562–1567.

Splawski, I., Tristani-Firouzi, M., Lehmann, M.H., Sanguinetti, M.C., Keating, M.T. (1997b) Mutations in the hminK gene cause long QT syndrome and suppress IKs function, *Nat. Genet.* **17**, 338–340.

Steinlein, O.K., Mulley, J.C., Propping, P., Wallace, R.H., Phillips, H.A., Sutherland, G.R., Scheffer, I.E., Berkovic, S.F. (1995) A missense mutation in the neuronal nicotinic acetylcholine receptor alpha-4 subunit is associated with autosomal dominant nocturnal frontal lobe epilepsy, *Nat. Genet.* **11**, 201–203.

Stoffel, M., Jan, L.Y. (1998) Epilepsy genes: excitement traced to potassium channels, *Nat. Genet.* **18**, 6–8.

Strom, T.M., Nyakatura, G., Apfelstedt-Sylla, E., Hellebrand, H., Lorenz, B., Weber, B.H.F., Wutz, K., Gutwillinger, N., Ruther, K., Drescher, B., Sauer, C., Zrenner, E., Meitinger, T., Rosenthal, A., Meindl, A. (1998) An L-type calcium-channel gene mutated in incomplete X-linked congenital stationary night blindness, *Nat. Genet.* **19**, 260–263.

Sugawara, T., Mazaki-Miyazaki, E., Fukushima, K., Shimomura, J., Fujiwara, T., Hamano, S., Inoue, Y., Yamakawa, K. (2002) Frequent mutations of SCN1A in severe myoclonic epilepsy in infancy, *Neurology* **58**, 1122–1124.

Sugawara, T., Tsurubuchi, Y., Agarwala, K.L., Ito, M., Fukuma, G., Mazaki-Miyazaki, E., Nagafuji, H., Noda, M., Imoto, K., Wada, K., Mitsudome, A., Kaneko, S., Montal, M., Nagata, K., Hirose, S., Yamakawa, K. (2001) A missense mutation of the Na^+ channel alpha II subunit gene Na(v)1.2 in a patient with febrile and afebrile seizures causes channel dysfunction, *Proc. Natl. Acad. Sci. U.S.A.* **98**, 6384–6389.

Sundin, O.H., Yang, J.M., Li, Y., Zhu, D., Hurd, J.N., Mitchell, T.N., Silva, E.D., Maumenee, I.H. (2000) Genetic basis of total colourblindness among the Pingelapese islanders, *Nat. Genet.* **25**, 289–293.

Tanabe, T., Beam, K.G., Powell, J.A., Numa, S. (1988) Restoration of excitation-contraction coupling and slow calcium current in dysgenic muscle by dihydropyridine receptor complementary DNA, *Nature* **336**, 134–139.

Tempel, B.L., Papazian, D.M., Schwarz, T.L., Jan, Y.N., Jan, L.Y. (1987) Sequence of a probable potassium channel component encoded at Shaker locus of Drosophila, *Science* **237**, 770–775.

Thomas, P., Ye, Y., Lightner, E. (1996) Mutation of the pancreatic islet inward rectifier Kir6.2 also leads to familial persistent hyperinsulinemic hypoglycemia of infancy, *Hum. Mol. Genet.* **5**, 1809–1812.

Thomas, P.M., Cote, G.J., Wohllk, N., Haddad, B., Mathew, P.M., Rabl, W., Aguilar-Bryan, L., Gagel, R.F., Bryan, J. (1995) Mutations in the sulfonylurea receptor gene in familial persistent hyperinsulinemic hypoglycemia of infancy, *Science* **268**, 426–429.

Tomita, H., Shakkittai, V.G., Gutman, G.A., Sun, G., Bunney, W.E., Cahalan, M.D., Chandy, K.G., Gargus, J.J. (2003) Novel truncated isoform of SK3 potassium channel is a potent dominant-negative regulator of SK currents: implications in schizophrenia, *Mol. Psychiatry.* **8**(5), 524–535,

Towbin, J.A., Vatta, M. (2001) Molecular biology and the prolonged QT syndromes, *Am. J. Med.* **110**, 385–398.

Tristani-Firouzi, M., Jensen, J.L., Donaldson, M.R., Sansone, V., Meola, G., Hahn, A., Bendahhou, S., Kwiecinski, H., Fidzianska, A., Plaster, N., Fu, Y.H., Ptacek, L.J., Tawil, R. (2002) Functional and clinical characterization of KCNJ2 mutations associated with LQT7 (Andersen syndrome), *J. Clin. Invest.* **110**, 381–388.

Turk, E., Zabel, B., Mundlos, S., Dyer, J., Wright, E.M. (1991) Glucose/galactose malabsorption caused by a defect in the Na(+)/glucose cotransporter, *Nature* **350**, 354–356.

van den Heuvel, L.P., Assink, K., Willemsen, M., Monnens, L. (2002) Autosomal recessive renal glucosuria attributable to a mutation in the sodium glucose cotransporter (SGLT2), *Hum. Genet.* **111**, 544–547.

Van Den Maagdenberg, A.M., Kors, E.E., Brunt, E.R., Van Paesschen, W., Pascual, J., Ravine, D., Keeling, S., Vanmolkot, K.R., Vermeulen, F.L., Terwindt, G.M., Haan, J., Frants, R.R., Ferrari, M.D. (2002) Episodic ataxia type 2 Three novel truncating mutations and one novel missense mutation in the CACNA1A gene, *J. Neurol.* **249**, 1515–1519.

Venter, J.C., Adams, M.D., Myers, E.W., Li, P.W., Mural, R.J., Sutton, G.G., Smith, H.O., et al. (2001) The sequence of the human genome, *Science* **291**, 1304–1351.

Walder, R.Y., Landau, D., Meyer, P., Shalev, H., Tsolia, M., Borochowitz, Z., Boettger, M.B., Beck, G.E., Englehardt, R.K., Carmi, R., Sheffield, V.C. (2002) Mutation of TRPM6 causes familial hypomagnesemia with secondary hypocalcemia, *Nat. Genet.* **31**(2), 171–174.

Wallace, R.H., Marini, C., Petrou, S., Harkin, L.A., Bowser, D.N., Panchal, R.G., Williams, D.A., Sutherland, G.R., Mulley, J.C., Scheffer, I.E., Berkovic, S.F. (2001a) Mutant

GABA(A) receptor gamma2-subunit in childhood absence epilepsy and febrile seizures, *Nat. Genet.* **28**, 49–52.

Wallace, R.H., Wang, D.W., Singh, R., Scheffer, I.E., George, A.L. Jr., Phillips, H.A., Saar, K., Reis, A., Johnson, E.W., Sutherland, G.R., Berkovic, S.F., Mulley, J.C. (1998) Febrile seizures and generalized epilepsy associated with a mutation in the Na(+)-channel beta-1 subunit gene SCN1B, *Nat. Genet.* **19**, 366–370.

Wallace, R.H., Scheffer, I.E., Barnett, S., Richards, M., Dibbens, L., Desai, R.R., Lerman-Sagie, T., Lev, D., Mazarib, A., Brand, N., Ben-Zeev, B., Goikhman, I., Singh, R., Kremmidiotis, G., Gardner, A., Sutherland, G.R., George, A.L. Jr., Mulley, J.C., Berkovic, S.F. (2001b) Neuronal sodium-channel alpha-1-subunit mutations in generalized epilepsy with febrile seizures plus, *Am. J. Hum. Genet.* **68**, 859–865.

Wang, H.-S., Pan, Z., Shi, W., Brown, B.S., Wymore, R.S., Cohen, I.S., Dixon, J.E., McKinnon, D. (1998) KCNQ2 and KCNQ3 potassium channel subunits: molecular correlates of the M-channel, *Science* **282**, 1890–1893.

Wang, Q., Shen, J., Splawski, I., Atkinson, D., Li, Z., Robinson, J.L., Moss, A.J., Towbin, J.A., Keating, M.T. (1995) SCN5A mutations associated with an inherited cardiac arrhythmia, long QT syndrome, *Cell* **80**, 805–811.

Wang, Q., Curren, M.E., Splawski, I., Burn, T.C., Millholland, J.M., VanRaay, T.J., Shen, J., Timothy, K.W., Vincent, G.M., de Jager, T., Schwartz, P.J., Towbin, J.A., Moss, A.J., Atkinson, D.L., Landes, G.M., Connors, T.D., Keating, M.T. (1996) Positional cloning of a novel potassium channel gene: KVLQT1 mutations cause cardiac arrhythmias, *Nat. Genet.* **12**, 17–23.

Warmke, J.W., Ganetzky, B. (1994) A family of potassium channel genes related to eag in Drosophila and mammals, *Proc. Natl. Acad. Sci. U.S.A.* **91**, 3438–3442.

Watanabe, H., Nagata, E., Kosakai, A., Nakamura, M., Yokoyama, M., Tanaka, K., Sasai, H. (2000) Disruption of the epilepsy KCNQ2 gene results in neural hyperexcitability, *J. Neurochem.* **75**(1), 28–33.

Watnick, T.J., Torres, V.E., Gandolph, M.A., Qian, F., Onuchic, L.F., Klinger, K.W., Landes, G., Germino, G.G. (1998) Somatic mutation in individual liver cysts supports a two-hit model of cystogenesis in autosomal dominant polycystic kidney disease, *Mol. Cell* **2**, 247–251.

Wilson, P.D. (2004) Polycystic kidney disease, *N. Engl. J. Med.* **350**(2), 151–164.

Wickenden, A.D., Yu, W., Zou, A., Jegla, T., Wagoner, P.K. (2000) Retigabine, a novel anticonvulsant, enhances activation of KCNQ2/Q3 potassium channels, *Mol. Pharmacol.* **58**, 591–600.

Wollnik, B., Schroeder, B.C., Kubisch, C., Esperer, H.D., Wieacker, P., Jentsch, T.J. (1997) Pathophysiological mechanisms of dominant and recessive KVLQT1 K^+ channel mutations found in inherited cardiac arrhythmias, *Hum. Mol. Genet.* **6**, 1943–1949.

Wright, E.M., Turk, E. (2004) The sodium/glucose cotransport family SLC5, *Pflugers Arch.* **447**(5), 510–518.

Yang, P., Kanki, H., Drolet, B., Yang, T., Wei, J., Viswanathan, P.C., Hohnloser, S.H., Shimizu, W., Schwartz, P.J., Stanton, M., Murray, K.T., Norris, K., George, A.L. Jr., Roden, D.M. (2002) Allelic variants in long-QT disease genes in patients with drug-associated torsades de pointes, *Circulation* **105**, 1943–1948.

Yue, Q., Jen, J.C., Nelson, S.F., Baloh, R.W. (1997) Progressive ataxia due to a missense mutation in a calcium-channel gene, *Am. J. Hum. Genet.* **61**, 1078–1087.

Zhuchenko, O., Bailey, J., Bonnen, P., Ashizawa, T., Stockton, D.W., Amos, C., Dobyns, W.B., Subramony, S.H., Zoghbi, H.Y., Lee, C.C. (1997) Autosomal dominant cerebellar ataxia (SCA6) associated with small polyglutamine expansions in the alpha(1A)-voltage-dependent calcium channel, *Nat. Genet.* **15**, 62–69.

Zuberi, S.M., Eunson, L.H., Spauschus, A., De Silva, R., Tolmie, J., Wood, N.W., McWilliam, R.C., Stephenson, J.P., Kullmann, D.M., Hanna, M.G. (1999) A novel mutation in the human voltage-gated potassium channel gene (Kv1.1) associates with episodic ataxia type 1 and sometimes with partial epilepsy, *Brain* **122**, 817–825.

Zwingman, T.A., Neumann, P.E., Noebels, J.L., Herrup, K. (2001) Rocker is a new variant of the voltage-dependent calcium channel gene Cacna1a, *J. Neurosci.* **21**, 1169–1178.

21
Pathophysiology of Stroke

Andreas Meisel, Konstantin Prass, Tilo Wolf, and Ulrich Dirnagl
Department of Experimental Neurology (Charité) Humboldt University, Berlin, Germany

Neurobiology. From Molecular Basis to Disease. Edited by Robert A. Meyers.

ISBN: 978-3-527-32293-0

Keywords

Agnosia
It is the inability to recognize and identify objects or persons despite having knowledge of the characteristics of those objects or persons.

Amaurosis
It is a partial or complete loss of sight.

Apraxia
It is the inability to execute a voluntary motor movement despite being able to demonstrate normal muscle function.

Ataxia
It is an incoordination and lack of balance.

Dyslexia
It is a difficulty (in learning) to read.

Dysphasia
It is a group of speech disorders with an impairment of expression by speech, or an impairment of the comprehension of spoken language. More severe forms of dysphasia are called aphasia.

Excitotoxicity
It is an excessive exposure to glutamate or related compounds killing brain neurons by overstimulating them.

Hemianopia
It is a blindness or reduction in vision in one half of the visual field.

Hemiparesis
It is a weakness on one side of the body.

Ischemic Tolerance
Ischemic tolerance or ischemic preconditioning is an intrinsic adaptive response to a mild ischemic stimulus protecting from an otherwise "lethal" ischemic event.

Penumbra
It is an area of constrained blood flow with partially preserved energy metabolism lying between the lethally damaged core and the normal brain.

SAGE
Serial analysis of gene expression, a technique for profiling gene expression.

The current understanding of stroke pathophysiology is basically founded on experimental models. Among these, *in vitro* models with primary cultures of cerebral cells permit stroke pathophysiology to be examined on a molecular level, while animal models (mainly mouse and rat) are used to evaluate medical intervention. There is a highly complex sequence of events leading to the eventual ischemic cerebral damage that follows a well-defined spatiotemporal pattern. Overwhelming excitotoxicity leads to early necrotic cell death in what is to become the core of the infarction, while the tissue damage in the surrounding zone called *penumbra* happens on a longer timescale. The excitotoxic or inflammatory mechanisms are milder, bearing the biochemical hallmarks of apoptosis. The brain cells, challenged by such a large-scale assault, activate endogenous protective programs. These have been studied by experimentally inducing ischemic tolerance (i.e. ischemic preconditioning). Importantly, cerebral ischemia not only affects the brain but also impacts other systems. For example, stroke induces dramatic immunodepression through overactivation of the sympathetic nervous system. As a result, severe bacterial infections such as pneumonia occur. The complex signaling cascades not only decide over cell survival in the brain and the neurological deficit but also over mortality after stroke from extracerebral complications. Their ability to govern not only the maturation of the eventual infarction but also the immune system make them a promising target for intervention and the development of neuroprotective drugs.

1 Introduction

In the United States, more than 600 000 people per year suffer from strokes, and, among acutely hospitalized neurological patients, stroke patients make up the largest share – about 50%. Currently, there are about 4 million stroke survivors in the United States. Mortality from stroke is an estimated 25%, which makes it the third major cause of death in industrialized countries. Because of the high rate of severe permanent disability, stroke is a burden not only for the affected patients and their families but also for national economies: In the United States, it is estimated that the annual costs caused by strokes ranges between 30 and 40 billion dollars. In the United Kingdom, the cost of one individual patient amounts to £30 000 over 5 years.

The term stroke accommodates a variety of different conditions. About 85% of all strokes are caused by cerebral ischemia due to vessel occlusion. Primary cerebral bleeding is rare in comparison – 15%. Of the ischemic strokes, 75% are caused by emboli, of either arterial or cardial origin, while microvascular occlusion, that is, hyalinosis or *in situ* thrombosis is responsible for 20% of cases. Hemodynamic ischemia, caused by stenoses of brain-supplying arteries, account for less than 5% of ischemic strokes.

The prospective risk of suffering an ischemic stroke is partially a function of social and behavioral (nutrition, tobacco use, stress) factors and long-standing disorders like hypertension, diabetes, disorders of cholesterol and lipid metabolism, and obesity. Atherosclerosis is not only the main underlying condition in ischemic stroke but also in coronary heart disease and peripheral vascular disease. Moreover, the genetic background of affected individuals is coming under increasing scientific scrutiny, as might be expected for familial stroke conditions like CADASIL (cerebral autosomal dominant arteriopathy with subcortical infarcts and leucencephalopathy), which is characterized by relapsing subcortical ischemia. The condition is caused by a mutation in the *notch3* gene on chromosome 19. MELAS (mitochondrial encephalopathy, lactate acidosis, stroke-like episodes) is another rare familial stroke disorder, characterized by migraine, grand-mal seizures, and recurrent cortical infarctions. The condition is caused by a range of mutations in the mitochondrial genome, which explains its maternal inheritance. The severity of the phenotype depends mainly on the mutation locus and on the proportion of mutated mitochondrial genomes in the cerebral mosaic. However, family and twin studies suggest that genetic factors are likely to be involved in common types of ischemic stroke, not just in rare and well-defined familial stroke disorders.

A locus on chromosome 5q12 described for the Icelandic population seemed to increase stroke susceptibility significantly. The responsible gene has been identified as phosphodiesterase 4D (pde4d). The risk-conveying polymorphisms are located in the gene-regulatory part. This hints at a defect in the regulation of the encoded cAMP degrading protein. Phosphodiesterase 4D belongs to a group of proteins that are drug targets in the treatment of asthma, erectile dysfunction, and inflammation.

Symptoms of focal cerebral ischemia depend on the individual affected vessel with its typical supply territory. Table 1 summarizes the frequency of affection and the typical clinical syndromes for the main brain arteries. Stroke mortality is rated between 20 and 30%. Unfavorable prognostic factors are old age, initial coma, papillary asymmetry, coronary heart disease, and heart failure. Pyrexia and infections (pneumonia, in particular) deteriorate the prognosis, particularly during the initial phase (see below). Of the surviving patients, one-third improve within a week, 40% remain unchanged in their disability, and 20% deteriorate further during the first week.

A great deal of our knowledge about the pathophysiology of stroke comes from experimental research. The experimental animal models follow two main paradigms. One is the model of focal cerebral ischemia as a model of ischemic stroke. The other is the model of global cerebral ischemia as model of circulatory arrest. For obvious

Tab. 1 Territorial stroke syndromes.

Anterior territory		
Middle cerebral artery (MCA)	(≈60%)	Hemiparesis mainly arm, hemihypaesthesia, hemianopia, dysphasia or neglect
Anterior cerebral artery (ACA)	(4%)	Hemiparesis mainly leg, urinary incontinence, apraxia
Anterior choroidal artery (AchA)	(8%)	Hemiparesis, hemihypaesthesia, hemianopia
Posterior territory		
Vertebral/basilar artery (VA/BA)	(10%)	Vertigo, diplopia, bilateral hypaesthesia, and paresis, crossed syndromes, amaurosis, ataxia, headache, coma
Cerebellum posterior/anterior inferior cerebellar artery (PICA/AICA)	(7%)	Headache, ataxia, vertigo, gaze palsy, facial weakness, deafness
Posterior cerebral artery (PCA)	(9%)	Hemianopia, dyslexia, visual agnosia

reasons, we will focus mainly at studies of focal cerebral ischemia. The bulk of experimental studies has been carried out with rats and mice, although some primates have been used. In rodents, ischemia is mostly induced by intravascular occlusion of the MCA, using a monofilament just exceeding the critical vessel diameter. Depending on the duration of occlusion, we distinguish permanent from transient ischemia models. The latter are also used as models for spontaneous reperfusion or for the state after successful lysis therapy with recombinant tissue plasminogen activator (rtPA) in humans, respectively.

In models of cerebral ischemia, perfusion to the brain is impeded by some form of manipulation. Since the essence of ischemic stroke is vascular occlusion, it is not surprising that in these models the extent of the eventual infarction can be predicted quite accurately from the degree of reduction in regional cerebral blood flow (rCBF). Thus, if the rCBF is reduced to less than 25% of normal, the likelihood of infarction in a given volume of brain tissue is greater than 95%. In contrast, the likelihood of infarction is less than 5% if rCBF does not fall below 50% of normal. These thresholds have been established in comparative human PET and MRI studies and correspond with experimental data from animal models. Thus, the initial reduced rCBF determines the extent of the anticipated infarction. This, however, holds only with the provision that a spontaneous or therapeutic reperfusion does not take place.

We now look at ischemic infarction as the result of a complex and prolonged process of infarct maturation rather than a simple result of reduced regional perfusion. As brain tissue has a high demand for oxygen and glucose, a disruption of perfusion leads to substrate depletion within few minutes, while toxic metabolites accumulate. The ensuing cellular energy deficit leads to a collapse of the established ion gradients and the membrane potential. Neurons and glial cells depolarize. Depending on the extent and the duration of this energy deficiency, the cells will suffer not only a functional but also a structural breakdown. The highly complex sequence of events within the ischemic area follows a fairly well-defined stereotypic spatiotemporal pattern, which we will discuss below in more detail. The concept of the ischemic *penumbra* is crucial to the understanding of these mechanisms. The cascade of ischemic damage begins with *excitotoxicity*, the formation of reactive *oxygen free radicals*, the increasing *tissue acidosis* and the occurrence of *periinfarct-depolarizations*. It is followed by the stages of *inflammation* and programmed cell death (*apoptosis*). This is associated with *DNA damage* that, in turn, induces *DNA repair programs*. Although the process is not yet fully understood, we know that chromatin remodeling, that is, *epigenetic mechanisms*, and the activation of transcription factors induce complex *gene programs*. These changes initiate the expression of destructive proteins involved in inflammation and apoptosis, as well as a host of protective genes that help repair the ischemic damage. It is the activation of these protective genes that builds up *ischemic tolerance*. Among the many newly discovered protective mechanisms, *endogenous* and *exogenous cell replacements* have met with a lot of interest. Apart from these autochthonous mechanisms of the brain tissue, there are other mechanisms on a systemic scale, which bear important clinical significance. For instance, the phenomenon of *stroke-induced immunodepression* can help understand why stroke patients are at such high risk of contracting serious bacterial infections. Neuroprotective treatment must be based on an understanding of these mechanisms.

2 The Penumbra Concept

In the ischemic brain, we commonly distinguish two tissue volumes – the core of the infarction and the surrounding zone, known as *ischemic penumbra* – the underperfused and metabolically compromised margin surrounding the irrevocably damaged core. Core and penumbra are characterized by two different kinds of cell death: necrosis and apoptosis (which is also called programmed cell death or delayed neuronal cell death). The severe perfusion deficit in the core causes a breakdown of metabolic processes, cellular energy supply, and ion homeostasis, which causes the cells to lose their integrity within minutes. Thus, acute necrosis of cell and tissue prevails in the core. In the penumbra, some residual perfusion is maintained by collateral vessels, which may be unable to maintain the full functional metabolism, but prevents immediate structural disintegration. However, over time, the alteration of cellular homeostasis causes more and more cells to die, and the volume of the infarction increases (Fig. 1). The penumbra has thus to be considered as tissue at risk during the maturation of the infarct. In this region, apoptosis and inflammatory signaling cascades play an important role. It may initially constitute 50% of the volume that will end up as infarction. The mechanisms that lead to delayed cell death within the penumbra are the subject of intense research, as they provide targets for a specific neuroprotective therapy in brain regions challenged by ischemia, but which are still viable.

3 Excitotoxicity

The depolarization of neurons and glia due to local energy deficit causes the activation of voltage-gated calcium channels and the release of excitatory amino

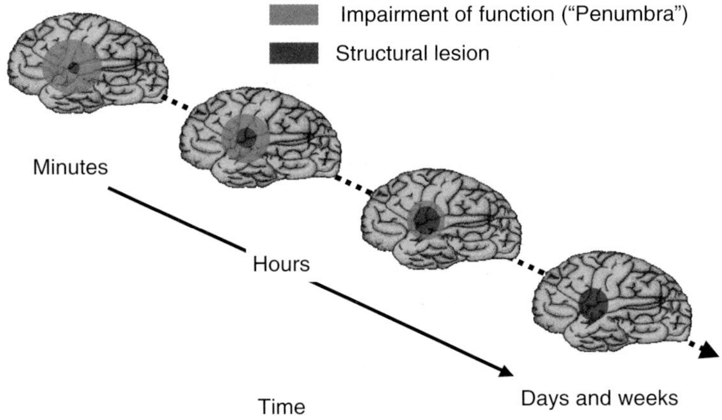

Fig. 1 Regression of the functional neurological deficit while the structural lesion grows. Early in the course of stroke, clinical symptoms mostly reflect an impairment of function (green) but not necessarily a structural lesion (blue). Over time, some areas either spontaneously, or because of therapy, recover function, which explains why symptoms in patients can regress, while the structural lesion actually grows (Reprinted from Dirnagl, U., Iadecola, C., Moskowitz, M.A. (1999) Pathobiology of ischaemic stroke: an integrated view, *Trends Neurosci.* **22**, 391–397, with permission from Elsevier) (See color plate p. xli).

acids into the extracellular space. Glutamate, in particular, which, under normal cellular energy conditions, would be immediately taken up presynaptically or through astrocytes, now remains in the extracellular space where it accumulates dramatically. Through the activation of NMDA and AMPA subtype glutamate receptors, the intracellular Ca^{2+} level rises (Fig. 2). Furthermore, metabotropic glutamate receptors are activated through the induction of phospholipase C (PLC) and inositol triphosphate (IP_3), and calcium is mobilized from intracellular stores. Additionally, the AMPA receptor induces an increase of intracellular Na^+ and Cl levels.

Altogether, the result is a massive disturbance of ion homeostasis, accompanied by passive water influx and cell edema. Ultimately, these massive changes in cell volume account for osmotic cell lysis. This lytic type of cell death, also referred to as *necrosis*, is primarily observed in the core of the infarction. Cells that escape this most dramatic form of disintegration, as they can not in the core but in the penumbra, excitotoxicity may be an initiator of molecular events that lead to apoptosis and inflammation.

4
Oxygen Free Radicals

As a consequence of ischemia and particularly of reperfusion, reactive oxygen free radicals such as superoxide, hydrogen peroxide, and hydroxyl radicals are generated. Nitric oxide is generated via the activation of the calcium calmodulin–dependent nitric oxide synthase (NOS); it reacts with superoxide radicals and forms thus the highly reactive peroxynitrite radical.

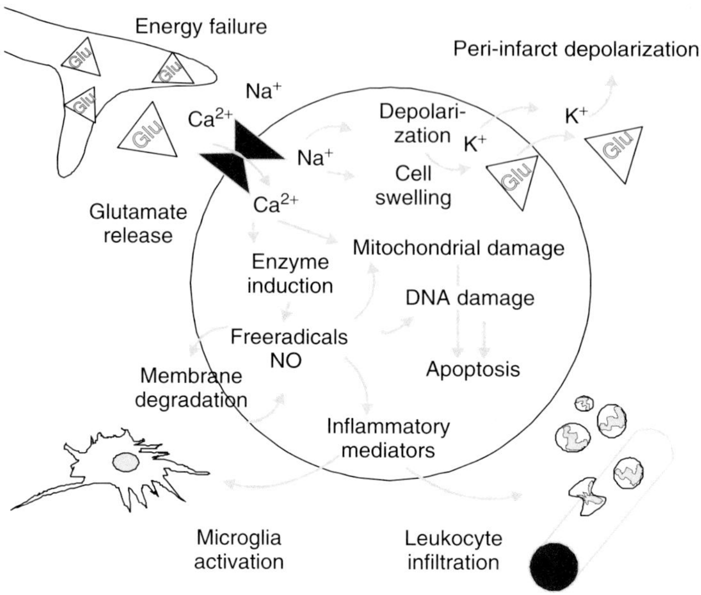

Fig. 2 Simplified overview of pathophysiological mechanisms in the focally ischemic brain. Energy failure leads to the depolarization of neurons. Activation of specific glutamate receptors dramatically increases intracellular Ca^{2+}, Na^{+}, Cl^{-} levels while K^{+} is released into the extracellular space. Diffusion of glutamate (Glu) and K^{+} in the extracellular space can propagate a series of spreading waves of depolarization (peri-infarct depolarizations). Water shifts to the intracellular space via osmotic gradients and cells swell (oedema). The universal intracellular messenger Ca^{2+} overactivates numerous enzyme systems (proteases, lipases, endonucleases, etc.). Free radicals are generated, which damage membranes (lipolysis), mitochondria, and DNA, in turn triggering caspase-mediated cell death (apoptosis). Free radicals also induce the formation of inflammatory mediators, which activate microglia and lead to the invasion of blood-borne inflammatory cells (leukocyte infiltration) via upregulation of endothelial adhesion molecules (Reprinted from Dirnagl, U., Iadecola, C., Moskowitz, M.A. (1999) Pathobiology of ischaemic stroke: an integrated view, *Trends Neurosci.* **22**, 391–397, with permission from Elsevier).

Further sources of oxygen free radicals in the damaged brain tissue are the breakdown products of the adenosine phosphates, which contribute to radical production via xanthine oxidase and the iron-catalyzed Haber–Weiss reaction. The many different radical species that are thus formed can react with virtually any cellular components (carbohydrates, amino acids, DNA, phospholipids) and damage them. The peroxidation of membrane lipids releases further radicals – and further glutamate (Fig. 2). Oxygen free radicals gain even more significance when new oxygen reaches the damaged tissue by virtue of reperfusion, or in the penumbra where oxygen supply has not ceased entirely.

Hypoxia itself as well as the elevated intracellular concentration of calcium ions and free radicals disrupt the function of neuronal mitochondria. Consequently, a so-called mitochondrial permeability transition pore (MPT) in the mitochondrial membrane may form. Besides impeding ATP production through loss of mitochondrial potential, the MPT leads to mitochondrial swelling, a burst of free oxygen radicals, and the release of proapoptotic molecules. Thus, a vicious cycle of further disintegration is fueled. This vicious cycle is counterbalanced in part by antioxidative enzymes like the manganese superoxide dismutase (Mn-SOD) and the cytosolic forms of the copper–zinc superoxide dismutase (CuZn-SOD). These may prevent the breakdown of the mitochondrial membrane and, thus, the release of cytochrome C, which would be the trigger of apoptosis.

5
Tissue Acidosis

In the context of stroke pathophysiology, the proton balance is intimately linked with the glucose metabolism. With reduced oxygen availability, anaerobic glycolysis, as the only remaining source of ATP production, leads to tissue acidosis. It has long been assumed that this acidosis was one of the main noxious mechanisms in ischemic stroke. This so-called "lactate-acidosis hypothesis" is often quoted as explanation for the "glucose paradox" of cerebral ischemia. This paradox refers to the observation that excessive supply of glucose, the most important source of energy of the brain, during focal cerebral ischemia does not reduce tissue damage as one should think but, instead, augments it. However, by which mechanism this happens and, in fact, whether levels of acidosis reached in brain ischemia can generate brain tissue damage at all is still far from being clear. Possibly, the pH-dependent transition of Fe(III) to Fe(II) and the release of iron from molecular storages lead to a facilitation of the Haber–Weiss reaction that forms toxic free oxygen radical species (see above). Besides the production of different species of oxygen free radicals, acidosis also interferes with intracellular protein synthesis. However, the lactate-acidosis hypothesis is not unchallenged. Particularly the fact that acidosis blocks the NMDA receptor and thus has an antiexcitotoxic effect indicates the complexity of the role that acidosis plays in cerebral ischemia.

A similarly hotly debated topic are the findings on hyperglycemia during stroke. Experimental data from animal models show a detrimental effect of hyperglycemia during focal cerebral ischemia. Clinical data also suggest that hyperglycemia during the acute phase of stroke worsens prognosis. Persisting hyperglycemia beyond the acute phase is also an independent prognostic factor for larger infarct volume and poorer functional outcome in stroke patients. It is a matter of debate whether these observations are due to a causal relationship or if hyperglycemia is a mere epiphenomenon, possibly due to a stress reaction (i.e. an effect of concomitant catecholamine and glucocorticoid release proportional to initial tissue damage). The controversial role of an associated lactate acidosis has already been discussed. An alternative concept is the glucocorticoid hypothesis. It suggests that hyperglycemia enhances the release of glucocorticoids, which are also released during ischemia in the sense of a stress reaction. Glucocorticoids have been shown to have a direct cytotoxic effect and the blockade of glucocorticoid receptors does reduce the damaging effect of hyperglycemia. On the other hand, hyperglycemia and acidosis have a protective effect *in vitro*, and some experimental studies demonstrate that hyperglycemia can also have a beneficial effect under special circumstances.

However, a recent study on humans seems to shed light on this controversy. Perfusion and diffusion MR imaging and MR spectroscopy shows that stroke patients with hyperglycemia have a higher lactate accumulation in the penumbra. While the perfusion–diffusion mismatch (as a measure of the size relation of penumbra and core) was similar in the normoglycemic and hyperglycemic groups, the eventual infarct size and disability scores were significantly higher in the hyperglycemic group. While it could still be argued that initially hyperglycemic patients were also likely to be longstanding or latent diabetics with an overall worse microvascular state, we take this study – which links in with much of the experimental animal data – as a strong indicator for the vulnerability of the penumbra to hyperglycemia. Indirectly, it also corroborates the lactacidosis theory, which, however, still lacks direct proof.

6 Peri-infarct Depolarizations

The anoxic as well as excitotoxic depolarization of neurons and glia increases the concentration of glutamate and potassium in the ischemic core. These diffuse into the penumbral zone where they induce the electrochemical depolarization of more neuronal and glial cells that are "just managing" under the circumstances of oligemia. Depolarization in this context is not to be confused with short-lived action potentials, but means sustained depolarization of the entire cell membrane, a breakdown of the membrane potential that has to be actively rectified. Depolarization also causes a further significant increase in extracellular glutamate and potassium (Fig. 2). Thus, waves of sustained electrochemical depolarization spread from the core of infarction into the penumbra, bearing similarity with Leao's spreading depression. These waves are called *peri-infarct depolarizations*. Peri-infarct depolarizations occur with a frequency of 1–4/h and have been detected by functional MRI and near infrared spectroscopy. Since active rectification of the membrane potential requires energy, these peri-infarct depolarizations contribute further to the metabolic compromise and each wave turns a larger area of the penumbra toward irrevocable infarction. In addition, peri-infarct depolarizations seem to compromise the impaired microcirculation within the penumbra even further. Peri-infarct depolarizations have been detected in patients with head trauma.

7 Inflammation

The early stage of inflammation, which starts a few hours after the onset of ischemia, is characterized by the expression of adhesion molecules in the vascular endothelium as well as on circulating leukocytes. Thus, leukocytes adhere to the endothelium and transmigrate from the blood into the brain parenchyma, which is of decisive importance for stroke-induced brain inflammation (Fig. 2).

Newly expressed endothelial adhesion molecules like ICAM-1 and VCAM-1 facilitate the interaction between endothelium and leucocytes as first step of transmigration of white blood cells from blood into the tissue. They interact with the β_2-integrins CD11b/CD18 (Mac-1) and CD11a/CD18

(LFA-1) that are, in turn, expressed on leucocytes. Neutrophil granulocytes accumulate in the capillaries of the penumbra and interfere with the already impaired microcirculation. Blocking this interaction with CD18, CD11, or ICAM-1 antibodies reduces not only the number of leucocytes but also the size of infarction. ICAM-1 knockouts have also smaller infarcts than wild-type mice. Next to the β_2-integrins, other integrins that are more specific for lymphocytes like VLA-4($\alpha_4\beta_1$) come into action, too. Consequently, blocking the α_4-component reduces also the volume of infarction.

A major part in inflammation is ascribed to the population of microglial cells. These cells are the primary immunoeffectors of the CNS and make up about 20% of the entire glial mass, about the same proportion as neurons. Activation of microglia is a striking feature in the penumbra. The cells change their shape from highly ramified cells to a more plump and amoeba-like form, and they proliferate. They also express a number of characteristic surface molecules (for instance, MHCI and MHCII) and are capable of antigen presentation. Similar to leukocytes, activated microglia cells are able to produce not only a variety of proinflammatory cytokines but also cytotoxic metabolites, especially oxygen free radicals (such as peroxynitrite and superoxide) and enzymes (for instance, cathepsins). In addition, microglia is phagocytically active. The inhibition of microglial activation turns out to be protective in experimental stroke models. On the other hand, activated microglia have also been shown to release anti-inflammatory cytokines (such as TGF-β1) and growth factors. Because of the Janus-faced nature of microglial products (destructive, for example, free radicals vs protective, for example, growth factors), the overall role of microglia in cerebral ischemia is not clear at present. It is very likely that microglia play different roles at different times, with protective or regenerative activities occurring days or even weeks after the onset of ischemia.

Activated leukocytes (granulocytes, monocytes/macrophages, lymphocytes) as well as neurons and glial cells (astrocytes, microglia) produce cytokines and chemokines. In particular, proinflammatory cytokines like TNFα, IL-1, and IL-6 play a major role as mediators of the inflammatory response. They are responsible for the transition from the early, excitotoxic phase to the inflammation phase. Their regulation depends on transcription factors like NFκB, which, in turn, are activated by oxygen free radicals. Cytokine-mRNA and also the protein of TNFα and IL-1 can be detected only few hours after induction of an experimental focal ischemia. Expression of IL-6 follows only after about 24 h. These cytokines are mainly released by microglial cells and macrophages. In stroke patients, it could be shown that the intrathecal concentration of IL-1 and IL-6 correlated with the infarct size. Cytokine receptor antagonists reduce the infarct volume in animal models. For example, blockade of TNFα, by TNF-binding proteins reduces brain injury after focal cerebral ischemia. The role of TNFα is not entirely clear, however, as mice lacking the TNFα-receptor (TNFR2) have larger infarctions. In part, these contrasting results may reflect a difference in signal transduction cascades activated by the two TNFα-receptors, TNFR1 and TNFR2. It is also arguable that an increased IL-6 induction may be responsible, since IL-6 knockouts do not have smaller infarctions than their litter mates. Further to the induction of adhesion molecules, the above-mentioned cytokines also render the blood–brain barrier more permeable and induce prothrombotic functions of the endothelium.

Not only can the proinflammatory cytokines be observed, but anti-inflammatory cytokines such as TGF-1β or IL-10 can also be induced. By downregulating the inflammation, these cytokines have a protective effect in the context of cerebral ischemia. The neuroprotective action of TGF-β1 is well known. Interestingly, this cytokine seems to play an important part in immunological tolerance. As in ischemic tolerance (see below), it is possible to induce a tolerant state toward ischemia. This protection is induced by immunogenetic proteins (for instance, myelin basic protein, MBP) and mediated by a certain lymphocyte population expressing TGF-β1. In contrast to ischemic tolerance, immunological tolerance is longer lasting. Protection lasts for months, whereas in ischemic tolerance it lasts no longer than 7 days.

Apart from cytokines, there are two other proteins that dominate the inflammatory phase: the inducible nitric oxide synthase (iNOS) and cyclooxygenase 2 (COX2). Focal cerebral ischemia models have shown that iNOS is induced on the mRNA as well as on the protein level. Protein expression reaches its height after 24 h. By inhibiting the expression of iNOS, the size of experimental infarcts can be reduced by about 30%, even if treatment is undertaken only 24 h after the onset of ischemia. Nitric oxide (NO), which is produced in much greater amounts by iNOS than by the constitutional NO synthases, acts via the formation of peroxynitrite highly cytotoxic. NO triggers apoptosis not only through peroxynitrite but also via the induction of p53, which probably causes direct DNA damage. Cox-2 is mainly generated in the penumbra. The enzyme has a destructive effect on the penumbral tissue, mainly through producing oxygen free radicals and toxic prostanoids. This seems a likely hypothesis, as both genetic and pharmacological inhibition of Cox-2 have a protective effect. Cox-2 as well as iNOS are promising targets for the treatment of stroke patients because blocking them is still effective even 6 to 24 h after ischemia.

Not only the activated residual microglia but also monocytes from the circulating blood migrate into the affected brain tissue. This transmigration is mediated by a chemokine, the monocyte-attractant protein 1 (MCP-1), which is induced 1 to 2 days after ischemia and causes immigration of monocytes. Moreover, about a third of these transmigrated cells differentiate within 14 days post stroke into microglial cells – virtually indistinguishable from autochthonous microglia. A transdifferentiation into astrocytes has also been observed, but our group has not been able to reproduce this finding. Irrespective of that, the important role of astrocytes during inflammation has been recognized, producing both proinflammatory as well as neuroprotective factors such as erythropoietin (EPO), TGF-β_1, or metallothionein 2.

In general, much further study needs to be conducted before the complex role of inflammation in cerebral ischemia is sufficiently understood to be able to specifically block the damaging proteins and foster protective elements during the inflammation process in stroke.

8
Damage to the Blood–Brain barrier

The function of the blood–brain barrier (BBB) depends on the integrity of its cellular matrix, which consists of endothelial cells, basal lamina, and astrocytes. Cerebral ischemia causes damage to this matrix and its intercellular signal exchange. Central to these damaging processes are proteases like cathepsins, plasminogen activators, and the matrix metalloproteinases (MMP). The action of the MMPs, a group of more than 20 zinc endopeptidases, has been extensively studied. Proteolytic destruction of the basal membrane by the MMPs permits the immigration of leukocytes and promotes vasogenic edema. MMP-2 and MMP-9 are expressed within 1 to 3 h after cerebral ischemia; the level of their expression correlates with the level of BBB breakdown, the risk of hemorrhagic transformation, and the extent of eventual neuronal damage. Pharmacological inhibition of the MMPs prevents this breakdown and reduces stroke volumes. Mice with target disruption of MMP-9 are partly protected from the effects of ischemia.

Of the plasminogen activators, tissue plasminogen activator (tPA) is best known. In the brain, tPA is expressed mainly in neurons, astrocytes, and microglia. Studies of mainly primary cortex neurons have shown that tPA acts also as neurotoxin, and experiments on tPA knockouts have produced conflicting data. Since recombinant tPA is the sole licensed drug for thrombolytic therapy, these findings reflect also on current therapeutic practice. tPA has furthermore been shown to induce MMP-9 and, thus, – at least indirectly – damaging the BBB. The same holds probably for the *recombinant* tPA. This may explain why such a high number of secondary hemorrhages have been observed after thrombolytic stroke therapy. Animal experiments showed that delayed rtPA treatment – mimicking real-life delays in treatment – lead to significantly greater damage to the BBB and a major increase in mortality. The pharmacological inhibition of MMPs can reverse this effect of rtPA. In addition, tPA enhances NMDA-mediated calcium signaling and subsequent excitotoxic neuronal injury. It is a matter of controversy whether this detrimental effect is due to a cleavage of the NR1 subunit of the NMDA receptor by tPA.

The example of rtPA demonstrates once again that in order to refine treatment of ischemia, a thorough understanding of the complex mechanisms is indispensable. The development of effective and low-risk thrombolytic treatment is, of course, still one of the main objectives of pharmacological stroke research. One promising approach is the development of novel plasminogen activators, such as recombinant *desmodus rotundus* salivary plasminogen activator (DSPA-α1; Desmoteplase). Experimental evidence indicates that Desmoteplase has pharmacological and toxicological properties superior to rtPA.

9 Programmed Cell Death and Apoptosis

Apoptosis follows a similar time schedule as inflammation. It sets in after a few hours and continues over days. It is the commonest form of cell damage in the penumbra, and is very distinct from the rapid neuronal and glial death in the adjacent ischemic core, where pannecrosis is complete within few hours. In necrosis, it is the cellular edema that leads to osmolysis (see above). The released cell contents are a potent inflammatory stimulus, as discussed above.

Caspases, a group of proteases, has a pivotal role in apoptosis. They facilitate the "orderly" destruction of the cell, which is characterized by a condensation of both nucleus and the cell as a whole. The cell contents are fragmented into so-called apoptotic bodies. Endonucleases fragment the cell's DNA into typical strands of 180 base pair length. Other morphologic criteria of apoptosis are membrane blebbing and shrunk cytoplasm with morphologically intact organelles. While the terms "delayed neuronal death" (DND) and "programmed cell death" (PCD) are often used synonymously with "apoptosis," the two former may well occur without the typical features of the latter. Although most experimental studies show some biochemical evidence for apoptosis to have taken place, histological features are often missing to substantiate the claim. This has led some researchers to question apoptosis as an actual feature of ischemia. A number of reviews have dealt extensively with the significance of apoptosis in the pathophysiology of ischemia. This is why we shall focus here on some of its mechanistic aspects in this context.

Essentially, apoptosis can be triggered in two ways, by intrinsic (i.e. basically mitochondrial) or extrinsic activation. Binding to the Fas-receptor, for example, by TNFα, triggers the extrinsic pathway. Intrinsic triggers of apoptosis in the context of ischemia are elevated concentrations of intracellular calcium, reactive oxygen species, and glutamate as well as increased DNA damage (Fig. 2). Through damage to the mitochondrial membranes, both pathways directly or indirectly lead to the activation of caspases.

The caspases are a hierarchic group of at least 14 cysteine-dependent and aspartate-specific proteases. Their inactive precursor proteins exist in every cell and can be activated through cleavage. Apoptosis is essentially an active and in itself energy-dependent cellular deconstruction process. In cerebral ischemia, the caspases 1, 3, 8, and 9 are involved, of which, the caspases 8 and 9 are the initiators of a signal cascade that activates the so-called execution caspase 3. Caspase 1 is predominantly engaged in cytokine activation. Caspase 3 is of predominant importance not only in the context of cerebral ischemia, and its genetic or pharmacological inhibition is not only in this context neuroprotective. The substrates degraded by caspase 3 are DNA repair enzymes like poly-(ADP-ribose)-polymerase (PARP) and the DNA dependent protein kinase (DNA-PK). The latter is important for the repair of double strand breaks. PARP recognizes single-strand disruptions and mediates self-repair of the DNA through swift automodification with synthesis of highly negatively charged, long, and branched ADP-ribose molecules. This process utilizes ATP and thus taps the cellular energy pool. Although not universally accepted, this ATP consumption by PARP is widely believed to be in itself cytotoxic. PARP is also capable of inducing caspase-independent cell suicide via the so-called *apoptosis-inducing factor* (AIF; see below). This also helps in explaining the partial failure of caspase inhibitors in some models of cerebral ischemia. The genetic or pharmacological blockade of PARP has, however, been demonstrated to be neuroprotective in experimental stroke. Of significance is the cleavage by caspase 3 of the inhibitor of caspase-activated DNase (ICAD). ICAD is thus activated and causes the characteristic feature of "laddering" that has been described in all stroke models. The term laddering draws on the regular appearance of those DNA fragments as ladderlike bands on a gel electrophoretic separation. The length of the resulting DNA fragments reflects the complex structure of eukaryotic genetic information in the chromatin. Chromatin is a structure, in which most of the DNA is enclosed by specific proteins, the histones (see epigenetics), leaving only few positions accessible for endonucleases. Complete apoptotic laddering leaves DNA-oligomers of about 140 to 180 base pairs in length, which do no longer contain any useful genetic information. Apart from destroying genetic information, caspases also degrade structure proteins of the nucleus and cell, such as laminin, actin, and gelsolin.

Cytochrome C is not only an essential part of the mitochondrial electron transfer chain and hence for the energy production in the cell but also a crucial mediator in the caspase activation cascade. On the extrinsic pathway of apoptosis, stimulation of the Fas-receptor leads to the formation of the "death-inducing signaling complex" (DISC). DISC activates caspase 8 from its precursor protein, which, in turn, activates Bid, a proapoptotic protein of the Bcl-2 family (see below). Bid then induces the liberation of cytochrome C from the mitochondrial membrane. Other members of the proapoptotic Bcl-2 family are Bad and Bax, important on the intrinsic pathway. Their activation leads to their translocation into the outer mitochondrial membrane and to the establishment of the mitochondrial permeability transition pore (MPTP), leading to the release of cytochrome C. Another way of its release from mitochondria is obviously direct membrane damage by free radicals. Cytochrome C, together with the cytosolic protein Apaf-1, forms the caspase precursor protein, and, using ATP, the so-called apoptosome. The apoptosome activates caspase 9 through cleavage from its precursor. Whether cytochrome C is eventually released depends, however, on the balance of proapoptotic members of the Bcl-2 family like Bid, Bax, Bad, and Bag, and their antiapoptotic counterparts Bcl-2 and Bcl-x_L. Prevailing expression of Bcl-2 and Bcl-x_L results in clearly smaller infarction after focal cerebral ischemia. Treatment with a Bcl-x_L fusion protein containing the TAT domain of the human immunodeficiency virus (HIV) for cell penetration acts also neuroprotectively in models of ischemia. Bcl-2 and Bcl-x_L seem to have a stabilizing effect on the mitochondrial membrane; they prevent the formation of MPTP, the ensuing cytochrome C liberation, and the formation of AIF. Not only the reinforcement of antiapoptotic factors but also the attenuation of proapoptotic proteins may be neuroprotective in cerebral ischemia. For instance, the phosphorylation of Bad leads to its sequestration by the protein 14-3-3 and thereby to its inactivation. The PI3K/Akt-pathway, utilized by many neurotrophic factors such as erythropoietin (see below: endogenous neuroprotection), plays an essential role in this. It has been reported that Akt has been activated during cerebral ischemia. Whether this has a functional impact in stroke remains unclear at the moment.

Other important protein groups involved in stroke are the members of the IAP-group (inhibitors of apoptosis proteins). The overexpression of the X-chromosome-linked IAP (XIAP) has been shown to be protective in stroke models. IAPs act either indirectly by inhibiting the apoptosome or directly inhibiting caspase 3, but can themselves be blocked by other proteins such as DIABLO (in mice, the human homologue is Smac) that is physiologically localized in mitochondria. DIABLO is released after focal ischemia and binds to XIAP. It is yet unclear what its function is during focal ischemia. The data discussed show, however, that apoptosis underlies a complex regulation process that is controlled at many levels.

10 Ischemia-induced DNA Damage, DNA Repair, and p53 as Genotoxic Sensor

The emergence of DNA damage and its repair functions in the context of focal cerebral ischemia are particularly interesting. DNA damage is caused by oxygen free radicals (see above). Particularly, the interaction of hydroxyl radicals with DNA results in strand breaks and base alterations. Hydroxyl radicals have a very short half-life – with the exception of peroxynitrite, formed from NO and superoxide – and are unlikely to reach the nuclear DNA from outside the cell. They are more likely to be produced in close proximity to the nucleus as a homolytic cleavage product of peroxynitrite, which has the ability to diffuse over longer distances.

Under physiological circumstances, DNA lesions are generated with a frequency of about 10 000 events per cell each day. They are recognized and repaired efficiently, provided the repair enzymes are intact and not overloaded. In the wake of ischemia and reperfusion, the tissue is flooded with oxygen free radicals, causing abundant DNA damage. Alterations that are typically caused by free radicals have been found in the mouse model of transient focal ischemia, where a massive increase in DNA damage was observed 10 to 20 min after reperfusion. Foremost, 8-hydroxy-deoxy guanosine becomes much more prevalent in both nuclear as well as mitochondrial DNA after ischemia. These alterations are repaired partially within 4 to 6 h, and the increase in repair activity is proportional to the degree of DNA damage.

There are three basic DNA repair mechanisms – base excision repair (BER), nucleotide excision repair (NER), and what is known as mismatch repair. BER seems to be of particular significance in ischemia. During BER, DNA glycosylases excise damaged or modified bases. The thus generated purine and pyrimidine-free sites (AP-sites) are eliminated from the damaged DNA strand by AP-endonucleases (apurinic/apyrimidinic endonucleases) and replaced by DNA polymerase β and DNA ligase with a short strand of 1 to 4 nucleotides based on the complementary blueprint of the remaining undamaged strand.

Apart from oxidative DNA damage, deamination of cytosine may produce uracil. In BER, such alterations are sorted out by uracil-DNA-glycosylase (UDG). We found that knockout mice deficient in UDG have substantially larger infarctions than the wild-type controls.

DNA repair is neither inexhaustible nor infallible. Persisting mutations are the result of erroneous DNA repair. In models of transient focal ischemia, a dramatic increase in the frequency of mutations has been observed. Half of them were transitions and transversions. The other half were deletions and occasional insertions, mostly resulting in a frame shift (frame shift mutations) of the open reading frame (ORF). It seems that not only the exhaustion of the repair capacity but also an impaired proofreading ability in the exonuclease-negative DNA polymerase β expressed in the brain is to blame for these mutations. The major role of DNA repair is to ensure transcription of mRNA from intact genes. An obvious consequence of mutations is instability of the genetic information that may lead to faulty proteins, eventually resulting in either cellular malfunction, disintegration, or spawning of tumors.

The maintenance of swift and effective DNA repair is therefore essential in postmitotic cells like neurons. In order to adapt to an increased demand, a cell may try to increase its repair efficiency. Apoptosis is another way of avoiding erroneous protein synthesis and the potential emergence of tumors (see above). Both mechanisms are found in focal cerebral ischemia, and it seems to be the extent of the damage that determines which scenario will prevail.

We have several reasons to believe that the tumor suppressor protein p53 represents the molecular switch on the crossroads of attempted repair and apoptosis. (1) p53 is mainly induced by DNA damage; (2) p53 can halt the cell cycle, and thus prevent the division of non-neuronal cells before complete repair; (3) p53 can induce apoptosis if the damage is extensive. In most models of ischemia, it could be shown that p53 had been induced. The fact that p53 knockouts have significantly lower infarct volumes accords with p53's proapoptotic function. However, it seems paradoxical that heterozygous $p53^{+/-}$ mice show even smaller infarctions. An attenuated expression of p53 seems to give better neuroprotection than a total lack of it, which hints at the dual function of the gene. It would be interesting to find out if longer-term neoplasias could be the downside of this short-term effect.

11 Epigenetic Mechanisms

Experimental work of the last years has shown that, apart from DNA repair, epigenetic mechanisms like DNA methylation and histone modification also play an important role in ischemic stroke. The transcriptional gene expression is not only regulated by transcription factors but also depends on epigenetic mechanisms that mainly modify the chromatin structure.

In the mammalian genome, DNA methylation is a covalent, postreplicative modification at the carbon 5 position of the pyrimidine ring of cytosine bases (m^5C); it typically occurs in CpG dinucleotides. DNA methylation is catalyzed by DNA methyltransferases (DNMT1, DNMT3a, DNMT3b). The main function of DNA methylation lies in the regulation of gene transcription. Methylated genes are barred from expression in a process called "gene silencing." First of all, many transcription activators are unable to bind to methylated loci at all. Furthermore, there is a group of methyl-CpG-binding domain proteins (MBD) that form a complex with histone deacetylases (HDAC), which, in turn, by deacetylation of histones, cause the chromatin structure to condense. Through this condensation, the gene-regulatory loci become inaccessible for transcription factors. Five proteins of the MBD family are known to date: MBD1 through 4 and MeCP2. HDACs deacetylate predominantly histones H3 and H4. Conversely, acetylated histones open up the chromatin structure, permitting the gene-regulatory regions (promoters, enhancers) to become active again, and, ultimately, enabling gene transcription through binding of transcriptional activator proteins. While the state of deacetylation or acetylation of a CpG hypomethylated promoter or enhancer region is comparatively volatile, DNA hypermethylation of these regulatory regions results in a more stable repression.

Initially, DNA methylation was thought to be of biological significance only for fundamental phenomena of cell biology like embryogenesis and differentiation, in X-chromosome inactivation, genomic imprinting, and in several neoplastic diseases. In more recent years, its pathophysiological role in a number of further conditions has been recognized, such as Fragile-X syndrome, ICF (*I*mmunodeficiency, *C*entromere instability, *F*acial anomalies) syndrome, or Rett syndrome.

For the first time, we were able to detect the involvement of DNA methyltransferase 1 and DNA methylation in an acute disease, such as ischemic stroke. In addition, a complex pattern of an altered expression of MBD-family proteins has been described in a model of transient cerebral ischemia. The inhibition of DNA methylation had a protective effect, as had manipulations to the acetylation state of histones. Pharmacological inhibition of a histone acetyltransferase with trichostatin A (TSA) caused enhanced histone acetylation *in vitro* as well as *in vivo*, and proved to be neuroprotective. A similar effect was shown recently in a mouse model of Huntington's disease. Thus, the epigenetic mechanisms of gene regulation may provide yet another axis of neuroprotective therapeutic intervention.

12 Gene Expression

The role of transcription factors such as AP-1, HIF-1, and NF-κB in the context of cerebral ischemia has been recognized for much longer. They initiate the transcriptional activation of complex programs of gene expression. The whole hierarchical system of such programs has been unraveled over more than 15 years. Within few hours, the transcription of immediate early genes such as c-fos is activated. Some of these, in turn, form transcription factors such as AP-1. They induce a variety of genes that may have protective or destructive effects. Cells that survive the ensuing biochemical negotiation finally express genes that contribute to a structural reorganization and thus to greater plasticity. A great many of the genes involved have been subject to pathophysiological studies in stroke. Their functional significance has been studied either in descriptive ways, that is, by describing candidate genes on RNA or protein level or using transgenic technology. Especially knockout mice have been extremely useful for this approach. The genetic elimination of specific proteins in mice used for these experiments has been particularly enlightening for the untangling of the signaling cascades involved in and after stroke.

In recent years, the advent of genetic screening methods has allowed for the systematic investigation of complex gene expression programs that are initiated by cerebral ischemia. In contrast to candidate-gene research, they permit a more unbiased approach. They are based on the assumption that focal cerebral ischemia does incite complex changes of gene expression. Not just a single, specifically targeted gene is examined for its role in the scenario of stroke but – ideally – all genes. For this new research, paradigm terms have been coined like discovery science, genomics, transcriptomics, proteomics, and cellomics. So-called high-throughput technologies as well as a highly evolved bioinformatic methodology are a prerequisite for acquiring and managing the vast bulk of information involved. DNA micro arrays or SAGE technology facilitate the examination of global changes in gene expression on mRNA level (transcriptomics). Global changes on the level of protein synthesis can be described with a combination of two-dimensional gel electrophoresis and identification techniques (proteomics). The first results of a number of studies using these techniques in the context of cerebral ischemia are now available.

What do these results imply? First of all, they confirm that the assumption of the existence of complex stroke-induced programs of gene expression is correct. Secondly, they demonstrate that they follow a robust temporal and spatial pattern. Not only do the expression patterns differ between core and penumbra but also within the penumbra there are many spatial and temporal differentiations. This has even given rise to a concept of "multiple penumbras." Thirdly, this approach has not only confirmed many findings of candidate gene research but also underpinned our pathophysiological understanding. Thus, a number of heat shock proteins, antioxidative enzymes, growth factors, and pro- as well as anti-inflammatory proteins have been detected. Not just transcription factors, but a variety of proteins have been found to play essential roles in the metabolism of RNA and DNA. Finally, proteins that contribute to structural and functional reorganization are thought to establish the molecular base of the functional healing potential after strokes. Not surprisingly, a number of hitherto unidentified genes have been brought forth by discovery science.

Although the results of these screening methods give a great boost to stroke research and open doors to new insights, a full functional characterization of the differentially expressed genes will take many more years. Our understanding of the emerging network of gene expression increasingly depends on not only biological but also mathematical models. However, there are things we do know already; for instance, that protein modifications are of great functional importance. Thus, the functional value of numerous proteins depends on their phosphorylation state. For example, the specific phosphorylation of the usually proapoptotic protein Bad prevents apoptosis in the ischemic penumbra. The differential function equals thus not simply differential gene expression, and genetic screening methods will never be sufficient to paint the full picture of stroke-induced molecular pathways. The same holds for proteomics, which, just like transcriptomics, neglects the functional importance of RNAs that do not code for proteins.

These molecular pathways, however, potentially lead to new, neuroprotective therapies and open up new perspectives for fundamental issues like neuronal plasticity and stem cell research.

13 Cell Replacement

In recent years, it has been shown in different models that ischemic brain injury induces neuron formation from neuronal stem cells. Such cells reside in the subgranular zone of the dentate gyrus, in the subventricular zone below the lateral ventricles, and periventricularly in the hippocampus. They can migrate into parts of ischemic lesions within the striatum, the dentate nucleus, and the hippocampal CA1 region. It is still a matter of debate as to what extent a similar neuroneogenesis also takes place within the penumbra in cortical regions. Neither is it entirely understood to what extent these migrating cells, which doubtlessly express neuronal markers, become functional. Observations so far indicate that it is not a very effective process and that most of the migrating cells perish. However, if these neuronal cells could be enabled or encouraged to assume the function of damaged neurons, this would imply another new therapeutic paradigm. Indeed, intrathecal infusion of growth factors like FGF-2 and EGF after global cerebral ischemia did not only stimulate the migration and the survival of these cells but also resulted in improved functional results both on a cellular and on a neurobehavioral level. Astrocytes are newly formed in a similar fashion. For future therapeutic evaluation, it should be kept in mind not only that growth factors such as erythropoietin, FGF-2, EGF, VEGF, SCF, and BDNF have this beneficial effect but also that their expression depends on the stimulation of the NMDA receptor by glutamate. A pharmacological blockade of this receptor with the intention of inducing neuroprotection could interfere with this effect and thus be counterproductive. This is yet another example of an incomplete understanding of the signaling mechanisms, leading to a paradox in the interpretation of our neuroprotective data. Bearing this in mind, it is needless to say that the ongoing experimental studies of cell replacement are still far from clinical implementation. However, the available experimental observations are very exciting: For instance, neural and also haematopoietic pluripotent cells differentiate into neuronal and glial cells when injected into infarction sites. On the other hand, this potentially helpful strategy may have detrimental effects too. In a mouse model of stroke, for example, murine embryonic stem cells did not migrate into the damaged tissue, but produced highly malignant teratocarcinomas at the site of implantation, independent of whether or not they had been predifferentiated *in vitro* into neural progenitor cells. As far as safety is concerned, an endogenous pathway of cell regeneration may be more promising. Bone marrow–derived cells (BMDC) introduced into the brain possess a differentiation potential capable not only of forming microglia but also neurons. This apparent plasticity has been attributed to transdifferentiation. However, more recent data suggest, in the case of neurons, that BMDCs do not differentiate into but rather fuse with Purkinje cells in the brain. These observations suggest that cell fusion contributes to the maintenance of neurons. One potential use for haematopoietic cells migrating from blood into ischemic brain tissue could be the therapeutic transfer of protective gene products such as growth factors.

14 Endogenous Neuroprotection – Ischemic Tolerance

In the wake of ischemic stroke, not only cascades of destruction are induced but also endogenous protective mechanisms are activated. These mechanisms are currently studied intensively. Central to this research are models of ischemic preconditioning. The idea behind ischemic preconditioning is to achieve a protected state of a cell, tissue, or whole organism by subliminal exposure to a noxious stimulus (trigger), which is applied just below the level at which damage would occur. The stimulus induces a protective state against insults that would normally be lethal. Subliminal ischemia ('ischemic tolerance'), hypoxia, reactive oxygen free radicals, inflammation, and so on can serve as tolerance-inducing stimuli. After induction, there are two time windows within which ischemic tolerance can be observed. Early tolerance occurs after about 5 to 120 min following stimulation (classical preconditioning), while delayed preconditioning sets in after a latency of about

24 to 72 h. Seven days after induction, the state of protection is no longer detectable. Studies of patients who had transient ischemic attacks (TIA) before suffering a persistent stroke indicate that ischemic tolerance may also exist in humans. Subsequent to a TIA, brain infarcts were smaller in comparison with patients without preceding TIA. As a TIA, by definition, does not lead to infarction, it may thus be considered as a stimulus that induces tolerance.

Unlike severe ischemia, leading to stroke, a preconditioning stimulus induces exclusively protective, but not destructive, signaling cascades. There are basically three stages of protective signaling that can be distinguished in delayed preconditioning. In the *induction phase*, molecular sensors (mostly receptors, channels, and regulators) are activated by transcription factors. During the *transduction phase*, certain protein kinases, transcription factors, and paracrine and autocrine mediators, such as growth factors, amplify the signal and thereby prepare the *effector phase*. In this phase, proteins with a direct protective impact (antiapoptotic, anti-inflammatory, antioxidative) are switched on. These mechanisms are the molecular basis for ischemic tolerance and have been reviewed in detail elsewhere. Here, we want to give only a brief outline of the concept. On the basis of animal models in mice and rats as well as cell culture experiments, we were able to describe the following signaling cascade: In astrocytes, a tolerance-inducing hypoxic stimulus induces the transcription factor HIF-1 (hypoxia-inducible-factor-1). HIF-1 is one of the central sensors of hypoxia and induces the expression of erythropoietin (Epo). Erythropoietin is released by astrocytes and binds to the neuronal Epo receptor. Via the activation of a protein kinase cascade, Epo prevents apoptosis, as shown in Fig. 3. This neuroprotective signaling cascade mediates approximately 50% of the protection observed in our model. The remaining 50% of protection that it does not account for suggests the presence of further mechanisms in ischemic preconditioning. As a matter of fact, beside this paracrine cascade, which is mediated by astrocytes and neurons, a number of other mechanisms have been described. A host of antiexcitotoxic, anti-inflammatory and antiapoptotic mechanisms, including programs for regeneration and repair, have been implicated so far (Fig. 4).

Understanding these mechanisms may enable us in the future to induce or boost endogenous protection in patients as a novel strategy to safeguard the brain against hypoxic/ischemic damage. For example, Epo is a promising drug in stroke therapy. There is solid evidence that exogenously applied Epois neuroprotective *in vivo* and *in vitro* models of cerebral ischemia. Clinical phase I and II trials suggest that Epo is not only safe but also beneficial in the therapy for acute stroke. In addition, pharmacological inducers of ischemic tolerance in brain could become a promising tool in circumstances when ischemia can be anticipated, for example, before cardiac or carotid surgery. For example, desferrioxamine, an approved drug for the treatment of hemosiderosis, induces robust ischemic tolerance by inducing erythropoietin in an HIF-1-dependent manner.

Combined, the above-mentioned results demonstrate that brain cells are not only challenged by deleterious mechanisms but also activate innate programs to protect themselves from ischemia. However, cerebral ischemia does not only affect the brain parenchyma but also other vital organs, in particular, the immune system.

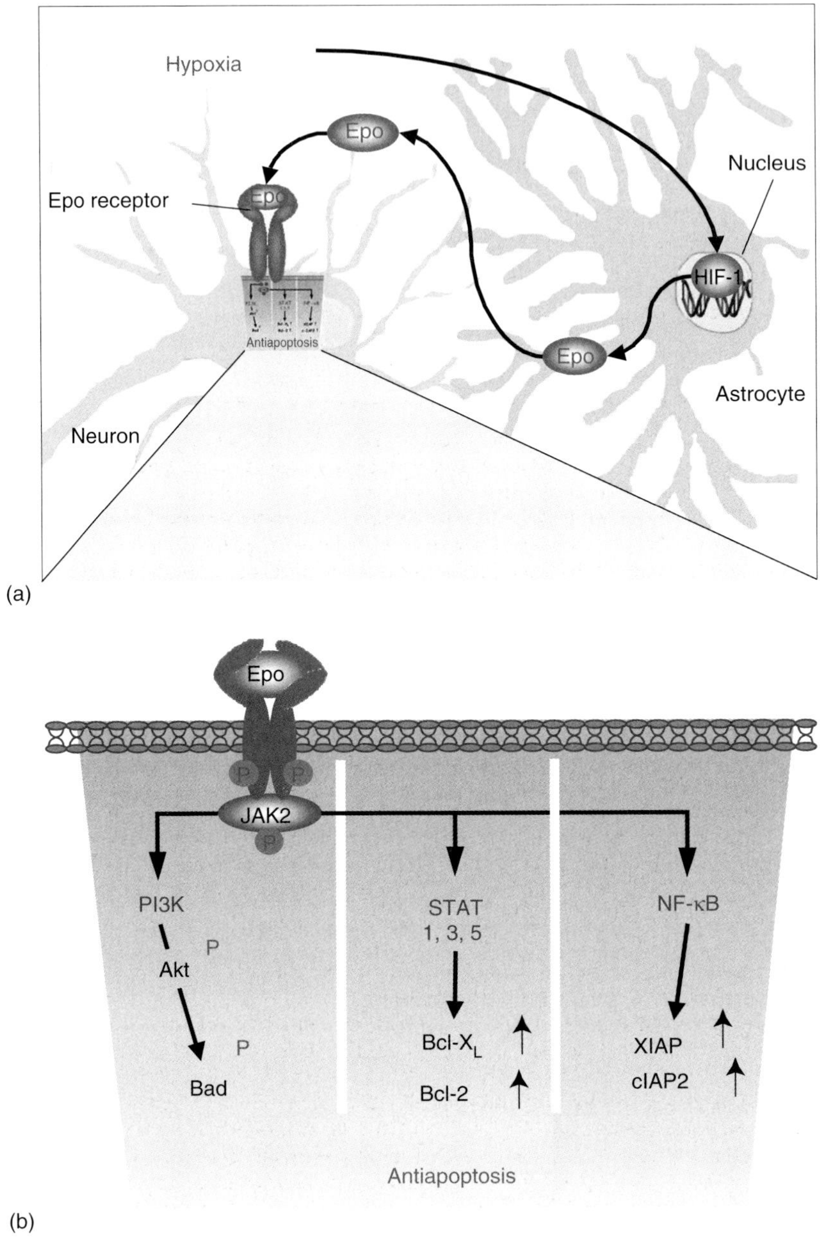
Hypoxia
Epo
Nucleus
Epo receptor
Epo
HIF-1
Antiapoptosis
Epo
Astrocyte
Neuron
(a)
Epo
JAK2
PI3K
P
Akt
P
Bad
STAT
1, 3, 5
Bcl-X_L ↑
Bcl-2 ↑
NF-κB
XIAP ↑
cIAP2 ↑
Antiapoptosis
(b)

15 Stroke-induced Immune Depression (SIDS)

Within three days after stroke, up to 61% of the stroke patients develop fever. Findings from animal experiments show that pyrexia of more than 1 °C does not only lead to a dramatic increase of the cytotoxic neurotransmitter glutamate but also to a significantly larger volume of the infarction. Clinical trials demonstrated that both mortality and morbidity rise with the occurrence of fever in stroke. The most common cause of fever in the acute course of stroke is infection. Infectious complications occur in 21 to 65% of stroke patients. Pneumonia is the most frequent infection in stroke patients and a major mortality risk factor.

Fig. 3 Epo-mediated endogenous neuroprotection (a) The transcriptional activator protein HIF-1 functions as a global regulator of oxygen homeostasis that facilitates both oxygen delivery and adaptation to oxygen deprivation. HIF-1 consists of two subunits, HIF-1α and HIF-1β. The HIF-1ß subunit is constitutively expressed, whereas the HIF-1α subunit is precisely regulated by the cellular oxygen concentration. Most relevant is the regulation of HIF-1α steady state protein levels. Under normoxic conditions, HIF-1α is rapidly hydroxylated by prolyl hydroxylase domain containing protein (PHD), subsequently ubiquitinated and degraded through a proteasomal pathway. Since oxygen is rate limiting for PH activity, hypoxia inhibits HIF-1α degradation and HIF-1α can heterodimerize with HIF-1ß. After dimerization, HIF-1 enters the nucleus, binds to its recognition site 5′-RCGTG-3′ in the hypoxia response element, and recruits coactivators as well as basal transcription factors. This transcription complex initiates the gene expression that mediates response to hypoxia. More than 40 HIF-1 target genes are known. One of them is Epo, a glycoprotein hormone that regulates hematopoiesis by promoting cell survival, proliferation, and differentiation of immature erythroid cells. Although Epo is mainly expressed in adult kidney and fetal liver tissue, it is also found in the brain. In ischemic preconditioning, low oxygen tension (hypoxia) induces Epo in astrocytes, depending on HIF-1. Epo is a key mediator of paracrine neuroprotection from astrocytes to neurons, which acts via its cognate receptor. (b) Epo exerts its neuroprotective effect by preventing neuronal apoptosis. The intracellular molecular pathways of Epo-induced neuroprotection involve tyrosine phosphorylation. Rather than containing an intrinsic protein tyrosine kinase activity, the relatively short cytoplasmatic domain of the Epo receptor (EpoR) interacts with nonreceptor tyrosine kinases. A key molecule is the janus kinase 2 (JAK2), which is constitutively associated with EpoR. The first step of this intracellular signaling is Epo-induced homodimerization of EpoR. Homodimerization of EpoR induces a conformational change of the whole protein complex, which place JAK2 in close proximity to EpoR, thus promoting transtyrosine phosphorylation and activation of JAK2. JAK2 induces phosphorylation of cytosolic tyrosines in the EpoR. This allows docking to the EpoR cytosolic domain of several signaling proteins, including phosphatidyl-inositol 3-kinase (PI3K) and Signal transducer and activator of transcription proteins (STAT1, STAT3, STAT5). These proteins, in turn, become tyrosine phosphorylated and thus activated. PI3K suppresses apoptotic cell death through its downstream effector protein kinase B (PKB)/Akt. Activated Akt kinase suppresses apoptosis by phosphorylation of the proapoptotic Bcl2-family protein Bad. Phosphorylated Bad prevents apoptosis, since unphosphorylated Bad is capable of forming heterodimers with the antiapoptotic proteins Bcl-X_L or Bcl-2 and thereby antagonizes their antiapoptotic function. STATs exert their antiapoptotic effects by transcriptional activation of antiapoptotic proteins from the bcl-2 gene family, in particular, bcl-2 and bcl-X_L. Activation of transcription factor NF-κB, a key event for ischemic preconditioning, is activated by phosphorylation of its inhibitor subunit IkappaB. NF-κB induces the expression of the inhibitor of apoptosis proteins cIAP2 and XIAP. Activating these three signaling pathways, Epo induces a fast and long-lasting neuroprotection by upregulation of antiapoptotic as well as downregulation of proapoptotic proteins.

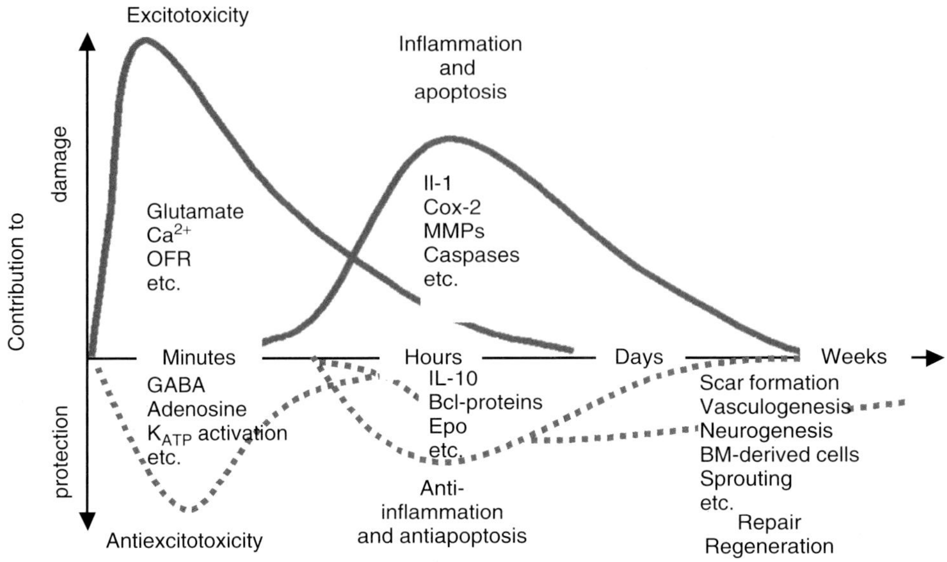

Fig. 4 Cell death–inducing and endogenous neuroprotective mechanisms in cerebral ischemia. Focal cerebral ischemia induces a complex series of mechanisms that damage brain cells. Brain tissue responds to most of the listed examples of noxious signals by inducing protective mechanisms. In infarcted tissue, destruction has overwhelmed protection, although tissue around the ischemic core might have been spared by restored substrate delivery (owing to delivery to collaterals, spontaneous or therapeutic reperfusion, or vasculogenesis) and cellular mechanisms of protection. In this cartoon, the *x*-axis reflects the evolution of the cascades over time and the *y*-axis illustrates the impact of each element of the destructive (top) and protective (bottom) cascades on the final outcome. The red dotted lines envelop major pathophysiological entities of tissue destruction in stroke (grouped for the acute mechanisms of excitotoxicity and the delayed mechanisms of inflammation and apoptosis) and the green broken lines envelop corresponding protective tissue responses. Abbreviations: BM, bone marrow; Cox-2, cyclooxygenase 2; Epo, erythropoietin; IL, interleukin; MMPs, matrix metalloproteinase; OFR, oxygen free radicals (Reprinted from Dirnagl, U., Simon, R.P., Hallenbeck, J.M. (2003) Ischemic tolerance and endogenous neuroprotection, *Trends Neurosci.* **26**, 248–254, with permission from Elsevier). (See color plate p. xlii).

Stroke patients have a much higher risk of developing severe bacterial infections. It is obvious that factors such as immobilization, reduced bulbar reflexes, drowsiness, and, subsequently, a higher risk of aspiration promote pulmonary infections. Although these are risk factors for bacterial infection, they cannot fully explain the high risk of infection observed. A large meta-analysis showed that other factors had to be equally important. In particular, a reduced function of the immune system was postulated. Very recently, we were able to prove that immunodepression following stroke causes severe bacterial infections. In a mouse model, we demonstrated that, three days after stroke, severe bacterial infections (mostly pneumonia and sepsis) develop spontaneously. Stroke induces an overactivation of the sympathetic nervous system and leads to a rapid, severe, and sustained lymphopenia as well as to a disturbed function of lymphocytes and monocytes.

Interestingly, there is also evidence for an overactivation of the sympathetic nervous system within the first two days after stroke in humans. Furthermore, we were able to show that a disturbed secretion of interferon-γ from T cells and NK cells is central to these severe infections. Both a cellular reconstitution of the immune system and the application of interferon-γ prevented infections. In addition, pharmacological inhibition of the sympathetic nervous system with the β-receptor-blocker propranolol not only prevented infections but also dramatically reduced the high rate of mortality in this model. Our studies are the first to provide a mechanistic explanation for the clinical phenomenon of increased susceptibility to infections after stroke.

Approaches to prevent infection after stroke seem to be desirable. Preventive anti-infective therapy is expected to reduce a number of negative prognostic factors such as fever, infection-induced arterial hypotension, and the systemic release of proinflammatory cytokines. In addition, with the prevention of severe infections, an earlier mobilization and rehabilitation becomes feasible. In an experimental stroke model, we were able to demonstrate that preventive anti-infective therapy with an antibiotic after focal cerebral ischemia not only reduces mortality and the infarct sizes but also decreases the functional deficit. These experimental results lead to the initiation of a Phase IIb trial for preventive antibacterial short-term therapy in patients with acute ischemic media infarction (Preventive AN*ti-infective* THER*apy* I*n* S*troke*; PANTHERIS, Berlin). Other groups initiated similar trials (A. Chamorro, Spain; M. Hennerici, Germany; personal communications).

16 Perspectives

Owing to intensive experimental and clinical research, our pathophysiological understanding of cerebral ischemia has greatly improved over the last 15 years. Complex pathways involving excitotoxicity, oxidative and nitrosative stress, peri-infarct depolarizations, inflammation, and apoptosis-like mechanisms have been described in cell culture as well as animal models of cerebral ischemia. Despite impressive results in animal experiments on neuroprotection, the transfer of preclinical results into human stroke therapy has been so far unsuccessful. Possible reasons for this dilemma include problems with the models used, the design of the clinical trials as well as the drugs, and they have been discussed intensively. However, we believe that although we have often not been able to transfer our experimental results into human stroke therapy, some cautious optimism is justified. More and more experimental data demonstrate that the division of molecular mechanisms in "good" and "bad guys" is an improper simplification, and we need to gather more data in order to identify more suitable molecular targets. Since in stroke there are multiple pathways that cause cell death, it seems reasonable to develop combination therapies targeting all these mechanisms. In addition, the advances in understanding endogenous neuroprotective mechanisms provide new therapeutic options to improve tissue survival after stroke. Targeting systemic complications like severe bacterial infections in a preventive manner may also improve outcome. Thus, an ever more refined understanding of complex damaging cascades as well as the development of more complex therapeutic strategies will continue to be the basis for more successful therapies of stroke in the future.

See also Brain Development.

Bibliography

Books and Reviews

Caplan, L.R. (2000) Caplan's stroke, *A Clinical Approach*, Butterworth-Heinemann, Boston, MA.

Chan, P.H. (2001) Reactive oxygen radicals in signaling and damage in the ischemic brain, *J. Cereb. Blood Flow Metab.* **21**, 2–14.

Dirnagl, U., Iadecola, C., Moskowitz, M.A. (1999) Pathobiology of ischaemic stroke: an integrated view, *Trends Neurosci.* **22**, 391–397.

Dirnagl, U., Simon, R.P., Hallenbeck, J.M. (2003) Ischemic tolerance and endogenous neuroprotection, *Trends Neurosci.* **26**, 248–254.

Emsley, H.C., Tyrrell, P.J. (2002) Inflammation and infection in clinical stroke, *J. Cereb. Blood Flow Metab.* **22**, 1399–1419.

Friedlander, R.M. (2003) Apoptosis and caspases in neurodegenerative diseases, *N. Engl. J. Med.* **348**, 1365–1375.

Kokaia, Z., Lindvall, O. (2003) Neurogenesis after ischaemic brain insults, *Curr. Opin. Neurobiol.* **13**, 127–132.

Lipton, P. (1999) Ischemic cell death in brain neurons, *Physiol. Rev.* **79**, 1431–1568.

Liu, P.K., Grossman, R.G., Hsu, C.Y., Robertson, C.S. (2001) Ischemic injury and faulty gene transcripts in the brain, *Trends Neurosci.* **24**, 581–588.

Lo, E.H., Dalkara, T., Moskowitz, M.A. (2003) Mechanisms, challenges and opportunities in stroke, *Nat. Rev. Neurosci.* **4**, 399–415.

Sharp, F.R., Bernaudin, M. (2004) HIF1 and oxygen sensing in the brain, *Nat. Rev. Neurosci.* **5**, 437–448.

Primary Literature

Arvidsson, A., Collin, T., Kirik, D., Kokaia, Z., Lindvall, O. (2002) Neuronal replacement from endogenous precursors in the adult brain after stroke, *Nat. Med.* **8**, 963–970.

Asahi, M., Wang, X., Mori, T., Sumii, T., Jung, J.C., Moskowitz, M.A., Fini, M.E., Lo, E.H. (2001) Effects of matrix metalloproteinase-9 gene knock-out on the proteolysis of blood-brain barrier and white matter components after cerebral ischemia, *J. Neurosci.* **21**, 7724–7732.

Azzimondi, G., Bassein, L., Nonino, F., Fiorani, L., Vignatelli, L., Re, G., D'Alessandro, R. (1995) Fever in acute stroke worsens prognosis. A prospective study, *Stroke* **26**, 2040–2043.

Back, T., Ginsberg, M.D., Dietrich, W.D., Watson, B.D. (1996) Induction of spreading depression in the ischemic hemisphere following experimental middle cerebral artery occlusion: effect on infarct morphology, *J. Cereb. Blood Flow Metab.* **16**, 202–213.

Baena, R.C., Busto, R., Dietrich, W.D., Globus, M.Y., Ginsberg, M.D. (1997) Hyperthermia delayed by 24 hours aggravates neuronal damage in rat hippocampus following global ischemia, *Neurology* **48**, 768–773.

Baird, T.A., Parsons, M.W., Phanh, T., Butcher, K.S., Desmond, P.M., Tress, B.M., Colman, P.G., Chambers, B.R., Davis, S.M. (2003) Persistent poststroke hyperglycemia is independently associated with infarct expansion and worse clinical outcome, *Stroke* **34**, 2208–2214.

Bak, S., Gaist, D., Sindrup, S.H., Skytthe, A., Christensen, K. (2002) Genetic liability in stroke: a long-term follow-up study of Danish twins, *Stroke* **33**, 769–774.

Becker, K., Kindrick, D., McCarron, R., Hallenbeck, J., Winn, R. (2003) Adoptive transfer of myelin basic protein-tolerized splenocytes to naive animals reduces infarct size: a role for lymphocytes in ischemic brain injury? *Stroke* **34**, 1809–1815.

Becker, K., Kindrick, D., Relton, J., Harlan, J., Winn, R. (2001) Antibody to the alpha4 integrin decreases infarct size in transient focal cerebral ischemia in rats, *Stroke* **32**, 206–211.

Becker, K.J., McCarron, R.M., Ruetzler, C., Laban, O., Sternberg, E., Flanders, K.C., Hallenbeck, J.M. (1997) Immunologic tolerance to myelin basic protein decreases stroke size after transient focal cerebral ischemia, *Proc. Natl. Acad. Sci. U. S. A.* **94**, 10873–10878.

Bonita, R. (1992) Epidemiology of stroke, *Lancet* **339**, 342–344.

Boysen, G., Christensen, H. (2001) Stroke severity determines body temperature in acute stroke, *Stroke* **32**, 413–417.

Bruce, A.J., Boling, W., Kindy, M.S., Peschon, J., Kraemer, P.J., Carpenter, M.K., Holtsberg, F.W., Mattson, M.P. (1996) Altered neuronal and microglial responses to excitotoxic and ischemic brain injury in mice lacking TNF receptors, *Nat. Med.* **2**, 788–794.

Busch, E., Gyngell, M.L., Eis, M., Hoehn-Berlage, M., Hossmann, K.A. (1996) Potassium-induced cortical spreading depressions during focal cerebral ischemia in rats: contribution to lesion growth assessed by diffusion-weighted NMR and biochemical imaging, *J. Cereb. Blood Flow Metab.* **16**, 1090–1099.

Cao, G., Pei, W., Ge, H., Liang, Q., Luo, Y., Sharp, F.R., Lu, A., Ran, R., Graham, S.H., Chen, J. (2002) In vivo delivery of a Bcl-xL fusion protein containing the TAT protein transduction domain protects against ischemic brain injury and neuronal apoptosis, *J. Neurosci.* **22**, 5423–5431.

Castillo, J., Davalos, A., Marrugat, J., Noya, M. (1998) Timing for fever-related brain damage in acute ischemic stroke, *Stroke* **29**, 2455–2460.

Chan, P.H. (2001) Reactive oxygen radicals in signaling and damage in the ischemic brain, *J. Cereb. Blood Flow Metab.* **21**, 2–14.

Chen, H., Chopp, M., Zhang, R.L., Bodzin, G., Chen, Q., Rusche, J.R., Todd, R.F. 3rd. (1994) Anti-CD11b monoclonal antibody reduces ischemic cell damage after transient focal cerebral ischemia in rat, *Ann. Neurol.* **35**, 458–463.

Chopp, M., Chan, P.H., Hsu, C.Y., Cheung, M.E., Jacobs, T.P. (1996) DNA damage and repair in central nervous system injury: National Institute of Neurological Disorders and Stroke Workshop Summary, *Stroke* **27**, 363–369.

Crumrine, R.C., Thomas, A.L., Morgan, P.F. (1994) Attenuation of p53 expression protects against focal ischemic damage in transgenic mice, *J. Cereb. Blood Flow Metab.* **14**, 887–891.

Davenport, R.J., Dennis, M.S., Wellwood, I., Warlow, C.P. (1996) Complications after acute stroke, *Stroke* **27**, 415–420.

del Zoppo, G.J. (1995) Why do all drugs work in animals but none in stroke patients? 1. Drugs promoting cerebral blood flow, *J. Intern. Med.* **237**, 79–88.

del Zoppo, G.J., Schmid-Schonbein, G.W., Mori, E., Copeland, B.R., Chang, C.M. (1991) Polymorphonuclear leukocytes occlude capillaries following middle cerebral artery occlusion and reperfusion in baboons, *Stroke* **22**, 1276–1283.

Digicaylioglu, M., Lipton, S.A. (2001) Erythropoietin-mediated neuroprotection involves cross-talk between Jak2 and NF-kappaB signalling cascades, *Nature* **412**, 641–647.

Eglitis, M.A., Mezey, E. (1997) Hematopoietic cells differentiate into both microglia and macroglia in the brains of adult mice, *Proc. Natl. Acad. Sci. U. S. A.* **94**, 4080–4085.

Ehrenreich, H., Hasselblatt, M., Dembowski, C., Cepek, L., Lewczuk, P., Stiefel, M., Rustenbeck, H.H., Breiter, N., Jacob, S., Knerlich, F., Bohn, M., Poser, W., Ruther, E., Kochen, M., Gefeller, O., Gleiter, C., Wessel, T.C., De Ryck, M., Itri, L., Prange, H., Cerami, A., Brines, M., Siren, A.L. (2002) Erythropoietin therapy for acute stroke is both safe and beneficial, *Mol. Med.* **8**, 495–505.

Eliasson, M.J., Sampei, K., Mandir, A.S., Hurn, P.D., Traystman, R.J., Bao, J., Pieper, A., Wang, Z.Q., Dawson, T.M., Snyder, S.H., Dawson, V.L. (1997) Poly(ADP-ribose) polymerase gene disruption renders mice resistant to cerebral ischemia, *Nat. Med.* **3**, 1089–1095.

Endres, M., Wang, Z.Q., Namura, S., Waeber, C., Moskowitz, M.A. (1997) Ischemic brain injury is mediated by the activation of poly(ADP-ribose)polymerase, *J. Cereb. Blood Flow Metab.* **17**, 1143–1151.

Endres, M., Fink, K., Zhu, J., Stagliano, N.E., Bondada, V., Geddes, J.W., Azuma, T., Mattson, M.P., Kwiatkowski, D.J., Moskowitz, M.A. (1999) Neuroprotective effects of gelsolin during murine stroke, *J. Clin. Invest.* **103**, 347–354.

Endres, M., Meisel, A., Biniszkiewicz, D., Namura, S., Prass, K., Ruscher, K., Lipski, A., Jaenisch, R., Moskowitz, M.A., Dirnagl, U. (2000) DNA methyltransferase contributes to delayed ischemic brain injury, *J. Neurosci.* **20**, 3175–3181.

Endres, M., Biniszkiewicz, D., Sobol, R.W., Harms, C., Ahmadi, M., Lipski, A., Katchanov, J., Mergenthaler, P., Dirnagl, U., Wilson, S.H., Meisel, A., Jaenisch, R. (2004) Increased postischemic brain injury in mice deficient in uracil-DNA glycosylase, *J. Clin. Invest.* **113**, 1711–1721.

Englander, E.W., Hu, Z., Sharma, A., Lee, H.M., Wu, Z.H., Greeley, G.H. (2002) Rat MYH, a glycosylase for repair of oxidatively damaged

DNA, has brain-specific isoforms that localize to neuronal mitochondria, *J. Neurochem.* **83**, 1471–1480.

Evans, A., Perez, I., Harraf, F., Melbourn, A., Steadman, J., Donaldson, N., Kalra, L. (2001) Can differences in management processes explain different outcomes between stroke unit and stroke-team care? *Lancet* **358**, 1586–1592.

Gasche, Y., Fujimura, M., Morita-Fujimura, Y., Copin, J.C., Kawase, M., Massengale, J., Chan, P.H. (1999) Early appearance of activated matrix metalloproteinase-9 after focal cerebral ischemia in mice: a possible role in blood-brain barrier dysfunction, *J. Cereb. Blood Flow Metab.* **19**, 1020–1028.

Ginsberg, M.D., Prado, R., Dietrich, W.D., Busto, R., Watson, B.D. (1987) Hyperglycemia reduces the extent of cerebral infarction in rats, *Stroke* **18**, 570–574.

Gladstone, D.J., Black, S.E., Hakim, A.M., Heart and Stroke Foundation of Ontario Centre of Excellence in Stroke Recovery. (2002) Toward wisdom from failure: lessons from neuroprotective stroke trials and new therapeutic directions, *Stroke* **33**, 2123–2136.

Gretarsdottir, S., Thorleifsson, G., Reynisdottir, S.T., Manolescu, A., Jonsdottir, S., Jonsdottir, T., Gudmundsdottir, T., Bjarnadottir, S.M., Einarsson, O.B., Gudjonsdottir, H.M., Hawkins, M., Gudmundsson, G., Gudmundsdottir, H., Andrason, H., Gudmundsdottir, A.S., Sigurdardottir, M., Chou, T.T., Nahmias, J., Goss, S., Sveinbjornsdottir, S., Valdimarsson, E.M., Jakobsson, F., Agnarsson, U., Gudnason, V., Thorgeirsson, G., Fingerle, J., Gurney, M., Gudbjartsson, D., Frigge, M.L., Kong, A., Stefansson, K., Gulcher, J.R. (2003) The gene encoding phosphodiesterase 4D confers risk of ischemic stroke, *Nat. Genet.* **35**, 131–138.

Hajat, C., Hajat, S., Sharma, P. (2000) Effects of poststroke pyrexia on stroke outcome: a meta-analysis of studies in patients, *Stroke* **31**, 410–414.

Hara, H., Friedlander, R.M., Gagliardini, V., Ayata, C., Fink, K., Huang, Z., Shimizu-Sasamata, M., Yuan, J., Moskowitz, M.A. (1997) Inhibition of interleukin 1beta converting enzyme family proteases reduces ischemic and excitotoxic neuronal damage, *Proc. Natl. Acad. Sci. U. S. A.* **94**, 2007–2012.

Heiss, W.D., Kracht, L.W., Thiel, A., Grond, M., Pawlik, G. (2001) Penumbral probability thresholds of cortical flumazenil binding and blood flow predicting tissue outcome in patients with cerebral ischaemia, *Brain* **124**, 20–29.

Hilker, R., Poetter, C., Findeisen, N., Sobesky, J., Jacobs, A., Neveling, M., Heiss, W.D. (2003) Nosocomial pneumonia after acute stroke: implications for neurological intensive care medicine, *Stroke* **34**, 975–981.

Hoehn-Berlage, M. (1995) Diffusion-weighted NMR imaging: application to experimental focal cerebral ischemia, *NMR Biomed.* **8**, 345–358.

Iadecola, C., Zhang, F., Casey, R., Nagayama, M., Ross, M.E. (1997) Delayed reduction of ischemic brain injury and neurological deficits in mice lacking the inducible nitric oxide synthase gene, *J. Neurosci.* **17**, 9157–9164.

Iadecola, C., Niwa, K., Nogawa, S., Zhao, X., Nagayama, M., Araki, E., Morham, S., Ross, M.E. (2001) Reduced susceptibility to ischemic brain injury and N-methyl-D-aspartate-mediated neurotoxicity in cyclooxygenase-2-deficient mice, *Proc. Natl. Acad. Sci. U. S. A.* **98**, 1294–1299.

Jin, K., Mao, X.O., Sun, Y., Xie, L., Greenberg, D.A. (2002) Stem cell factor stimulates neurogenesis in vitro and in vivo, *J. Clin. Invest.* **110**, 311–319.

Jin, K., Mao, X.O., Eshoo, M.W., Nagayama, T., Minami, M., Simon, R.P., Greenberg, D.A. (2001) Microarray analysis of hippocampal gene expression in global cerebral ischemia, *Ann. Neurol.* **50**, 93–103.

Joutel, A., Corpechot, C., Ducros, A., Vahedi, K., Chabriat, H., Mouton, P., Alamowitch, S., Domenga, V., Cecillion, M., Marechal, E., Maciazek, J., Vayssiere, C., Cruaud, C., Cabanis, E.A., Ruchoux, M.M., Weissenbach, J., Bach, J.F., Bousser, M.G., Tournier-Lasserve, E. (1996) Notch3 mutations in CADASIL, a hereditary adult-onset condition causing stroke and dementia, *Nature* **383**, 707–710.

Katzan, I.L., Cebul, R.D., Husak, S.H., Dawson, N.V., Baker, D.W. (2003) The effect of pneumonia on mortality among patients hospitalized for acute stroke, *Neurology* **60**, 620–625.

Keyvani, K., Witte, O.W., Paulus, W. (2002) Gene expression profiling in perilesional and contralateral areas after ischemia in rat brain, *J. Cereb. Blood Flow Metab.* **22**, 153–160.

Kilic, E., Dietz, G.P., Hermann, D.M., Bahr, M. (2002) Intravenous TAT-Bcl-Xl is protective after middle cerebral artery occlusion in mice, *Ann. Neurol.* **52**, 617–622.

Lapchak, P.A., Chapman, D.F., Zivin, J.A. (2000) Metalloproteinase inhibition reduces thrombolytic (tissue plasminogen activator)-induced hemorrhage after thromboembolic stroke, *Stroke* **31**, 3034–3040.

Le, D.A., Wu, Y., Huang, Z., Matsushita, K., Plesnila, N., Augustinack, J.C., Hyman, B.T., Yuan, J., Kuida, K., Flavell, R.A., Moskowitz, M.A. (2003) Caspase activation and neuroprotection in caspase-3- deficient mice after in vivo cerebral ischemia and in vitro oxygen glucose deprivation, *Proc. Natl. Acad. Sci. U. S. A.* **99**, 15188–15193.

Lindsberg, P.J., Carpen, O., Paetau, A., Karjalainen-Lindsberg, M.L., Kaste, M. (1996) Endothelial ICAM-1 expression associated with inflammatory cell response in human ischemic stroke, *Circulation* **94**, 939–945.

Linnik, M.D., Zahos, P., Geschwind, M.D., Federoff, H.J. (1995) Expression of bcl-2 from a defective herpes simplex virus-1 vector limits neuronal death in focal cerebral ischemia, *Stroke* **26**, 1670–1674.

Liu, P.K., Hsu, C.Y., Dizdaroglu, M., Floyd, R.A., Kow, Y.W., Karakaya, A., Rabow, L.E., Cui, J.K. (1996) Damage repair, and mutagenesis in nuclear genes after mouse forebrain ischemia-reperfusion, *J. Neurosci.* **16**, 6795–6806.

Lu, A., Tang, Y., Ran, R., Clark, J.F., Aronow, B.J., Sharp, F.R. (2003) Genomics of the periinfarction cortex after focal cerebral ischemia, *J. Cereb. Blood Flow Metab.* **23**, 786–810.

Martinou, J.C., Dubois-Dauphin, M., Staple, J.K., Rodriguez, P., Frankowski, H., Missotten, M., Albertini, P., Talabot, D., Catsicas, S., Pietra, C., Huarte, J. (1994) Overexpression of bcl-2 in transgenic mice protects neurons from naturally occurring cell death and experimental ischemia, *Neuron* **13**, 1017–1030.

Meisel, C., Prass, K., Braun, J., Victorov, I., Wolf, T., Megow, D., Halle, E., Volk, H.D., Dirnagl, U., Meisel, A. (2004) Preventive antibacterial treatment improves the general medical and neurological outcome in a mouse model of stroke, *Stroke* **35**, 2–6.

Moncayo, J., de Freitas, G.R., Bogousslavsky, J., Altieri, M., van Melle, G. (2000) Do transient ischemic attacks have a neuroprotective effect? *Neurology* **54**, 2089–2094.

Mori, E., del Zoppo, G.J., Chambers, J.D., Copeland, B.R., Arfors, K.E. (1992) Inhibition of polymorphonuclear leukocyte adherence suppresses no-reflow after focal cerebral ischemia in baboons, *Stroke* **23**, 712–718.

Nakatomi, H., Kuriu, T., Okabe, S., Yamamoto, S., Hatano, O., Kawahara, N., Tamura, A., Kirino, T., Nakafuku, M. (2002) Regeneration of hippocampal pyramidal neurons after ischemic brain injury by recruitment of endogenous neural progenitors, *Cell* **110**, 429–441.

Namura, S., Zhu, J., Fink, K., Endres, M., Srinivasan, A., Tomaselli, K.J., Yuan, J., Moskowitz, M.A. (1998) Activation and cleavage of caspase-3 in apoptosis induced by experimental cerebral ischemia, *J. Neurosci.* **18**, 3659–3668.

Nedergaard, M., Hansen, A.J. (1993) Characterization of cortical depolarizations evoked in focal cerebral ischemia, *J. Cereb. Blood Flow Metab.* **13**, 568–574.

Parsons, M.W., Barber, P.A., Desmond, P.M., Baird, T.A., Darby, D.G., Byrnes, G., Tress, B.M., Davis, S.M. (2002) Acute hyperglycemia adversely affects stroke outcome: a magnetic resonance imaging and spectroscopy study, *Ann. Neurol.* **52**, 20–28.

Perry, L., Love, C.P. (2001) Screening for dysphagia and aspiration in acute stroke: a systematic review, *Dysphagia* **16**, 7–18.

Pfefferkorn, T., Rosenberg, G.A. (2003) Closure of the blood-brain barrier by matrix metalloproteinase inhibition reduces rtPA-mediated mortality in cerebral ischemia with delayed reperfusion, *Stroke* **34**, 2025–2030.

Prass, K., Scharff, A., Ruscher, K., Lowl, D., Muselmann, C., Victorov, I., Kapinya, K., Dirnagl, U., Meisel, A. (2003) Hypoxia-induced stroke tolerance in the mouse is mediated by erythropoietin, *Stroke* **34**, 1981–1986.

Prass, K., Meisel, C., Hoflich, C., Braun, J., Halle, E., Wolf, T., Ruscher, K., Victorov, I.V., Priller, J., Dirnagl, U., Volk, H.D., Meisel, A. (2003) Stroke-induced immunodeficiency promotes spontaneous bacterial infections and is mediated by sympathetic activation reversal by poststroke T helper cell Type 1-like immunostimulation, *J. Exp. Med.* **198**, 725–736.

Priller, J., Flugel, A., Wehner, T., Boentert, M., Haas, C.A., Prinz, M., Fernandez-Klett, F., Prass, K., Bechmann, I., de Boer, B.A., Frotscher, M., Kreutzberg, G.W., Persons, D.A., Dirnagl, U. (2001) Targeting gene-modified hematopoietic cells to the central nervous system: use of green fluorescent protein uncovers microglial engraftment, *Nat. Med.* **7**, 1356–1361.

Pulsinelli, W.A., Waldman, S., Rawlinson, D., Plum, F. (1982) Moderate hyperglycemia augments ischemic brain damage: a neuropathologic study in the rat, *Neurology* **32**, 1239–1246.

Raghavendra Rao, V.L., Bowen, K.K., Dhodda, V.K., Song, G., Franklin, J.L., Gavva, N.R., Dempsey, R.J. (2002) Gene expression analysis of spontaneously hypertensive rat cerebral cortex following transient focal cerebral ischemia, *J. Neurochem.* **83**, 1072–1086.

Reith, J., Jorgensen, H.S., Pedersen, P.M., Nakayama, H., Raaschou, H.O., Jeppesen, L.L., Olsen, T.S. (1996) Body temperature in acute stroke: relation to stroke severity, infarct size, mortality, and outcome, *Lancet* **347**, 422–425.

Rosenberg, G.A., Estrada, E.Y., Dencoff, J.E. (1998) Matrix metalloproteinases and TIMPs are associated with blood-brain barrier opening after reperfusion in rat brain, *Stroke* **29**, 2189–2195.

Ruscher, K., Isaev, N., Trendelenburg, G., Weih, M., Iurato, L., Meisel, A., Dirnagl, U. (1998) Induction of hypoxia inducible factor 1 by oxygen glucose deprivation is attenuated by hypoxic preconditioning in rat cultured neurons, *Neurosci. Lett.* **254**, 117–120.

Ruscher, K., Freyer, D., Karsch, M., Isaev, N., Megow, D., Sawitzki, B., Priller, J., Dirnagl, U., Meisel, A. (2002) Erythropoietin is a paracrine mediator of ischemic tolerance in the brain: evidence from an in vitro model, *J. Neurosci.* **22**, 10291–10301.

Sairanen, T., Ristimaki, A., Karjalainen-Lindsberg, M.L., Paetau, A., Kaste, M., Lindsberg, P.J. (1998) Cyclooxygenase-2 is induced globally in infarcted human brain, *Ann. Neurol.* **43**, 738–747.

Saito, A., Hayashi, T., Okuno, S., Ferrand-Drake, M., Chan, P.H. (2003) Interaction between XIAP and Smac/DIABLO in the mouse brain after transient focal cerebral ischemia, *J. Cereb. Blood Flow Metab.* **23**, 1010–1019.

Schurr, A., Payne, R.S., Miller, J.J., Tseng, M.T. (2001) Preischemic hyperglycemia-aggravated damage: evidence that lactate utilization is beneficial and glucose-induced corticosterone release is detrimental, *J. Neurosci. Res.* **66**, 782–789.

Seo, S.Y., Kim, E.Y., Kim, H., Gwag, B.J. (1999) Neuroprotective effect of high glucose against NMDA, free radical, and oxygen-glucose deprivation through enhanced mitochondrial potentials, *J. Neurosci.* **19**, 8849–8855.

Sharp, F.R., Lu, A., Tang, Y., Millhorn, D.E. (2000) Multiple molecular penumbras after focal cerebral ischemia, *J. Cereb. Blood Flow Metab.* **20**, 1011–1032.

Siren, A.L., Fratelli, M., Brines, M., Goemans, C., Casagrande, S., Lewczuk, P., Keenan, S., Gleiter, C., Pasquali, C., Capobianco, A., Mennini, T., Heumann, R., Cerami, A., Ehrenreich, H., Ghezzi, P. (2001) Erythropoietin prevents neuronal apoptosis after cerebral ischemia and metabolic stress, *Proc. Natl. Acad. Sci. U. S. A.* **98**, 4044–4049.

Soriano, S.G., Lipton, S.A., Wang, Y.F., Xia, M., Springer, T.A., Gutierrez-Ramos, J.C., Hickey, P.R. (1996) Intercellular adhesion molecule-1-deficient mice are less susceptible to cerebral ischemia-reperfusion injury, *Ann. Neurol.* **39**, 618–624.

Strong, A.J., Fabricius, M., Boutelle, M.G., Hibbins, S.J., Hopwood, S.E., Jones, R., Parkin, M.C., Lauritzen, M. (2002) Spreading and synchronous depressions of cortical activity in acutely injured human brain, *Stroke* **33**, 2738–2743.

Takeda, H., Spatz, M., Ruetzler, C., McCarron, R., Becker, K., Hallenbeck, J. (2002) Induction of mucosal tolerance to E-selectin prevents ischemic and hemorrhagic stroke in spontaneously hypertensive genetically stroke-prone rats, *Stroke* **33**, 2156–2163.

Tarkowski, E., Rosengren, L., Blomstrand, C., Wikkelso, C., Jensen, C., Ekholm, S., Tarkowski, A. (1995) Early intrathecal production of interleukin-6 predicts the size of brain lesion in stroke, *Stroke* **26**, 1393–1398.

Trapp, T., Korhonen, L., Besselmann, M., Martinez, R., Mercer, E.A., Lindholm, D. (2003) Transgenic mice overexpressing XIAP in neurons show better outcome after transient cerebral ischemia, *Mol. Cell. Neurosci.* **23**, 302–313.

Trendelenburg, G., Prass, K., Priller, J., Kapinya, K., Polley, A., Muselmann, C., Ruscher, K., Kannbley, U., Schmitt, A.O., Castell, S., Wiegand, F., Meisel, A., Rosenthal, A., Dirnagl, U. (2002) Serial analysis of gene expression identifies metallothionein-II as major neuroprotective gene in mouse focal cerebral ischemia, *J. Neurosci.* **22**, 5879–5888.

Wang, H., Zhan, Y., Xu, L., Feuerstein, G.Z., Wang, X. (2001) Use of suppression subtractive hybridization for differential gene expression in stroke: discovery of CD44 gene expression and localization in permanent focal stroke in rats, *Stroke* **32**, 1020–1027.

Wang, X., Lee, S.R., Arai, K., Lee, S.R., Tsuji, K., Rebeck, G.W., Lo, E.H. (2003) Lipoprotein receptor-mediated induction of matrix metalloproteinase by tissue plasminogen activator, *Nat. Med.* **9**, 1313–1317.

Wang, Y.F., Tsirka, S.E., Strickland, S., Stieg, P.E., Soriano, S.G., Lipton, S.A. (1998) Tissue plasminogen activator (tPA) increases neuronal damage after focal cerebral ischemia in wild-type and tPA-deficient mice, *Nat. Med.* **4**, 228–231.

Wass, C.T., Lanier, W.L., Hofer, R.E., Scheithauer, B.W., Andrews, A.G. (1995) Temperature changes of > or = 1 degree C alter functional neurologic outcome and histopathology in a canine model of complete cerebral ischemia, *Anesthesiology* **83**, 325–335.

Wehner, T., Boentert, M., Eyupoglu, I., Prass, K., Prinz, M., Klett, F.F., Heinze, M., Bechmann, I., Nitsch, R., Kirchhoff, F., Kettenmann, H., Dirnagl, U., Priller, J. (2003) Bone marrow-derived cells expressing green fluorescent protein under the control of the glial fibrillary acidic protein promoter do not differentiate into astrocytes in vitro and in vivo, *J. Neurosci.* **23**, 5004–5011.

Weih, M., Kallenberg, K., Bergk, A., Dirnagl, U., Harms, L., Wernecke, K.D., Einhaupl, K.M. (1999) Attenuated stroke severity after prodromal TIA: a role for ischemic tolerance in the brain? *Stroke* **30**, 1851–1854.

Yrjänheikki, J., Tikka, T., Keinanen, R., Goldsteins, G., Chan, P.H., Koistinaho, J. (1999) A tetracycline derivative, minocycline, reduces inflammation and protects against focal cerebral ischemia with a wide therapeutic window, *Proc. Natl. Acad. Sci. U. S. A.* **96**, 13496–13500.

Yu, S.W., Wang, H., Poitras, M.F., Coombs, C., Bowers, W.J., Federoff, H.J., Poirier, G.G., Dawson, T.M., Dawson, V.L. (2002) Mediation of poly(ADP-ribose) polymerase-1-dependent cell death by apoptosis-inducing factor, *Science* **297**, 259–263.

Index

Neurobiology. From Molecular Basis to Disease. Edited by Robert A. Meyers.

ISBN: 978-3-527-32293-0

i

q